AI-Guided Design and Property Prediction for Zeolites and Nanoporous Materials

AI-Guided Design and Property Prediction for Zeolites and Nanoporous Materials

Edited by

German Sastre
Instituto de Tecnología Química UPV-CSIC
Universidad Politecnica de Valencia
Valencia, Spain

Frits Daeyaert
SynopsisDeNovoDesign
Beerse, Belgium

This edition first published 2023

Registered Offices
John Wiley & Sons, Inc., 111 River Street, Hoboken, NJ 07030, USA
John Wiley & Sons Ltd, The Atrium, Southern Gate, Chichester, West Sussex, PO19 8SQ, UK

Editorial Office
The Atrium, Southern Gate, Chichester, West Sussex, PO19 8SQ, UK

For details of our global editorial offices, customer services, and more information about Wiley products visit us at www.wiley.com.

Wiley also publishes its books in a variety of electronic formats and by print-on-demand. Some content that appears in standard print versions of this book may not be available in other formats.

Library of Congress Cataloging-in-Publication Data
Names: Sastre, German, editor. | Daeyaert, Frits, editor.
Title: AI-guided design and property prediction for zeolites and nanoporous materials / edited by German Sastre, Frits Daeyaert.
Description: Hoboken : John Wiley and Sons Ltd, 2023. | Includes bibliographical references and index.
Identifiers: LCCN 2022042565 (print) | LCCN 2022042566 (ebook) | ISBN 9781119819752 (hardback) | ISBN 9781119819769 (pdf) | ISBN 9781119819776 (epub) | ISBN 9781119819783 (ebook)
Subjects: LCSH: Zeolites--Analysis | Nanostructured materials--Analysis | Molecules--Models | Artificial intelligence
Classification: LCC QE391.Z5 A44 2023 (print) | LCC QE391.Z5 (ebook) | DDC 549/.68--dc23/eng20221121
LC record available at https://lccn.loc.gov/2022042565
LC ebook record available at https://lccn.loc.gov/2022042566

Cover design by Wiley
Cover Images: Binary Code: © fotograzia/Getty Images; Lightning: © John Sirlin/Getty Images; Zeolite image: Courtesy of Frits Daeyaert; Server: © cybrain/Getty Images; Parthenon: © Cliff Wassmann/Getty Images

Set in 9.5/12.5pt STIXTwoText by Integra Software Services Pvt. Ltd, Pondicherry, India

Contents

List of Contributors

Estefania Argente
Valencian Research Institute for Artificial Intelligence (VRAIN), Universitat Politècnica de València, Valencia, Spain

Laurent Baumes
ExxonMobil, Clinton, New Jersey, USA

Steven Bennett
Department of Chemistry, Molecular Sciences Research Hub, Imperial College London, White City Campus, London, United Kingdom

Rocio Bueno-Perez
Adsorption & Advanced Materials Laboratory (AAML), Department of Chemical Engineering & Biotechnology, University of Cambridge, Cambridge, United Kingdom

Watcharop Chaikittisilp
Research and Services Division of Materials Data and Integrated System (MaDIS), National Institute for Materials Science (NIMS), Tsukuba, Ibaraki, Japan

Ruxandra G. Chitac
EaStCHEM School of Chemistry, University of St Andrews, St Andrews, United Kingdom

Yamil J. Colon
Department of Chemical and Biomolecular Engineering, University of Notre Dame, Notre Dame, Indiana, USA

Paul A. Cox
School of Pharmacy and Biomedical Sciences, University of Portsmouth, Portsmouth, United Kingdom

Frits Daeyaert
SynopsisDeNovoDesign, Beerse, Belgium

Archit Datar
William G. Lowrie Department of Chemical and Biomolecular Engineering, The Ohio State University, Columbus, Ohio, USA

Tracy M. Davis
Chevron Technical Center, Richmond, California, USA

Michael Deem
Certus LLC, Houston, Texas, USA

Giulia Lo Dico
IMDEA Materials Institute, Madrid, Spain

Tolsa Group, Madrid, Spain

Saleh Elomari
Chevron Technical Center, Richmond, California, USA

David Fairen-Jimenez
Adsorption & Advanced Materials Laboratory (AAML), Department of Chemical Engineering & Biotechnology, University of Cambridge, Cambridge, United Kingdom

Lawson. T. Glasby
Department of Chemical and Biological Engineering, University of Sheffield, Sheffield, United Kingdom

Rafael Gómez-Bombarelli
Department of Materials Science and Engineering, Massachusetts Institute of Technology, Massachusetts, USA

Diego A. Gómez-Gualdrón
Chemical and Biological Engineering, Colorado School of Mines, Colorado, USA

Maciej Haranczyk
IMDEA Materials Institute, Madrid, Spain

Kim E. Jelfs
Department of Chemistry, Molecular Sciences Research Hub, Imperial College London, White City Campus, London, United Kingdom

Christopher M. Lew
Chevron Technical Center, Richmond, California, USA

Aurelia Li
Adsorption & Advanced Materials Laboratory (AAML), Department of Chemical Engineering & Biotechnology, University of Cambridge, Cambridge, United Kingdom

Li-Chiang Lin
William G. Lowrie Department of Chemical and Biomolecular Engineering, The Ohio State University, Columbus, Ohio, USA

Department of Chemical Engineering, National Taiwan University, Taipei, Taiwan

María Gálvez-Llompart
Molecular Topology and Drug Design Unit, Department of Physical Chemistry, University of Valencia, Valencia, Spain

Instituto de Tecnologia Quimica (UPV-CSIC), Universidad Politecnica de Valencia, Valencia, Spain

Qiang Lyu
School of Materials Science and Engineering, China University of Petroleum (East China), Qingdao, Shandong, China

James Mattock
School of Pharmacy and Biomedical Sciences, University of Portsmouth, Portsmouth, United Kingdom

Alechania Misturini
Instituto de Tecnología Química UPV-CSIC; Universitat Politècnica de València, Valencia, Spain

Peyman Z. Moghadam
Department of Chemical and Biological Engineering, University of Sheffield, Sheffield, United Kingdom

Department of Chemical Engineering, University College London (UCL), London, United Kingdom

Seyed Mohamad Moosavi
Department of Mathematics and Computer Science/Mathematics, Artificial Intelligence for the Sciences, Freie Universität Berlin, Berlin, Germany

Gyoung S. Na
Korea Research Institute of Chemical Technology (KRICT). Daejeon, Republic of Korea

German Sastre
Instituto de Tecnologia Quimica (UPV-CSIC), Universidad Politecnica de Valencia, Valencia, Spain

Joel E. Schmidt
Chevron Technical Center, Richmond, California, USA

Daniel Schwalbe-Koda
Department of Materials Science and Engineering, Massachusetts Institute of Technology, Massachusetts, USA

Mervyn D. Shannon
Johnson Matthey Technology Centre, Billingham, United Kingdom

Cory M. Simon
School of Chemical, Biological, and Environmental Engineering, Oregon State University, Oregon, USA

Michael M.J. Treacy
Department of Physics, Arizona State University, Tempe, Arizona, USA

Alessandro Turrina
Johnson Matthey Technology Centre, Billingham, United Kingdom

Soledad Valero
Valencian Research Institute for Artificial Intelligence (VRAIN), Universitat Politècnica de València, Valencia, Spain

Emily H. Whaites
Department of Chemical and Biological Engineering, University of Sheffield, Sheffield, United Kingdom

Paul A. Wright
EaStCHEM School of Chemistry, University of St Andrews, St Andrews, United Kingdom

Dan Xie
Chevron Technical Center, Richmond, California, USA

Stacey I. Zones
Chevron Technical Center, Richmond, California, USA

Preface

There will never be a better video game than Pac-Man.[1] Well, yes, this is a matter of opinion, but it suggests a level of creativity that might never be reached by AI (Artificial Intelligence). But, if AI could ever reach a similar level, would we be happy with an AI equaling human's creativity? Limits to AI are necessary. AI is becoming more and more integrated into an ever-increasing number of aspects of human life and ethical concerns must be discussed along with or even prior to these developments. The ethics applied to AI will mirror our own ethical beliefs, hence concerns about AI are also concerns about ourselves.[2] Justice and transparency are perennial needs of our society and will be implemented in future developments of AI.[3] But power also attracts particular interests, with the risk of a globally biased AI.[4]

With our task in this book being rather modest with respect to the big trends outlined above, it is nevertheless our duty not to produce useless data as well as to follow recommendations and protocols that will make information and results understandable, reproducible, and widely available.[5] AI has made an enormous impact recently in the field of materials chemistry, and in particular nanoporous materials. Books may not contribute actively to impact, but they do provide an invaluable service to the scientific community by presenting a more careful approach to the foundations of our knowledge. This is our task with this book, with the help of the invaluable commitment of all the authors, who analyze how AI in its various flavors is pervading our way to do research and, hopefully, gain knowledge.

1 Molecular Modeling and New Zeolite Structures

The book opens with an account of the expertise collected by Stacey Zones and coworkers at the Chevron zeolite group in the development of new zeolites and the role therein of OSDAs (organic structure directing agents) in combination with other synthesis parameters. For more than 40 years this world leading group have been pioneers which, from the industry, have played an outstanding role in the elucidation and interpretation of the mechanisms of zeolite synthesis. The chapter describes in particular how molecular modeling techniques have provided insights into the working mechanism of OSDAs, and how these insights are becoming instrumental in the search for new zeolites. The role of high-throughput and automation of collecting, validating, and analyzing data in order to further develop machine learning and AI tools is discussed. The ultimate goal is to gain a deeper insight into the zeolite synthesis, as well as to use these tools to open up the large structural space of hypothetical zeolites, paving the way for future prediction of on-demand synthesis of zeolite materials for target applications.

2 *De Novo* Design of OSDAs

Given the progress in molecular modeling and machine learning techniques to accurately model the templating effect of OSDAs (organic structure directing agents), the second chapter by Frits Daeyaert and Michael Deem describes *de novo* design of OSDAs as an effective approach to exploit this ability. *De novo* design originated in drug design as an automated tool to generate molecules interacting with a biological target. In the zeolite-OSDA setting, the aim is to design molecules that closely interact with a target zeolite structure. It is discussed how issues like the synthesizability of the designed compounds and the simultaneous optimization of multiple additional properties and constraints of OSDAs can be addressed. The application to several OSDA design projects of a *de novo* algorithm developed by the authors is described.

3 ML Search for SDAs Using Molecular Topology and MC Techniques

In Chapter 3, María Gálvez-Llompart and German Sastre present the application of 3D-QSAR algorithms to the modeling and design of SDAs for zeolite synthesis. 3D-QSAR algorithms have proven their merit in drug design. They rely on the principle that the activity of a drug molecule relies on its binding energy with a biological target, and that this binding energy can be modeled using a combination of well-chosen molecular descriptors and appropriate machine learning algorithms. This situation resembles zeolite synthesis with the aid of SDAs, where the interaction energy between an SDA and its target zeolite is a determining factor in the outcome of a zeolite synthesis. As in drug design, accurate prediction of this interaction energy using atomistic modeling is very computationally expensive. It is shown how the use of molecular topology descriptors in combination with appropriate statistical and machine learning techniques allows an orders of magnitude speed-up in predicting zeolite–SDA interaction energies. This in turn enables virtual screening of large numbers of available molecules for use as OSDAs for zeolite synthesis. The method is demonstrated for BEA zeolite and, importantly, uses carefully selected experimental data.

4 Big Data in Zeolite Simulations

Application of data science methods to zeolite discovery has been hindered by the diversity of synthesis routes and conditions and by the absence of machine-readable experimental data. In Chapter 4, Daniel Schwalbe-Koda and Rafael Gómez-Bombarelli discuss how combining simulations and data-driven methods lead to the understanding and prediction of the role of OSDAs in zeolite synthesis. The concept of the FAIR, findable–accessible–interoperable–reusable, principle in the context of computational material databases is discussed. An overview is given of existing zeolite databases of experimental and predicted structures, and of efforts to create curated datasets of zeolite syntheses based upon the FAIR principle. The creation of a web-based platform to interactively explore OSDA-templated synthesis routes by simultaneously mining both experimental and computational data is described. Details of how the computational data have been compiled are discussed. These include efficient algorithms for OSDA-zeolite docking and binding energy calculations, the set-up and analysis of high-throughput computational screening, and consideration of additional OSDA descriptors such as synthetic accessibility, volume, and shape and charge descriptors.

5 Co-Templating for Small-Pore Zeolite Catalysts

Small-pore zeolites, which are of importance for their use in catalytic reactions and selective adsorption of small molecules, often contain more than one type of cavity. Thus, improved synthesis of these materials might be envisaged by the use of more than one template. This is discussed in Chapter 5 by Alessandro Turrina and coworkers forming a solid cluster of collaborators, including several UK universities and the leadership of Johnson Matthey who has worked on zeolite synthesis for more than three decades since the pioneering work by Paul Wright, John Casci, and Paul Cox. Synthesis of nanoporous solids using co-templating, aided by computational simulations, has been described for SAPO zeotypes. For Al-Si zeolites, the situation is complicated by the presence of metal cations that interact with both the crystallizing frameworks and OSDAs that are present in reaction mixtures. In the case of mixed inorganic/organic templating, it is discussed how computational modeling is applied in both synthesis design and structure elucidation. Co-templating has been successfully exploited in the synthesis and structural control of disordered intergrowth zeolite structures containing more than one cavity type.

6 Computer Generation of Hypothetical Zeolites

In Chapter 6, German Sastre and coworkers discuss the algorithmically closely related fields of computer generation of hypothetical zeolite structure prediction (ZSP) and determination (ZSD). Both ZSP and ZSD present complex, combinatorial problems, and one approach to solve these is the use of genetic algorithms. Thus, a separate paragraph is devoted to this subject. An overview of ZSP and ZSD algorithms and their software implementation is given and one particular method, the zeoGAsolver program, is discussed in detail. ZSP and ZSD are very computationally expensive, and therefore can benefit from the advent of high-speed GPUs, requiring efficient parallelization of existing algorithms, as has been implemented by Laurent Baumes and coworkers in the code Parallel Genetic Hybrid Algorithm for Zeolites. Also, the groundbreaking algorithms included in the SCIBS (Symmetry-Constrained Intersite Bonding Search) software are described in detail by Michael Treacy. This method enumerates all possible four-valent networks within each space group given the number of unique tetrahedral atoms. Using this program, in collaboration with other authors, the first database of hypothetical zeolites was created and has been freely available since 2004.

7 Numerical Representations of Chemical Data for Machine Learning

In Chapter 7, Gyoung Na discusses how molecular structures are represented in a way that can be input into machine learning algorithms. In chemistry, a natural representation of molecules is as molecular graphs, which can be input into graph neural networks. The general architecture of graph neural networks is presented, and practical implementations including graph convolutional networks, graph attention networks, message passing neural networks, and crystal graph convolutional neural networks are summarized. Many problems in chemistry involve interactions between multiple molecules, and therefore the adaptation of graph neural networks to these problems is also discussed. While graph-based representations of molecular structures have proven to be very useful, they do not always capture all aspects of molecular structure. Representation learning can be applied to generate latent representations of molecular graphs that can subsequently be used as input for classification and regression algorithms. In a separate paragraph, Python implementations for generating molecular graphs and using these as input in the graph-based machine learning algorithms presented in this chapter are discussed. Finally, the application of graph-based machine learning for a number of chemical applications is demonstrated.

8 Extracting Metal–Organic Framework Data from the Cambridge Structural Database

To successfully apply data-based learning to the design and property prediction of nanoporous materials, data have to be collected and carefully curated. Chapter 8, by Aurelia Li, Rocio Bueno-Perez, and David Fairen-Jimenez, presents how the MOF subset of the Cambridge Structural Database, CSD MOF, has been extracted. It describes how the experimental data have been curated in terms of disorders, solvent removal, and addition of missing hydrogens. It is shown how the CSD MOF subset can be used to analyze the textural properties of MOFs and to classify MOFs in terms of chemical families, functional groups, chirality, porosity, and crystal quality.

9 Data-Driven Approach for Rational Synthesis of Zeolites and Other Nanoporous Materials

Searching for and optimizing nanoporous materials using traditional trial and error experiments is time-consuming and cost-inefficient. In Chapter 9, Watcharop Chaikittisilp illustrates through a number of example studies how data-driven techniques can significantly speed up these processes at different stages. In one study, it is shown how organic free zeolite syntheses can be modeled and predicted by applying machine learning on data collected from the literature. Property prediction of nanoporous materials is illustrated by setting up and validating models to predict the electrochemical catalytic oxygen reduction reaction of doped nanoporous carbons. This particular case is characterized by the occurrence of missing data, and appropriate ways to address this issue are discussed. Optimization of experimental conditions to generate better performing materials is illustrated by the use of active learning to design nanoporous metal alloys for electrochemical applications. It is suggested that data science techniques as discussed in this chapter can be combined with automated high-throughput experiments to significantly enhance the discovery of new and better performing nanoporous materials.

10 Porous Molecular Materials: Exploring Structure and Property Space With Software and Artificial Intelligence

Porous molecular materials, or PMMs, differ from other materials such as zeolites or MOFs in that they are not formed by direct covalent or coordination bonds, but instead consist of molecular units that are held together by non-covalent interactions. In Chapter 10, Steve Bennett and Kim Jelfs describe how computational methods are playing an increasing role in rationalizing the structure and property of PMMS and in the discovery of new PMMs. The chapter starts with an overview of computational modeling of PMMs. Next, the generation of data and the construction of appropriate descriptors required for the application of machine learning techniques is discussed. It is then shown how these ML techniques can be combined with generative algorithms to explore chemical space in search for novel PMM materials. The importance of combining these efforts with the consideration of synthetic accessibility is emphasized.

11 ML-Aided Discovery of Nanoporous Materials for Energy- and Environmental-Related Applications

An important application of nanoporous materials involves gas adsorption for gas storage and separation. Traditionally, computational methods based upon molecular simulations and first principle calculations have been deployed for the discovery of novel materials in this field. In Chapter 11, Archit Datar, Wiang Lyu, and Li-Chiang Lin describe how data-driven approaches have led to significant improvements in this area, especially in terms of computational efficiency. First, an introduction to commonly used statistical and machine learning techniques is provided. It is then described how these techniques are deployed in data-driven approaches for studying gas adsorption and separation in nanoporous materials. Topics discussed are the modeling of interatomic and intermolecular interactions, the development of reliable nanoporous structure datasets, the selection of appropriate molecular descriptors, and methods to search chemical space for the discovery of optimal materials. Deployment of the methods described is illustrated in a number of case studies, including the screening of materials for post-combustion CO_2 capture as well as methane and hydrogen storage.

12 Big Data Science in Nanoporous Materials: Datasets and Descriptors

The digital revolution brought about by the widespread availability of digital computers and electronic databases has enabled the application of big data science to accelerate materials discovery. In Chapter 12, Maciej Haranczyk and Giulia Lo Dico describe this evolution in the context of nanoporous materials research. Topics discussed are repositories of experimental and predicted crystal structures, appropriate descriptors of these structures, the collection and representation of materials properties, and the prediction of properties from structure descriptors using machine learning models. The chapter concludes with an overview of current applications and a reflection on future possibilities.

13 Efficient Data Utilization in Training ML Models for Nanoporous Materials Screening

Machine learning approaches have become important tools in materials science and will remain so in the future. However, machine learning is a data hungry approach and available data in the field of nanoporous materials remains relatively scarce. It is therefore important to effectively use these available data. This is discussed in depth by Diego Gómez-Gualdrón, Cory Simon, and Yamil Colón in Chapter 13, which covers descriptor, material, and model selection for the prediction of material properties, and methods to effectively deploy these in the search of novel materials. While these topics are discussed with examples in the context of modeling of adsorption in MOFs, the conclusions and lessons learned are generally applicable in the prediction of other properties and of nanoporous materials.

14 ML and Digital Manufacturing Approaches for Solid-State Materials Development

In Chapter 14, Lawson Glasby, Emily Whaites, and Peyman Moghadam discuss how the introduction of data science and digital technologies paves the way to more efficient synthesis and development of novel solid materials. An important challenge here is the collection and use of appropriate data. This is illustrated in the case of MOFs by discussing existing databases and the application of natural language processing to extend and maintain these data, and by giving an overview of machine learning algorithms to mine these. The state-of-the -art in automated solid-state synthesis is illustrated through a number of examples. The emergence of open-source data repositories of consistent and reliable reports of synthesis results and conditions, including those of unsuccessful experiments, is identified as an important evolution that holds promise to overcome current limitations.

15 Overview of AI in the Understanding and Design of Nanoporous Materials

Throughout the chapters of this book, the contributing authors have described the latest developments and state-of-the-art of the application of AI methods in nanoporous materials development. Topics covered included the development and maintenance of databases of structures, properties, and synthesis protocols, and the mining of these using big data science. Current applications of machine learning were discussed, including generative algorithms and data-driven synthesis approaches, prediction of adsorption properties, and OSDA modeling. The authors of the concluding chapter of this book summarize these contributions, identify remaining challenges, and give their view on future developments.

Frits Daeyaert
July 2022 *German Sastre*

Notes

1 Cohen, D.S. "'Pac-Man' – the Most Important and Iconic Video Game of All Time." https://www.lifewire.com/pac-man-video-game-729560.
2 Lauer, D. (2021). You cannot have AI ethics without ethics. *AI Ethics* 1: 21–25. https://doi.org/10.1007/s43681-020-00013-4.
3 Benjamins, R. (2021). A choices framework for the responsible use of AI. *AI Ethics* 1: 49–53. https://doi.org/10.1007/s43681-020-00012-5.
4 Asimov, I. (1950). *I Robot*. Gnome Press.
5 Wang, A.Y.-T., Murdock, R.J., Kauwe, S.K. et al. (2020). Sparks: Machine learning for materials scientists: an introductory guide toward best practices. *Chem. Mater.* 32: 4954–4965.

About the Cover

The storm over the Parthenon symbolizes how the advent of artificial intelligence is having a profound impact on our traditional approach of building up human knowledge. STW is one of the few known zeolite frameworks exhibiting chirality, which opens up possibilities for enantioselective absorption, separation and catalysis. BHEHPI is an organic linker molecule used to build the metal-organic framework with the highest specific surface area known. The AI revolution is made possible by high performance computing hardware.

Acknowledgments

First of all we would like to thank Wiley for the call to identify an area of materials chemistry where a new book might be of wide and general interest, and in particular to Sarah Higginbotham who actively supported our initial suggestions in the months following September 2019 when Emma Strickland knocked at our door with her kind and challenging invitation. During this time of almost three years that it took to prepare this book, we have also received active support from other members of the team whom we also would like to thank. Gnanapriya Pattel, who took care of the CTA Agreements, and the other co-editors of the book, Sakeena Quraishi and Stefanie Volk, who showed a continued and enthusiastic support that contributed to keep the ball rolling almost without interruption during the hard pandemic times. And not to forget the hard work carried by Durgadevi Shanmugasundaram, Hemalathaa Krishnamoourthy, and Jo Egre, editing bits and pieces here and there, always with exemplary professionality and good sense of humor.

We were also happily surprised by an invitation by Sakeena to contribute to the cover design and were allowed to explore a huge in-house store of drawings and photographs. After hours, one of them immediately caught our attention, “Storm over Parthenon.” Seldom could a better symbol of human knowledge than the Parthenon be found. And then, the thunderstorm resembles how this traditional paradigm may be jeopardized by a new way of acquiring knowledge, which is Artificial Intelligence. Hopefully the storm will transform into water that will irrigate a new fertile ground for the benefit of mankind. Jennifer Cossham helped so much throughout this process of cover design and allowed us, incompetent amateurs, to put our two cents into the marvelous work made by the designers.

Well, this shows the atmosphere of working hand-in-hand with the Wiley team. We exchanged warm greetings at Christmas and felt almost part of the 100-years Celebration at Wiley VCH[1] in March 2021. E-mails indicated that the Wiley team was mostly located in Germany but during all this time we had a feeling that they were next door to us.

We are especially indebted to authors who took over replacements of chapters that were scheduled but could not be finished and very kindly adapted their schedule to facilitate a timely update. And hence we come to the most important acknowledgment, that due to the authors of this book. Thanks for contributing generously with the best of your research to build a coherent, complete. and overall high-quality view of how Artificial Intelligence is taking the science of materials chemistry by storm.

Finally, thanks to all the people who contributed by revising chapters and suggesting additions and improvements to the book, and also those who gave us confidence and support: Jeffrey D. Rimer (University of Houston), Wei Fan (University of Massachusetts), Rob G. Bell (University College London), Koki Muraoka (The University of Tokyo), Zach Jensen (MIT), Kenta Iyoki (The University of Tokyo), Takahiko Moteki (The University of Tokyo), Yi Li (Jilin University), Hiromasa

Kaneko (Meiji University), Seda Keskin (Koç University), Angeles Pulido (Cambridge Crystallographic Data Centre), Dewi W. Lewis (University College London), Daniele Ongari (Swiss Federal Institute of Technology Lausanne), Aditi Krishnapriyan (Lawrence Berkeley National Laboratory), Andrew Rosen (Lawrence Berkeley National Laboratory), Anubhav Jain (Lawrence Berkeley National Laboratory), Teng Zhou (Max-Planck-Institut für Dynamik Komplexer Technischer Systeme), Andrew M. Beale (University College London), François-Xavier Coudert (Paris Sciences et Lettres University), Valeria Molinero (University of Utah), Johannes Hachmann (University at Buffalo), Steven K. Kauwe (University of Utah), Anthony Wang (University of Waterloo), Mark E. Davis (Caltech), Chris Wilmer (University of Pittsburgh), Keisuke Takahashi (Hokkaido University), and Amanda S. Barnard (Australian National University).

Frits Daeyaert

July 2022 *German Sastre*

Note

1 The Wiley Network; Celebrate with Wiley-VCH – 100 Years of Growing Knowledge; https://www.wiley.com/network/latest-content/celebrate-with-wiley-vch-100-years-of-growing-knowledge

1

The Confluence of Organo-Cations, Inorganic Species, and Molecular Modeling on the Discovery of New Zeolite Structures and Compositions

Christopher M. Lew, Dan Xie, Joel E. Schmidt, Saleh Elomari, Tracy M. Davis, and Stacey I. Zones

Chevron Technical Center, Richmond, California, USA

1.1 Introduction

The use of quaternary ammonium cations as organic structure-directing agents (OSDAs) in zeolite synthesis dates back to well over half a century. They are perhaps the greatest driver in discovering new zeolite frameworks and compositions, but at the same time are limited by their cost, which can prevent economic commercial-scale production of new materials [1]. There are already many fine reviews on zeolite synthesis and structure direction, and Table 1.1 contains a part of a non-exhaustive list of these. The purpose of this review is to highlight a subset of zeolite synthesis reactions where subtle changes in the synthesis variables lead to new structures and compositions. Moreover, we elucidate the role of molecular modeling to help us understand the results and to guide new experiments. While many materials found with these minor perturbations tend to be difficult to access synthetically, their unique properties can give them superior performance in numerous applications, which can catapult them from a laboratory curiosity to a high-demand commercial product. Perhaps there is no better modern example of this than the use of SSZ-13 as an automotive deNO_x catalyst, and the story of that material is well-known in the zeolite community (the SSZ-designation is used for novel materials prepared by Chevron) [2].

A timeline of zeolite development is given in Table 1.2, and the first synthetic zeolites date back to around 1950 with the pioneering inorganic work of Richard Barrer and Robert Milton, the founding fathers of synthetic zeolites, and a little later with Donald Breck and Edith Flanigen [3]. Over 20 synthetic zeolites were quickly discovered, but the first zeolite patent was not granted until 1959, even though the application was filed in 1953 [4]. The first reported organics for zeolite synthesis were simple methylammoniums used to produce higher silica versions of zeolites A and X, as was first reported in 1961 and followed later by patents [5–9]. Once the genie was out of the bottle with regard to this innovation, it was quickly followed by reports of zeolite beta in 1967 [10] and ZSM-5 in 1972 [11]. To this day, these zeolites remain two of the high-silica zeolites with the largest industrial applications and have been the subject of innumerable patents and research papers. Some of the most notable early organizations in the zeolite field were Linde, Mobil, Grace, and Union Carbide, and their labs discovered and commercialized a remarkable number of materials in a short time. A great amount of progress has been made in the field in the past 50 years, and herein we will give an overview of zeolite synthesis variables, with special attention paid to OSDAs, to highlight how subtle perturbations to these variables can lead to significant differences in the crystalline product.

AI-Guided Design and Property Prediction for Zeolites and Nanoporous Materials, First Edition. Edited by German Sastre and Frits Daeyaert.

Table 1.1 A non-exhaustive listing of review papers covering zeolite synthesis and structure direction.

Title	Authors	Year	References
Zeolite and Molecular Sieve Synthesis	Davis, M.E. and Lobo, R.F.	1992	[35]
Synthesis of Porous Silicates	Helmkamp, M.M. and Davis, M.E.	1995	[127]
Searching for New High-Silica Zeolites Through a Synergy of Organic Templates and Novel Inorganic Conditions	Zones, S.I., Nakagawa, Y., Lee, G.S. et al.	1998	[16]
Synthesis of All-Silica and High-Silica Molecular Sieves in Fluoride Media	Camblor, M.A., Villaescusa, L.A., and Diaz-Cabanas, M.J.	1999	[128]
Ordered Porous Materials for Emerging Applications	Davis, M.E.	2002	[129]
Towards the Rational Design of Zeolite Frameworks	Wagner, P. and Davis, M.E.	2002	[130]
The Hydrothermal Synthesis of Zeolites: History and Development from the Earliest Days to the Present Time	Cundy, C.S. and Cox, P.A	2003	[3]
The Hydrothermal Synthesis of Zeolites: Precursors, Intermediates, and Reaction Mechanism	Cundy, C.S. and Cox, P.A.	2005	[131]
The Chemistry of Phase Selectivity in the Synthesis of High-Silica Zeolites	Burton, A.W., Zones, S.I., and Elomari, S.	2005	[80]
The Fluoride-Based Route to All-Silica Molecular Sieves: A Strategy for Synthesis of New Materials Based Upon Close-Packing of Guest–Host Products	Zones, S.I., Hwang, S.J., Elomari, S. et al.	2005	[23]
Zeolite Molecular Sieves: Preparation and Scale-Up	Casci, J.L.	2005	[132]
Organic Molecules in Zeolite Synthesis: Their Preparation and Structure-Directing Effects	Burton, A.W. and Zones, S. I.	2007	[31]
Present and Future Synthesis Challenges for Zeolites	Coronas, J.	2010	[133]
Inorganic Molecular Sieves: Preparation, Modification and Industrial Application in Catalytic Processes	Martínez, C. and Corma, A.	2011	[134]
Zeolites: From Curiosity to Cornerstone	Masters, A.F. and Maschmeyer, T.	2011	[135]
Needs and Trends in Rational Synthesis of Zeolitic Materials	Wang, Z., Yu, J., and Xu, R.	2012	[136]
Towards the Rational Design of Efficient Organic Structure-Directing Agents for Zeolite Synthesis	Moliner, M., Rey, F., and Corma, A.	2013	[137]
New Trends in the Synthesis of Crystalline Microporous Materials	Bellussi, G., Carati, A., Rizzo, C. et al.	2013	[138]
New Stories of Zeolite Structures: Their Descriptions, Determinations, Predictions, and Evaluations	Li, Y. and Yu, J.	2014	[139]
Zeolites From a Materials Chemistry Perspective	Davis, M.E.	2014	[140]
Synthesis Strategies for Preparing Useful Small Pore Zeolites and Zeotypes for Gas Separations and Catalysis	Moliner, M., Martinez, C., and Corma, A.	2014	[141]
Synthesis of New Zeolite Structures	Li, J., Corma, A., and Yu, J.	2015	[32]
Introduction to the Zeolite Structure-Directing Phenomenon by Organic Species: General Aspects	Gómez-Hortigüela, L. and Camblor, M.A.	2017	[142]
Small-Pore Zeolites: Synthesis and Catalysis	Dusselier, M. and Davis, M.E.	2018	[2]

Table 1.2 Zeolite timeline.

Year	Discovery
Antiquity	Used in Roman aqueducts and Mayan filtration systems
1756	Axel Fredrick Cronstedt coins the term "zeolite," meaning boiling stone
1905	Used commercially to soften water
1930s	Pioneering work of Richard Barrer on adsorption and synthesis (**MOR**, **KFI**)
1950s	Milton and Breck discover A, X, Y
1954	Union Carbide commercializes synthetic zeolites for separations
1962	Mobil Oil uses zeolite X for cracking
1967–1969	Mobil reports beta and ZSM-5
1969	Grace reports USY
1974	Henkel introduces zeolite A in detergents
1977	Union Carbide commercializes ion-exchange separations
1980s	AlPOs discovered by Union Carbide. Heteroatoms: Fe, Ga, Ti, Ge. Secondary modifications
1990s	MeAPOs, MOFs, mesoporous materials
2000s	Charge density mismatch aluminosilicates, germanosilicate compositions, fluoride media
2010s	ADOR, 2D materials, chirality, dry gel conversion, computational advances, small-pore $deNO_x$
2020s	Machine learning, data mining of historical data, high-throughput experiments

Source: Adapted with permission from [143].

1.2 Inorganic Studies

In the preface of his recent monograph, "Insights into the Chemistry of Organic-Structure Directing Agents in the Synthesis of Zeolite Materials," Luis Gomez-Hortiguela states that the activity in this area has largely been based upon empirical observations and then decisions, but that the field is emerging into new opportunities with advanced computer modeling [12]. This comparison is, in essence, the relationship of our writing this first chapter in this collection, which then moves on to a number of chapters that emphasize these computational advances. Our experience in looking for ways to discover new zeolite materials spans this time range from about 1980 to the present. While our team at Chevron, over these decades, has discovered a number of novel zeolite materials, either structures, compositions, or other physical properties that have centered on the creation and use of novel organo-cations (which we will give some history of later in this chapter), we have also been influenced by some fundamental observations that emerged in other research studies. In particular, a study by Navrotsky et al. enlightened us that while there are great differences in the three-dimensional (3D) structures of zeolites, their differences in enthalpies of formation are remarkably small [13]. That tells us that how you arrive at stabilizing a structure is going to be a key factor in which a zeolite can be recovered in a synthesis. In fact, an important assessment was made by Navrotsky herself when she stated, "Rather than stabilizing an otherwise unstable structure, the template selects one framework from a variety of possibilities of similar energy" [14]. In fact, over the years, we have found that a novel and unusual OSDA, the N,N,N-trimethyladamantammonium molecule shown in Figure 1.1, can make a half-dozen zeolites as the reaction conditions change [15–17]. In addition, the same OSDA produces MOR when used in conjunction with 1,2-hexanediol [18]. What are some of these factors that result in producing different zeolite structures?

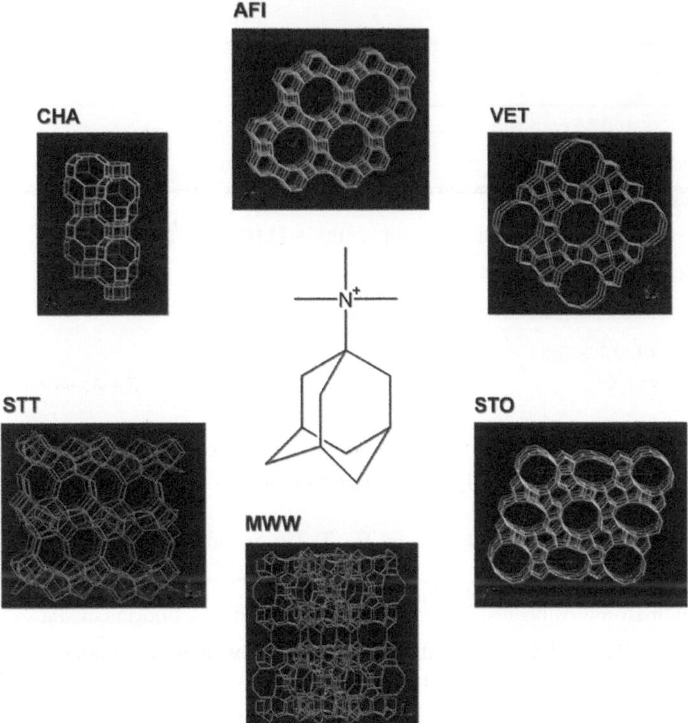

Figure 1.1 Zeolites made from N,N,N-trimethyladamantammonium and its structure [16, 126].

A key experiment for our work was the findings shown in Figure 1.2. Here we see that two different paths, each with the same net overall stoichiometry, will produce two different zeolites. Parenthetically, these are examples with the use of the N,N,N-trimethyladamantammonium OSDA just mentioned. This discovery would be in line with the Navrotsky statement above. Using soluble boron, alkali, and a silica source did not lead to the same zeolite structure as supplying the silica and boron via calcined boron beta. The boron beta source does not dissolve at the outset of the reaction but may provide an anchoring site for the organo-cation to then begin to nucleate a new zeolite, SSZ-24 (**AFI**) in this instance. We then extended this synthesis approach to make a number of interesting novel borosilicate zeolites [16].

If we conclude that in many instances a variety of zeolites can be made from a single OSDA, then what we are considering here is the context of the synthetic reaction. What are the events surrounding the use of the OSDA? This will then include considerations like the rest of the synthesis reaction, particularly the other inorganic chemical contributors. For the synthesis of many zeolite materials, basic media is needed. We need to contribute the hydroxide, which helps to drive the synthesis as a component known as a mineralizing reagent, and this can be done with contributions from either alkali cation hydroxides or from the OSDA itself. On the one hand, enough hydroxide is needed to be able to dissolve some of the silica, while keeping the system under conditions of supersaturation. If our ratio of hydroxide to silica were too high, all the silica would dissolve, be depolymerized from its contributing source, and there would be no drive toward nucleation and crystallization as the silicates would remain nicely dissolved and solvated. On the other hand, too much alkali can cause issues in the synthesis of zeolites with a desired SiO_2/Al_2O_3 ratio of 10 or beyond. Over a range of OH to SiO_2 ratio, zeolite mordenite or sodium silicate

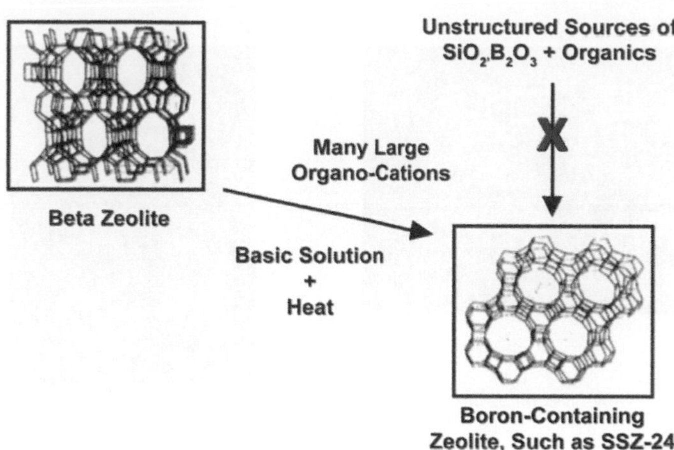

Figure 1.2 The graphic shows that the use of boron beta as a reagent for both B and Si leads to the novel zeolite, B-SSZ-24, whereas a synthesis with the same reagent ratios and solubilized reagents would lead to a different product. Thus, the reaction path matters. *Source:* Reproduced from [16], with permission from Elsevier.

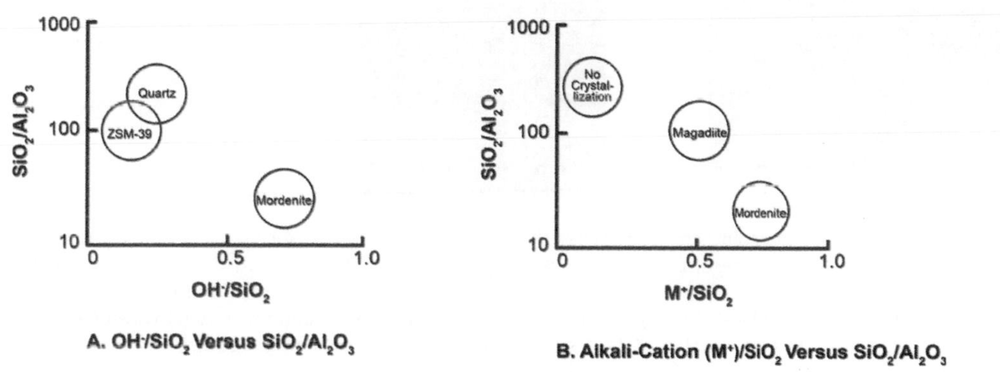

Figure 1.3 Magadiite/mordenite diagrams. Here we can see the gap region of OH/Si around 0.3 where the use of the OSDA to provide much of the OH content prevents either magadiite or mordenite, neither of which will use the OSDA, from dominating the crystallization. *Source:* Reproduced from [19] with permission of Taylor & Francis Group LLC – Books; permission conveyed through Copyright Clearance Center, Inc.

Magadiite (on the lower OH end) become favored where the OSDA is not even used in the synthesis [19]. It is in dealing with these undesirable competing phases in zeolite synthesis (Figure 1.3) that our work moved to using the OSDA as a hydroxide source and typically providing the bulk of it, with the alkali contributors a minor reagent. This provided a successful pathway to a number of new zeolite structures with higher SiO_2:Al_2O_3 (silica-to-alumina ratio or SAR) values that were nonetheless still very active catalysts. This demarcation into the world of higher silica zeolites also brought an increase in products whose sub-unit structures were rich in 5-rings. In fact, a number of the earliest discovered materials were termed pentasils for the preponderance of this type of structure. We will return to this type of change in structure, seen by the sub-ring construction of the zeolite, when we discuss the influence of aluminum on zeolite structure formation.

Figure 1.4 shows four zeolite materials, and they are representative of the general trends observed when we make the following adjustments to zeolite synthesis:

1) Increasing the amount of Al in the synthesis leads away from pentasil-type structures to multi-dimensional pore systems more often seen for either large-pore or small-pore zeolite products. One factor here is that the length of the bond for Al–O versus Si–O may alter the preferred bond

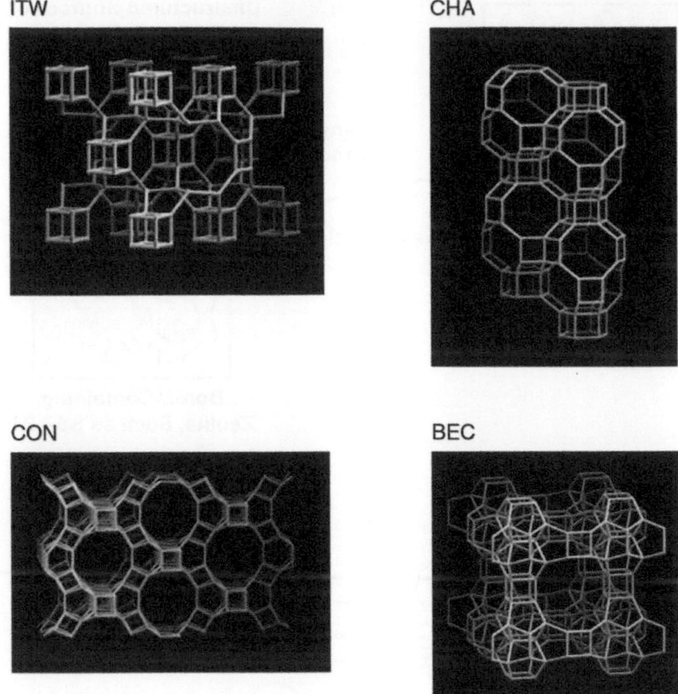

Figure 1.4 Four different zeolite structure types that are influenced by the inorganic chemistry. **ITW** is made when there is fluoride present to create these double 4-rings. **CHA** is a multi-dimensional small-pore zeolite favored by higher Al and OH contents in the synthesis. **CON** is a more typical high-silica zeolite, as the Al and OH values are reduced in a synthesis with the OSDA. **BEC** is a frequent product when Ge is introduced into the syntheses, which also favors the formation of double 4-rings [126].

angles for its siting in the zeolite structure and that may impact the type of zeolite structure that develops around the OSDA.

2) The introduction of boron also has consequences if enough is incorporated into the structure. The boron bond length is short, and it is not unusual for the boron to revert to trigonal in the structure. It is common for boron to make this reversion once the OSDA is removed under calcination.
3) A very large impact on the product selectivity emerged when the Corma group at ITQ developed a reaction with sizable amounts of mineralizing reagent based upon fluoride ions instead of hydroxide, and the key parameter of working with very low water concentrated reaction conditions [20,21]. What we were struck by is the ability to generate five-coordinate Si in these reactions (which sometimes maintained its presence right into the final products [22]), which then also changed the bond angles in the emerging tetrahedra. For the first time, this allowed the development of multi-dimensional zeolites of a variety of structures that were either all-silica or contained very low amounts of Al. Prior to that, the researchers just had ZSM-5 or -11 to work with for separations or catalysis.
4) After the fluoride breakthrough from the Corma group, they discovered that a large series of novel structures (labeled as ITQ zeolites) could be found by introducing germanium into the synthesis where it could substitute for Si. With time, it was recognized that the optimal synthesis systems here also contained low water content and a sizable Ge contribution to the reactants. Upon closer analysis of the novel zeolite products, it was also recognized that the Ge

preferred to reside in a double 4-ring type of subunit. This, in turn, contributed to the preponderance of multi-dimensional zeolite products. A good example of these events is the **BEC** structure (ITQ-17) that is shown in Figure 1.4. With time, it has been found that a large number of sizable OSDAs ($C/N^+ > 14$) were very good at producing **BEC** (polymorph C) when Ge was present. Without Ge, the same OSDA then could lead to either ***BEA** (polymorph A) itself (with enough Al in the synthesis) or **MTW** if it could fit into the one-dimensional (1D) 12-ring [23].

We have mentioned that the alkali cation content could alter the zeolite synthesis in such a way that the OSDA could be left a spectator as either Magadiite might dominate (lower OH to Si ratio) or mordenite (higher OH to Si ratio). However, there is a more subtle zone of activity that involves the OSDA when we move to synthesis conditions that include OSDA, Na, and additional cations like K. As seen in Table 1.3, Xie found some interesting intergrowth effects in the ABC-6 family of small-pore zeolites [24]. The ratio of Na to K then changes the intergrowth frequency in such important zeolite products like **LEV** to **ERI**.

A further reaction parameter needs mentioning at this point when we consider the use of the OSDA, which we imagine will be key to the formation of a stable nuclei, based upon a good fit in a host lattice, leading to the eventual crystal growth. In a recent publication, we described how over a number of years we have used **FAU** zeolites as a source of first just Al, and then later, the combined Si and Al contributors as the **FAU** SAR became high enough [25]. The implementation of these materials in a synthesis combined with the OSDA and a mineralizing source (either OH or F) has led to the discovery of some novel materials for us. Take, for example, our highly unusual small-pore, large-cavity zeolite SSZ-27 that best forms when the particular isomer of a diquaternary ammonium compound is used (Figure 1.5). If we think back to the reaction comparison described above with the conversion of boron beta to borosilicate zeolites, perhaps there is a commonality. In a recent study that focused on the synthesis rates of converting the **FAU** zeolite to zeolite SSZ-13 (**CHA**) by comparing a series of increasingly larger OSDAs, the rates correlated well with the favorable fit calculated for each OSDA in the product host [26]. However, we also noticed that we could find some OSDA present in the **FAU** zeolite prior to its conversion to SSZ-13. So did the **FAU** provide an anchoring point for the nucleation to begin? This is a subject of current study for our group.

What we have seen over time is that the OSDA fills the void space of the eventual zeolite product that forms. The better the fit (and the energetics are dominated by vdW interactions), the more likely the OSDA will be selective for a given structure over some number of possibilities that the

Table 1.3 Table of discoveries, cations, and intergrowths by the Chevron zeolite group.

Zeolite	Intergrowth	Key alkali cations	Patent
SSZ-98	**ERI**	K	US Patent 9,409,786
SSZ-99	**CHA/GME**	Na	US Patent 9,192,924
SSZ-101	**CHA/GME**	Na	US Patent 9,505,626
SSZ-102	**ESV**	Na	US Patent 9,598,289
SSZ-103	**AFX**	Na	US Patent 9,908,108
SSZ-104	**CHA/GME**	Na	US Patent 9,725,328
SSZ-105	**ERI/LEV**	Na + K	US Patent 9,663,380
SSZ-107	**CHA/AFX**	Na	US Patent 9,981,853
SSZ-108	**CHA/AFX**	Na	US Patent 10,221,073

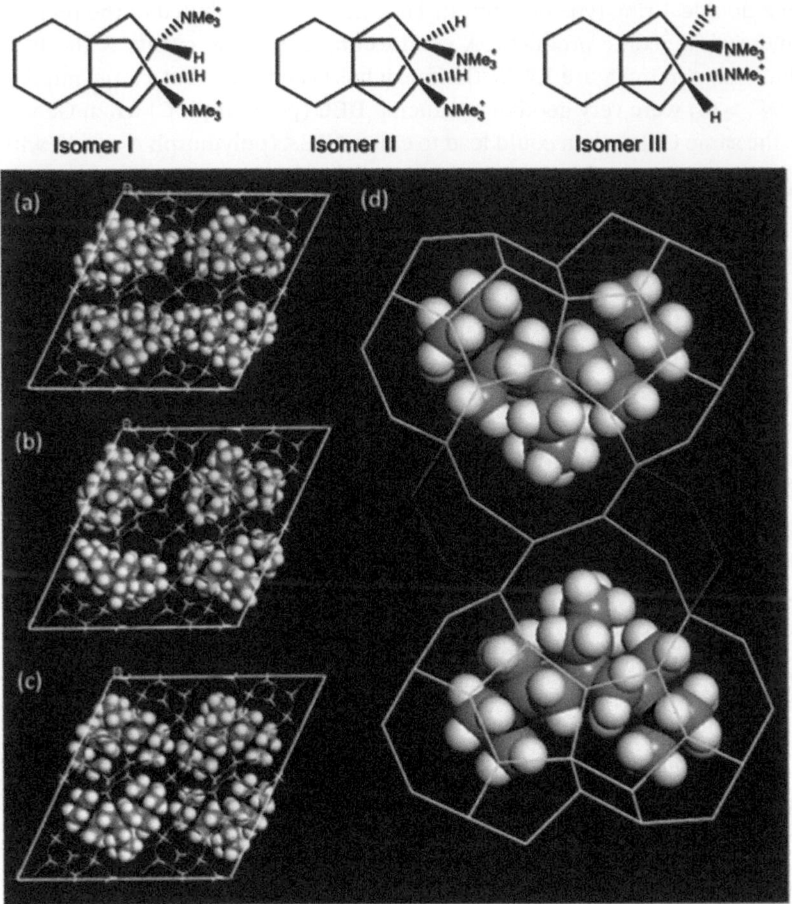

Figure 1.5 OSDA and structure for SSZ-27. The choice of isomer matters in specifying the small-pore, large-cage SSZ-27, over the large-pore, multi-dimensional SSZ-26 (**CON**). (a)–(c) = I–III. (d) is the isomer III in the cage of SSZ-27 [89].

inorganic conditions could make favorable [26]. In extreme cases, we might find a given OSDA that makes only one zeolite structure (as has been seen for the use of the OSDA to make SSZ-52: Figure 1.6) [27,28]. More often, we find that a given OSDA can make more than one product as the reaction context changes. In the recent study mentioned on the conversion of **FAU** to SSZ-13, the trimethyl adamantylammonium cation gives the fastest formation rates and is among the highest favorable energetics for fit from the calculations, and yet even this OSDA can make six zeolite structures [15–17]. As we show later in this chapter, we have developed a number of series of OSDAs where an organic synthesis methodology has been employed. In each series, we may create a central feature, such as the imidazole ring, and then modifications of the substituents of the key central component lead to differences in the eventual zeolite product found, if one will crystallize. However, in each case, once we know the structure, we can rationalize how the OSDA fits in the zeolite. Of course, we can then be aided by increasingly better modeling methods and the emerging use of newer diffraction methods that can localize the details of the OSDA [29].

The interplay of kinetics and thermodynamics in zeolite synthesis is complex, and the studies are made more challenging by the system contexts. pH values above 11 and temperatures high enough

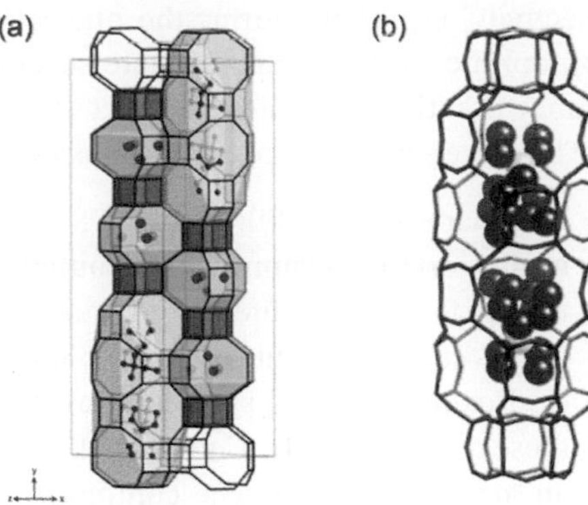

Figure 1.6 The arrangement of OSDA in the cages of the SSZ-52 structure (a) and then space-filling by the OSDA in the large cage (b). One of the surprises in this study was that the very large cage of SSZ-52 actually contains two large OSDAs. *Source:* Reproduced with permission from [28]; © 2013, American Chemical Society.

to create pressurized closed reactors are inhospitable to real-time reaction measurements. In high-silica zeolite syntheses, the bulk of the reaction time is an induction period that generates stable nuclei for crystal growth to proceed. If we take into account the experimental data and observations from Navrotsky (which we have pointed out above), zeolite synthesis becomes a much greater kinetics problem, more than arriving at a "most stable zeolite product" with the OSDA. From this angle, we can then note the following outstanding issues, whose answers are mostly unknown to us:

1) Is having an anchor point for the charged OSDA a key point in the nucleation?
2) How are water molecules in the primary solvation shell around the OSDA eventually replaced by silica as the nuclei get created and stabilized? In theory, this would be measured by solution viscosity changes if we had a sensitive enough technique.
3) Why are elevated temperatures required to drive these events? It is reasonably clear that mineralizer hydrolysis has already reached an equilibrium before nucleation is seen.
4) How is the charge of the OSDA compensated through the sequence of transformations from solvated organo-cation to eventual, tight-fitting OSDA in a host lattice? Remember that accompanying water molecules are rarely seen as part of the guest–host final crystalline product.

1.3 Organic Structure-Directing Agents (OSDAs)

Numerous research groups have written review articles on OSDAs, and we refer the reader to several of these [30–34]. In this section we focus on aspects of OSDA synthesis and application that we have found useful in our zeolite program at Chevron.

1.3.1 Purpose and Important Properties

Organic molecules serve a variety of functions during zeolite synthesis. Lobo and Davis categorized organics into three main roles: 1) templates; 2) structure-directing agents; and 3) space fillers [35]. As a template, the geometric shape of the void space in the zeolite framework typically matches well with that of the organic. In these cases, the non-bonded energy (primarily van der Waal's interactions) between the framework and the organic is at a minimum. A classic example of templating is the synthesis of ZSM-18 (**MEI**) using triquaternary ammonium cations that have the same three-fold rotational symmetry as the ZSM-18 cage [36]. Interesting examples from Stucky et al. and Davis et al. illustrate cases where the framework symmetry and chirality, respectively, are determined by the organic [37,38]. On the other hand, when a specific organic plays a role in phase selectivity, especially during the nucleation process, the concept of structure direction is more appropriate. In other cases, where several organics can be used in the synthesis of one zeolite framework, these organics likely act as space fillers.

There are several properties to consider when choosing an OSDA for zeolite synthesis:

1) The shape and flexibility of the OSDA often determine various structural features of the zeolite framework. For example, large, bulky organics often form cage structures, while linear OSDAs tend to form 1D zeolites. Alkylamines and other organics with multi-dimensional hydrocarbon chains can form frameworks with intersecting channels [39,40].
2) Hydroxide-mediated syntheses result in high pH gels, and the OSDAs must be stable under these conditions. The Hofmann elimination reaction of a quaternary ammonium compound to an amine and an alkene commonly occurs under basic and high temperature conditions. However, degradation products can sometimes act as the effective OSDA, which can lead to either surprising discoveries or unwanted impurities.
3) An intermediate hydrophobicity, which is well-correlated to the C/N^+ ratio, maximizes the effectiveness of the OSDA, especially when synthesizing high-silica zeolites. Davis et al. found that a C/N^+ ratio between 11 and 16 strikes a good balance between solubility in the aqueous synthesis solution and favorable interactions with silicate species [41]. Zones et al. found that the C/N^+ ratio correlates with the kinetics of crystallization, although certain conditions that allow for better interactions between the OSDA and the inorganic species can overcome limitations to OSDAs that fall outside of the optimum C/N^+ range [26].

1.3.2 Classes of Ammonium-based OSDAs

Zeolite researchers have had success with OSDAs of a wide variety of shapes and sizes, and we refer the reader to another reference that discusses recent innovations [33]. Here, we highlight several classes of ammonium-based OSDAs that we have worked with over the years at Chevron:

1) Rigid Polycyclic OSDAs
   As first outlined by Liebau and Gies [39,40], empirical observations about the types of organic molecules used and the resulting zeolite frameworks reveal several domains that are broadly governed by the amount of heteroatom substitution and the size and shape of the organic [42]. High-silica gels with small organic molecules tend to form clathrates. As the organic becomes too large for a clathrate, 1D-, medium-, and large-pore zeolites become the next class of frameworks to form. Increased amounts of lattice substitution with these organics or the use of larger organic molecules often results in multi-dimensional zeolites. At even higher amounts of heteroatom substitution, small-pore, cage-based frameworks form. Because large organics with a high degree of flexibility can sometimes lead to 1D zeolites (e.g., ZSM-12 [**MTW**]), we purposefully developed classes of rigid, spherical-shaped OSDAs for the purpose of forming cage-based zeolites. Studies by Zones et al. and Davis et al. used polycyclic OSDAs which resulted in SSZ-13, SSZ-35 (**STF**), SSZ-36 (**RTH/ITE** intergrowth), and SSZ-39 (**AEI**) [42]. Zones et al. also used a family of $[5.2.1.0^{2.6}]$tricyclodecane derivatives to make SSZ-31 (***STO**), SSZ-33 (**CON**), and SSZ-37 (**NES**) [43]. Finally, we used a series of fused bicyclo [l.m.0] cations that similarly produced SSZ-13, SSZ-31, SSZ-35, SSZ-36, SSZ-43, and SSZ-48 (**SFE**) [44]. A few of these examples are shown in Table 1.4. As previously discussed, one of the OSDAs studied, N,N,N-trimethyladamantammonium, can be used to make SSZ-13, SSZ-23 (**STT**), SSZ-24, SSZ-25 (**MWW**), SSZ-31, and VPI-8 (**VET**) [16]. Of the structures listed in this section, five of them are cage-based, small-pore zeolites (SSZ-13, SSZ-16 [**AFX**], SSZ-17 [**LEV**], SSZ-36, and SSZ-39), and four are 1D, 12 MR (SSZ-24, SSZ-31, SSZ-48, and VPI-8). An unusual exception is SSZ-23, which contains 9 MR.

Table 1.4 Examples of rigid polycyclic OSDAs and some of their resulting products.

OSDA	Zeolite product(s)	References
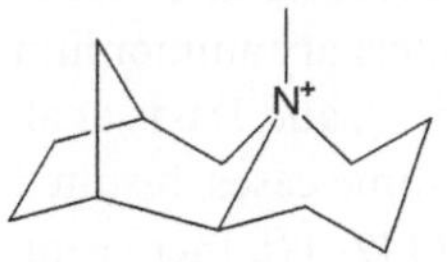 N⁺	SSZ-13, SSZ-31, SSZ-35	[42]
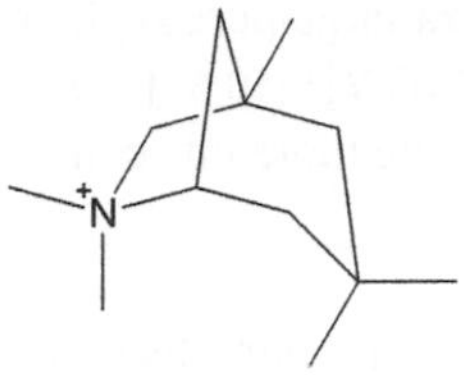 N⁺	SSZ-31, SSZ-35, SSZ-36, SSZ-39	[42]
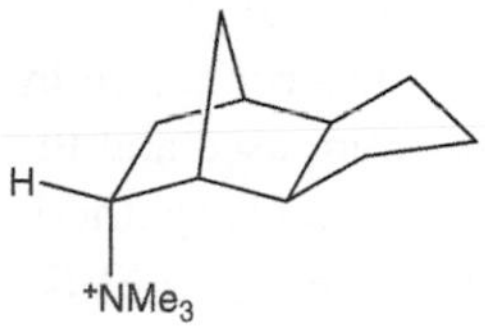 H, ⁺NMe₃	SSZ-31, SSZ-33	[43]
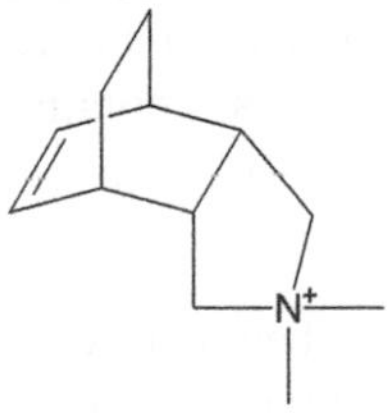 N⁺	SSZ-31, SSZ-33, SSZ-37	[43]
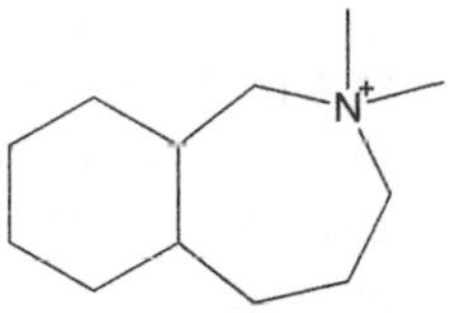 N⁺	SSZ-13, SSZ-31, SSZ-37	[44]
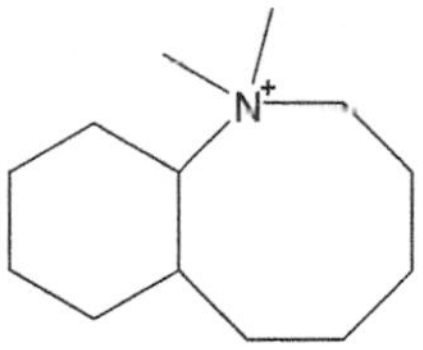 N⁺	SSZ-31, SSZ-36	[44]
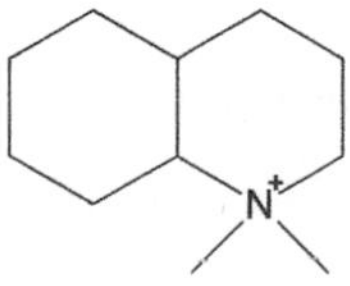 N⁺	SSZ-13, SSZ-35, SSZ-36, SSZ-43	[44]
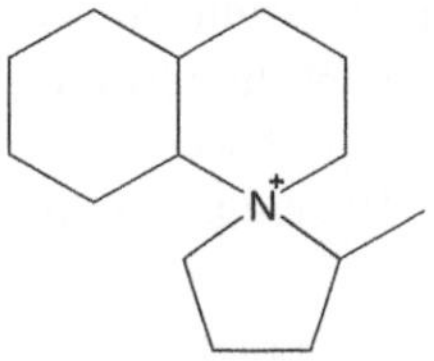 N⁺	SSZ-31, SSZ-48	[44]

2) Imidazoles
   Many research groups have looked at imidazole derivatives as OSDAs. Our early work on imidazoles with isopropyl substituents or smaller resulted in primarily 1D zeolites (ZSM-12, ZSM-22 [**TON**], ZSM-23 [**MTT**], and ZSM-48 [***MRE**]) [45]. Curiously, by connecting these same imidazole derivatives with alkyl chains of varying length and creating diquaternary ammonium OSDAs, the same zeolites were produced (except ZSM-23) [46]. Camblor et al. and Davis et al. found that different inorganic conditions, including the use of fluoride in some cases, resulted in small- and medium-pore zeolites, primarily **ITW**, **RTH**, **STF**, and **STW** [47–49]. Increasing the size of the imidazole substituent resulted in a number of large- and extra-large-pore zeolites (CIT-5 [**CFI**], CIT-13 [***CTH**], ITQ-26 [**IWS**], ***BEA**, **BEC**, and SSZ-70 [***-SVY**]) [50,51]. As is often the case, phase selectivity was governed by both the OSDA type and the inorganic conditions.
3) Diquaternary Ammonium Compounds (Diquats)
   Diquats have been used by the zeolite community since their original use in the 1990s by Casci at Imperial Chemical Industries and Moini et al. at Mobil [52–54]. Several of our studies have focused on the use of symmetric OSDAs consisting of heterocyclic end groups connected by alkyl chains of varying length. In one such investigation, five end groups were used and the chain lengths were varied from C4 to C6. These OSDAs were used in 9 inorganic synthesis conditions, and 18 zeolite structures were formed, including 3 novel materials, SSZ-74 [**-SVR**], SSZ-75, and SSZ-81 [55,56]. With the exception of a couple of clathrates and SSZ-16, the products were 10 and/or 12 MR zeolites. Another study using end groups made through an enamine chemistry (see Section 1.3.3) resulted in a novel material, SSZ-82 (**SEW**) [57]. SSZ-82 contains intersecting 12 MR and 10 MR channels, and the enamine end groups are located in the 12 MR channels, while the bridging C6 chain is located along the 10 MR channels.

The synthesis of SSZ-16 presents a noteworthy case study for the relationship between templating effects and charge density of the organics. The original patent by Zones for the aluminosilicate version of SSZ-16 used a diquat consisting of two DABCO molecules connected by a C4 alkyl chain (1,4-di(1-azoniabicyclo[2.2.2]octane)butyl dibromide) [58]. SSZ-16, however, could not be made with an Si/Al ratio higher than ~8. Davis used an imidazolium derivative with two adamantane substituents on either end; the length and shape of the new OSDA closely resembled that of the original DABCO OSDA but only had one charge instead of two [59]. With fewer charges requiring compensation by aluminum in the framework, the Si/Al could be increased to ~45.

1.3.3 Methods of Making

We have investigated a number of chemistries to synthesize quaternary ammonium compounds. Most of the methods we have studied involve some sort of addition or transformation to make an amine that is then quaternized using a Menshutkin reaction. Several of these chemistries were used to make OSDAs that were subsequently studied under fluoride conditions [23]:

1) Diels-Alder
   Nakagawa et al. used the Diels-Alder reaction to combine monocyclic compounds into polycyclic amines by reacting a diene with a substituted alkene (dienophile). Many of the resulting OSDAs were of the rigid, polycyclic class of organics as discussed above and produced either cage-based or 1D zeolite structures. The novel phase, SSZ-35, is a 1D, 10 MR zeolite that resulted from this chemistry (Scheme 1.1).

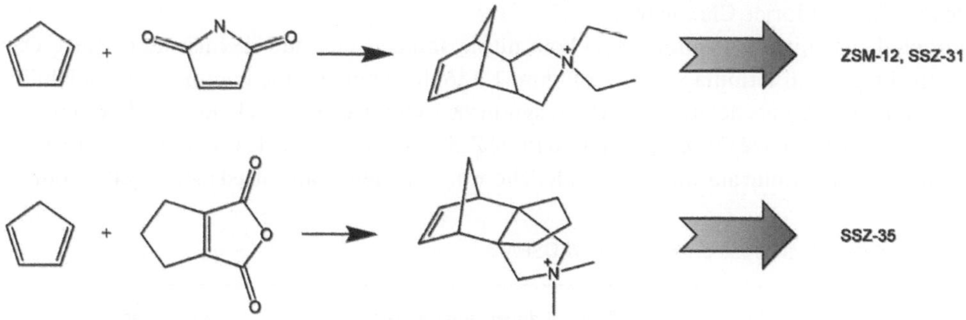

ZSM-12, SSZ-31

SSZ-35

Scheme 1.1 Two OSDAs made using the Diels-Alder chemistry and their resulting zeolite phases (not all the organic synthesis steps are shown). *Source:* Adapted with permission from [42]; © 2000, American Chemical Society.

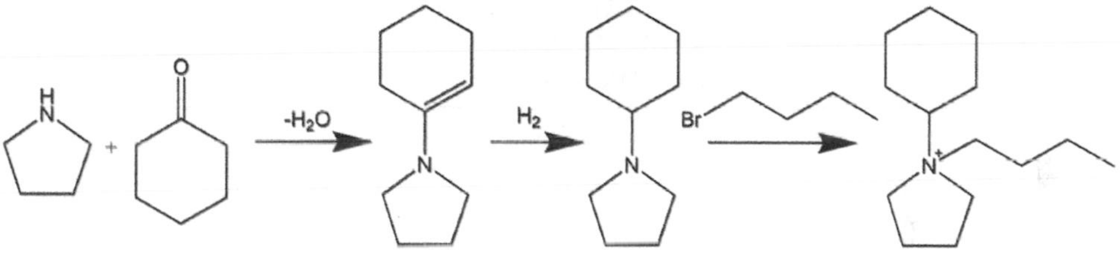

Scheme 1.2 An example of using enamine chemistry to make the OSDA for SSZ-57.

2) Michael Addition and Beckmann Rearrangement
   Polycyclic ketones were created through the use of the Michael addition reaction, and nitrogen was introduced by way of a Beckmann rearrangement. Similar to the Diels-Alder products, this chemistry also produced rigid, polycyclic OSDAs. Several of these OSDAs were selective for SSZ-35 and SSZ-36 [42]. Furthermore, when used in minor amounts with isobutylamine as the major organic component, they exhibited structure-direction capabilities for SSZ-25 [60]. The SSZ-25 study was a nice example of a method we have championed to introduce a cheaper amine to reduce the overall costs of the organics required to make a zeolite.

   We also used the Beckmann rearrangement (without the Michael Addition) to study a series of bicyclo OSDAs that resulted in several phases, including novel zeolites SSZ-43 and SSZ-48 [44]. In these studies, the position of the nitrogen in the ring and the resulting symmetry or asymmetry of the OSDA had an influence on the phase selectivity: symmetric OSDAs resulted in ***BEA** or ZSM-12, while less symmetry produced cage-based zeolites. In addition, the *cis* and *trans* isomers of the OSDAs, which resulted from products of the Beckmann rearrangement, were often extremely phase-specific. Careful separation of the isomers needed to be performed in order to get a pure-phase zeolite product.
3) Enamines
   Elomari et al. explored the use of enamine chemistry to primarily combine cyclic amines with cyclic ketones. Scheme 1.2 shows the synthesis of the OSDA used for making SSZ-57 (***SFV**). First, pyrrolidine is combined with cyclohexanone to form the unsaturated enamine. The enamine is then hydrogenated, and the tertiary amine is quaternized with 1-bromobutane. Table 1.5 shows the novel SSZ materials that have been made using the enamine chemistry.

4) Nitrile and Acyl Chloride Chemistries
Elomari et al. also investigated the use of both nitrile and acyl chloride chemistries to make OSDAs with the general formula shown in Figure 1.7 [61]. Generally, the commercial availability of the precursor reagents determined which synthesis route was used. Scheme 1.3 describes the synthesis of a nitrile-based OSDA that is used for SSZ-55 (**ATS**). The nitrile is first reduced to a primary amine using lithium aluminum hydride. The amine is then quaternized using methyl iodide.

Table 1.5 SSZ materials that use enamine-based OSDAs.

Material	IZA framework type code	Channel dimensionality (Pore size)	References
SSZ-56	**SFS**	2D (12 MR × 10 MR)	[144]
SSZ-57	***SFV**	3D (12 MR × 10 MR × 10 MR)	[145]
SSZ-58	**SFG**	2D (10 MR)	[29,146]
SSZ-60	**SSY**	1D (12 MR)	[29,147]
SSZ-63	–	likely 3D (12 MR)	[148]
SSZ-82	**SEW**	2D (12 MR × 10 MR)	[149]

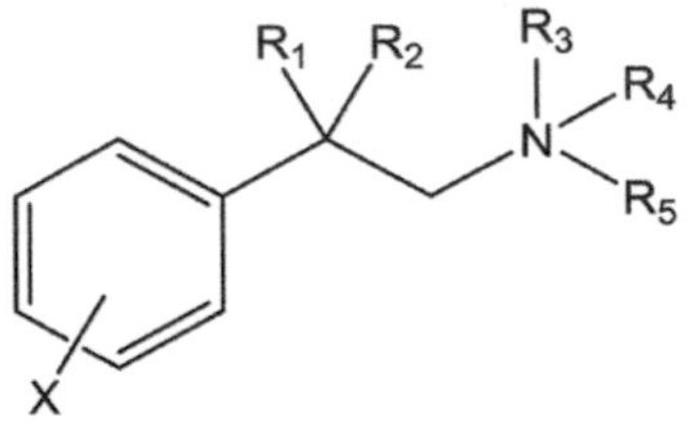

Figure 1.7 General formula for OSDAs generated from nitrile and acyl chloride chemistries. X is -H, methyl, -F, -Cl, -F, and -Cl, or methoxy; R_1 and R_2 are each methyl or ethyl; R_1 and R_2 together are $-(CH_2)_x-$, where x is 2, 3, 4, or 5; or R_1 and R_2 together are methylated or dimethylated $-(CH_2)_y-$, where y is 3, 4, or 5; and R_3, R_4, and R_5 are each methyl or ethyl, or one of R_3, R_4, or R_5 is methyl and the other two together are $-(CH_2)_z-$, where z is 4, 5, 6, or 7.

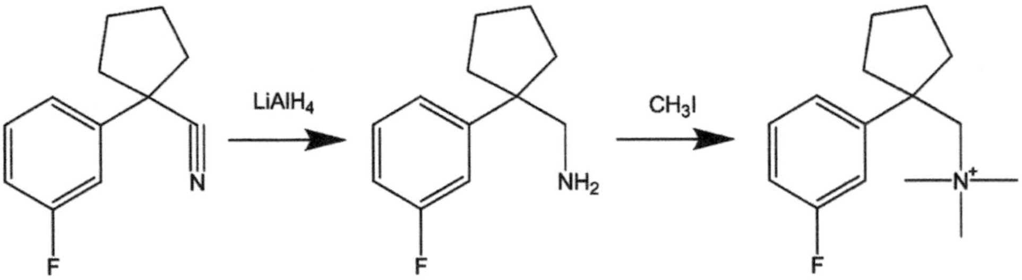

Scheme 1.3 An example of using nitrile chemistry to make the OSDA for SSZ-55.

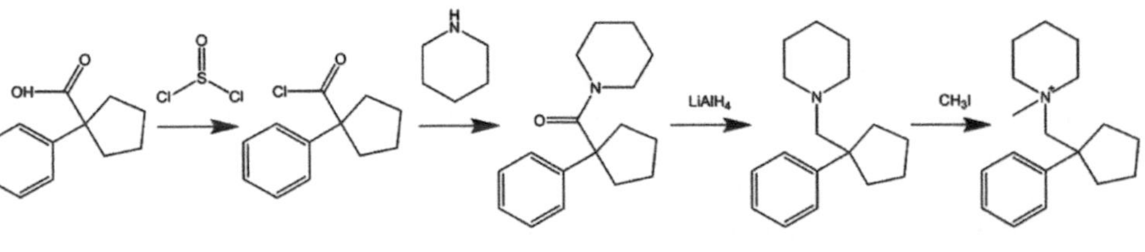

Scheme 1.4 An example of using acyl chloride chemistry to make the OSDA for SSZ-59.

Table 1.6 SSZ materials that use nitrile or acyl chloride-based OSDAs.

Material	IZA Framework type code	Channel dimensionality (Pore size)	References
SSZ-53	**SFH**	1D (14 MR)	[29,71]
SSZ-55	**ATS**	1D (12 MR)	[150]
SSZ-59	**SFN**	1D (14 MR)	[29,71]
SSZ-64	–	likely 3D (12 MR)	[151]
SSZ-65	**SSF**	2D (12 MR)	[152]

Scheme 1.4 details an example of an acyl chloride-based OSDA that is used to make SSZ-59 (**SFN**). A carboxylic acid is first turned into an acyl chloride by reaction with thionyl chloride. Here, the chemistry of interest is the reaction of the acyl chloride with a secondary amine, followed by the reduction and removal of the carbonyl group with lithium aluminum hydride. Finally, the tertiary amine is alkylated with methyl iodide.

These nitrile and acyl chloride chemistries create a class of OSDAs that are large but flexible with several degrees of freedom. The resulting products are primarily large-pore (e.g., SSZ-48 and SSZ-55) and extra-large-pore (e.g., SSZ-53 [**SFH**] and SSZ-59) zeolites or clathrates (ZSM-39 [**MTN**] and ZSM-51 [**NON**]). Several novel SSZ materials using these OSDAs are summarized in Table 1.6.

1.4 OSDA–Zeolite Energetics and Rational Synthesis

Computational approaches have been applied in the zeolite community since the 1980s, largely for modeling adsorption and diffusion in zeolites [62–66]. However, researchers quickly realized that guest–host interaction simulations that were used for locating sorption sites and docking energy calculations (or sorption enthalpy) could also be used to estimate the templating effect of organics in zeolite synthesis [67–70]. The hydrocarbon compounds were replaced by quaternary ammonium or amine OSDAs in the calculations. Early studies were focused on investigating the non-bonded interaction between experimentally proven OSDA–zeolite pairs in order to understand the correlation between the geometry of OSDAs and the zeolite pore systems they form. These results were used to further explain synthesis outcomes and establish rules for OSDA design [15,36,42,43,46,71–83]. Nowadays, with the establishment of better simulation processes, more reliable interatomic potential parameters, and stronger computational power allowing comprehensive screening, molecular modeling techniques are more frequently used to predict alternative OSDAs for synthesizing existing target zeolite structures for cost reduction and/or property modification (e.g., composition, acid site location, crystal size, and morphology), and even to design new OSDAs for synthesizing hypothetical zeolites [27,38,49,84–88].

Most calculations involving OSDA–zeolite interactions are based on Molecular Mechanics, where a variety of well-developed force fields (e.g., CVFF, Dreiding, COMPASS, and Universal) have been successfully applied to these materials. Although the non-bonding parameters of the above-mentioned force fields differ from one another, it was found that the minimized OSDA–zeolite configuration is close regardless of which one is chosen. Moreover, the general rankings of guest–host interactions are similar even though the absolute energy values calculated from different force fields are somewhat different [74]. The non-bonded interaction energy contains two parts: the vdW term and the electrostatic term. The vdW term is found to be the dominant factor in determining OSDA geometry and templating ability. The Lennard–Jones potential is highly repulsive for even small steric overlap between the guest molecule and host framework, making the vdW interaction energy very sensitive to the OSDA–zeolite configuration. The electrostatic term is usually hard to derive due to the difficulty of the assignments of charges for the OSDA–zeolite system. This is particularly true when the location of Al atoms (or other non-tetravalent heteroatoms) in the silicate framework is not known, and the charge balancing cations are omitted in the calculation. Therefore, in many cases, the host zeolites are assumed to be all-silica so that the Coulombic interactions are neglected for simplification. Local charges are removed from the lattice even though the silicate structure may actually contain significant amounts of non-tetravalent heteroatoms. Exclusion of electrostatics in the calculations causes deviations in the simulations from reality since most OSDAs are charged molecules. However, early studies from Lewis et al. showed that the final OSDA–zeolite configuration and the trend of templating ability is about the same with and without the electrostatic term being included in the calculation [73,74]. Sastre et al. studied guest–host interactions on all-silica frameworks synthesized in fluoride-media (assuming the fluoride anion was sitting in the center of double 4-ring units), and their calculations showed that the electrostatic contributions were more than an order of magnitude smaller than the contributions from non-Coulombic interactions [87]. In addition, the differences in the electrostatic terms of different OSDA–framework pairs were negligible compared to the overall calculated energies.

The simulation process can be done by simply docking the OSDA molecules to the void space of the zeolite framework manually through direct visualization and then performing geometry optimization to get the lowest possible interaction energy. However, if the starting position of the docked OSDA molecules is away from the "ideal" position, a local rather than global, minimum energy configuration could be easily reached. A better way to search for global minimum and avoid bias from prior assumption of the docking site is to involve Molecular Dynamics (MD) and Monte Carlo (MC) in the process [67,68,70,72–76]. In general, a high-temperature MD trajectory for the guest OSDA molecule is calculated and used to sample a range of conformational possibilities of the molecule, each of which is then inserted into the zeolite framework using an MC procedure. The MC procedure is based on a random selection of docking positions and molecular orientations inside the zeolite framework. If the interaction energy of such guest–host configurations is better than a certain threshold value (that indicates an excessive steric contact between guest and host), the docked conformation is accepted and saved for further analysis. If the energy is worse than the threshold value, the random choice of position and orientation is made again, and the energy evaluation is repeated until a new configuration is accepted or a predefined number of trials has been reached. As a result, each conformation extracted from the MD trajectory of the isolated OSDA molecule generates a docked configuration for final energy minimization, which then allows the docked molecule to optimize its interaction with the zeolite framework and give the representative best energy configuration for the guest–host system. Sometimes an additional Simulated Annealing (SA) procedure is applied after the MD docking and before the final energy minimization step. The annealing procedure allows the system to dynamically search the energy surface to find the global minimum. SA involves first heating the system, followed by a slow cooling stage (to allow the system to fall into a more relaxed configuration) before using conventional energy minimization to find the final optimized structure [68,76]. More details on the background and recent developments of simulation theory and computation processes are described in Chapter 2.

Early studies on modeling OSDA–zeolite interactions demonstrated an impressive correlation between simulation and experimental outcomes in predicting both OSDA location and phase selectivity. For instance, Bell et al. were able to derive the minimum energy tetrapropylammonium location in the **MFI** structure that matched well with the experimental structural data obtained from single-crystal XRD (X-ray diffraction) analysis [67]. Catlow et al. demonstrated that the non-bonded energy calculation could distinguish small changes in the unit cell parameters of host structures [72]. The example they used was zeolite NU-3 (**LEV**), whose unit cell parameters were noticeably different (by ~1.5%) when synthesized with two different cyclic templates. Calculations showed that the most stable OSDA–unit cell combinations were those that were found experimentally. Njo et al. showed that the interaction energy of N,N-diethyl-3,5-dimethylpiperidinium (both *cis* and *trans*) ions with all-silica zeolite **MEL** was better than with **MFI** [75]. Consequently, they were able to rationalize why pure-phase **MEL** was synthesized instead of the more commonly-observed **MFI** phase. Schmitt and Kennedy successfully identified alternative OSDA molecules through molecular modeling that could replace the triquaternary aromatic molecule used in the discovery of ZSM-18 (**MEI**) [36]. This was the first example of a priori phase prediction in zeolite synthesis through computation.

In the same year that Schmitt and Kennedy published their work on ZSM-18, Harris and Zones at Chevron also contributed an exciting finding when they discovered that the vdW interaction was the most important contributor toward favorable guest–host pairs in the systems of all-silica nonasil (**NON**) and aluminosilicate SSZ-13 (**CHA**) [70]. More importantly, the interaction energy values of different OSDAs for the same zeolite correlated well with the crystallization rates of that particular zeolite made by the different OSDAs. A favorable guest–host interaction may lower the activation energy for generating viable nuclei that lead to crystal growth. The importance of nucleation as the strongest determinant of overall crystallization was further supported by experiments where less favorable OSDAs (based on interaction energies) have crystallization rates greatly enhanced by seeding or increasing stirring rates during synthesis. The same research group led by Zones then carried out a series of studies on OSDA–zeolite energetics for a variety of novel OSDAs and SSZ materials, investigating not only how different OSDAs crystallize a specific zeolite (like the earlier work from Harris and Zones [70]), but also how a specific OSDA crystallizes a list of competing zeolites [15,42,43,46,71,77–83]. These studies provided insight into the relationship between OSDA–zeolite interactions and final product phase selectivity.

The zeolite community has long known that both the organic component and inorganic gel chemistry are important in determining what structure will be crystallized. Because the crystallization process is mainly kinetically driven, the best OSDA–zeolite pair (based on calculated interaction energy) that represents the thermodynamic minimum, when compared to the same OSDA with other zeolite frameworks, may not necessarily guarantee crystallization for the given pair. This is particularly true when the system lacks a sterically-demanding conformation, such as zeolite beta. Zones et al. performed an interesting experiment on the phase selectivity of beta versus SSZ-33 using mixed isomers (*endo*- and *exo*-) of tricyclo[5.2.1.0^{2,6}]decane [43]. Molecular modeling showed that the *endo*-isomer favors SSZ-33 while the *exo*-isomer favors beta. This indeed agreed with the synthesis results that demonstrated a 95/5 ratio of *endo-/exo*-isomers crystallized pure-phase SSZ-33 and a 5/95 ratio produced pure-phase beta. What is more interesting was that the use of SSZ-33 seeds resulted in the successful crystallization of pure-phase SSZ-33 for conditions where the *endo-/exo*-isomer ratios in the synthesis gel would normally give pure-phase beta. This experiment offered insight on the relationship between the stability of the final OSDA-containing zeolite phase (which was explicitly modeled through interaction energy calculations) and the stability of the OSDA-containing phases at or preceding nucleation in determining which structure will be preferred. A similar situation to this example was a more recent modeling study on SSZ-27 where the guest–host interaction energy calculation correctly predicted the different stabilization preference between SSZ-26 and SSZ-27 using different isomers of the same OSDA [89]. This work highlighted how a small conformational difference can lead to the formation of a different zeolite framework (Figure 1.5).

In studying the guest–host relationships in the synthesis of cage-based SSZ-materials, molecular modeling was used to understand how the intergrowth zeolite, SSZ-36 (**RTH/ITE** intergrowth), could be formed from the same OSDA [42,79]. It was suggested that the main reason lies in the very similar vdW interaction energies of the OSDA in the cages of the **RTH** and **ITE** frameworks, along with the very similar structural features between **RTH** and **ITE** (i.e., both are cage-based frameworks having the same compositional building units). Another intergrowth pair, **STF/SFF**, can also be synthesized, and this system offers the same scenario where both structural and vdW interaction energy similarities are present when N,N-diethyl-2,5-*cis*-dimethylpiperidinium is used in the calculation [90]. Given such knowledge, it was then no surprise to predict a priori that the pair of **ERI** and **LEV** could also form an intergrowth, and indeed, by using N,N-dimethylpiperidinium as the OSDA, the novel zeolite, SSZ-105, covering the full range of **ERI/LEV** intergrowth ratios, was recently synthesized [88,91]. Figure 1.8 shows the energy-minimized configuration of N,N-dimethylpiperidinium in the *eri* and *lev* cages, and Figure 1.9 shows the PXRD (Powder X-ray diffraction) and corresponding SEM images at various **ERI/LEV** intergrowth ratios.

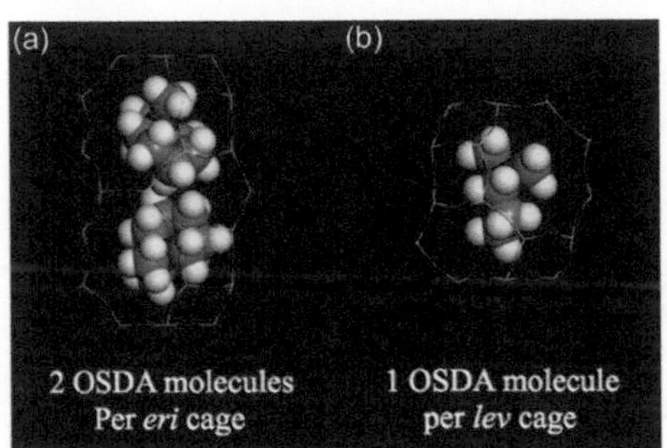

Figure 1.8 Predicted energy-minimized configurations of N,N-dimethylpiperidinium in the *eri* cage (a) and *lev* cage (b). The OSDA has a loading of two molecules per *eri* cage and one per *lev* cage.

Eventually, molecular modeling should be used to predict the phase crystallized in a synthesis with a given OSDA molecule. The structure-directing capability of a molecule for a target zeolite

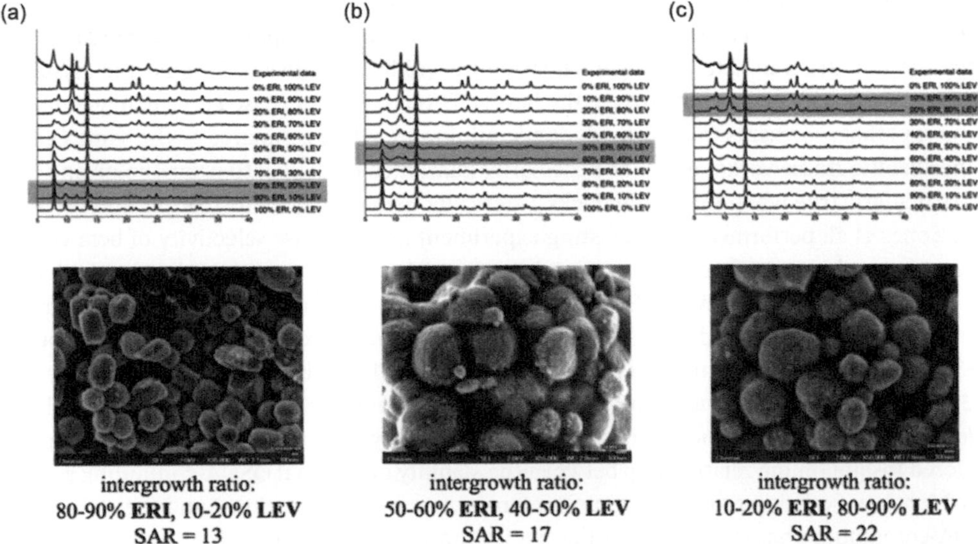

Figure 1.9 The PXRD pattern and the corresponding SEM image of SSZ-105 with different **ERI/LEV** intergrowth ratios are shown in (a), (b), and (c). The DIFFaX simulated PXRD patterns for the intergrowth are also included for comparison.

framework is typically evaluated by comparing its calculated interaction energy with those of other OSDA molecules that successfully made that framework. Indeed, sometimes such a relatively simple strategy works well for designing and identifying thermodynamically more favorable OSDA molecules for target zeolite syntheses. This is especially true if kinetically favorable synthesis conditions for the target zeolite are already well-established. One good example is from Davis and Deem et al., where they successfully used molecular modeling to identify several OSDA molecules for SSZ-52 (**SFW**), which were subsequently verified by lab syntheses [27] (see Chapter 2 for more details).

Similar to the case of SSZ-52, we recently conducted a rational synthesis on the high-silica **ERI** zeolite, SSZ-98 [88,92–96]. Both **SFW** and **ERI** structures belong to the ABC-6 family of 8 MR frameworks, and the well-defined cages (different length with about the same cross-section of the 12 MR) in this family of structures make them ideal for molecular modeling studies. **ERI** has two different types of cages: a small *can*-cage and a longer *eri*-cage. From previous synthesis experience on SSZ-98 using 1,4-dimethyl-DABCO as the OSDA, we found that the crystallization of **ERI** was only possible if K^+ ion was used in the synthesis gel, which is likely because K^+ ion favors the formation of *can*-cages. In addition, we found that using dealuminated Y zeolite as a reagent for both aluminum and silicon provided tremendous advantages to the synthesis. In addition to enhancing the crystallization rate and phase selectivity, starting with Y zeolite also increased the range of the product SAR. After such kinetically favorable synthesis conditions were established, we began a search for more energetically favorable OSDAs for this framework by screening an internal library of existing quaternary ammonium compounds. The vdW interaction energy between each OSDA in the library and the **ERI** structure was then optimized and calculated. Such minimized energy values were then used to rank the OSDAs from most-to-least energetically favorable. More than 20 candidates were found to give interaction energies better than 1,4-dimethyl-DABCO, and eventually 7 of them were able to successfully crystalize pure-phase SSZ-98 using the kinetically favorable synthesis conditions established earlier. The interaction energies and the optimum loading for these seven molecules (denoted OSDA-2 to OSDA-8) are listed in Table 1.7, and the predicted energy-minimized configurations of these OSDA molecules in the *eri*-cage are shown in Figure 1.10. It was also interesting to note that different crystal sizes and morphologies of the **ERI** structure were observed when different OSDAs and starting gel SAR were used. Here, molecular modeling was very powerful in identifying thermodynamically favorable OSDA molecules for this particular structure.

If the target zeolite: i) can only be crystallized in a narrow range of synthesis conditions; ii) has competing phases that can easily form from similar gel composition and synthesis conditions; or iii) if the target zeolite is a hypothetical framework that has not yet been synthesized, rational synthesis through molecular modeling is more challenging. It is then necessary to evaluate not only how well an OSDA candidate fits within the target framework but also how well it fits within other competing frameworks. Burton et al. have explicitly demonstrated that both evaluations are important by comparing experimental results with molecular modeling results in the synthesis of cage-based zeolite frameworks using a large database of piperidinium derivatives and polycyclic quaternary compounds [79]. While energy comparisons among different OSDAs in the same zeolite framework are straightforward, energy comparisons among different zeolite frameworks for the same OSDA are usually not. Different zeolite frameworks may have different framework densities (FD) and/or a different percentage of framework atoms that truly participate in vdW interactions with OSDA molecules. As a result, vdW interaction energy normalized either by the number of T atoms or by the number of OSDA molecules per unit cell is not ideal, especially when evaluating a cage-based lower FD phase versus a channel-based higher FD phase. In addition, two other

Table 1.7 Calculated vdW interaction energies for various OSDAs in the **ERI** framework structure.

OSDA molecule	Number of Torsion Angles	Number of molecules per unit cell	Interaction energy kJ (mol Si)-1
1	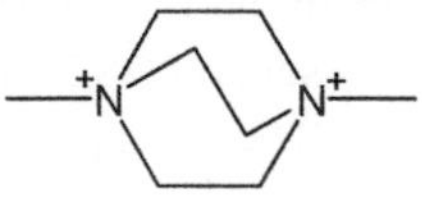 0	4	−7.8
2	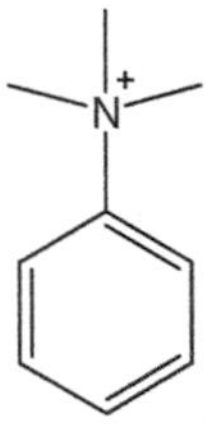 1	4	−14.5
3	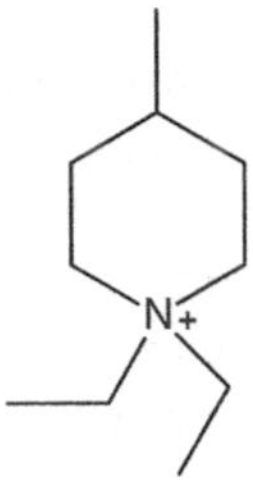 2	2	-9.0
4	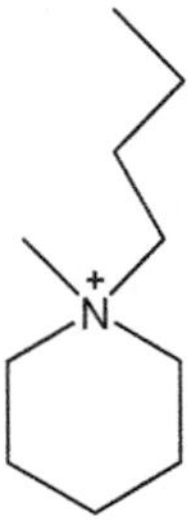 3	2	-8.6
5	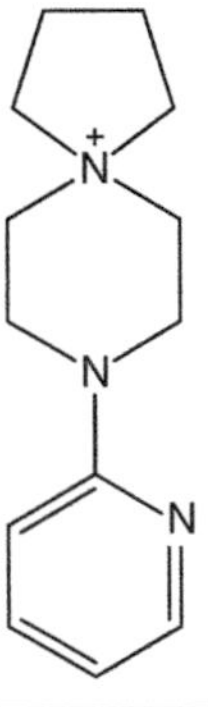 1	2	−14.1

(*Continued*)

Table 1.7 (*Continued*)

OSDA molecule	Number of Torsion Angles	Number of molecules per unit cell	Interaction energy kJ (mol Si)-1
6	2	2	−11.3
7	0	4	−13.5
8	2	4	−13.2

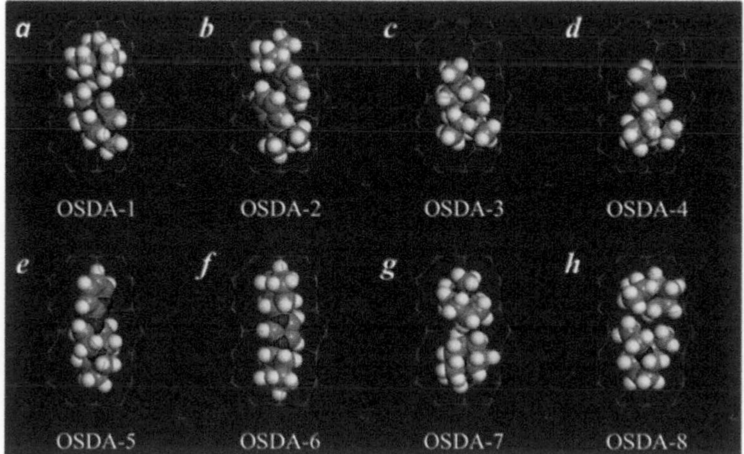

Figure 1.10 Predicted energy-minimized configurations of the OSDA molecules in the *eri* cage. OSDA-1, -2, -7, and -8 have a loading of two molecules per *eri* cage, while the rest have one per *eri* cage.

contributions to the energy of an OSDA–zeolite pair also need to be considered when evaluating phase selectivity, as discussed by Burton et al. [77,79,80]: i) the inherent energy of the defect-free framework, and ii) the concentration of silanol/siloxy defects in the final product. Piccione et al. conducted calorimetry experiments to measure the enthalpies of defect-free, silica zeolites relative to quartz [97]. Their work suggested that there is an approximately linear correlation between FD and the enthalpies of defect-free silica frameworks, and such enthalpies are about 6–12 kJ/mol T atom greater than quartz. For example, the enthalpy of 3D, cage-based, 8 MR, all-silica **CHA** (FD = 15.4) is ~2.6 KJ/mol T atom better (i.e., lower) than that of 1D, channel-based, 14 MR, all-silica **CFI** (FD = 18.3). For aluminosilicate zeolites that have more OSDA cations than aluminum atoms, the framework will need negatively charged siloxy defects for charge balance, and this creates additional defect energy penalties in zeolite structures. For example, Piccione et al. reported that there is a ~1.2 KJ/mol T atom defect energy penalty in all-silica **MFI** prepared by hydroxide media compared to the one prepared by fluoride media [97]. If the differences in the framework enthalpies (the denser the framework, the lower its enthalpy) is of similar magnitude to the differences in stabilization provided by a given OSDA molecule for each of the different frameworks (in general, the denser framework has a higher value of absolute interaction energy), the interaction energy may not be as sensitive to phase selectivity. Of course, the dream goal of molecular modeling is to identify a specific OSDA that offers both the greatest interaction energy for a particular framework and the greatest difference in the interaction energy between the target framework and potential competing frameworks.

1.5 Role of High Throughput and Automation

Combinatorial synthesis methods that have become commonplace in industries, such as the pharmaceutical industry, are becoming increasingly relevant to zeolite science. High initial capital costs can hamper their full-scale use in small academic settings, although smaller versions of high-throughput equipment can still be adopted at many levels. For example, well plates with tens of microreactor wells are relatively easy to machine and can replace individual autoclave reactors. Depending on the desired goals, there is not necessarily a need to employ multi-million dollar robotics and automation options. Several reviews on high-throughput and zeolite synthesis automation have been written, and we refer the reader to these papers for further information [98–104].

There are several reasons to use automation and high-throughput techniques, and we discuss a few of them here. While the zeolite community has made great strides in understanding aspects of synthesis, (such as nucleation and crystallization, the interaction of organic and inorganic species, and the interplay between kinetics and thermodynamics), reliable and comprehensive a priori knowledge of input versus output remains an elusive goal, and experimental zeolite synthesis largely remains a trial-and-error art. High-throughput techniques can speed up the discovery of new zeolite phases, and groups at ITQ, ExxonMobil, Jilin University, and UOP, among many others, have found several materials using these methods [105–111].

High-throughput can also optimize synthesis conditions. The structure of ZSM-43 (**MRT**) remained unsolved for nearly 40 years after its initial discovery because the samples had poor crystallinity and small crystal sizes. A sample with improved crystallinity was developed through high-throughput screening, and the structure was able to be solved [112]. In the case of ITQ-43, a hierarchical morphology was targeted and developed using high-throughput techniques [113]. Combinatorial methods have also been used for developing new polymorphs [114] and modifying crystal sizes and porosity [115]. Finally, successful commercialization of zeolites often requires using the least amount of water and organics as possible in order to reduce production costs, and high-throughput techniques can quickly find an optimized recipe and decrease the time to market.

In a more explicit way, using many synthesis experiments to develop new materials and optimize their properties is, in actuality, exploring and expanding the synthesis space around which the best material can be found. For example, the Sintef group reproduced and expanded on Breck's early work on sodium aluminosilicate syntheses [116], and Bein et al. mapped out the synthesis space of aluminophosphates using two different organic OSDAs [117]. The ability to quickly generate ternary diagrams and study the kinetic stability of a particular zeolite phase is extremely useful in understanding the synthesis system of interest, and high throughput allows these experiments to be done in a practical amount of time with a reasonable amount of effort.

As described in more detail in later chapters, artificial intelligence (AI) models require datasets for training, validation, and testing. Mining the literature is one way to generate a large dataset [118]. High-throughput experimentation can also provide datasets in a relatively small amount of time. Corma et al. have been able to use experimental datasets created from high-throughput screening and have shown very good prediction performance from their AI models [106,119,120]. At present, datasets are typically limited to one or two particular synthesis systems within one research institution. The increased use of high throughput and the sharing of data within the zeolite community will improve AI models and expand their range of predictive capabilities.

While the use of high throughput and automation has vast potential for increasing scientific knowledge, automation has practical benefits from the standpoint of safety and reproducibility. Repeated pipetting and weighing of chemicals expose the researcher to a variety of hazards, such as repetitive stress injuries and chemical contact. In addition, these tasks are fairly low value-added jobs, and highly-trained researchers can better use their time analyzing data or designing experiments. Finally, automation can eliminate human error and issues associated with reproducibility.

We feel that it is important to note that data analysis is a component of high throughput that is often overlooked but presents serious challenges if not handled properly. Since XRD is typically the primary screening tool for zeolite synthesis, we talk about data analysis in terms of XRD data. However, any type of data that is generated in large volumes with high-throughput tools needs to be carefully considered. Manual analysis of XRD data is time-consuming and can easily become a bottleneck. As such, XRD data analysis presents opportunities for automation and machine learning. In the ideal case, the researcher could input multiple XRD patterns and the output would be the zeolite phase(s) for each sample. However, complicating factors (such as unknown phases, impurities, mixtures, amorphous content, differences in crystal size and composition, and intergrowths) make XRD pattern automation difficult, and there is currently no off-the-shelf software that is capable of performing these tasks. Corma et al. applied a methodology called adaptable time warping that attempts to tackle some of these challenges [121,122]. This learning algorithm was able to correctly identify single phases and mixtures from a complex dataset of eight zeolite phases, amorphous material, and mixtures thereof. Expansion of algorithms such as these to all known zeolite phases and common impurities/minerals would allow for truly automated analysis of any set of zeolite XRD patterns.

A typical zeolite synthesis workflow includes several places where automation and high-throughput tools can be employed. These include dispensing of chemicals, well plates for crystallization, sample work-up, and high-throughput XRD. Dispensing robots can range from simple lab balances outfitted with dispensing pumps to highly sophisticated robotic platforms. Several commercial companies, such as Chemspeed Technologies, Labman Automation, Zinsser NA, Tecan, and Mettler Toledo, offer both customized and off-the-shelf products. These companies typically also offer filtration and centrifugation options for sample work-up. The literature contains several examples of well plates and synthesis tools that are often machined in-house [116,119,123,124]. In some cases, these well plates can be directly transferred to an XRD with a high-throughput sample stage.

1.6 Cataloguing, Archiving, Harvesting, and Mining Years of Historical Data

Given that we are currently in what seems to be an exponentially accelerating portion of the information age, it is no surprise that zeolite science is not any different. Figure 1.11 is a Web of Science search for the term "zeolite" over time, which returned ~350,000 results through the year 2020. There is certainly an immense amount of data on zeolite synthesis available in publications and patents, and this likely is an order of magnitude less data than what is in dust-collecting lab notebooks. Making use of historical zeolite synthesis data has already been considered in several publications [103,118], and there are a number of key steps to making this a reality:

1) Extraction: This is by far the most challenging aspect in utilizing historical data. Even in modern publications and patents that were designed to be in digital form, there is no common format for reporting zeolite synthesis parameters, so extracting this data into a common format will require clever software and rigorous verification. The difficulty increases in older patents and publications as optical character recognition (OCR) must be applied to first convert them into digital form, and there will inevitably be errors in this process. While patents are found in the public domain, many publications are behind pay walls that may prevent access to the data. The most challenging data sources are lab notebooks and other hard copy, handwritten records, and extracting data from these sources was the subject of a 2020 *Nature* article that concluded that OCR was not very effective with handwritten records [125]. Therefore, an organization

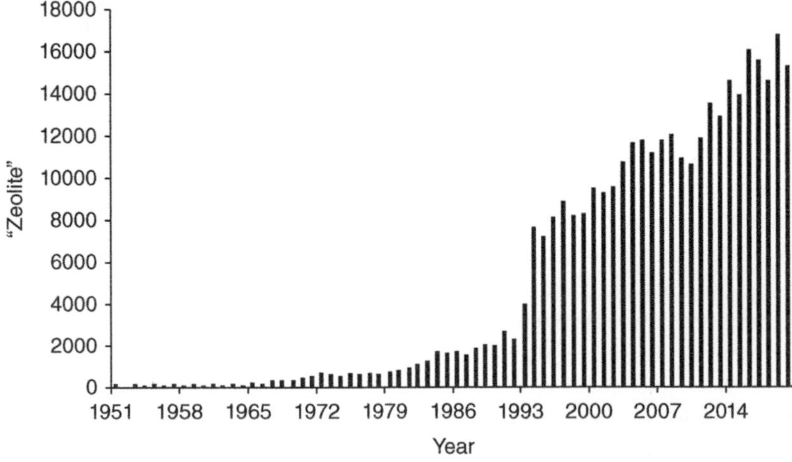

Figure 1.11 Web of Science search for the term "zeolite" showing the number of publications per year with the search term through the year 2020.

would have to decide if obtaining this data is worth the cost of manual extraction, which would also likely need to be performed by an expert in the topic to avoid errors. One critical aspect motivating the painstaking effort that will be required to extract data from handwritten records is that publications and patents generally only include experiments that give the desired product. However, knowing which conditions did not do this is vital for any complete analysis; this is especially true for predictive capabilities.

2) Validation: When OCR is applied, there will inevitably be errors that will need to be found. The best case would be automated error analysis to flag potential errors for manual inspection.
3) Analysis: Many tools are presently available for analyzing large amounts of data and extracting useful information from this data. This part of the process normally draws the most attention, but in this case, it will likely require the least amount of manual labor, as the extraction and validation steps will have created high-quality data that is ready for analysis.

Overall, extracting, validating, and analyzing large amounts of data across disparate sources is a monumental challenge. However, we believe that the correct way to approach this problem is to consider the cost of reproducing all the experiments, even using a high-throughput synthesis tool. Given the high cost of laboratory work, it is likely that even manual extraction of results from lab notebooks would be a tremendous bargain compared to reproducing all the work.

1.7 Concluding Remarks

In this first chapter of this compilation of contributions, we brought together the considerable amount of experimental experience we have with synthesizing zeolites and the progression of attempts to model and rationalize synthesis results. In the chapters to come, a wealth of approaches to computationally understand and predict zeolite synthesis will be presented. We have brought along our efforts in the modeling world, much of it focused on energetically favorable OSDA interactions with the zeolite frameworks. Where the future for modeling then needs to go is how to predict the successful crystallization of a zeolite in a sea of possibilities. In this chapter, we have attempted to elucidate some of the factors that impact the likelihood of the type of zeolite product crystallizing. With progress, we can then imagine a future where the design of a synthesis of the considerable number of hypothetical zeolite structures computationally predicted for a demanded application, but not yet realized, can take off.

References

1 Zones, S.I. (2011). Translating new materials discoveries in zeolite research to commercial manufacture. *Microporous Mesoporous Mater.* 144 (1–3): 1–8.

2 Dusselier, M. and Davis, M.E. (2018). Small-pore zeolites: synthesis and catalysis. *Chem. Rev.* 118 (11): 5265–5329.

3 Cundy, C.S. and Cox, P.A. (2003). The hydrothermal synthesis of zeolites: history and development from the earliest days to the present time. *Chem. Rev.* 103 (3): 663–702.

4 Milton, R.M. (1959). Molecular sieve adsorbents. US Patent 2,882,243, 14th April.

5 Barrer, R.M. and Denny, P.J. (1961). Hydrothermal chemistry of the silicates. Part IX: Nitrogenous aluminosilicates. *J. Chem. Soc.* 0: 971–982.

6 Barrer, R.M., Denny, P.J., and Flanigen, E.M. (1967). Molecular sieve adsorbents. US Patent 3,306,922, 28th February.

7 Kerr, G.T. (1966). Chemistry of crystalline aluminosilicates. Part II: The synthesis and properties of zeolite ZK-41. *Inorg. Chem.* 5 (9): 1537–1539.

8 Kerr, G.T. (1967). Synthetic zeolite. US Patent 3,314,752, 18th April.

9 Kerr, G.T. and Kokotailo, G.T. (1961). Sodium zeolite ZK-4: a new synthetic crystalline aluminosilicate. *J. Am. Chem. Soc.* 83 (22): 4675.

10 Wadlinger, R.L., Kerr, G.T., and Rosinski, E.J. (1967). Catalytic composition of a crystalline zeolite. US Patent 3,308,069, 7th March.

11 Argauer, R.J. and Landolt, G.R. (1972). Crystalline zeolite ZSM-5 and method of preparing the same. US Patent 3,702,886, 14th November.

12 Gómez-Hortigüela, L. (2017). Preface. In: *Insights into the Chemistry of Organic Structure-Directing Agents in the Synthesis of Zeolitic Materials. Structure and Bonding* (ed. L. Gómez-Hortigüela), 175. Cham: Springer.

13 Petrovic, I., Navrotsky, A., Davis, M.E. et al. (1993). Thermochemical study of the stability of frameworks in high silica zeolites. *Chem. Mater.* 5: 1805–1813.

14 Navrotsky, A. (2005). Calorimetric insights into the synthesis of templated materials. *Curr. Opin. Colloid. Interface Sci.* 10 (5–6): 195–202.

15 Nakagawa, Y., Lee, G.S., Harris, T.V. et al. (1998). Guest/host relationships in zeolite synthesis: ring-substituted piperidines and the remarkable adamantane mimicry by 1-azonio spiro [5.5] undecanes. *Microporous Mesoporous Mater.* 22: 69–85.

16 Zones, S.I., Nakagawa, Y., Lee, G.S. et al. (1998). Searching for new high silica zeolites through a synergy of organic templates and novel inorganic conditions. *Microporous Mesoporous Mater.* 21: 199–211.

17 Zones, S.I., Nordstrand, R.A.V., Santilli, D.S. et al. (1989). Sequence of high silica zeolites found during synthesis experiments in the presence of a quaternary adamantammonium cation. *Stud. Surf. Sci. Catal.* 49: 299–309.

18 Kumar, M., Berkson, Z.J., Clark, R.J. et al. (2019). Crystallization of mordenite platelets using cooperative organic structure-directing agents. *J. Am. Chem. Soc.* 141 (51): 20155–20165.

19 Davis, M.E. and Zones, S.I. (1997). A perspective on zeolite synthesis: how do you know what you'll get? In: *Synthesis of Porous Materials: Zeolites, Clays, and Nanostructures* (ed. M.L. Occelli and H. Kessler), 1–34. New York: Marcel Dekker.

20 Barrett, P.A., Camblor, M.A., Corma, A. et al. (1998). Synthesis and structure of as-prepared ITQ-4, a large pore pure silica zeolite: the role and location of fluoride anions and organic cations. *J. Phys. Chem. B* 102 (21): 4147–4155.

21 Camblor, M.A., Corma, A., and Valencia, S. (1996). Spontaneous nucleation and growth of pure silica zeolite-β free of connectivity defects. *ChemComm.* 20: 2365–2366.

22 Zones, S.I., Darton, R.J., Morris, R.E. et al. (2005). Studies on the role of fluoride ion vs. reaction concentration in zeolite synthesis. *J. Phys. Chem. B* 109: 652–661.

23 Zones, S.I., Hwang, S.-J., Elomari, S. et al. (2005). The fluoride-based route to all-silica molecular sieves: a strategy for synthesis of new materials based upon close-packing of guest–host products. *C R Chim.* 8: 267–282.

24 Xie, D. and Zones, S.I. (2017). Method for making molecular sieve SSZ-102. US Patent 9,573,819, 21st February.

25 Zones, S.I., Lew, C.M., Xie, D. et al. (2020). Studies on the use of faujasite as a reagent to deliver silica and alumina in building new zeolite structures with organo-cations. *Microporous Mesoporous Mater.* 300: 110162.

26 Zones, S.I., Jayanthi, K., Pascual, J. et al. (2021). Energetics of the local environment of structure-directing agents influence zeolite synthesis. *Chem. Mater.* 33 (6): 2126–2138.

27 Davis, T.M., Liu, A.T., Lew, C.M. et al. (2016). Computationally guided synthesis of SSZ-52: a zeolite for engine exhaust clean-up. *Chem. Mater.* 28 (3): 708–711.

28 Xie, D., McCusker, L.B., Baerlocher, C. et al. (2013). SSZ-52, a zeolite with an 18-layer aluminosilicate framework structure related to that of the $DeNO_x$ catalyst Cu-SSZ-13. *J. Am. Chem. Soc.* 135 (28): 10519–10524.

29 Smeets, S., McCusker, L.B., Baerlocher, C. et al. (2016). Locating organic guests in inorganic host materials from X-ray powder diffraction data. *J. Am. Chem. Soc.* 138: 7099–7106.

30 Lobo, R.F., Zones, S.I., and Davis, M.E. (1995). Structure-direction in zeolite synthesis. *J. Lncl. Phenom. Mol. Recogni. Chem.* 21: 47–78.

31 Burton, A.W. and Zones, S.I. (2007). Organic molecules in zeolite synthesis: their preparation and structure-directing effects. In: *Introduction to Zeolite Science and Practice*, 3e (ed. J. Čejka, H.V. Bekkum, A. Corma, et al.). Elsevier.

32 Li, J., Corma, A., and Yu, J. (2015). Synthesis of new zeolite structures. *Chem. Soc. Rev.* 44: 7112–7127.

33 Lew, C.M., Davis, T.M., and Elomari, S. (2016). Synthesis of new molecular sieves using novel structure-directing agents. In: *Verified Syntheses of Zeolitic Materials*, 3e (ed. S. Mintova). Elsevier.

34 Burton, A.W. (2018). Recent trends in the synthesis of high-silica zeolites. *Catal. Rev.* 60 (1): 132–175.

35 Davis, M.E. and Lobo, R.F. (1992). Zeolite and molecular sieve synthesis. *Chem. Mater.* 4: 756–768.

36 Schmitt, K.D. and Kennedy, G.J. (1994). Toward the rational design of zeolite synthesis: the synthesis of zeolite ZSM-18. *Zeolites* 14: 635–642.

37 Bu, X., Feng, P., and Stucky, G.D. (2000). Host-guest symmetry and charge matching in two germanates with intersecting three-dimensional channels. *Chem. Mater.* 12: 1505–1507.

38 Brand, S.K., Schmidt, J.E., Deem, M.W. et al. (2017). Enantiomerically enriched, polycrystalline molecular sieves. *Proc. Natl. Acad. Sci. USA* 114 (20): 5101–5106.

39 Gies, H. and Marler, B. (1992). The structure-controlling role of organic templates for the synthesis of porosils in the system SiO_2/template/H_2O. *Zeolites* 12 (1): 42–49.

40 Liebau, F. (1985). *Structural Chemistry of Silicates*. Berlin Heidelberg: Springer-Verlag.

41 Kubota, Y., Helmkamp, M.M., Zones, S.I. et al. (1996). Properties of organic cations that lead to the structure-direction of high-silica molecular sieves. *Microporous Mater.* 6: 213–229.

42 Wagner, P., Nakagawa, Y., Lee, G.S. et al. (2000). Guest/host relationships in the synthesis of the novel cage-based zeolites SSZ-35, SSZ-36, and SSZ-39. *J. Am. Chem. Soc.* 122: 263–273.

43 Zones, S.I., Nakagawa, Y., Yuen, L.-T. et al. (1996). Guest/host interactions in high silica zeolite synthesis: [5.2.1.0$^{2.6}$]tricyclodecanes as template molecule. *J. Am. Chem.Soc.* 118: 7558–7567.

44 Lee, G.S., Nakagawa, Y., Hwang, S.-J. et al. (2002). Organocations in zeolite synthesis: fused bicyclo [l.m.0] cations and the discovery of zeolite SSZ-48. *J. Am. Chem. Soc.* 124: 7024–7034.

45 Zones, S.I. (1989). Synthesis of pentasil zeolites from sodium silicate solutions in the presence of quaternary imidazole compounds. *Zeolites* 9: 458–467.

46 Zones, S.I. and Burton, A.W. (2005). Diquaternary structure-directing agents built upon charged imidazolium ring centers and their use in synthesis of one-dimensional pore zeolites. *J. Mater. Chem.* 15: 4215–4223.

47 Rojas, A., Martínez-Morales, E., Zicovich-Wilson, C.M. et al. (2012). Zeolite synthesis in fluoride media: structure direction toward ITW by small methylimidazolium cations. *J. Am. Chem. Soc.* 134: 2255–2263.

48 Schmidt, J.E., Deimund, M.A., Xie, D. et al. (2015). Synthesis of RTH-type zeolites using a diverse library of imidazolium cations. *Chem. Mater.* 27: 3756–3762.
49 Schmidt, J.E., Deem, M.W., and Davis, M.E. (2014). Synthesis of a specified, silica molecular sieve by using computationally predicted organic structure-directing agents. *Angew. Chem. Int. Ed.* 126 (32): 8512–8514.
50 Archer, R.H., Zones, S.I., and Davis, M.E. (2010). Imidazolium structure directing agents in zeolite synthesis: exploring guest/host relationships in the synthesis of SSZ-70. *Microporous Mesoporous Mater.* 130 (1–3): 255–265.
51 Boal, B.W., Deem, M.W., Xie, D. et al. (2016). Synthesis of germanosilicate molecular sieves from Mono- and Di-quaternary ammonium OSDAs constructed from benzyl imidazolium derivatives: stabilization of large micropore volumes including new molecular sieve CIT-13. *Chem. Mater.* 28 (7): 2158–2164.
52 Casci, J.L. (1992). Zeolites. US Patent 5,108,579, 28th April.
53 Casci, J.L. (1992). Zeolite NU-87. US Patent 5,102,641, 7th April.
54 Moini, A., Schmitt, K.D., Valyocsik, E.W. et al. (1994). The role of diquaternary cations as directing agents in zeolite synthesis. *Zeolites* 14 (7): 504–511.
55 Jackowski, A., Zones, S.I., Chaudhuri, K. et al. (2014). A hydrothermal synthesis of a 2-dimensional layered silicate followed by a transition to a 3-dimensional aluminosilicate zeolite. *Microporous Mesoporous Mater.* 197: 33–39.
56 Jackowski, A., Zones, S.I., Hwang, S.-J. et al. (2009). Diquaternary ammonium compounds in zeolite synthesis: cyclic and polycyclic N-heterocycles connected by methylene chains. *J. Am. Chem. Soc.* 131: 1092–1100.
57 Burton, A.W. (2010). Molecular sieve SSZ-82 composition of matter and synthesis thereof. US Patent 7,820,141, 26th October.
58 Zones, S.I. (1985). Zeolite SSZ-16. US Patent 4,508,837, 2nd April.
59 Davis, T.M. (2018). High-silica AFX framework type zeolites. US Patent 9,908,108, 6th March.
60 Zones, S.I., Hwang, S.-J., and Davis, M.E. (2001). Studies of the synthesis of SSZ-25 zeolite in a “Mixed-Template” system. *Chem.-Eur. J.* 7 (9): 1990–2001.
61 Elomari, S. (2003). Process for preparing zeolites. US Patent 6,632,417, 14th October.
62 Freeman, C.M., Catlow, C.R.A., Thomas, J.M. et al. 1991). Computing the location and energetics of organic molecules in microporous adsorbents and catalysts: a hybrid approach applied to isometric butenes in a model zeolite. *Chem. Phys. Lett.* 186 (2–3): 137–142.
63 Nowak, A.K. and Cheetham, A.K. (1986). The use of computer graphics to study adsorption, diffusion and catalysis in zeolites. In: *New Developments in Zeolite Science and Technology, Proceedings of the 7th International Zeolite Conference*, 28 (ed. Y. Murakami, A. Iijima, and J.W. Ward), 475–479. Elsevier Science.
64 Pickett, S.D., Nowak, A.K., Thomas, J.M. et al. (1989). Computer simulation of the adsorption and diffusion of benzene in silicalite, theta-1, and a new zeolite, EU-1. *Zeolites* 9 (2): 123–128.
65 Santilli, D.S., Harris, T.V., and Zones, S.I. (1993). Inverse shape selectivity in molecular sieves: observations, modeling, and predictions. *Microporous Mater.* 1 (5): 329–341.
66 Yashonath, S., Demontis, P., and Klein, M.L. (1988). A molecular dynamics study of methane in zeolite NaY. *Chem. Phys. Lett.* 153 (6): 551–556.
67 Bell, R.G., Lewis, D.W., Voigt, P. et al. (1994). Computer modelling of sorbates and templates in microporous materials. *Stud. Surf. Sci. Catal.* 84: 2075–2082.
68 Cox, P.A., Stevens, A.P., Banting, L. et al. (1994). Molecular modelling studies of zeolite synthesis. *Stud. Surf. Sci. Catal.* 84: 2115–2122.

69 de Vos Burchart, E., Jansen, J.C., van de Graaf, B. et al. (1993). Molecular mechanics studies on MFI-type zeolites. Part 4: Energetics of crystal growth directing agents. *Zeolites* 13 (3): 216–221.

70 Harris, T.V. and Zones, S.I. (1994). A study of guest/host energetics for the synthesis of cage structures NON and CHA. *Stud. Surf. Sci. Catal.* 84: 29–36.

71 Burton, A.W., Elomari, S., Chen, C.-Y. et al. (2003). SSZ-53 and SSZ-59: two novel extra-large pore zeolites. *Chem.- Eur. J.* 9: 5737–5748.

72 Catlow, C.R.A., Bell, R.G., Gale, J.D. et al. (1995). Modelling of structure and reactivity in zeolites. *Stud. Surf. Sci. Catal.* 97: 87–100.

73 Lewis, D.W., Catlow, C.R.A., and Thomas, J.M. (1996). Influence of organic templates on the structure and on the concentration of framework metal ions in microporous aluminophosphate catalysts. *Chem. Mater.* 8 (5): 1112–1118.

74 Lewis, D.W., Freeman, C.M., and Catlow, C.R.A. (1995). Predicting the templating ability of organic additives for the synthesis of microporous materials. *J. Phys. Chem.* 99 (28): 11194–11202.

75 Njo, S.L., Koegler, J.H., van Koningsveld, H. et al. (1997). Molecular mechanics calculations on the N,N-diethyl-3,5-dimethylpiperidinium ions in MEL and MFI. *Microporous Mater.* 8 (5–6): 223–229.

76 Stevens, A.P., Gorman, A.M., Freeman, C.M. et al. (1996). Prediction of template location via a combined Monte Carlo–simulated annealing approach. *J. Chem. Soc. Faraday Trans.* 92 (12): 2065–2073.

77 Burton, A.W. (2007). A priori phase prediction of zeolites: case study of the structure-directing effects in the synthesis of MTT-type zeolites. *J. Am. Chem. Soc.* 129: 7627–7637.

78 Burton, A.W., Darton, R.J., Davis, M.E. et al. (2006). Structure-directing agent location and non-centrosymmetric structure of fluoride-containing zeolite SSZ-55. *J. Phys. Chem. B* 110: 5273–5278.

79 Burton, A.W., Lee, G.S., and Zones, S.I. (2006). Phase selectivity in the syntheses of cage-based zeolite structures: an investigation of thermodynamic interactions between zeolite hosts and structure directing agents by molecular modeling. *Microporous Mesoporous Mater.* 90 (1–3): 129–144.

80 Burton, A.W., Zones, S.I., and Elomari, S. (2005). The chemistry of phase selectivity in the synthesis of high-silica zeolites. *Curr. Opin. Colloid Interface Sci.* 10: 211–219.

81 Wragg, D.S., Morris, R., Burton, A.W. et al. (2007). The synthesis and structure of SSZ-73: an all-silica zeolite with an unusual framework topology. *Chem. Mater.* 19 (16): 3924–3932.

82 Zones, S.I., Burton, A.W., Lee, G.S. et al. (2007). A study of piperidinium structure-directing agents in the synthesis of silica molecular sieves under fluoride-based conditions. *J. Am. Chem. Soc.* 129: 9066–9079.

83 Zones, S.I., Hwang, S.-J., Olmstead, M.M. et al. (2010). A most unusual zeolite templating: cage to cage connection of one guest molecule. *J. Phys. Chem. C* 114: 8899–8904.

84 Gallego, E.M., Portilla, M.T., Paris, C. et al. (2017). "Ab initio" synthesis of zeolites for preestablished catalytic reactions. *Science* 355 (6329): 1051–1054.

85 Jo, D. and Hong, S.B. (2019). Targeted synthesis of a zeolite with pre-established framework topology. *Angew. Chem. Int. Ed.* 58 (39): 13845–13848.

86 Pophale, R., Daeyaert, F., and Deem, M.W. (2013). Computational prediction of chemically synthesizable organic structure directing agents for zeolites. *J. Mat. Chem. A* 1 (23): 6750–6760.

87 Sastre, G., Cantin, A., Díaz-Cabañas, M.J. et al. (2005). Searching organic structure directing agents for the synthesis of specific zeolitic structures: an experimentally tested computational study. *Chem. Mater.* 17 (3): 545–552.

88 Xie, D. (2021). Rational design and targeted synthesis of small-pore zeolites with the assistance of molecular modeling, structural analysis, and synthetic chemistry. *Ind. Eng. Chem. Res.* 60 (43): 15403–15415.

89 Smeets, S., Zones, S.I., Xie, D. et al. (2019). SSZ-27: a small-pore zeolite with large heart-shaped cavities determined by using multi-crystal electron diffraction. *Angew. Chem. Int. Ed.* 58 (37): 13080–13086.

90 Villaescusa, L.A., Zhou, W., Morris, R.E. et al. (2004). Synthesis, characterization and control of faulting in STF/SFF topologies, a new family of intergrowth zeolites. *J. Mat. Chem.* 14 (13): 1982–1987.

91 Xie, D. and Lew, C.M. (2017). Molecular sieve SSZ-105. US Patent 9,663,380, 30th May.

92 Xie, D. (2017). Method for preparing zeolite SSZ-98. US Patent 9,815,704, 14th November.

93 Xie, D. (2017). Synthesis of molecular sieve SSZ-98. US Patent 9,815,705, 14th November.

94 Xie, D. and Chen, C.-Y. (2017). Synthesis of aluminosilicate zeolite SSZ-98. US Patent 9,662,642, 30th May.

95 Xie, D. and Lew, C.M. (2017). Method for preparing zeolite SSZ-98. US Patent 9,663,379, 30th May.

96 Xie, D., Zones, S.I., Lew, C.M. et al. (2016). Molecular sieve SSZ-98. US Patent 9,409,786, 9th August.

97 Piccione, P.M., Laberty, C., Yang, S. et al. (2000). Thermochemistry of pure-silica zeolites. *J. Phys. Chem. B* 104: 10001–10011.

98 Newsam, J.M., Bein, T., Klein, J. et al. (2001). High throughput experimentation for the synthesis of new crystalline microporous solids. *Microporous Mesoporous Mater.* 48 (1–3): 355–365.

99 Schüth, F. (2005). High-throughput experiments for synthesis and applications of zeolites. In: *Studies in Surface Science and Catalysis* (ed. J. Cejka and H.V. Bekkum), 161–180. Elsevier.

100 Maier, W.F., Stöwe, K., and Sieg, S. (2007). Combinatorial and high-throughput materials science. *Angew. Chem. Int. Ed.* 46: 6016–6067.

101 Stock, N. (2010). High-throughput investigations employing solvothermal syntheses. *Microporous Mesoporous Mater.* 129: 287–295.

102 Potyrailo, R., Rajan, K., Stoewe, K. et al. (2011). Combinatorial and high-throughput screening of materials libraries: review of state of the art. *ACS Comb. Sci.* 13(6): 579–633.

103 Moliner, M., Román-Leshkov, Y. et al. (2019). Machine learning applied to zeolite synthesis: the missing link for realizing high-throughput discovery. *Acc. Chem. Res.* 52: 2971–2980.

104 Clayson, I.G., Hewitt, D., Hutereau, M. et al. (2020). High throughput methods in the synthesis, characterization, and optimization of porous materials. *Adv. Mater.* 32(44): 2002780.

105 Corma, A., Díaz-Cabañas, M.J., Moliner, M. et al. (2006). Discovery of a new catalytically active and selective zeolite (ITQ-30) by high-throughput synthesis techniques. *J. Catal.* 241 (2): 312–318.

106 Corma, A., Moliner, M., Serra, J.M. et al. (2006). A new mapping/exploration approach for HT synthesis of zeolites. *Chem. Mater.* 18 (14): 3287–3296.

107 Corma, A., Díaz-Cabañas, M.J., Jordá, J.L. et al. (2006). High-throughput synthesis and catalytic properties of a molecular sieve with 18- and 10-member rings. *Nature* 443: 842–845.

108 Jiang, J., Xu, Y., Cheng, P. et al. (2011). Investigation of extra-large pore zeolite synthesis by a high-throughput approach. *Chem. Mater.* 23 (21): 4709–4715.

109 Schmitt, K.D., Burton, A.W., Vroman, H.B. et al. (2020). EMM-31 materials and processes and uses thereof. US Patent 10.710,889, 14th July.

110 Mabon, R., Burton, A.W., Vroman, H.B. et al. (2021). Zeolite synthesis using diquaternary structure directing agents. WO 2021/040915 A1, 4th March 2021.

111 Blackwell, C.S., Broach, R.W., Gatter, M.G. et al. (2003). Open-framework materials synthesized in the TMA^+/TEA^+ mixed-template system: the new low Si/Al ratio zeolites UZM-4 and UZM-5. *Angew. Chem. Int. Ed.* 42 (15): 1737–1740.

112 Willhammar, T., Su, J., Yun, Y. et al. (2017). High-throughput synthesis and structure of zeolite ZSM-43 with two-directional 8-ring channels. *Inorg. Chem.* 56 (15): 8856–8864.

113 Jiang, J., Jorda, J.L., Yu, J. et al. (2011). Synthesis and structure determination of the hierarchical meso-microporous zeolite ITQ-43. *Science* 333 (6046): 1131–1134.

114 Cantín, A., Corma, A., Díaz-Cabañas, M.J. et al. (2006). Rational design and HT techniques allow the synthesis of new IWR zeolite polymorphs. *J. Am. Chem. Soc.* 128 (13): 4216–4217.

115 Dhainaut, J., Daou, T.J., Bats, N. et al. (2013). The influence of L-lysine and PDADMA on the crystal size and porosity of zeolite Y material. *Microporous Mesoporous Mater.* 170: 346–351.

116 Akporiaye, D.E., Dahl, I.M., Karlsson, A. et al. (1998). Combinatorial approach to the hydrothermal synthesis of zeolites. *Angew. Chem. Int. Ed.* 37 (5): 609–611.

117 Choi, K., Gardner, D., Hilbrandt, N. et al. (1999). Combinatorial methods for the synthesis of aluminophosphate molecular sieves. *Angew. Chem. Int. Ed.* 38 (19): 2891–2894.

118 Jensen, Z., Kim, E., Kwon, S. et al. (2019). A machine learning approach to zeolite synthesis enabled by automatic literature data extraction. *ACS Cent. Sci.* 5: 892–899.

119 Moliner, M., Serra, J.M., Corma, A. et al. (2005). Application of artificial neural networks to high-throughput synthesis of zeolites. *Microporous Mesoporous Mater.* 78 (1): 73–81.

120 Serra, J.M., Baumes, L.A., Moliner, M. et al. (2007). Zeolite synthesis modelling with support vector machines: a combinatorial approach. *Comb. Chem. High Throughput Screen.* 10 (1): 13–24.

121 Baumes, L.A., Moliner, M., and Corma, A. (2009). Design of a full-profile-matching solution for high-throughput analysis of multiphase samples through powder X-ray diffraction. *Chem. Eur. J.* 15 (17): 4258–4269.

122 Baumes, L.A., Moliner, M., Nicoloyannis, N. et al. (2008). A reliable methodology for high throughput identification of a mixture of crystallographic phases from powder X-ray diffraction data. *CrystEngComm* 10: 1321–1324.

123 Caremans, T.P., Kirschhock, C.E.A., Verlooy, P. et al. (2006). Prototype high-throughput system for hydrothermal synthesis and X-ray diffraction of microporous and mesoporous materials. *Microporous Mesoporous Mater.* 90 (1–3): 62–68.

124 Klein, J., Lehmann, C.W., Schmidt, H.-W. et al. (1998). Combinatorial material libraries on the microgram scale with an example of hydrothermal synthesis. *Angew. Chem. Int. Ed.* 37 (24): 3369–3372.

125 Nowogrodzki, A. (2020). How to digitize your lab notebooks. *Nature* 586: 159–160.

126 Baerlocher, C. and McCusker, L.B. (2021). Database of zeolite structures. http://www.iza-structure.org/databases (accessed 4th August 2021).

127 Helmkamp, M.M. and Davis, M.E. (1995). Synthesis of porous silicates. *Annu. Rev. Mater. Sci.* 25: 161–192.

128 Camblor, M.A., Villaescusa, L.A., and Díaz-Cabañas, M.J. (1999). Synthesis of all-silica and high-silica molecular sieves in fluoride media. *Top. Catal.* 9: 59–76.

129 Davis, M.E. (2002). Ordered porous materials for emerging applications. *Nature* 417: 813–821.

130 Wagner, P. and Davis, M.E. (2002). Towards the rational design of zeolite frameworks. In: *Supramolecular Organization and Material Design* (ed. W. Jones and C.N.R. Rao). Cambridge, UK: Cambridge University Press.

131 Cundy, C.S. and Cox, P.A. (2005). The hydrothermal synthesis of zeolites: precursors, intermediates and reaction mechanism. *Microporous Mesoporous Mater.* 82: 1–78.

132 Casci, J.L. (2005). Zeolite molecular sieves: preparation and scale-up. *Microporous Mesoporous Mater.* 82: 217–226.

133 Coronas, J. (2010). Present and future synthesis challenges for zeolites. *Chem. Eng. J.* 156 (2): 236–242.

134 Martínez, C. and Corma, A. (2011). Inorganic molecular sieves: preparation, modification and industrial application in catalytic processes. *Coord. Chem. Rev.* 255 (13–14): 1558–1580.

135 Masters, A.F. and Maschmeyer, T. (2011). Zeolites: from curiosity to cornerstone. *Microporous Mesoporous Mater.* 142: 423–438.

136 Wang, Z., Yu, J., and Xu, R. (2012). Needs and trends in rational synthesis of zeolitic materials. *Chem. Soc. Rev.* 2012 (41): 1729–1741.

137 Moliner, M., Rey, F., and Corma, A. (2013). Towards the rational design of efficient organic structure-directing agents for zeolite synthesis. *Angew. Chem. Int. Ed.* 52: 2–12.

138 Bellussi, G., Carati, A., Rizzo, C. et al. (2013). New trends in the synthesis of crystalline microporous materials. *Catal. Sci. Technol.* 3: 833–857.

139 Li, Y. and Yu, J. (2014). New stories of zeolite structures: their descriptions, determinations, predictions, and evaluations. *Chem. Rev.* 114: 7268–7316.

140 Davis, M.E. (2014). Zeolites from a materials chemistry perspective. *Chem. Mater.* 26 (1): 239–245.

141 Moliner, M., Martínez, C., and Corma, A. (2014). Synthesis strategies for preparing useful small pore zeolites and zeotypes for gas separations and catalysis. *Chem. Mater.* 26 (1): 246–258.

142 Gómez-Hortigüela, L. and Camblor, M.Á. (2017). Introduction to the zeolite structure-directing phenomenon by organic species: general aspects. In: *Insights into the Chemistry of Organic Structure-Directing Agents in the Synthesis of Zeolitic Materials. Structure and Bonding*, 175 (ed. L. Gómez-Hortigüela), 175. Cham: Springer.

143 Flanigen, E.M., Broach, R.W., and Wilson, S.T. (2010). Introduction. In: *Zeolites in Industrial Separation and Catalysis* (ed.S. Kulprathipanja), 5. Weinheim: WILEY-VCH: Verlag GmbH & Co. KGaA.

144 Elomari, S.A. and Singa, T.P. (2014). Method for making aluminosilicate zeolite SSZ-56. US Patent 8,647,602, 11th February.

145 Baerlocher, C., Weber, T., McCusker, L.B. et al. (2011). Unraveling the perplexing structure of the zeolite SSZ-57. *Science* 333: 1134–1137.

146 Burton, A., Elomari, S., Medrud, R.C. et al. (2003). The synthesis, characterization, and structure solution of SSZ-58: a novel two-dimensional 10-ring pore zeolite with previously unseen double 5-ring subunits. *J. Am. Chem. Soc.* 125: 1633–1642.

147 Burton, A. and Elomari, S. (2004). SSZ-60: a new large-pore zeolite related to ZSM-23. *Chem. Comm.* 22: 2618–2619.

148 Burton, A.W., Elomari, S., Chan, I. et al. (2005). Structure and synthesis of SSZ-63: toward an ordered form of zeolite beta. *J. Phys. Chem. B* 109: 20266–20275.

149 Xie, D., McCusker, L.B., and Baerlocher, C. (2011). Structure of the borosilicate zeolite catalyst SSZ-82 solved using 2D-XPD charge flipping. *J. Am. Chem. Soc.* 133: 20604–20610.

150 Wu, M.G., Deem, M.W., Elomari, S.A. et al. (2002). Synthesis and structure determination by ZEFSAII of SSZ-55: a new high-silica, large-pore zeolite. *J. Phys. Chem. B* 106: 264–270.

151 Elomari, S. (2003). Zeolite SSZ-64 composition of matter and synthesis thereof. US Patent 6,569,401, 27th May.

152 Elomari, S., Burton, A.W., Ong, K. et al. (2007). Synthesis and structure solution of zeolite SSZ-65. *Chem. Mater.* 19: 5485–5492.

2

De Novo Design of Organic Structure Directing Agents for the Synthesis of Zeolites

Frits Daeyaert[1] *and Michael Deem*[2]

[1] *Synopsisdenovodesign, Beerse, Belgium*
[2] *Certus LLC, Houston, Texas, USA*

2.1 Introduction

Zeolites are typically synthesized from crystallization in a gel medium under hydrothermal conditions. The outcome of a synthesis depends on the reaction conditions and the composition in terms of inorganic species of the gel. In addition, the presence of small organic molecules can direct the synthesis toward the formation of a specific framework. The mechanism of action of this structure direction is believed to be a templating effect whereby the structure directing agent is complementary in terms of size and charge distribution to the pores of the zeolite [1]. The efficiency of the templating has been quantified in terms of the non-bond interaction between the "organic structure directing agent (OSDA)" and the zeolite structure [2]. The interaction energy between an OSDA and a zeolite, also termed the stabilization energy, can be calculated by fitting the OSDA into the zeolite structure through molecular modeling, typically by combining several techniques such as molecular docking, Monte Carlo simulated annealing, molecular mechanics, and molecular dynamics [3]. Thus, in a considerable number of combined theoretical and experimental studies, the interaction energies between known OSDAs and the zeolite frameworks obtained with these have been shown to be correlated [2, 4–13]. This observation in turn has led to the targeted rational design of novel OSDAs for zeolites [14].

In general, when molecular properties can be predicted using a computational protocol or scoring function, novel molecules that exhibit desirable values of these properties can be designed by several approaches.

One approach is to manually design two-dimensional (2D) chemical structures and predict their properties by submitting them to the scoring function. In this manner, good scoring structures can be generated by trial and error. Promising molecules are then synthesized and experimentally tested. This approach requires both profound insight into the field of application and knowledge of synthetic chemistry to ensure that the proposed molecules can be synthesized. While this approach can be successful, it is time-consuming and requires a team of highly skilled professionals from different specialties working closely together.

Another approach to the discovery of molecules with desirable properties is virtual screening. The structures of molecules in a database of compounds that can be purchased from commercial vendors or are otherwise available is submitted to the scoring function, and good scoring molecules are selected for testing. A widely used non-commercial database of purchasable molecules is the ZINC database [15], which (as of May 2021) contained over 750 million purchasable compounds (http://zinc15.docking.org). Other examples are eMolecules (https://www.emolecules.

AI-Guided Design and Property Prediction for Zeolites and Nanoporous Materials, First Edition. Edited by German Sastre and Frits Daeyaert.

com) and Chemspace (https://chem-space.com). Depending on the computational complexity of the scoring function, exhaustive screening of all molecules present in these databases can be unfeasible in practice, even with high speed and massively parallel computers. This problem can be addressed by first filtering the compounds by calculating computationally inexpensive properties, or by screening a structurally diverse subset of a database [16]. At the same time, while the number of molecules in these databases seems high, they nevertheless comprise a minute fraction of the number of hypothetical chemical structures that are chemically feasible, which is estimated to be of the order of 10^{60} [17]. Thus, virtual screening avoids the high cost and uncertainty of chemical synthesis, but at the same time it is basically a random search of a small subset of the available chemical space, which limits its effectiveness.

A third approach to computational molecular design is *de novo* design. A *de novo* design program or algorithm automatically generates molecular structures with desirable predicted properties. *De novo* design is fully automated and therefore requires less human input than manual design. In contrast to virtual screening, its chemical search space is not limited to existing compounds.

A *de novo* design algorithm specifically directed toward OSDAs for zeolites was first developed in the seminal work by Lewis et al. [18]. Template molecules were "grown" in a target zeolite structure by adding fragments to an initial random or user specified template. The putative OSDA was evolved in a genetic algorithm by operators including the addition of fragments, rotation along newly added or existing rotatable bonds, by translation and rotation in space, by ring formation, and by local minimization. The scoring function was based upon the overlap of the van der Waals spheres of the template molecule and the host, and periodic boundary conditions were implemented to allow for the crystalline and symmetric nature of the target zeolite. A close derivative of a *de novo* designed molecule was shown to form a microporous cobalt-aminophosphate (DAF-4, LEV framework [19]). In addition, the algorithm generated multiple analogs of known templates used for the synthesis of ZSM-5 (MFI framework). The method also led to the discovery of a more effective OSDA for the synthesis of the Chabazitic cobalt aluminophosphate (DAF-5 [20]).

A problem with the "growing" of molecules by combining chemical fragments is that, while the molecules can be perfectly feasible with respect to chemical valence rules, the chance that they can be synthesized in the laboratory is very slim. This issue of synthetic accessibility of the compounds generated by a *de novo* design algorithm has long been recognized and addressed in the field of computational drug design. This has led the authors to adapt and apply a *de novo* design program originally developed for drug design to the design of OSDAs for the synthesis of zeolites. Other important features of a *de novo* design program or algorithm are the methods with which the molecular properties to be optimized are calculated and the possibility to simultaneously optimize multiple molecular properties. In this chapter, we describe how these features have been implemented in the "*Synopsis*" *de novo* design program [21, 22], which may serve as a blueprint for future, and improved, programs and algorithms.

2.2 *De Novo* Design

A *de novo* design algorithm generally consists of a molecular structure generator, a scoring function, and an optimization algorithm. Molecule structures are generated by the structure generator and are scored by the scoring function. The optimization algorithm ensures that eventually good scoring molecular structures are obtained.

2.2.1 Molecular Structure Generator

An important challenge with *de novo* design is the synthesizability of the designed molecules. A randomly generated molecular structure may be perfectly valid with respect to chemical valency rules, but the chances are high that the corresponding chemical compound cannot be synthesized in the laboratory with acceptable effort and cost. One approach to address this issue is to, a posteriori, assess the synthesizability of the generated molecular structures by a retro-synthesis program [23]. An alternative approach is to not directly generate molecules, but instead to generate synthesis routes to molecules, where molecules are obtained *in silico* by applying well-documented organic chemistry reactions to available chemical reagents. Other *de novo* algorithms that address synthesizability in a similar way are the DOGs [24] and AutoGrow [25] programs and the multistep reaction-based algorithm described in [26]. Our program uses a list of 96 organic chemistry reactions. These have been compiled from chemistry textbooks, from suggestions from organic chemists, and online sources such as the Organic Chemistry Portal (https://www.organic-chemistry.org). Both zero-order reactions, requiring a single reagent, and first-order reactions, requiring two reagents, have been implemented. Each reaction has a descriptive mnemonic that is used to describe a reaction step in the synthesis route to a *de novo* generated molecule. As an illustration, Table 2.1 lists two of these 96 reactions.

In order for a reagent to participate in a given reaction, both the presence of necessary functional groups and the absence of interfering functional groups are considered. For instance, the ALCOHOLTOACID reaction requires a reagent that contains an alcohol function. At the same time, the reagent should not contain an aldehyde function as this would be oxidized first, as illustrated in Figure 2.1.

Table 2.1 Two reactions available to the *de novo* design program.

Mnemonic	Reactant(s)	Product	Description
REDUCECYAN	1) Alkyl cyanide ($RC \equiv N$)	Alkyl amine (RNH_2)	Reduction of nitrile to amine
MENSHUTKIN	1) Alkyl halide (R_1X) 2) Tertiary amine ($NR_2R_3R_4$)	Quaternary ammonium ($N^+R_1R_2R_3R_4$)	Menshutkin reaction

The first reaction is a zero-order reaction, requiring a single starting material. The second reaction is a first order reaction, requiring two reagents.

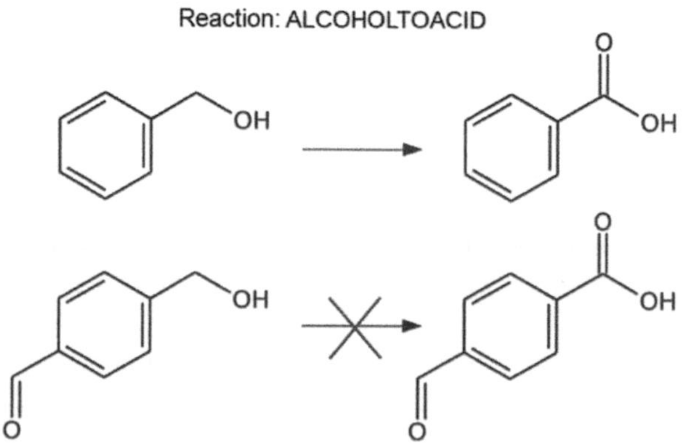

Figure 2.1 Example of a reaction and a restriction to the presence of interfering functional groups.

The reactions and the requirements for their reagents are hard-coded in the software. A list of reagents is to be supplied by the user and typically contains commercially available products that are filtered by cost price and in-stock availability. The list of reagents is pre-processed by a separate program to determine which compounds can participate in each of the 96 reactions. To generate a molecule, a reaction or a sequence of reactions is picked randomly together with a set of suitable reagents to create a synthesis route. The molecular structure of the resulting virtual end product is then submitted to the scoring function. During a *de novo* design run, the program will search for combinations of reagents and reactions that lead to high scoring molecules, as described further below. To reduce the complexity of a synthesis route, the number of reaction steps in a synthesis route is limited. It was found that the success rate of chemical synthesis of a molecule rapidly decreases with the number of reaction steps in a synthesis route. This is illustrated in figure 6 of [21]. Hence, the number of reaction steps in a synthesis route is typically limited to three.

2.2.2 Scoring Function

The scoring function used in a *de novo* design algorithm is often an integral part of the program code [24, 25]. However, we argue that an important asset of a *de novo* design program is that the scoring function is independent of the molecular structure generator and the optimization algorithm. We therefore use a scoring function that is user implemented as a program or script that reads in a molecular structure and outputs a score. This allows the *de novo* design program to be used for very diverse molecular design tasks such as drug design and the design of OSDAs for zeolite synthesis. Additionally, latest insights into molecular property calculations can be easily incorporated into the scoring function during a design effort, independently from the *de novo* design program.

The score that rates a molecule is generally difficult to express as a single number. For OSDA design, an important property is the interaction energy with the zeolite, but other properties such as molecular flexibility, the presence of charge centers, and the absence of undesirable functional groups also determine the quality of a proposed molecule. Thus, the output of the scoring function is a vector of score components that has to be optimized. One approach to this multi-objective optimization problem is the application of a weighing scheme that condenses the score vector into a single number that must be optimized. This requires that for each design effort an appropriate weighing scheme be set up a priori. Alternatively, one can use a Pareto optimization algorithm that generates molecules that are optimal in a Pareto sense. A molecule is Pareto optimal when no other molecules exist that score better on all individual score components. We have preferred to adopt this Pareto approach, which is described in Section 2.2.3.

To score a molecule, a number of easy to calculate properties are often applied as filters. Only for molecules that pass these filters, are more computationally demanding properties that have to be minimized or maximized calculated. Examples of filter-like properties are molecular weight, molecular flexibility, and presence or absence of certain functional groups. To handle score vectors with different types of filters, the score types listed in Table 2.2 have been defined, where filter score types include scores that are binary, have to be below or above a threshold value, or are to be bracketed within an interval. These score types are used to compare score vectors when ranking the score of two molecules, as explained in Section 2.2.3.

Table 2.2 Score types that can be output by the scoring function.

Score type
Binary
Greater than a threshold (>)
Less than a threshold (<)
Greater than or equal to a threshold (≥)
Less than or equal to a threshold (≤)
Bracketed between a minimum and a maximum value
Minimize
Maximize

2.2.3 Optimization Algorithm

The optimization algorithm at the heart of the *Synopsis de novo* design program is a Pareto-driven genetic algorithm (GA). Genetic or evolutionary algorithms mimic biological evolution. Putative solutions to a given optimization problem are coded as chromosomes that evolve in a population by applying genetic operators. At the start of a GA, an initial population of individual solutions with randomly generated chromosomes is created. For each individual, a fitness is determined, which is the quantity to be optimized in the optimization problem at hand. This initial population is evolved by generating child solutions. A child solution is generated by selecting one or more parent solutions from the population, based on their fitness, and combining and mutating their chromosomes. The fitness of the child solution is then determined and the child is inserted into the population by replacing an existing solution. By repeating this cycle of selection, combination, mutation, and replacement, the quality of the solutions gradually improves and eventually individuals with a very high fitness appear in the population. Thus, GAs have been shown to be effective in the solution of hard optimization problems, and have found intensive application in molecular design [27]. Various flavors of GAs exist, differing in the coding into chromosomes of the optimization problem, the implementation of the genetic operators, and the flow of the evolution cycle. Our algorithm is an adaptation of the implementation of a multi-objective GA by Deb et al. [28].

The first step in the design of a GA to a given optimization problem is the coding of the problem into chromosomes on which genetic operators can be applied. As explained above, the problem we want to solve is to find combinations of organic chemistry reactions and reagents that form reaction products that score well in a given scoring function. Thus, a chromosome of our GA is a virtual synthesis route containing a single reaction step or a sequence of reaction steps in which reagents participate. Each reaction step is one of the set of 96 well-documented organic chemistry reactions implemented in the program. Each reagent is a compound from a user supplied database of available compounds, having the correct chemical functionalities. A graphical illustration of a chromosome representing a synthesis route is shown in Figure 2.2.

At the start of a GA run, a number N, typically 100, of synthesis routes are generated by randomly choosing reactions and reagents. The number of reaction steps in a synthesis route is limited to reduce synthetic complexity. The end products of the synthesis routes are submitted to the scoring function to obtain their fitness, which consists of a score vector of molecular properties, as discussed above. When the initial population of synthesis routes is complete and the

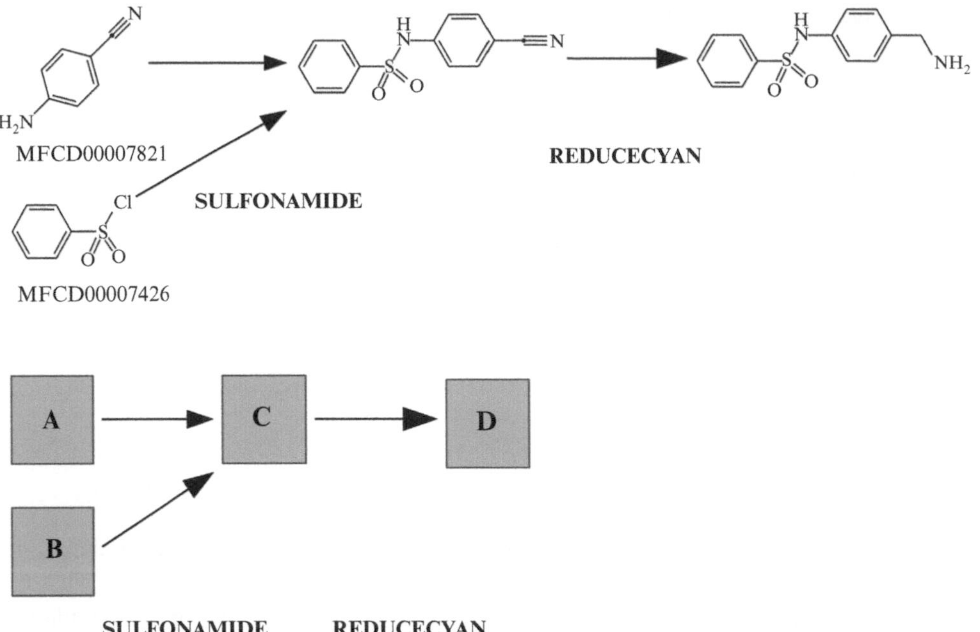

Figure 2.2 Example of a synthesis route and its schematic representation as a chromosome in the GA. The synthesis route consists of two steps. The first step is the SULFONAMIDE reaction and requires two reagents. These are the randomly chosen commercially available compounds MFCD0000782 and MFCD00007426. The second step in the synthesis route is the REDUCECYAN reaction, which is applied to the virtual reaction product of the first step. The end product, in the schematic representation marked as "D," is to be submitted to the scoring function to obtain its fitness.

fitness of the end products has been obtained, the evolutionary cycle illustrated in Figure 2.3 is started by applying the genetic operators. The first operator is the selection operator to select parent synthesis routes that will produce child synthesis routes. The GA uses tournament selection: the population of synthesis routes is ranked, *n* synthesis routes are picked at random, and the highest ranked synthesis route is selected as the winner. The fitness of a synthesis route is determined by applying the scoring function to the molecule corresponding to its virtual end product. As this fitness is a vector of multiple score components, the multi-objective Pareto sorting method described in [28] is applied to rank the synthesis routes in the population. This non-dominated Pareto sorting is based upon the concept of dominance: when two score vectors are compared, one of the two dominates the other, when for each of the score components it performs better than the other. When this is not the case, there is a tie and the two score vectors belong to the same Pareto front. To compare the score vectors of two molecules, both the values of the score components and the score types defined in Table 2.2 are taken into account. For filter

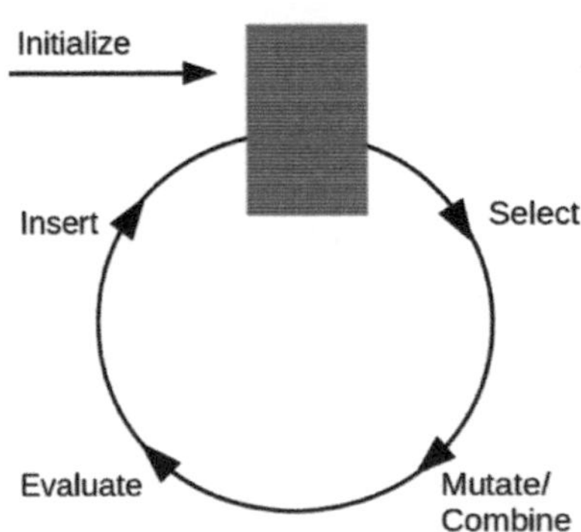

Figure 2.3 Evolutionary cycle of the genetic algorithm at the heart of the *de novo* design program.

type score components, if the score component of one molecule passes the filter and the other does not, the former is the winner for that score component. If the score components of both molecules fall within a threshold or bracket, there is a tie. If both molecules fail to pass the filter, the one closest to the threshold or bracket is the winner for that score component. To determine whether one molecule and its corresponding synthesis route dominates the other, all score components in the score vector are considered and dominance occurs when one molecule wins on all score components. As mentioned above, for molecules that do not pass all filter-type score components, the calculation of the more computationally demanding score components can be skipped to save CPU time. To perform the Pareto ranking, the values of these score components are assigned default values that are bad.

To rank the population of the GA, the different Pareto fronts are determined and the synthesis routes are sorted according to their Pareto front number. Within each Pareto front the synthesis routes are ranked using a secondary sorting method. The original algorithm uses a crowding distance, which is the average difference between the values of the score components of an individual and those of the individuals in the population [28]. Within a Pareto front, the individuals are ranked according to their decreasing crowding distance in order to evenly spread the individuals over the front. This crowding distance is straightforwardly defined for score components that have to be minimized or maximized, but cannot be calculated for individuals that do not pass all filter-type score components. For these synthesis routes, a different secondary ranking method based upon the order in which they have been generated and have entered the population is applied. Within a Pareto front, "younger" synthesis routes are ranked higher than "older" ones. The rationale for this is that the preference for younger synthesis routes will drive evolution and explore new areas of the molecular search space.

When the initial population of synthesis routes has been generated and scored and ranked, child synthesis routes are generated by applying one of six genetic operators:

1) *Add*: extend the synthesis route with a randomly chosen reaction step
2) *Cut*: delete the last reaction step of a synthesis route
3) *Replace Random*: replace one reagent by another, randomly selected reagent in a synthesis route
4) *Replace Like*: replace one reagent by another, chemically similar reagent in a synthesis route
5) *Combine*: combine the reaction sequences of two synthesis routes
6) *Random*: randomly generate a new synthesis route

The parent synthesis routes required by these operators are selected by tournament selection as explained above. The operators are illustrated in Figures 2.4 to 2.9.

When applying the "Add" operator, an additional reaction step is applied in which the end product of a parent synthesis route is a reagent. The additional reaction step is picked at random, under the constraint that the end product of the parent synthesis route has the correct chemical functionality. If the newly added reaction step requires an additional reagent, it is picked randomly from the reagent list. The "Add" operator can only be applied when the maximum number of reaction steps in the parent

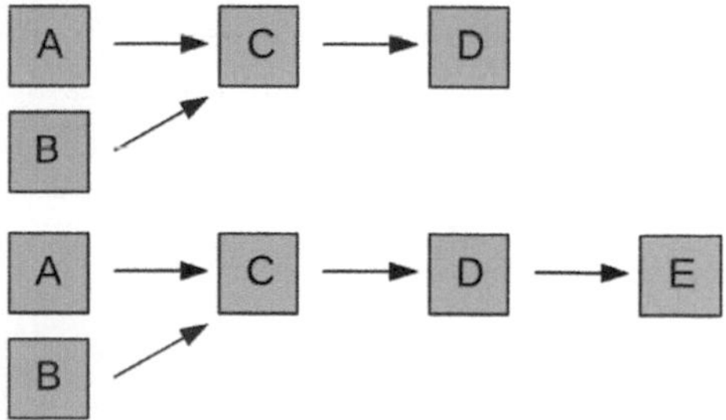

Figure 2.4 Schematic illustration of the ADD operator. A parent synthesis route, represented by the blue squares, is extended by an additional reaction step that is randomly chosen out of the 96 available reactions. The end product of the child synthesis route is represented as "E," which is to be submitted to the scoring function to obtain its fitness.

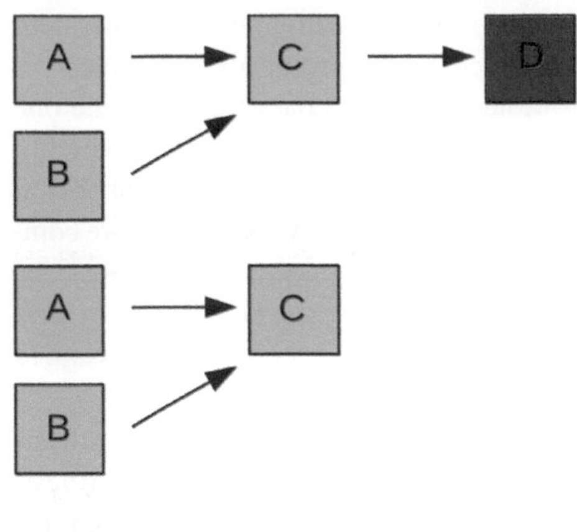

Figure 2.5 Schematic illustration of the CUT operator. The last synthesis step of a parent synthesis route, represented by the upper blue and red squares, is removed. The end product of the child synthesis route is the square represented as "C" in the lower scheme, which is to be submitted to the scoring function to obtain its fitness.

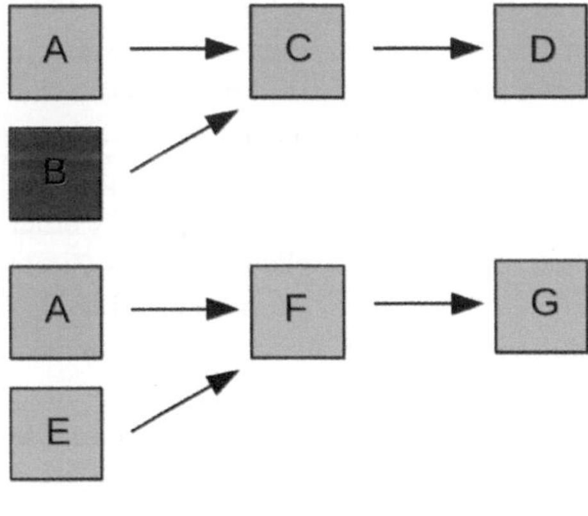

Figure 2.6 Schematic illustration of the REPLACERANDOM operator. The starting material represented as "B" in the red square in the parent synthesis route (top) is replaced by another starting material represented as "E" in the green square in the child synthesis route (bottom). The only restriction to the new starting material is that it has the correct chemical functionality to participate in the parent synthesis route. The end product of the child synthesis route is the square represented as "G" in the lower scheme, which is to be submitted to the scoring function to obtain its fitness.

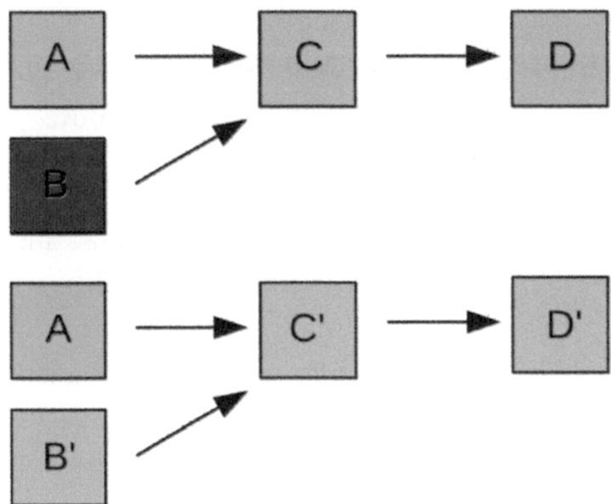

Figure 2.7 Schematic illustration of the REPLACELIKE operator. The starting material represented as "B" in the red square in the parent synthesis route (top) is replaced by another starting material represented as "B′" in the green square′ in the child synthesis route (bottom). The new starting material must have the correct chemical functionality to participate in the parent synthesis route, and must have a 2D chemical similarity larger than a given threshold to the original starting material "B." The end product of the child synthesis route is the square represented as "D′" in the lower scheme, which is to be submitted to the scoring function to obtain its fitness.

synthesis route has not been reached. The "Cut" operator removes the last reaction step from a parent synthesis route. When applying the "Replace Random" operator, one reagent in a parent synthesis route is replaced by another, randomly picked, reagent that has the required chemical functionality. In the "Replace Like" operator, this newly picked reagent also has to have a minimal 2D similarity to the original reagent. The similarity measure used is a Tanimoto coefficient based upon a binary string of 1024 descriptors, and the similarity threshold is typically 0.75. The

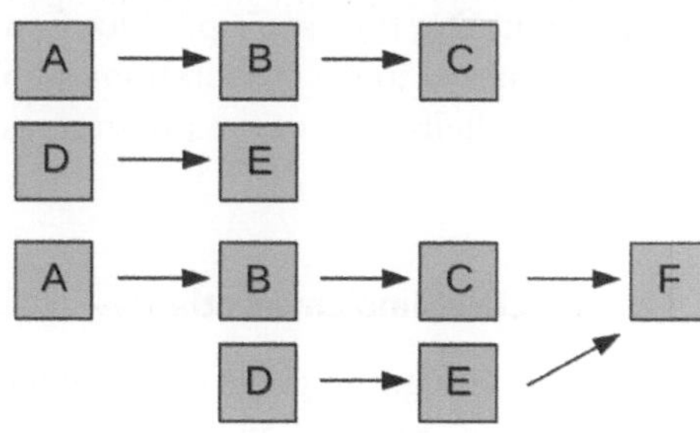

Figure 2.8 Schematic illustration of the COMBINE operator. The end products of two parent synthesis routes (top), represented as "C" and "E," participate in an additional reaction step to form the new end product "F." The added reaction step is a randomly chosen reaction from the list of 96 available reactions for which compounds "C" and "E" have the correct chemical functionality. The end product of the child synthesis route is the square represented as "F" in the lower scheme, which is to be submitted to the scoring function to obtain its fitness.

"Combine" operator searches two parent synthesis routes for end products or intermediates that can participate in one of the available reactions. Care is taken that the maximum number of reaction steps is not exceeded in the resulting child synthesis route. The "Random" operator generates a random synthesis route in the same way that the synthesis routes of the initial population were generated. The mutation and combination operators are summarized in Table 2.3.

Figure 2.9 Schematic illustration of the NEWRANDOM operator. A new synthesis route is generated from scratch, using a randomly chosen set of reactions and reagents available in the reagent database. No parent synthesis route is needed. The end product of the new child synthesis route is the square represented as "H," which is to be submitted to the scoring function to obtain its fitness.

When a child synthesis route has been generated and its end product is not already present in the population, its fitness is calculated and it is inserted into the population by replacing another synthesis route according to one of two replacement schemes. In one scheme, the fitness of the child synthesis route is compared to the fitness of the lowest ranked synthesis route in the population. When the fitness of the child synthesis route is better, it replaces the lowest ranked synthesis route. Otherwise, the child synthesis route is discarded. In an alternative replacement scheme, the child synthesis route replaces the oldest synthesis route in the population, regardless of its fitness. In both replacement schemes, an elitist mechanism is applied such that the m highest scoring synthesis routes of the population are excluded from replacement. After the child synthesis route has been inserted into the population, the population is re-ranked as described above.

Table 2.3 Mutation and combination genetic operators.

Operator	Action	Parents
Add	Add reaction step	1
Cut	Remove reaction step	1
Replace Random	Randomly replace reagent	1
Replace Like	Replace reagent with similar one	1
Combine	Combine two synthesis routes	2
Random	Generate new synthesis route	0

Child synthesis routes are produced and evaluated one by one. This "steady-state" mechanism, as opposed to the "generational replacement" mechanism often used in GAs, allows an efficient parallelization of the program as it requires no synchronization points in the course of a run.

2.2.4 Practical Implementation

The *de novo* design program is implemented as a main program that performs the structure generation and optimization, and an independent daemon program that calls the user-supplied scoring function. Multiple instances of the daemon program can be run on multiple CPUs and computers. Communication between the main program and the daemons is through disk file semaphores.

The GA uses a number of adjustable parameters such as the population size, the tournament size of the selection operator, the number of elitist solutions, the relative probabilities of the mutation and combination operators, the Tanimoto similarity threshold for the Replace Like operator, and the replacement scheme. In practice, the performance of the algorithm turns out to be rather insensitive to the precise values of these parameters as long as they are reasonable [22].

The output of a *de novo* design run is a list of molecular structures together with their score vectors and the synthesis routes by which they were generated. It can be converted to an .html document for examination with a web browser. The synthesis route of each entry is provided through a link that is activated by clicking on the entries in the .html document. An example of the raw output of a synthesis route is shown in Figure 2.10.

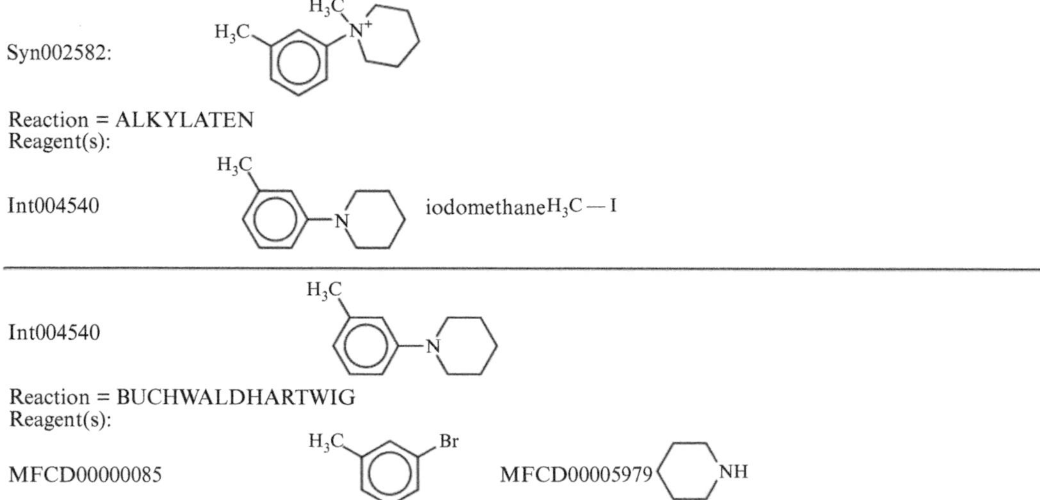

Figure 2.10 Raw output as .html of the synthesis route to a *de novo* designed molecule. The synthesis route consists of two reaction steps picked from the set of 96 available reactions: a Buchwald–Hartwig coupling followed by methylation of an sp3 nitrogen. The compounds with MFCD numbers are commercially available starting materials.

2.3 Scoring Functions for OSDAs

As discussed in the Introduction, the efficiency of an OSDA in templating a given zeolite framework depends on the stabilization energy when the OSDA is fitted into the zeolite pores. Thus, typically, an OSDA scoring function calculates the stabilization energy of a molecule, and the *de novo* design engine generates molecules for which this stabilization energy is most negative. However, other molecular properties such as flexibility, size, or charge also determine whether a molecule is fit to act as an OSDA. In addition, for some OSDA design projects, the stabilization energy in multiple zeolites needs to be calculated.

2.3.1 Stabilization Energy

Computational design of OSDAs relies on the paradigm that a successful OSDA acts as a template molecule that fits into the pores of its target zeolite [1]. This templating effect can be quantized by the stabilization energy of an OSDA docked into the pores of a target zeolite [2]. The stabilization energy is generally obtained by first fitting a number of OSDA copies into the pores of the target zeolite, and then performing an atomistic simulation using either force field-based or quantum mechanical methods. An overview of methodologies is given in [3], and specific approaches are detailed in [29–31]. We here describe the procedure we have used in our own OSDA design work.

The stabilization energy of an OSDA is defined as the van der Waals term of the interaction energy between the OSDA and the zeolite. To calculate the stabilization energy, first the structure of the free OSDA has to be determined. For OSDAs with rotatable bonds, this requires a conformational search. For molecules with multiple rotatable bonds, this is a challenging problem that cannot be solved in a deterministic way with reasonable computational effort. In our work we use an in-house developed genetic algorithm-based program. The chromosomes in this GA are the decimally encoded values of the torsion angles of the rotatable bonds of the molecule. At the start of the program, a population of conformations with randomized torsion angles is generated. The default maximal population size is $20 \times N$, where N is the number of rotatable bonds. It is ensured that all conformations in the population are unique, which means that the actual population size may be less than this maximal number. The fitness of a conformation is the molecular mechanics (MM) energy obtained by initializing the torsion angles with the values in the chromosome and then applying a full local MM optimization. The force field used is MMFF [32]. After the MM minimization, the torsion angle values in the chromosome are replaced by the torsion angle values in the minimized conformation. When the initial population is complete, child conformations are generated by applying selection, random mutation, and combination operators. By default, $50 \times N$ children are generated in a run. Due to the stochastic nature of the algorithm, it cannot be guaranteed that the globally optimal conformation will be present in the final population. However, it has been our experience that for molecules with up to 10 rotatable bonds, this procedure consistently generates the same low-energy conformation, and we assume that this is the global optimum on the MM potential energy surface. The CPU time needed for a conformational analysis depends on the number of rotatable bonds and the number of atoms in the OSDA, and typically varies between 1 and 10 minutes.

For molecules with unsaturated rings, a ring-flipping algorithm is used to generate multiple ring conformations before the conformational analysis. These conformations are used to seed the initial population of the conformational analysis program.

When the lowest energy conformation of an OSDA has been obtained, a number of copies of this conformation are successively fitted into the zeolite. The optimal translation of each copy is determined by fast Fourier transform from the convolution shift on a van der Waals grid. The optimal rotation is determined by repeating this procedure for 100 random orientations. The number of copies fitted into a zeolite depends on the zeolite and is determined by trial and error. The stabilization energy per copy of the OSDA is then determined by molecular dynamics (MD). First, four rounds of MM minimization are used with alternate BFGS (Broyden–Fletcher–Goldfarb–Shannon) and conjugate gradient minimizers. Next, three MD runs are performed with different time steps and simulation times of 0.1, 1.0, and 30 ps, respectively. The stabilization energy is defined as $E = E_{system} - E_{zeolite} - nE_{OSDA}$. E_{system} is the energy of the zeolite with *n* copies of the OSDA in the unit cell, $E_{zeolite}$ is the energy of the zeolite, and E_{OSDA} is the energy of a single OSDA. The energies are averages taken from the last 5 ps of the third MD run. They are reported as energies per silicon atom in units of kJ/(mol Si), allowing comparison of different OSDAs in the same zeolite. The minimization and MD calculations are performed with the GULP program [33] using the Dreiding force field [34]. A dynamics run requires between one and several hours on a single CPU, depending on the number of atoms in the zeolite unit cell, the number of atoms in the OSDA, and the number of OSDA copies.

2.3.2 Other Constraints

To design OSDAs targeted toward a given zeolite, the *de novo* design program can be run in combination with a scoring function that minimizes the stabilization energy as calculated with the procedure described in the previous paragraph. However, besides a highly negative stabilization energy, there are other molecular properties that are required for a molecule to act as a successful OSDA. These properties can generally be formulated as constraints, and a number of these are summarized in Table 2.4.

A first constraint is molecular flexibility, as more rigid molecules have been found to be more efficient OSDAs [35]. Molecular flexibility can be reduced by limiting both the overall number of rotatable bonds and the highest number of sequential sp3–sp3 atoms in a designed molecule. We define a rotatable bond as a bond between any atom group and an sp3 hybridized atom, with the exception of end-standing methyl and CX_3 groups. Limiting the total number of rotatable bonds can still lead to molecules with highly flexible chains of sp3 hybridized atoms such as *n*-alkyl

Table 2.4 Molecular properties and constraints for OSDAs.

Property	Score type
Flexibility: number of rotatable bonds	Less than a threshold (<)
Flexibility: number of sequential sp3-sp3 bonds	Less than a threshold (<)
Stability: elements other than C, N, H	Less than a threshold (<)
Stability: no $C \equiv C$, $C \equiv N$	Less than a threshold (<)
Composition: C to N^+ ratio	Less than a threshold (<)
Charge: positively charged N	Bracketed between a minimum and maximum value
Size: volume	Bracketed between a minimum and maximum value

substituents on an otherwise rigid scaffold. We therefore impose a second flexibility constraint by explicitly limiting the number of adjacent sp3–sp3 atoms in a chain of atoms.

Chemical stability of OSDAs under the hydrothermal reaction conditions generally required to synthesize zeolites precludes the presence of reactive functional groups such as esters, amides, lactams, ketones, aldehydes, alcohols, and acids in an OSDA. Also, the presence of certain elements such as sulfur and halogens in zeolites to be used as catalysts after removal of OSDAs by calcination is often unwanted. A simple and straightforward way to ensure both the chemical stability of an OSDA and the absence of unwanted elements is to limit the elements in a designed OSDA molecule to C, N and, H atoms. An additional constraint to ensure chemical stability under hydrothermal reaction conditions is the absence of triply bound C atoms, as in alkynes and cyan groups.

Another property that determines the ability of a molecule to be effective as an OSDA is its hydrophobicity. Molecules with moderate hydrophobicity are better suited to structure-direct zeolites. A simple measure for hydrophobicity is the ratio of C atoms to charged N atoms [35]. Thus, the presence of one or more positively charged N atoms and the C to N^+ ratio are introduced as additional constraints.

Molecular volume is another constraint that can be used as a filter, as molecules that are too large will not fit into the zeolite pores, and molecules that are too small will not be able to effectively fill the pores. Obviously, the threshold values for the molecular volume depend on the target zeolite.

The above-mentioned molecular properties can be obtained in a trivial way from just the molecular connectivity or from the 2D or 3D structure of the putative OSDA. They are used as filters, and any molecules that do not pass the threshold values are not subjected to the more elaborate calculation of the stabilization energy in the target zeolite. These properties and constraints have been introduced by us in our own design work to increase the probability that a designed molecule will be a successful OSDA. We emphasize that their introduction and implementation is totally independent from the *de novo* design engine. Any other or additional properties or constraints can be added to a scoring function together with their desired score types and threshold values without altering the *de novo* design program.

2.3.3 Multiple Objectives

To design OSDAs for a single zeolite, the *de novo* design engine can be run in combination with a scoring function that first calculates a number of easily obtained properties that are used as filters. When a molecule passes all the filters, the stabilization energy of the molecule when fitted into the target zeolite is calculated using the MD protocol. At the start of a *de novo* design run, the population of the GA generally contains molecules that do not pass many of the filters, and the Pareto front contain multiple molecules. The Pareto front is multidimensional, with one dimension for each constraint that has not been met by all molecules in the front. When molecules appear that pass all the filters, the Pareto front condenses into one dimension, with the molecule having the most favorable stabilization energy forming the first Pareto "front." The end result of a *de novo* design run is a list of molecules that meet all the constraints and are ranked according to their stabilization energy. Even when the aim of a design run is the optimization of a single final objective, we have observed that the Pareto sorting is important to effectively generate molecules that pass the multiple constraints in the initial stages of the GA [22].

In addition, the Pareto approach allows the optimization of multiple final objectives. One example is the design of OSDAs that template one particular zeolite, but do not template a competing zeolite. In this case, the stabilization energy of OSDA in the target zeolite is to be minimized, and the stabilization energy in the competing zeolite is to be maximized. The result of a design run is then a set of molecules that meet all imposed constraints and that form a 2D Pareto optimal front. Within the front, no molecule has a lower stabilization energy toward the target zeolite without at the same time having a lower stabilization energy toward the competing zeolite. An example is the design of OSDAs that stabilize polymorph A of zeolite beta (BEA), but not the competing polymorph B (BEB), as described in [36]. This is illustrated in Figure 2.11, which shows the three first Pareto fronts in the final population of a *de novo* design run. Each data point corresponds to a designed OSDA. The data points are connected with solid lines to show the Pareto fronts. The x-axis in this figure presents the stabilization energy in BEA, and the y-axis presents the stabilization energy in BEB. The blue, green, and red lines show the first, second, and third Pareto fronts in the final population of the design run, respectively. An ideal OSDA would show up in the upper left part of the Pareto plot, which corresponds to a molecule having a very low stabilization energy in BEA and at the same time a very high stabilization energy in BEB. In Pareto terms, this part of the plot is called the "Utopia." Some molecules with a favorable very low BEA stabilization energy, like the one corresponding to the left-most lower blue data point, also have an unfavorable low BEB stabilization energy. This can be attributed to the close similarity between the two polymorphs. Other molecules, with a favorable very high BEB stabilization energy, like the one corresponding to the right-most upper blue data point, have at the same time an unfavorable higher BEA stabilization energy. The Pareto optimization algorithm tries to find molecules that score well on both criteria, and it is up to the designer to manually select molecules from the Pareto front for OSDA and zeolite synthesis. In this particular example, it can be seen that in the first Pareto front, several molecules are present with a BEA stabilization energy below 15 kJ/(mol Si) and a BEB stabilization energy that is positive, so these molecules would be a first choice.

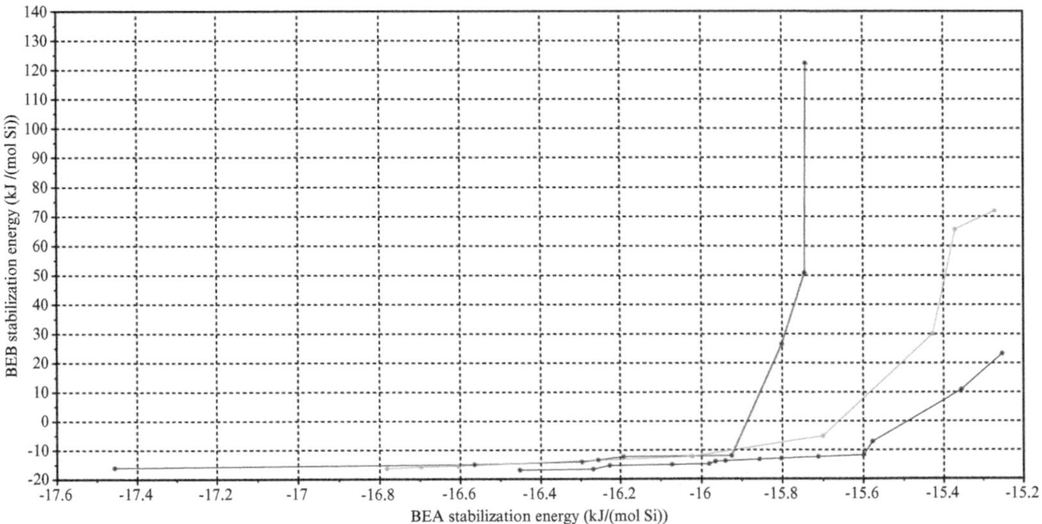

Figure 2.11 First three Pareto fronts in the final population of a *de novo* design run to generate OSDAs that target BEA but not BEB. The first front is formed by the blue data points. The data points have been linked with solid lines for clarity. The data are taken from [36].

Another example where the final objective of an OSDA design effort is dual is in the search for chiral OSDAs that selectively target one enantiomer of a chiral zeolite. Indeed, it has been shown that chiral, stereochemically pure OSDAs can lead to enantiomerically-enriched chiral zeolites [37, 38], which offers perspectives for chiral separation and catalysis. An example design study is the design of chiral OSDAs to template the $P4_122$ enantiomorph of zeolite BEA [39]. The scoring function to be used by the Pareto-driven *de novo* design algorithm is adapted by first adding the constraint that requires the presence of at least one chiral center in a putative OSDA. The number of chiral centers is limited to one or two in order to reduce the effort that will be required to eventually synthesize the enantiomerically pure OSDAs. For molecules that pass all the other imposed constraints, all enantiomers are generated by repeated inversion through all chiral centers. Meso forms of symmetrical molecules and mirror image pairs that are too energetically stressed, as in some fused-cycle molecules, are rejected. All chiral enantiomers are then passed to the MD protocol to calculate their stabilization energy in the $P4_122$ enantiomorph of zeolite BEA. The stabilization energy of the lowest scoring enantiomer is the first Pareto criterion, which must be minimized. The second criterion is the chiral energy gap. This is the difference in stabilization energy in the $P4_122$ enantiomorph of zeolite BEA of the best fitting enantiomer and its mirror image. This energy gap has to be maximized. Figure 2.12 shows the three first Pareto fronts in the final population of a *de novo* design run. As in Figure 2.11, the data points are connected by blue, green, and red solid lines to illustrate the first, second, and third Pareto fronts, respectively. The OSDAs with the very lowest stabilization energy of the fittest enantiomer, corresponding to the lower left data points, have a very small energy gap and therefore can be expected to not selectively target a single enantiomorph of BEA. The upper right data points correspond to OSDAs that are predicted to exhibit a very large energy gap, but their templating capability toward BEA is reduced. The data points in the center of the front approach the Pareto Utopia region in the upper left of the plot. They correspond to molecules of which one enantiomer is predicted to have a high templating capacity toward $P4_122$ BEA, while the other enantiomer is predicted to be ineffective in templating this enantiomorph. Thus, these are the molecules that are most promising as OSDAs for producing enantiomerically-enriched BEA.

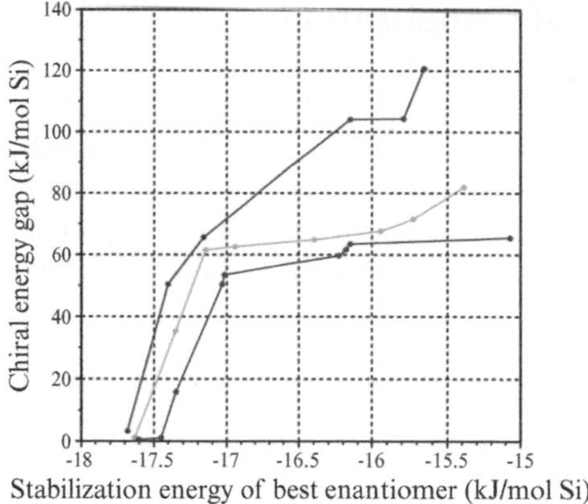

Figure 2.12 First three Pareto fronts in the final population of a *de novo* design run to generate chiral OSDAs that selectively target the $P4_122$ enantiomorph of zeolite BEA. The first front is formed by the blue data points. The data points have been linked with solid lines for clarity. The data are taken from [39].

In the two examples discussed above, the final objective of a *de novo* design run is two-dimensional. Extension to higher-dimensional final objectives is independent of the *de novo* design engine and only requires the setting up of an appropriate scoring function. An example of a three-dimensional (3D) design problem would be the design of chiral OSDAs that target one enantiomorph of zeolite BEA, but not the BEB polymorph. For the selection of good candidate molecules from a higher-dimensional Pareto front appropriate algorithms can be invoked [40].

2.4 Applications

Over the years, we have deployed our *de novo* design engine in various OSDA design projects. These are summarized in Table 2.5, together with references to the corresponding publications. In this section we briefly summarize and comment upon these projects. We also discuss the recently published work by Muraoka et al. [41] on multi-objective *de novo* design of OSDAs using an ant colony optimization algorithm.

2.4.1 From Drug Design to the Design of OSDAs for Zeolites [42]

As mentioned at the beginning of this chapter, the *Synopsis* program was originally developed and applied in a drug design setting. In this project, we have for the first time applied it to the design of OSDAs for zeolites. As the *de novo* design engine and the scoring function are fully separate programs, this did not require any alterations to the former. We did, however, expand the list of organic chemistry reactions available at that time with the Menshutkin reaction. This reaction is of special interest to OSDA design, as it is an important route to quaternary ammonium ions [52]. In later projects we added more reactions directed specifically to OSDA design, adding up to the 96 reactions available at present. A scoring function was set up that calculated a score vector comprised of four score components: the absence of unwanted chemical groups, molecular flexibility, molecular volume, and the stabilization energy in the target zeolite. The first three score components were used as filters, while the stabilization energy had to be minimized and was only computed for molecules passing the filters. To test the method, we applied it to three zeolites for which effective OSDAs have been documented: AEI, ITE, and STF. For all three zeolites, we were able to design molecules with predicted stabilization energies well below the predicted stabilization energies of known state-of-the-art OSDAs.

Table 2.5 Overview of *de novo* OSDA design projects.

References	Target Zeolite(s)	Synthesized
[42]*	AEI, ITE, STF	
[43]*	STW	V
[44]*	AEI	V
[45]*	SFW	V
[46]*	STW	
[36]*	BEA vs. BAB	
[39]*	chiral BEA	
[47]*	BEA	
[48]*	WEI, JBW, GIS, SIV, DAC, 8124767, 8277563	
[49]*	GIS, ABW, 8186909, 8198030	
[50]*	DFT, ACO, NAT, JRY	
[51]*	Organic linkers for MOFs	
[41]**	CHA, AEI, CON	

*By the authors.
**Multi-objective *de novo* design of OSDAs using an ant colony optimization algorithm by K. Muraoka et al. [41].

2.4.2 Experimental Confirmation: Pure Silica STW [43]

This project resulted in the first experimentally confirmed zeolite synthesis with the aid of an OSDA designed using our *de novo* design engine, thus delivering a proof-of-principle of the approach. The zeolite synthesized was STW (HPM-1). A pure-silica, fluoride-mediated inorganic synthesis system was used to limit the number of parameters influencing the outcome of the synthesis. The designed OSDA, the monoquaternary pentamethylimidazolium, in a later study served as a basis for a chiral diquaternary imidazolium OSDA that was used to prepare enantiomerically-enriched STW zeolite [37].

2.4.3 Experimental Confirmation: Zeolite AEI [44]

In this study, we designed a novel OSDA that was used to synthesize zeolite AEI (SSZ-39) in a hydroxy-mediated synthesis mixture in the presence of aluminum and other inorganic cations other than silica. This study demonstrates that *de novo* design can be successful even when a larger number of synthetic variables come into play, as is typically the case for the synthesis of commercially interesting microporous materials.

2.4.4 Practical Application: SSZ-52 (SFW) [45]

In this study, *de novo* design was used in combination with virtual screening of a library of known quaternary ammonium compounds to design effective OSDAs for the synthesis of SSZ-52 (SFW). This is a small-pore zeolite that shows promise for application in engine exhaust clean-up but for which the one known effective OSDA is difficult to synthesize and therefore cost prohibitive. After minor modification, a *de novo* designed OSDA, in combination with a reduced amount of the original known OSDA, was shown to promote the synthesis of SSZ-52, thus opening a pathway to economically feasible application of this zeolite.

2.4.5 Design of Chiral OSDAs to Direct the Synthesis of Chiral STW [46]

In previous work mentioned above, we designed a monoquaternary imidazolium directing the synthesis of STW [43]. This OSDA was subsequently used as a monomer in a dimerization reaction using a chiral linker molecule to form a chiral OSDA leading to enantiomatically-enriched STW [37]. The monoquaternary imidazolium and the derived chiral dimer are depicted in Figure 2.13. The synthesis route leading to the chiral dimer can be generalized, as in Scheme 1.

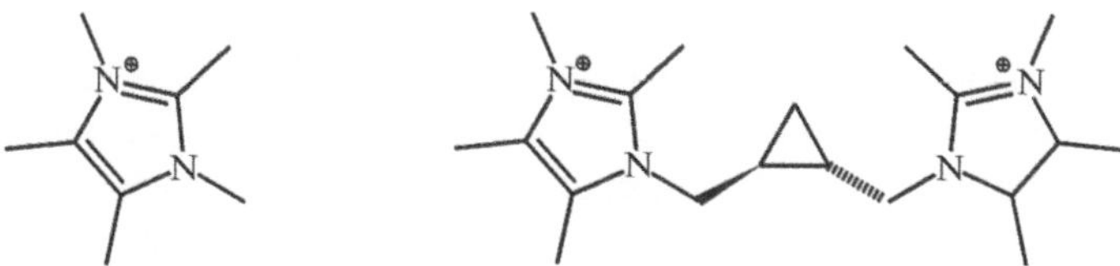

Figure 2.13 Monoquaternary imidazolium and its derived chiral dimer as OSDA for enantiomerically-enriched STW.

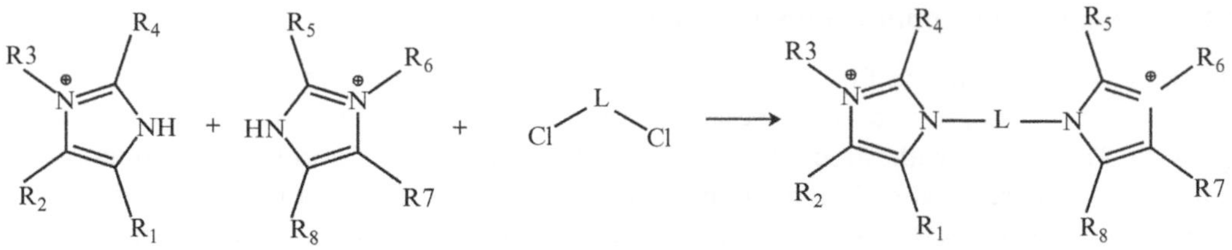

Scheme 1: Reaction scheme leading to chiral imidazolium dimers as OSDAs directed toward enantiomerially-enriched STW. The linker molecule L must contain at least one chiral center.

Using this reaction scheme, virtual combinatorial chemistry was applied to identify novel chiral OSDAs directed toward STW. Combinations of monomers and linkers were explored to identify dimers of which one enantiomer has a favorable stabilization energy in STW, while its mirror image has a less favorable stabilization energy. The set of imidazolium structures was initially obtained from a substructure search on a database of available compounds. The resulting structures were fit into the STW zeolite, and those molecules with a stabilization energy below −14 kJ/(mol Si) were retained. To this set, 23 imidazolium molecules resulting from a *de novo* design run were added, resulting in a set of 139 monomers to participate in the reaction scheme. As linker molecules, a set of 17 commercially available chiral dichlorides were obtained from a vendor database. Except for the two reactive chlorine atoms, no elements other than C, N, or H were allowed in the linker molecules. With these monomer and linker sets, a total of $139 \times 139 \times 17 = 328{,}457$ dimers could be formed, a number that is not amenable to an exhaustive calculation of their stabilization energies. We therefore devised a stochastic virtual combinatorial chemistry algorithm to search this chemical space. This algorithm is a GA closely analogous to the *de novo* design program. The input to the algorithm is a single, fixed synthesis route that consists of a number of reaction steps chosen from the set of 96 reactions available to the *de novo* design engine, and a user supplied database of reagents. A chromosome of the GA consists of a list of references to reagents in the reagent database, and the aim of the GA is to find combinations of reagents for which the synthesis route generates products that score favorably in an externally provided scoring function. The working of the algorithm, including the use of different score types and Pareto-based ranking and selection to allow multi-objective optimization is identical to that of the *de novo* design engine. However, as the synthesis route coded by a chromosome of the GA is fixed, the “Add” and “Cut” operators to generate child chromosomes are not defined, and the “Combine” and “Random” operators apply to the reagents only. The virtual combinatorial chemistry problem addressed here with a GA is very similar to the one encountered in the recent *de novo* work of Muraoka et al. [41], where it is addressed using ant colony optimization as a stochastic optimization algorithm. This work is described below Section 2.4.11.

In the STW study, the synthesis route was the scheme depicted in Scheme 1 and the reagent database contained the 139 imidazolium compounds and the 17 dichloride linkers. The scoring function first calculated a number of 1D and 2D molecular properties that were used as filters, including the presence of at least one but no more than two chiral centers. For molecules passing these filters, the stabilization energies in STW of the lowest energy conformation of each enantiomer and its mirror image were calculated. The GA was used to generate molecules with a low stabilization energy for one enantiomer, and a high chiral stabilization energy gap with respect to its mirror image. A total of 146 molecules were identified with a predicted stabilization energy in STW below −15 kJ/(mol Si), and chiral energy gaps between a few tenths up to ~100 kJ/(mol Si).

The final Pareto front, depicted in figure 3 of [46], consisted of 11 molecules with stabilization energies and energy gap ranging from −17.5 kJ/(mol Si) and 4.0 kJ/(mol Si) to −15.5 kJ/(mol Si) and 99.7 kJ/(mol Si).

2.4.6 Design of Selective OSDAs Directed Toward BEA vs. BEB [36]

In this study, OSDAs were designed that selectively stabilize the A polymorph of zeolite beta. This industrially important zeolite is typically obtained as an intergrowth between two polymorphs, BEA and BEB. Availability of pure polymorph zeolite beta BEA has the potential to lead to more efficient catalysis and separation processes. Also, the synthesis of the pure polymorph of chiral BEA would be a first step toward the enantiomerically pure or enantiomerically-enriched zeolite. Thus, the aim of the study was to design OSDAs that have a favorable stabilization energy toward BEA, and a less favorable stabilization energy toward BEB. Several design methods were applied: *de novo* design, virtual screening, and exhaustive and stochastic virtual combinatorial chemistry.

The structure of BEA can accommodate 8 smaller or 4 larger guest molecules that can act as OSDAs. We decided to first design a set of small "monomer" molecules that in a later stage were linked into "dimer" molecules using an appropriate linker molecule. The *de novo* design engine was therefore first applied with a scoring function that fits 8 OSDA copies into BEA, leading to a set of ~160 molecules with a stabilization energy below −15 kJ/(mol Si). Additionally, a set of commercially available compounds that were close analogs to the *de novo* designed molecules were screened. The result of these two design rounds was a set of ~350 small molecules with a stabilization energy toward BEA below −15 kJ/(mol Si). Of these molecules, a subset of 271 had the correct chemical functionality for virtual dimerization according to the reaction scheme depicted in Scheme 2.

For screening the larger dimer molecules, the scoring function was adapted to fit only 4 OSDA copies in BEA. Additionally, for dimer molecules with a stabilization energy below −15 kJ/(mol Si) toward BEA, the stabilization energy toward BEB was also calculated.

A set of 78 linker molecules was obtained from commercially available compound databases. With these, a set of ~20,000 unique symmetrical dimers consisting of a linker and two identical monomers could be generated and exhaustively screened.

The number of asymmetrical dimers that could be generated from a linker and two different monomers was no longer amenable to exhaustive computational screening. Therefore, the stochastic virtual combinatorial chemistry algorithm described in Section 2.4.5 was used to search this chemical search space. Two runs of the algorithm were carried out.

In addition to the exhaustive and stochastic combinatorial chemistry approaches, we performed two *de novo* design runs using the *Synopsis* program to generate molecules of which four copies were fitted into the zeolite.

Combining the results of the total of 5 design runs, we identified 212 molecules with a predicted stabilization energy toward BEA lower than −15 kJ/(mol Si), and a stabilization energy toward BEB that was at least 2 kJ/(mol Si) higher. We observed that, in terms of number of high scoring molecules per number of MD calculations, *de novo* design is most successful in searching the chemical search space, followed closely by the stochastic combinatorial chemistry algorithm.

ArN(H) + Ar'N(H) + XLX → Ar-L-Ar'

Scheme 2: Dimerization reaction to generate OSDAs selectively directed toward BEA. ArN is an aromatic 5- or 6-membered N-hetero cycle. X is a halogen. L is a linker with 1 up to 6 C atoms between the halogens.

2.4.7 Design of OSDAs for Chiral Zeolite BEA [39]

In this study, chiral OSDAs directed toward the $P4_122$ enantiomer of zeolite BEA were designed. Using a combination of *de novo* design, exhaustive virtual combinatorial chemistry, and stochastic virtual combinatorial chemistry, we identified 175 putative OSDAs with a predicted stabilization energy in BEA of one enantiomer below −15 kJ/(mol Si), and a stabilization energy for its mirror image enantiomer that is predicted to be more than 2 kJ/(mol Si) higher. For these 175 molecules, we also predicted the stabilization energy in the competing polymorph BEB and found that 42 also have a stabilization energy gap with respect to BEA that is larger than 2 kJ/(mol Si). Thus, these compounds are promising putative OSDAs for zeolite beta that is both enantiomerically enriched and enriched in the A polymorph.

2.4.8 Application of a Machine-Learning Scoring Function in the *De Novo* Design of OSDAs for Zeolite Beta [47]

While performing the studies discussed in the two previous paragraphs, a large number of molecular dynamics calculations of putative OSDAs fitted into the structure of zeolite beta were carried out. To tap this large body of information, a machine-learning (ML) method was applied to speed up these computationally expensive energy calculations by two orders of magnitude. The 3D molecular structures of a set of 4781 putative OSDAs with a range of calculated stabilization energies in BEA were encoded as linear sets of intensities in a calculated diffraction pattern [53]. These were used as input into a neural network that was trained, tested, and validated to output the MD calculated stabilization energies. A scoring function was set up in which the molecular dynamics call to predict the stabilization energy of an OSDA in BEA was replaced by a call to the trained and validated neural network. This scoring function was used in combination with the *de novo* design engine and led to the discovery of 469 putative OSDAs with a verified stabilization energy below −17 kJ/(mol Si), which is comparable to or better than the stabilization energies of known OSDAs for zeolite beta.

2.4.9 Design of OSDAs for Zeolites for Gas Adsorption and Separation [48–50]

In the search for novel and better performing nanoporous materials for gas separation and storage applications, considerable effort has been directed to the computational screening of databases of existing and hypothetical zeolites [54]. In a series of studies, we have illustrated how targeted *de novo* design can be deployed to design OSDAs directed toward zeolite structures that have been identified as hits in such computational screens.

2.4.9.1 Carbon Capture and Storage: WEI, JBW, GIS, SIV, DAC, 8124767, 8277563 [48]

In a large-scale computational screening effort, Lin et al. [55] identified a number of existing and theoretical zeolite frameworks that were predicted to effectively absorb CO_2. These zeolites have potential application in fossil fuel power plants for carbon capture and storage (CSS), which is considered an important strategy for mitigation of global warming by CO_2 emission. In selecting promising frameworks, the parasitic energy of the CSS process was used as a criterion. The parasitic energy is the total energy cost to extract and pressurize the CO_2 generated in a fossil fuel power plant for transport and storage for further use. By screening of the known zeolite frameworks in the IZA database [56], the WEI, JBW, GIS, SIV, and DAC frameworks were predicted to have the lowest parasitic energy. None of these IZA frameworks have been synthesized in the all-Si form, which is the composition in which they were predicted to be most promising for CSS. We therefore applied *de novo* design to design OSDAs targeted toward their synthesis. In addition, a considerable number of hypothetical zeolite structures [57] were predicted to have an even lower parasitic energy than the six best scoring IZA structures. From these, we selected two structures, with frameworks 8124767 and 8277563, to design OSDAs. In addition to a low parasitic energy, these structures also have a large maximum free sphere diameter. The latter is important for rapid transport and therefore efficient diffusion of CO_2 in the zeolite pores. For these two hypothetical structures, we checked their position on the framework energy vs. framework density plot. It was found that both structures lie close to the line along which the majority of known zeolites are positioned [58]. We observed that the ease with which the *de novo* design engine identified high scoring putative OSDAs for these seven targeted zeolites is highly dependent on the structural characteristics of each framework. For one framework, JBW, only one favorably scoring OSDA was generated in three runs of the program, while for the hypothetical 8277563, a large number of molecules with a low predicted stabilization energy were identified. For each target, the optimal number of OSDA copies in the zeolite unit cell had to be carefully examined. For the JBW and 8124767 frameworks, the unit cell had to be expanded along one crystallographic axis to allow the fitting of an OSDA.

2.4.9.2 Carbon Dioxide/Methane Separation: GIS, ABW, 8186909, 8198030 [49]

An alternative strategy for CSS is the separation of CO_2 from post-combustion flows or from natural gas through membrane separation. This has motivated a computational screening study of a large number of zeolite topologies to identify frameworks with optimal adsorption and diffusion properties for CO_2/CH_4 mixtures [59]. In this study, two known zeolites frameworks, GIS and ABW, and two hypothetical structures from the PCOD database, PCOD8186909 and PCOD8198030, were found to be the most promising. PCOD8186909 was predicted to have a higher CO_2 permeability than the two known frameworks, and PCOD8198030 was predicted to be a candidate material for the inverse CO_2/CH_4 separation process, in which CH_4 is the absorbed species. We therefore set out to design OSDAs targeting these four frameworks. For GIS and PCOD8186909, numerous putative OSDAs with low predicted stabilization energies were identified, with different numbers of OSDA copies fitted into the unit cell. For the ABW framework, no molecules with a stabilization energy below −8.2 KJ/(mol Si) could be generated, indicating that this framework does not easily accommodate guest molecules. The favorably scoring OSDAs for PCOD8198030 also scored well, or sometimes better, in the PCOD8186909 framework, which is problematic as their eventual use in zeolite synthesis may give rise to framework mixtures. Application of the *de novo* engine with a dual objective, minimization of the stabilization energy toward PCOD8198030, and maximization of the stabilization energy toward PCOD8186909, might alleviate this problem. The result of this study was a set of promising, synthetically accessible OSDAs that are a step forward along the way to the practical implementation of zeolite membrane separation of CO_2 from CH_4.

2.4.9.3 Separation of Ethylene-Ethane: DFT, ACO, NAT, JRY [50]

The separation of ethane from ethylene produced by industrial cracking is highly energy consuming, and therefore alternative separation techniques including the application of zeolites are the subject of intensive research. In this context, Shah et al. [60] performed a computational screening of 214 known zeolite structures and identified several promising frameworks. Two zeolites, with frameworks DFT and ACO, were predicted to selectively adsorb ethylene, and two zeolites, with frameworks NAT and JRY, were predicted to selectively adsorb ethane.

As no all-Si structures with these frameworks have been reported, we set out to apply *de novo* design to design OSDAs directed toward these. For all four zeolites, we designed putative OSDAs with favorable stabilization energies, comparable to or lower than reported stabilization energies for known OSDA/zeolite pairs. The number of favorable scoring hits and the value of the best stabilization energies obtained varied largely for the four target zeolites, again indicating the varying capacity of different zeolite frameworks to accommodate guest molecules.

2.4.10 Design of MOFs for Methane Storage and Delivery [51]

As an illustration of how *de novo* design can be applied to the design of nanoporous materials other than zeolites, we here mention our work on the design of organic linkers for metal organic frameworks (MOFS) with high methane deliverable capacity. A set of nine well-known MOF networks, each with a given type of secondary building unit (SBU), were selected. The aim of the study was to design linkers that are capable to form MOFs of the given frameworks that are predicted to have a high methane storage capacity. To do this, a scoring function that calculates five molecular properties was set up to be run in combination with the *de novo* design engine. As mentioned and emphasized before, this scoring function is totally independent of the *de novo* design program, requiring no adaptation of the latter. The score components comprising the scoring function are listed in Table 2.6.

The first four molecular properties must ensure that the designed linker had the correct chemical functionality and geometry to be able to form a MOF with a given framework and SBU. These properties were used as filters. Linker molecules that passed all the filters were used to build a MOF structure, for which the methane deliverable capacity was predicted using grand canonical Monte Carlo simulation. Running the *de novo* design program in combination with this scoring function, a large number of high scoring putative linkers were generated. The synthesizability of the linkers was inherently addressed by the *de novo* design program, and the scoring function ensured their rigidity and correct geometry. For each of the network topologies considered, we used an SBU with which MOFs have been synthesized in practice. Interestingly, the known MOF-5, a material with one of the highest deliverable capacities reported, was among the results generated. In addition, MOFs consisting of 48 predicted linkers in 4 of the 9 frameworks considered were predicted to perform even better than MOF-5.

Table 2.6 Components of the score vector used for the design of MOFs with high methane deliverable capacity.

Score	Score type
Number of rotatable bonds	Less than a threshold
Number of carboxylic acid groups	Bracketed
Angular orientation of the carboxylic acid groups	Greater than a threshold (>)
Distance between the carboxylic acid groups	Less than a threshold (<)
Methane deliverable capacity	Maximize

2.4.11 Multi-Objective *De Novo* Design of OSDAs for Zeolites Using an Ant Colony Optimization Algorithm [41]

Recently, a novel *de novo* algorithm to design OSDAs was developed by Muraoka et al. [41]. The chemical search space of the algorithm consisted of the reaction products resulting from the alkylation reaction between commercially available amines and organic halides, thus ensuring the synthesizability of the designed compounds. To effectively search this chemical space, consisting of ~6×10^6 molecules, an ant colony optimization (ACO) algorithm was used. The multiple molecular properties to be optimized were the hydrophobicity, the flexibility, and size, and finally the stabilization energy of a designed molecule in a target zeolite as calculated with molecular dynamics. An important motivation to develop the algorithm was to lower the cost of OSDAs for the production of industrially interesting zeolites. Therefore, the ACO algorithm was adapted by defining a cost parameter that was used to steer the algorithm toward the generation of cheaper OSDAs. The *de novo* design algorithm was shown to generate putative OSDAs similar to known OSDAs that template the CHA, AEI, and CON frameworks, and that can be synthesized at a significantly lower cost.

2.5 Conclusions and Outlook

De novo design has been explored, both in theory and in practice, as a viable approach toward the discovery of OSDAs for zeolites. This has been illustrated in the seminal work of Lewis et al. [18], by our own efforts in the field as detailed in this contribution, and more recently by Muraoka et al. [41]. Important features of a *de novo* design algorithm, in our eyes, are the appropriate address of synthetic accessibility of the designed molecules, the separation of molecule generation and scoring methods, and the ability to perform multi-objective optimization.

The generation of novel chemical entities by *de novo* design with no or little regard with respect to the possibility of their eventual synthesis, although sometimes successful [18], is bound to be limited to idea generation. We, and others [24–26, 41], have addressed this issue by deploying a virtual synthetic chemistry engine as part of a *de novo* design algorithm. The alternative of using retro-synthesis algorithms as an a-posteriori assessment of synthetic accessibility has for a long time been hampered by the limited success of the latter [62]. Recent progress in the field based upon deep learning methodology is impressive [63] and may form the basis for a structure generation paradigm as part of a *de novo* design engine. Another fast-growing field is deep learning for molecular design [64]. However, this is still limited by the need for very fast molecular property prediction methods, which precludes the use of computationally intensive physics-based simulations such as molecular dynamics and quantum mechanics. Replacement of these direct methods by machine learning methods may provide an outcome to this issue [65, p. 20277].

The separation of the scoring and molecule generation and optimization in a *de novo* design engine allows for maximal flexibility in the use of the methodology used to predict the desired molecular properties and in the actual design goal. In our work in OSDA design, we have focused on the van der Waals stabilization energy of putative OSDAs in all-Si target zeolites. Introduction of an appropriate charge model can be used to steer Al siting in acid zeolites [66] and should allow the design of zeolites with desired acid proton positions [67, 68]. Inclusion of less computationally demanding protocols such as the neural network described in the Section 2.4.8 or molecular topology-based QSAR methods explored in [69] can significantly speed up the design process without altering the *de novo* design engine.

The ability to perform multi-objective optimization has proven to be an important asset to the *de novo* design of OSDAs for zeolites. As we have illustrated in Sections 2.4.5 through 2.4.7, it allows the design of stereo selective OSDAs directed toward the synthesis of enantiomerically-enriched chiral zeolites, and the design of OSDAs that selectively template one particular zeolite framework and not a competing one. This may be particularly useful for the design of OSDAs for hypothetical zeolites. This was shown in [70], where an OSDA for a hypothetical zeolite identified by computational screening experimentally led to the synthesis of another, known zeolite. Importantly, this result could be, at least in part, rationalized by the zeolite–OSDA interaction energies of the target zeolite and the competing zeolite framework.

References

1 Lok, B.M., Cannan, T.R., and Messina, C.A. (1983). The role of organic molecules in molecular sieve synthesis. *Zeolites* 3: 282–291.

2 Lewis, D.W., Freeman, C.M., and Catlow, C.R.A. (1995). Predicting the templating ability of organic additives for the synthesis of microporous materials. *J. Phys. Chem. C* 99: 11194–11202.

3 Turrina, A. and Cox, P.A. (2017). Molecular modelling of structure direction phenomena. In: *Insights into the Chemistry of Organic Structure-Directing Agents in the Synthesis of Zeolitic Materials. Structure and Bonding* (ed. L. Gómez-Hortigüela), 75–102. Cham: Springer. doi: 10.1007/430_2017_16.

4 Li, J., Yu, J., Yan, W. et al. (1999). Structures and templating effect in the formation of 2D layered aluminophosphates with Al_3P_4O163-stoichiometry. *Chem. Mater.* 11: 2600–2606.

5 Rollmann, L.D., Schlenker, J.L., Kennedy, C.L. et al. (2000). On the role of small amines in zeolite synthesis 2. *J. Phys. Chem. B* 104: 721–726.

6 Sun, P., Jin, Q., Wang, L., Li, B. et al. (2003). A study on the templating ability of diquaternary cations for the zeolite synthesis in terms of energetics. *J. Porous Mater.* 10: 145–150.

7 Sastre, G., Leiva, S., Sabater, M.J. et al. (2003). Computational and experimental approach to the role of structure-directing agents in the synthesis of zeolites: the case of cyclohexyl alkyl pyrrolidinium salts in the synthesis of β, EU-1, ZSM-11, and ZSM-12 zeolites. *J. Phys. Chem. B* 107: 5432–5440.

8 Sastre, G., Cantin, A., Diaz-Cabañas,M.J. et al. (2005). Searching organic structure directing agents for the synthesis of specific zeolitic structures: an experimentally tested computational study. *Chem. Mater.* 17: 545–552.

9 Gómez-Hortigüela, L., Pérez-Pariente, J., Corà,F. et al. (2005). Structure-directing role of molecules containing benzyl rings in the synthesis of a large-pore aluminophosphate molecular sieve: an experimental and computational study. *J. Phys. Chem. B* 109: 21539–21548.

10 Burton, A.W., Lee, G.S., and Zones, S.I. (2006). Phase selectivity in the syntheses of cage-based zeolite structures: an investigation of thermodynamic interactions between zeolite hosts and structure directing agents by molecular modeling. *Micropor. Mesopor. Mater.* 90: 129–144.

11 Elanany, M., Su, B.L., and Vercauteren, D.P. (2007). Strong templating effect of TEAOH in the hydrothermal genesis of the AlPO4-5 molecular sieve: experimental and computational investigations. *J. Mol. Catal. A Chem.* 270: 295–301.

12 Zones, S.I., Burton, A.W., Lee, G.S. et al. (2007). A study of piperidinium structure-directing agents in the synthesis of silica molecular sieves under fluoride-based conditions. *J. Am. Chem. Soc.* 129: 9066–9079.

13 O'Brien, M.G., Sanchez-Sanchez, M., Beale A.M. et al. (2007). Effect of organic templates on the kinetics and crystallization of microporous metal-substituted aluminophosphates. *J. Phys. Chem. C* 111: 16951–16961.

14 Moliner, M., Rey, F., and Corma, A. (2013). Towards the rational design of efficient organic structure-directing agents for zeolite synthesis. *Angew. Chemie – Int. Ed.* 52: 13880–13889.

15 Sterling, T. and Irwin, J.J. (2015). ZINC 15: ligand discovery for everyone. *J. Chem. Inf. Model.* 55: 2324–2337.

16 Dubois, J., Bourg, S., Vrain, C. et al.(2008). Collections of compounds. How to deal with them? *Curr. Comput. Aided. Drug Des.* 4: 156–168.

17 Virshup, A.M., Contreras-García, J., Wipf, P. et al. (2013). Stochastic voyages into uncharted chemical space produce a representative library of all possible drug-like compounds. *J. Am. Chem. Soc.* 135: 7296–7303.

18 Lewis, D.W., Willock, D.J., Catlow, C.R.A. et al. 1996). *De novo* design of structure-directing agents for the synthesis of microporous solids. *Nature* 382: 604–607.

19 Barrett, P.A., Jones, R.H., Thomas, J.M. et al. (1996). Rational design of a solid acid catalyst for the conversion of methanol to light alkenes: synthesis, structure and performance of DAF-4. *Chem. Commun.* 17: 2001–2002.

20 Lewis, D.W., Sankar, G., Wyles, J. et al. (1997). Synthesis of a small-pore microporous material using a computationally designed template. *Angew. Chemie (International Ed. English)* 36: 2675–2677.

21 Vinkers, H.M., De Jonge, M., Daeyaert, F. et al. (2003). SYNOPSIS: SYNthesize and OPtimize system *in Silico*. *J. Med. Chem.* 46: 2765–2773.

22 Daeyaert, F., and Deem, M.W. (2017). A Pareto Algorithm for efficient *de novo* design of multi-functional molecules. *Mol. Inform.* 36: 1–14.

23 Warr, W.A. and Short, A. (2014). Review of chemical reaction database systems, computer-aided synthesis design, reaction prediction and synthetic feasibility. *Mol. Inform.* 33: 469–476.

24 Hartenfeller, M., Zettl., Walter, M. et al. (2012). DOGS: reaction-driven *de novo* design of bioactive compounds. *PLoS Comput. Biol.* 8 (2): e1002380. doi:10.1371/journal.pcbi.1002380.

25 Durrant, J.D., Lindert, S., and McCammon, J.A. (2013). AutoGrow 3.0: an improved algorithm for chemically tractable, semi-automated protein inhibitor design. *J. Mol. Graph. Model.* 44: 104–112.

26 Masek, B.B., Baker, D.S., Dorfman, R.J. et al. (2016). Multistep reaction based *de novo* drug design: generating synthetically feasible design ideas. *J. Chem. Inf. Model.* 56: 605–620.

27 Maddalena, D.J. and Snowdon, G.M. (1997). Applications of genetic algorithms to drug design. *Expert Opin. Ther. Pat.* 7: 247–254.

28 Deb, K., Pratap, A., Agarwal, S. et al. (2002). A fast and elitist multiobjective genetic algorithm: NSGA-II. *IEEE Trans. Evol. Comput.* 6: 182–197.

29 Gálvez-Llompart, M., Cantín, A., Rey, F. et al. (2019). Computational screening of structure directing agents for the synthesis of zeolites. A simplified model. *Zeitschrift fur Krist. – Cryst. Mater.* 234: 451–460.

30 Schwalbe-Koda, D. and Gómez-Bombarelli, R. (2021). Supramolecular recognition in crystalline nanocavities through Monte Carlo and Voronoi network algorithms. *J. Phys. Chem. C* 125: 3009–3017.

31 Schwalbe-Koda, D. and Gómez-Bombarelli, R. (2021). Benchmarking binding energy calculations for organic structure-directing agents in pure-silica zeolites. *Journal of Chemical Physics* 154. https://doi.org/10.1063/5.0044927.

32 Halgren, T.A. (1996). Merck molecular force field. I: Basis, form, scope, parameterization, and performance of MMFF94. *J. Comput. Chem.* 17: 490–519.

33 Gale, J.D. and Rohl, A.L. (2003). The General Utility Lattice Program (GULP). *Mol. Simul.* 29: 291–341.

34 Mayo, S.L., Olafson, B.D., and Goddard, W.A. (1990). DREIDING: a generic force field for molecular simulations. *J. Phys. Chem.* 94: 8897–8909.

35 Kubota, Y., Helmkamp, M.M., Zones, S.I. et al. (1996). Properties of organic cations that lead to the structure-direction of high-silica molecular sieves. *Microporous Mater* 6: 213–229.

36 Daeyaert, F. and Deem, M.W. (2019). Design of organic structure directing agents for polymorph A zeolite beta. *J. Mater. Chem. A* 7: 9854–9866.

37 Brand, S.K., Schmidt, J.E., Deem, M.W. et al. (2017). Enantiomerically-enriched, polycrystalline molecular sieves. *Proc. Natl. Acad. Sci. USA* 114: 5101–5106.

38 de la Serna, R., Nieto, D., Sainz, R. et al. (2022). GTM-3, an extra-large pore enantioselective chiral zeolitic catalyst. *J. Am. Chem. Soc.* 144: 8249–8256.

39 Daeyaert, F. and Deem, M.W. (2019). Design of organic structure directing agents for chiral zeolite beta A. *ChemistrySelect* 4: 3531–3537.

40 Arora, J.S. (2017). Multi-objective optimum design concepts and methods. In: *Introduction to Optimum Design* (ed. J. Singh Arora), 771–794. Elsevier Inc. doi: 10.1016/B978-0-12-800806-5/00018-4.

41 Muraoka, K., Chaikittisilp, W., and Okubo, T. (2020). Multi-objective *de novo* molecular design of organic structure-directing agents for zeolites using nature-inspired ant colony optimization. *Chem. Sci.* 11: 8214–8223.

42 Pophale, R., Daeyaert, F., and Deem, M.W. (2013). Computational prediction of chemically synthesizable organic structure directing agents for zeolites. *J. Mater. Chem. A* 1: 6750–6760.

43 Schmidt, J.E., Deem, M.W., and Davis, M.E. (2014). Synthesis of a specified, silica molecular sieve by using computationally predicted organic structure-directing agents. *Angew. Chemie – Int. Ed.* 53: 8372–8374.

44 Schmidt, J.E., Deem, M.W., Lew, C. et al. (2015). Computationally-guided synthesis of the 8-ring zeolite AEI. *Top. Catal.* 58: 410–415.

45 Davis, T.M., Liu, A.T., Lew, C.M. et al. (2016). Computationally guided synthesis of SSZ-52: a zeolite for engine exhaust clean-up. *Chem. Mater.* 28: 708–711.

46 Daeyaert, F. and Deem, M.W. (2018). *In silico* design of chiral dimers to direct the synthesis of a chiral zeolite. *Mol. Phys.* 116: 2836–2855.

47 Daeyaert, F., Ye, F., and Deem, M.W. (2019). Machine-learning approach to the design of OSDAs for zeolite beta. *Proc. Natl. Acad. Sci. USA* 116: 3413–3418.

48 Daeyaert, F. and Deem, M.W. (2019). Design of organic structure directing agents to control the synthesis of zeolites for carbon capture and storage. *RSC Adv.* 9: 41934–41942.

49 Daeyaert, F. and Deem, M.W. (2020). Design of organic structure-directing agents for the controlled synthesis of zeolites for use in carbon dioxide/methane membrane separations. *Chempluschem* 85: 277–284.

50 Daeyaert, F. and Deem, M.W. (2020). Design of organic structure directing agents to guide the synthesis of zeolites for the separation of ethylene-ethane mixtures. *RSC Adv.* 10: 20313–20321.

51 Bao, Y., Martin, R.L., Simon, C.M. et al. (2015). *In silico* discovery of high deliverable capacity metal-organic frameworks. *J. Phys. Chem. C* 119: 186–195.

52 Smith, M.B. and March, J. (2007). *March's Advanced Organic Chemistry: Reactions, Mechanisms and Structure*. Wiley Interscience.

53 Schuur, J.H., Selzer, P., and Gasteiger, J. (1996). The coding of the three-dimensional structure of molecules by molecular transforms and its application to structure-spectra correlations and studies of biological activity. *J. Chem. Inf. Comput. Sci.* 36: 334–344.

54 Boyd, P.G., Lee, Y., and Smit, B. (2017). Computational development of the nanoporous materials genome. *Nat. Rev. Mater.* 2.

55 Lin, L.-C., Berger, A.H., Martin, R.L. et al. (2012). *In silico* screening of carbon-capture materials. *Nat. Mater.* 11: 633–641.

56 Baerlocher, C. and McCuscker, L.B. (2022). Database of zeolite structures. www.iza-structure.org/databases.

57 Deem, M.W., Pophale, R., Cheeseman, P.A. et al. (2009). Computational discovery of new zeolite-like materials. *J. Phys. Chem. C* 113: 21353–21360.

58 Henson, N.J., Cheetham, A.K., and Gale, J.D. (1996). Computational studies of aluminum phosphate polymorphs. *Chem. Mater.* 8: 664–670.

59 Kim, J., Abouelnasr, M., Lin, L.C. et al. (2013). Large-scale screening of zeolite structures for CO_2 membrane separations. *J. Am. Chem. Soc.* 135: 7545–7552.

60 Shah, M.S., Fetisov, E.O., Tsapatsis, M. et al. (2018). C_2 adsorption in zeolites: *in silico* screening and sensitivity to molecular models. *Mol. Syst. Des. Eng.* 3: 619–626.

61 Bao, Y., Martin, R.L., Haranczyk, M. et al. (2015). *In silico* prediction of MOFs with high deliverable capacity or internal surface area. *Phys. Chem. Chem. Phys.* 17: 11962–11973.

62 Szymkuc, S., Gajewska, E., Klucznik, T. et al. (2016). Computer-assisted synthetic planning: the end of the beginning. *Angew. Chemie – Int. Ed.* 55: 5904–5937.

63 Segler, M.H.S., Preuss, M., and Waller, M.P. (2018). Learning to plan chemical syntheses. *Nature* 555: 604–610.

64 Elton, D.C., Boukouvalas, Z., Fuge, M.D. et al. (2019). Deep learning for molecular design: a review of the state of the art. *Mol. Syst. Des. Eng.* 4: 828–849.

65 Moosavi, S.M., Jablonka, K.M., and Smit, B. (2020). The role of machine learning in the understanding and design of materials. *J. Am. Chem. Soc.* 142: 20273–20287.

66 Sastre, G., Fornes, V., and Corma, A. (2002). On the preferential location of Al and proton siting in zeolites: a computational and infrared study. *J. Phys. Chem. B* 106: 701–708.

67 Gómez-Hortigüela, L., Pinar, A.B., Corà, F. et al. (2010). Dopant-siting selectivity in nanoporous catalysts: control of proton accessibility in zeolite catalysts through the rational use of templates. *Chem. Commun.* 46: 2073–2075.

68 Pinar, A.B., Gómez-Hortigüela, L., McCusker, L.B. et al. (2013). Controlling the aluminum distribution in the zeolite ferrierite via the organic structure directing agent. *Chem. Mater.* 25: 3654–3661.

69 Gálvez-Llompart, M., Gálvez, J., Rey, F. et al. (2020). Identification of new templates for the synthesis of BEA, BEC, and ISV zeolites using molecular topology and Monte Carlo techniques. *J. Chem. Inf. Model.* 60: 2819–2829.

70 Boruntea, C.R., Sastre, G., Lundegaard, L.F. et al. (2019). Synthesis of high-silica erionite driven by computational screening of hypothetical zeolites. *Chem. Mater.* 31: 9268–9276.

3

Machine Learning Search for Suitable Structure Directing Agents for the Synthesis of Beta (BEA) Zeolite Using Molecular Topology and Monte Carlo Techniques

María Gálvez-Llompart[1,2] and German Sastre[1]

[1]*Instituto de Tecnologia Quimica (UPV-CSIC), Universidad Politecnica de Valencia; Valencia, Spain*
[2]*Molecular Topology and Drug Design Unit, Department of Physical Chemistry, University of Valencia, Valencia, Spain*

3.1 Introduction

Computer-aided material design is experiencing exponential growth with the introduction of high-performance computations. Brand new big data techniques can be used for the rational design of target materials, with machine learning (ML) being one of the most innovative and promising techniques [1–3]. Its strong predictive capability has been observed in areas such as molecular design, property optimization, and synthesis prediction [4–6]. Now is the time for materials science to unveil the potential of machine learning techniques [7].

ML techniques have already been used for the design of new materials, for instance, zeolites and related zeotype materials. Zeolites are crystalline, microporous materials extensively used in a variety of industrial applications such as catalysis, gas separation, and ion exchange. The industrial application of zeolites is determined by their structural and topological properties. As in many other areas of material design, the "trial-and-error" approach was common when searching for novel zeolites because the mechanisms associated with the nucleation and crystallization of zeolites are still not fully understood. Consequently, rational design of zeolites may be considered a chimera. Nevertheless, different properties such as types and amounts of framework atoms, mineralizing agents, and inorganic/organic structure directing agents, are known to have an impact on zeolite formation type and can be considered a solid starting point [8–12].

Among these factors, organic structure directing agent (OSDA) molecules play an important role in the selective formation of a given zeolite polymorph [3,10,12]. These molecules are responsible for charge balancing, space filling, template, and even acting like a lock-and-key relationship, similar to drug–receptor interactions. In fact, depending on the characteristics of the OSDA, it may show greater or less specificity when favoring the synthesis of one or more zeolites [10,13,14].

Different authors have recently applied ML for the identification of new OSDA, such as Jensen et al. [9,15], Moliner et al. [16], Kwak et al. [17], and Daeyaert et al. [18]. Unlike other computational approaches such as density functional theory or molecular dynamics [19,20], which try to identify candidate OSDAs for specific zeolite structures, ML techniques are computationally fast, focused not only on pure silica materials, and are not limited to a single zeolite system.

Among all the different ML techniques available, artificial neural networks (ANN) is one of the most relevant. ANN consist of an information-processing method that shares a structural architecture similar to that of the biological nervous system. The main component of every ANN is called a layer. Most of the ANN contain at least three layers: input, hidden, and output (Figure 3.1). The input layers collect the data from the input files, while the output layer sends the processed information directly to the outside world [21]. Between input and output layers there may be one or

AI-Guided Design and Property Prediction for Zeolites and Nanoporous Materials, First Edition. Edited by German Sastre and Frits Daeyaert.

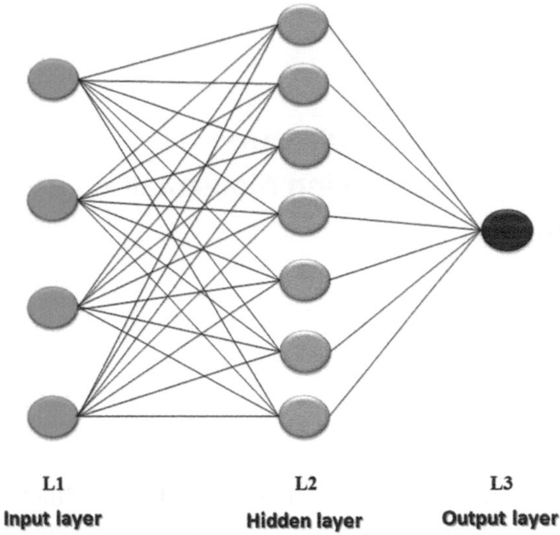

Figure 3.1 Types of layers from an ANN.

many hidden layers. These internal layers contain units or neurons, which can be interconnected in various structures to send messages and share information that will be relevant to determine the corresponding output [21]. The more the number of hidden layers, the more the complexity of the ANN and, consequently, more resources such as time and computational power will be needed. The output layer can have one (i.e., when performing regression analysis) or several neurons (classification analysis), depending on the number of characteristics to be found.

ANN build up a number of networks to solve the problem and then selects those networks which best represent the relationship between the input and target variables. ANN are able to classify a category (classification analysis) or predict a property (regression analysis).

Neurons communicate by means of activation functions, that is, algorithms will have all the inputs or training data, and will assign weights expressing their importance. Different types of activation functions are available: linear (identity function) and non-linear (logistic sigmoid function, hyperbolic tangent function, negative exponential function, and standard sine activation function). Each neural network can have different activation functions in its internal and output layers [21].

The objective of a neural network is to learn behaviors that are capable of highlighting characteristics of the input objects, a process called training. Still, it is almost impossible to obtain 100% correct predictions, so at the penalty of an incorrect prediction, we have what is known as a loss [21]. The presence of a loss, which may be present in an ANN model, is not negative per se, since a 100% correct prediction would mean overfitting and the model's inability to predict external data. Thanks to the training, we should be able to measure the loss and then iterate until the algorithm discovers the parameters of the model with the lowest possible loss. Once this occurs, the model has converged. One of the most effective methods to determine the parameters related with the least loss is to calculate the loss with each of the possible weights and then determine the point at which the loss is minimal. The problem is that this method (gradient descent algorithm) requires a very long time and too much computational power, so modifications of the gradient descent algorithm are preferably used such as the Broyden-Fletcher-Goldfarb-Shanno (BFGS) optimization algorithm [22] or conjugate gradient descent [23]. Differences between gradient descent (steepest descent), conjugate gradient descent, and BFGS are related to the method by which the descent direction is determined (from the gradient). However, the gradient descent only considers descent in the negative gradient direction, Steepest Gradient Descent, descents in the direction of the largest directional derivative, and BFGS modifies gradient descent by introducing a Hessian approximation matrix computed from finite gradient differences. In summary, these modified gradient descent algorithms are used to minimize loss functions by iteratively adjusting the parameters of a network in such a way that its output deviation is minimized.

Although models using ML techniques usually have a high predictive capacity, their Achilles heel seems to be linked with a very complex and non-linear relation between the descriptors (or features) and the desired information; therefore, it is difficult to give an interpretation of the resulting model [1,24].

Hence, the use of complementary statistical techniques for data analysis such as pattern recognition techniques (linear discriminant analysis, minimum distance classifier, cluster analysis), and regression analysis techniques (linear regression analysis, polynomial regression analysis, multilinear regression analysis, logistic regression analysis) are of great interest. Several examples are available in the literature, combining either ANN to other statistical techniques for material design purposes; for example, Mukherjee and Colón [25] employed ML techniques and linear discriminant analysis for MOF discovery and Wanyonyi et al. [26] applied ANN combined with linear discriminant analysis and regression analysis techniques to model the role of pore chemistry and topology in the heavy metal sorption by zeolites.

Variable selection is a key step in modeling, and different approaches can be adopted in either linear or non-linear models. Forward stepwise variable selection [27], partial least squares (PLS) [28], and principal component analysis (PCA) [28] are techniques widely employed to perform a dimensionality reduction of variables before modeling.

Forward stepwise selection is a variable selection method which adds the most significant variables (i.e., the smallest p-value) one by one until all variables under consideration are included in the model [27]. Partial least squares (PLS) regression is a technique that reduces the number of variables by simultaneously modeling the variance in X-space and the covariance between the X- and Y-spaces [28], whereas PCA describes a dataset in terms of new uncorrelated variables ("components"). The components are ordered by the amount of original variance they describe, so the technique is useful for reducing the dimensionality of a dataset [28].

All the methods and techniques described until now are capable of rapidly reducing the dimensionality of the chemical space, using few computational resources and represent an important first step toward developing a high-throughput zeolite research pipeline. However, additional atomistic simulations on a target zeolite can improve the final selection of potential OSDAs [12,16].

Once the statistical techniques have been introduced, it is time to determine how to describe the OSDAs. To define an OSDA, different approaches can be employed, but in this chapter the topological descriptors (TD) are proposed. The methodology which employs TDs to define molecular structures is called Molecular Topology (MT) [29–31]. Its basic building relies on graph theory to describe molecules by means of topological and topo-chemical descriptors [32]. As shown in Figure 3.2, the chemical structure of a molecule is transformed into a mathematical graph, which can be translated into different types of matrices, such as the adjacency or distance matrix. After manipulating these matrices by means of different mathematical operations, we obtain different topological or topo-chemical indices to allow us to describe our molecules.

The first research group to establish the term "Molecular Topology" describing this methodology was "The Molecular Topology and Drug Design Unit" at the University of Valencia. MT has been

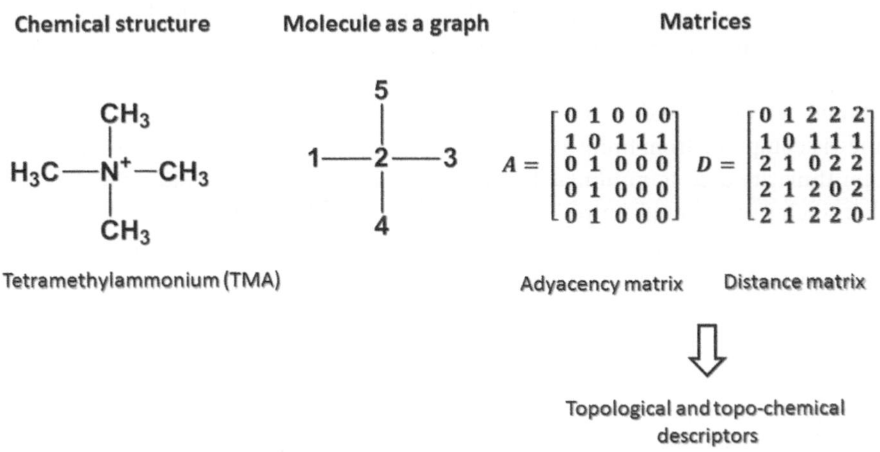

$$A = \begin{bmatrix} 0&1&0&0&0 \\ 1&0&1&1&1 \\ 0&1&0&0&0 \\ 0&1&0&0&0 \\ 0&1&0&0&0 \end{bmatrix} \quad D = \begin{bmatrix} 0&1&2&2&2 \\ 1&0&1&1&1 \\ 2&1&0&2&2 \\ 2&1&2&0&2 \\ 2&1&2&2&0 \end{bmatrix}$$

Figure 3.2 Example of transforming OSDA into a set of topological descriptors. This OSDA is tetramethylammonium (TMA).

applied in a wide range of areas, such as agro-chemistry [33,34], sustainable chemistry [35], biological chemistry [36–38], and physico-chemical properties [39].

Models employing descriptors to predict a biological activity or physico-chemical property are called QSAR (quantitative structure activity relationships) [40] or QSPR (quantitative structure property relationship) models [41]. Since the introduction of Comparative Molecular Field Analysis by Cramer et al. in 1988 [42], new 3D-QSAR algorithms allowed a significant advance in the identification of new drugs and established the concept relating the effects of shape on binding of steroids to carrier proteins [42]. Zeolites do share this "common principle" by the increasing structure directing effects when occluded organic molecules fit conveniently into the zeolite micropores being formed during the synthesis. Host–guest van der Waals interactions are of crucial importance in this process and their calculation allows to test numerically the concept above.

In what follows, readers can find an example of the MT methodology combined with machine learning (ANN) and regression (MLR) techniques for modeling zeolite-SDA van der Waals energy applied to BEA zeolite.

3.2 Artificial Neural Networks for Modeling Zeolite-SDA van der Waals Energy Applied to BEA Zeolite

The importance of zeolite beta (with a large 3D 12-ring channel system) appears in different industrial processes, including the alkylation of benzene and the separation of organics from water. Several organic structure directing agents (OSDAs) have been reported for the syntheses of zeolite beta [43], with the main synthetic route using tetraethyl amine [44]. Polymorph A (BEA), polymorph B (BEB), and polymorph C (BEC) conform zeolite beta and so far no synthetic route to pure BEA has been found [18] and instead BEA/BEB intergrowths are obtained. Nevertheless, different computational techniques (*de novo* design, virtual screening, exhaustive virtual combinatorial screening, and stochastic virtual combinatorial screening) have been applied to identify synthesizable organic structure directing agents that may aid the nucleation and growth of pure zeolite beta A, excluding the competing product zeolite beta B. An example of this is found in the work published by Daeyaert and Deem [45], who identify putative OSDAs that stabilize BEA (stabilization energy in zeolite beta A lower than −15 kJ per (mol Si)) but not BEB (energy gap greater than 2 kJ per (mol Si) with respect to zeolite BEA).

The BEA polymorph is chiral, and therefore an enantiomerically-enriched form of pure BEA would be of great interest for enantiospecific catalysis and separation [46]. In 2019, Deem and Daeyaert [47] published an interesting computational study applying a combination of a multiobjective scoring function with exhaustive and stochastic virtual combinatorial chemistry and *de novo* design algorithms to design chiral OSDAs to direct the formation of enantiomerically-enriched zeolite BEA. In this case, they identified putative OSDAs of which one enantiomer has a lower stabilization energy than the mirror image enantiomer and also, a significantly less favorable stabilization energy in the competing BEB zeolite. So, both aspects have been addressed in this study.

Finally, Daeyaert et al. apply a machine-learning strategy for designing OSDAs for zeolite beta. By doing so, they replace a computationally expensive molecular dynamics evaluation of the stabilization energy of the OSDA [18] inside zeolite beta with a neural network prediction. They build a neural network able to predict the OSDA-BEA stabilization energies calculated from molecular dynamics, and then apply an evolutionary design algorithm that screens the space of chemically feasible OSDAs (stabilization energies below −17 kJ/(mol Si)).

Once commercial importance and challenges of obtaining pure and chiral BEA zeolite have been addressed, we present the reader with a practical example in which ANN, combined with multilinear regression (MLR) models, are used to predict BEA–SDA van der Waals interaction energy.

The study presented below is based on previous data published by Galvez-Llompart et al. [11], in which different organic compounds were proposed as structure directing agents to favor the synthesis of BEA, BEC, and ISV zeolites (identified through the use of discriminant linear analysis and applying topological descriptors). Monte Carlo and lattice energy minimization calculation of zeolite-SDA interactions for these organics were calculated (table S10, [11]). Now, we intend to take a step forward and build models able to predict BEA–SDA van der Waals energies by applying two different statistical techniques: one based on machine learning (artificial neural networks) and the other on regression analysis (multi-linear regression analysis). These models should be able to replace the Monte Carlo and lattice energy minimization calculation of the van der Waals energy with a trained neural network and multi-linear regression model. After obtaining our quantitative structure–property relationship (QSPR) models, we will use this approach to identify candidate OSDAs for the synthesis of BEA zeolite.

The available data is made of 29 organics (see table S10, [11]), whose $E_{ZEO\text{-}SDA}$ were calculated using Monte Carlo and lattice energy minimization. van der Waals zeolite-SDA energy is indicated for the maximum loading calculated by zeoTsda and divided by the number of SiO_2 units in the unit cell (see reference [11] for more information). The unit cell of BEA (and also BEC, ISV) zeolite contains 64 SiO_2 units.

The set of available data (29 compounds) was handled in a slightly different way according to the statistical technique applied for building the models, even if all the available data was used either in MLR or ANN models. Construction of the neural networks was made using 70% of the data as a training set and the remaining 30% as an internal validation group (test set). This way, all available data has been used to train the model using both statistical techniques.

Once the data used to build the prediction models has been defined, it is possible to characterize the organics whose energy values of interaction with the zeolites to study are available [11]. To this end, we will use topological and topochemical descriptors calculated with the commercial software AlvaDesc [48]; descriptors that consider both 2D and 3D properties of the compounds to be calculated.

A forward stepwise variable selection procedure was used, in which variables are sequentially entered into the model depending on the "p" statistic value selected (threshold: $p < 0.05$) [27]. The p value shows us the probability of having obtained the result, if we assume that the null hypothesis is true. That is, if the value of p is lower than the level of significance, it is possible to test the null hypothesis that the multivariate means of the groups (active and inactive) are equal. Subsequently, the best subseries of 4 or 5 descriptors with respect to the property $E_{BEA\text{-}SDA}$ are identified. Therefore, from almost 3000 descriptors calculated, just 4 or 5 descriptors were selected for modeling $E_{BEA\text{-}SDA}$. Statistical software was used for modeling $E_{BEA\text{-}SDA}$, employing multilinear regression analysis and artificial neural networks (regression analysis) [21]. Default parameters for MLR techniques were employed. Two types of neural networks were tested: MLP (multilayer perceptron) and RBF (radial basis function). The number of hidden layers in the neural networks has been determined by the equation below, with I being input dimension of ANN, $H1$ is the hidden number of neurons in layer 1, and O is the number of output layers in the ANN.

$$N_w = (I+1) \times H1 + (H1+1) \times O$$

This equation determines the total number of synaptic connections in a neural network for avoiding over-fitting phenomena. When applying this equation to calculate the total number of weights in the neural network, the minimum and maximum of hidden layers were set as 1 and 3 respectively for our ANN model.

Table 3.1 shows the different models for the prediction of $E_{BEA\text{-}SDA}$. For both statistical techniques, within the training group, at least 88% of the values of zeo-SDA energy can be considered to be well-described (fitted) by the regression models, with 88% being the variance. In addition, the use of artificial neural networks (machine learning technique) allows us to achieve greater precision when predicting $E_{BEA\text{-}SDA}$ with respect to the multi-linear regression analysis technique, as confirmed by the higher value of the correlation coefficient (r^2) and the root mean square error (RMSE): a parameter useful for comparing the fit of different regression models [49,50].

For the MLR model we highlight how descriptors with a negative value in the equation favor lower BEA–SDA interaction, such as SpMin2_Bh (i), RDF105p, and R2e. SpMin2_Bh(i) is a topochemical Burden eigenvalue derived from the Burden matrix, weighted by its ionization potential [51,52]. From these descriptors, special attention is deserved by those considering specific information of 3D structure, such as RDF105p and R2e. RDF105p, a molecular descriptor obtained by radial basis functions centered on interatomic distances (10.5 Å) considers molecular

Table 3.1 Summary of predictive models built using MLR and ANN to predict BEA–SDA van der Waals interaction energy.

Multilinear Regression Analysis		Automated Neural Networks	
MLR Model	**Stat. Param.**	**ANN Model**	**Stat. Param.**
$E_{BEA-SDA} = 16.398 - (7.661 \times SpMin2_{Bh(i)}) + (0.033 \times P_{VSA_{s6}}) + (0.164 \times RDF085u) - (1.840 \times RDF105p) - (4.932 \times R2e)$	N = 29 $r^2 = 0.88$ F = 34.72 $p < 0.000001$ RMSE = 0.36	MLP 4[a]-3-1	N = 29 $r^2 = 0.98$ RMSE = 0.20

[a]SpMin2_Bh(i), RDF085u, RDF105p, and R4s+.

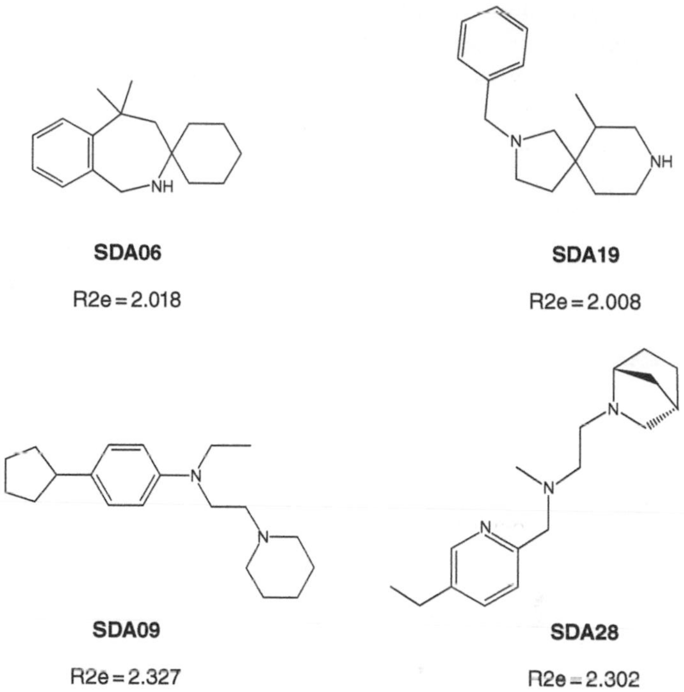

Figure 3.3 R2e descriptor values from training set compounds of MLR model.

SDA06

R2e = 2.018

SDA19

R2e = 2.008

SDA09

R2e = 2.327

SDA28

R2e = 2.302

conformation and is weighted by atom polarizability [52]. R2e is a GEometry, Topology, and Atom-Weights AssemblY (GETAWAY) descriptor based on spatial autocorrelation, encoding information on the effective position of substituents and fragments in molecular space [53,54]. These descriptors contemplate Sanderson electronegativity of atoms. As observed in Figure 3.3, SDA09 (R2e = 2.327) and SDA28 (R2e = 2.302) present a higher value for this descriptor, which is not surprising since they are the compounds that have a greater number of atoms at distance 2 from each other (added to a greater presence of atoms with larger Sanderson electronegativity). This, as a general rule, is valid for molecules with a more elongated and less spherical conformation. On the contrary, SDA19 (R2e = 2.008) and SDA06 (R2e = 2.018) show a lower presence of electronegative atoms at distance 2 from each other, and have a more spherical conformation.

The architecture of the neural networks constructed to predict the $E_{BEA\text{-}SDA}$ (Figure 3.4) shows that the neurons of the input layers are made up of the top-4 subseries of descriptors that best predict the property under study ($E_{ZEO\text{-}SDA}$).

Additional information regarding the construction of neural networks is shown in Table 3.2, such as the activation functions used by the neurons of the inner and output layers, the algorithm used to train the network, the type of function used to calculate the error associated with each function and, finally, the number of training cycles that each network has required. In each training cycle the entire training set is passed through the networks and the network error is calculated. This information is then used to adjust the weights so that the error is further reduced.

The topo-chemical descriptors used in the construction of the different models presented are reflected in Table 3.3. These descriptors determine 2D and 3D characteristics of the different SDAs studied.

In Figure 3.5, we can see the predicted and calculated BEA–SDA van der Waals energy (MLR and ANN models) for all available data (29 SDAs).

After the construction and analysis of the models, validation is required. As the number of dataset available is small, an internal validation procedure is preferred over an external one [55].

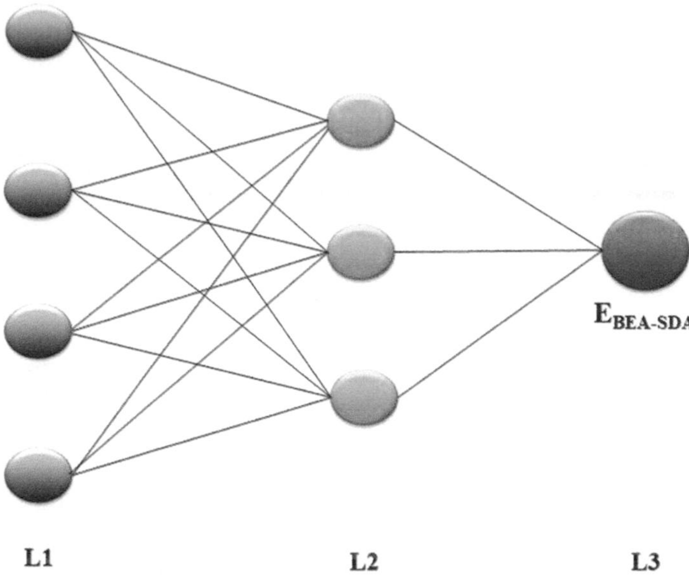

Figure 3.4 Layout of neural network predicting $E_{BEA-SDA}$: input layer with the 4 descriptors selected for each ANN, hidden layer with 3 neurons, and output layer corresponding to the dependent variable value ($E_{BEA-SDA}$).

Table 3.2 Features of ANN build up for predicting $E_{BEA-SDA}$.

ANN	Activation function for hidden layers	Activation function for output layer	Algorithm to train networks	Error function	Number of training cycles
MLP 4-3-1	Tangential function	Identity function	BFGS	SOS*	42

*Sum of Squares

Table 3.3 Topo-chemical descriptors used in the construction of MLR and ANN models for predicting BEA zeolite-SDA dispersion energy are presented below.

Descriptor type	Descriptor name	Descriptor definition
Burden eigenvalues	SpMin2_Bh(i)	Smallest eigenvalue n. 2 of Burden matrix weighted by ionization potential
GETAWAY descriptors	R2e	R autocorrelation of lag 2/weighted by Sanderson electronegativity
GETAWAY descriptors	R4s+	R maximal autocorrelation of lag 4/weighted by I-state
P_VSA-like descriptors	P_VSA_s_6	P_VSA-like on I-state, bin 6
RDF descriptors	RDF085u	Radial Distribution Function – 085/unweighted
RDF descriptors	RDF105p	Distribution Function – 105/weighted by polarizability

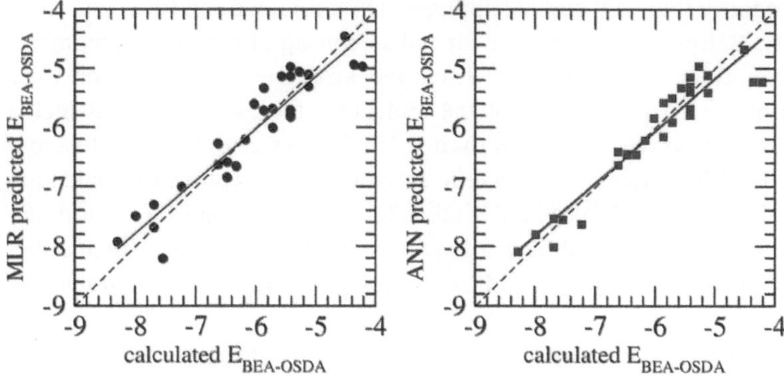

Figure 3.5 Scatterplots of the calculated and predicted BEA–SDA energy for the training set of each model (MLR and ANN models). Validation and training points in blue and red respectively for the ANN model.

Table 3.4 Parameters resulting of validation techniques for MLR and ANN models predicting BEA-SDA van der Waals energy.

Model		Internal validation technique		
Type	**Predictive capability**	**L-O-O**	**L-S-O**	**Cross-validation**
MLR	$r^2 = 0.88$	$Q^2 = 0.82$	$Q^2 = 0.92$	
ANN	$r^2 = 0.98$			$Q^2 = 0.89$

Cross-validation, leave-one-out and leave-some-out methods are used. Leave-one-out (LOO) [56], is a technique in which one compound at a time of the training group is left out and the model is built with the rest of the data from the training set. After obtaining the new model, it is asked to predict the value of the property for the compound left out that would act as a test. This is repeated for all dataset compounds. Another internal validation technique is the leave-some-out (LSO) call [57], in which the same process mentioned above is repeated in the LOO, only instead of leaving a compound out, we will leave out approximately 25% of the data (6 out of 29). LOO and LSO were used for internal validation of the MLR model. Cross-validation was used to validate the ANN model; 30% of the data was used for testing the model and the rest (70%) were used for developing the model. Table 3.4 shows the different validation procedures applied to the MLR model and its correlation coefficient data obtained together with external validation of ANN model.

Table 3.4 shows the robustness (internal validation) of the models. Both MLR and ANN models show internal validation correlation coefficient (Q^2) similar to those of the model correlation coefficient (r^2). Therefore, neither a single nor a group of compounds are determining the predictive capability of the model.

3.3 Virtual Screening: Identifying Novel SDA with Favorable $E_{ZEO\text{-}SDA}$ for the Synthesis of BEA Zeolite

Once BEA–SDA van der Waals energy prediction models have been constructed and validated, a database of commercial compounds is screened in order to select potential SDAs for the synthesis of BEA zeolite. The Emolecules database [58] provides over 32 million unique chemical structures from a network of more than 100 global suppliers. Therefore, we can find structures with high chemical diversity. Before performing the virtual screening of this database, molecules containing atoms other than C and N were discarded and 1360468 molecules were selected. Bash and awk scripts to extract those 1360468 molecules took a few days. Selected molecules were stored in directories containing less than 2^{16} molecules each in order to avoid handling errors. From the resulting directories, specific software was employed to calculate the "shoebox" distances of all molecules [14]. Since it takes less than 1 second for each molecule, it took less than 18 hours to obtain the shoebox distances for each directory. Since directories are independent this can be run in parallel and can be done in 18 hours for all molecules of our selection.

Next, we use MLR and ANN models to identify molecules showing a low $E_{BEA\text{-}SDA}$. Table 3.5 and Figure 3.6 shows 19 candidate OSDAs for the synthesis of BEA.

Once a virtual screening and selection of candidate BEA OSDAs has been performed on the basis of ANN and MLR algorithms to model BEA–OSDA van der Waals energy, it is necessary to validate the predictive models using our atomistic approach, based on Monte Carlo and lattice energy minimization, for determining BEA–OSDA energies. A potential SDA for BEA would ideally exhibit a low value of E(BEA-SDA) and should give a larger stability in BEA over the other competing phases (BEC and ISV) [11].

Table 3.5 E_{BEA_SDA} (kJ/mol Si) prediction value for pre-selected organic compound from Emolecules database as potential OSDA for BEA zeolites synthesis.

EMol id	MLR $E_{BEA\text{-}SDA}$	ANN $E_{BEA\text{-}SDA}$
14312311	−13.3	-7.0
50961168	-8.3	-3.8
106927669	−12.6	-4.2
177523224	−13.5	-7.2
179716557	−13.7	-6.8
233016343	−14.3	-4.5
259818077	−11.4	-5.2
259947421	−11.8	-7.5
260117959	−12.3	-5.3
260301480	−11.4	-8.3
260351136	−11.5	-8.5
260460703	−11.3	-5.9
260702877	−13.7	-7.6
260714801	−11.7	-4.3
260972836	−14.0	-5.8
301308413	−13.4	-5.2
301311920	−11.4	-6.9
301326260	−11.5	-6.3
302639806	−11.6	-6.2

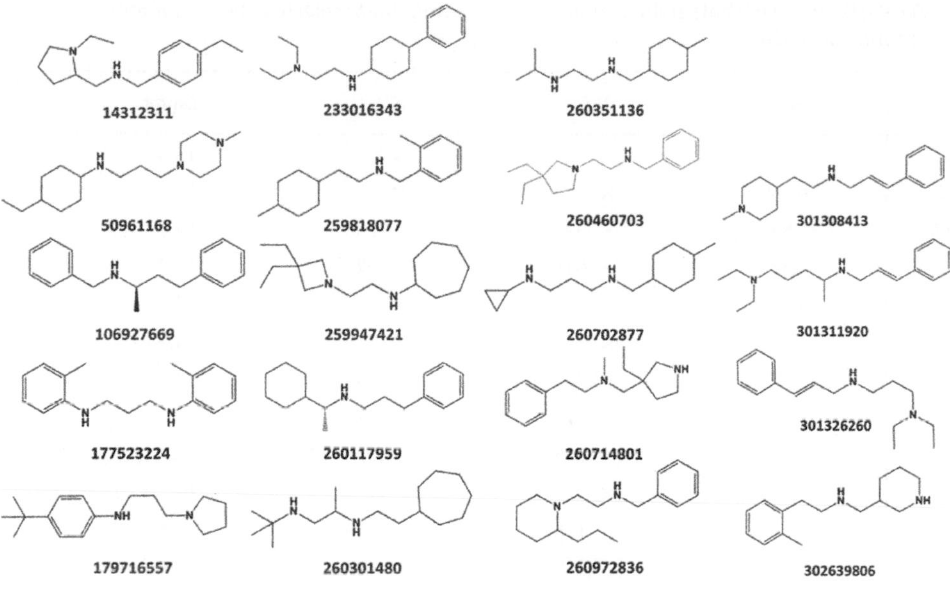

Figure 3.6 Potential OSDAs for BEA: virtual screening of Emolecules database.

3.4 Zeo-SDA Energy Calculation Using Atomic Models

Taking into account the predictions of the MLR and ANN models in Table 3.6, regarding the suitable OSDAs for the synthesis of BEA, we now calculate explicitly the zeo-OSDA energetic stabilization using the combination of force fields and algorithms (Monte Carlo and lattice energy minimization) employed in our previous work [11]. The maximum interatomic distance (Dmax) of the SDA molecule will be useful for the interpretation of the results. The results are indicated in Table 3.6 and can be divided in three groups:

a) Those OSDA molecules (14312311, 179716557, 233016343, 260301480, 260702877, 301311920, 50961168) that do not fit in BEC and ISV, whilst they fit in BEA. This corresponds to OSDA molecules whose Dmax is in the interval [13.3,15.0]. BEA, BEC, and ISV contain two of their crystallographic cell lengths in the interval [12.6,13.0] Å and the other cell length in the interval [25.5,26.2] Å. But BEA is the unit cell with a larger channel extension (Figure 3.7) through the longest cell length and so the OSDA can be aligned across this direction, achieving a loading of four molecules per unit cell. Contrarily, BEC and ISV need to distribute at least two OSDA molecules along the shortest channel axis, limiting the loading of molecules larger than ca. 13.5 Å. For these cases, BEA will be particularly selective by not only the zeo(BEA)-SDA stabilization, but also for the impossibility to achieve a loading of four OSDA molecules per unit cell in the cases of BEC and ISV.
b) Those OSDA molecules (301308413, 301326260) that do not fit in any of the three zeolites, BEA, BEC, ISV. This corresponds to very large OSDA molecules, whose Dmax (largest interatomic distance according to the shoebox algorithm) is larger than 15.8 Å, precluding the fit in the unit cells. These cases fall within the expected shortcoming of the MLR and ANN models,

Table 3.6 Zeo-OSDA van der Waals stabilization energies (kJ/mol Si) of selected OSDAs suitable candidates to synthesize BEA.

SDA	BEA	BEC	ISV	Dmax
14312311	−5.623	×	×	13.3
50961168	−6.769	×	×	14.6
106927669	−6.995	−3.528	−3.588	12.6
177523224	−7.342	−6.045	−5.970	13.7
179716557	−4.764	×	×	14.7
233016343	−6.498	×	×	14.7
259818077	−6.362	−5.397	−5.608	13.1
259947421	−6.422	−5.276	−5.518	13.3
260117959	−6.196	−5.065	−5.096	12.8
260301480	−4.854	×	×	15
260351136	−6.196	−5.156	−4.779	13.5
260460703	−6.663	−5.819	−5.985	13.4
260702877	−6.528	×	×	14.6
260714801	−6.844	−5.563	−5.201	12.4
260972836	−7.387	−3.407	−4.869	13.8
301308413	×	×	×	15.9
301311920	−3.724	×	×	14.5
301326260	×	×	×	15.9
302639806	−7.101	−3.060	−4.869	12.6

Loading in BEA is always four OSDA molecules per u.c., except in the case of 301311920 (2 OSDAs/u.c.). "Dmax" values correspond to the maximum interatomic distance within the OSDA molecule (Å), as calculated from the "shoebox" algorithm. "×" means the OSDA could not be fitted in the corresponding micropore.

giving some "false positive" molecules, but they are a minority and have no effect in allowing identification of valuable OSDA candidates in cases (a) and (c).

c) OSDA molecules in the interval [12.4, 13.8] Å, which can fit in the three zeolites (106927669, 177523224, 259818077, 259947421, 260117959, 260351136, 260460703, 260714801, 260972836, 302639806): although with the possibility to achieve the loading of four OSDA/u.c. only in BEA. Then, although zeo-SDA stabilization energies are provided for BEA, BEC, and ISV, there is a significant energy difference, with BEA giving always the most stable zeo-SDA interaction and hence BEA being the predicted phase.

Fitting of OSDAs in BEA is shown in Figures 3.8 and 3.9. Apart from the useless two OSDA molecules found in (b) above, case (a) gives seven OSDA molecules that, by not fitting in BEC and ISV, can be good candidates to synthesize BEA. Also, some of the molecules in case (c) show much larger stabilization (>1.0 eV) for BEA than the other two zeolites (BEC, ISV), such as 106927669, 260972836, and 302639806. These OSDA molecules have a non-linear shape, with a marked L-shape for 106927669 and 260972836, which allow a very good fit in BEA (Figure 3.10), while precluding a good fit in BEC and ISV. These 10 OSDA molecules (14312311, 179716557, 233016343, 260301480, 260702877, 301311920, 50961168, 106927669, 260972836, and 302639806) are the best candidates to give BEA zeolite according to the present study.

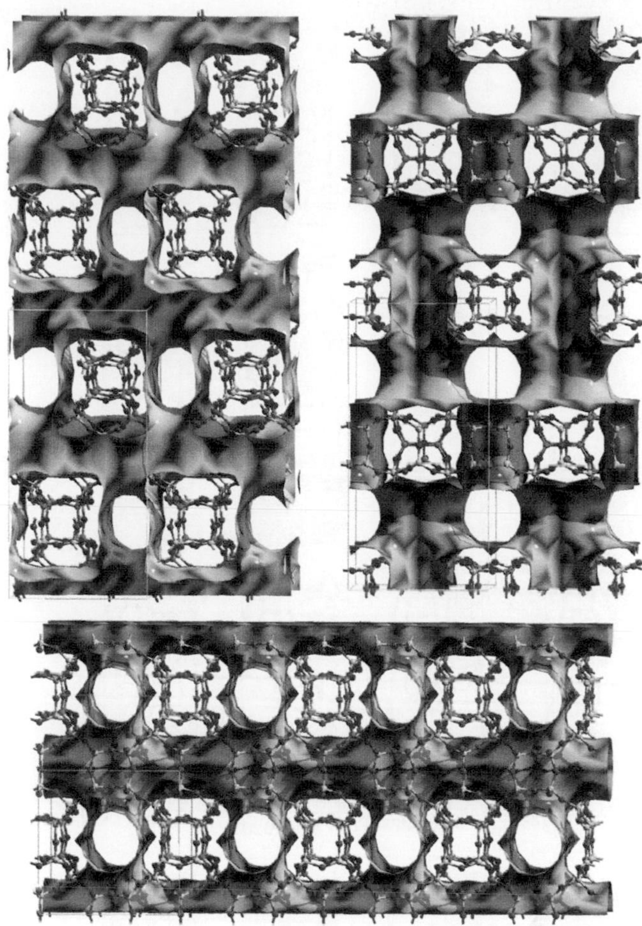

Figure 3.7 Channels (blue shaded) of BEA (top left): ISV (top right) and BEC (bottom).

3.5 Comparing Zeo-SDA Energy Calculation Using MLR, ANN, and Atomic Models

An ultimate validation of the MLR and ANN model can be performed by comparing zeo-SDA stabilization energy calculation predicted by the models and Monte Carlo and lattice energy minimization calculations.

As seen in Table 3.7, the range of van der Waals stabilization energy values predicted by the ANN model (−4 to −8 kJ/mol Si) is more accurate than the MLR model (−8 to −14 kJ/mol Si) if compared with Monte Carlo and lattice energy minimization calculations (−4 to −7 kJ/mol Si). In addition, most of residual values from ANN predictions to atomic model calculations differ less or equal to 2 kJ/mol Si. Therefore, we can state that the ANN model is more accurate than the MLR model in the stabilization energy prediction comparing with reported Monte Carlo and lattice energy minimization calculations.

Nevertheless, the MLR model was able to rank correctly the compound with a more favorable stabilization energy to synthesize BEA (260972836). Therefore, when applying the MLR model, compounds should be selected with stabilization energy of −14 kJ/mol Si or lower. Whereas when applying the ANN model, we should select compounds with predicted stabilization energy smaller than −7 kJ/mol Si.

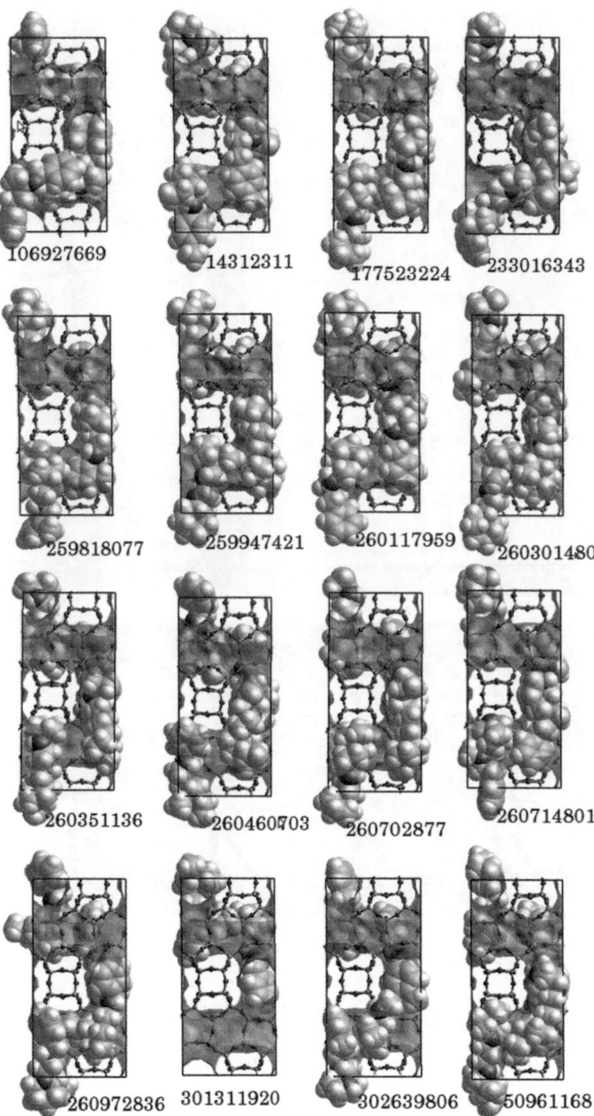

Figure 3.8 Projection along [100] of some OSDAs from Table 3.6 in BEA zeolite.

3.6 Conclusions

A combination of Molecular Topology (MT) and Monte Carlo (MC) techniques has been used in order to find suitable OSDAs for a particular zeolite, all-silica Beta (BEA). Based on a previous study in which 29 OSDAs were tested for the selective synthesis of BEA, BEC, and ISV zeolites, the calculated zeo-OSDA energies were used as a training set. MLR and ANN models were obtained. These models were applied to the emolecules database and 19 new OSDAs for the selective synthesis of BEA were selected. MC calculations in BEA, BEC, and ISV zeolites showed that, in agreement with the MT models, these 19 OSDAs are selective for BEA.

The combination of MT (artificial intelligence) and MC (atomistic modeling) methods has been improved using a larger training set and a larger database than in previous work. The application

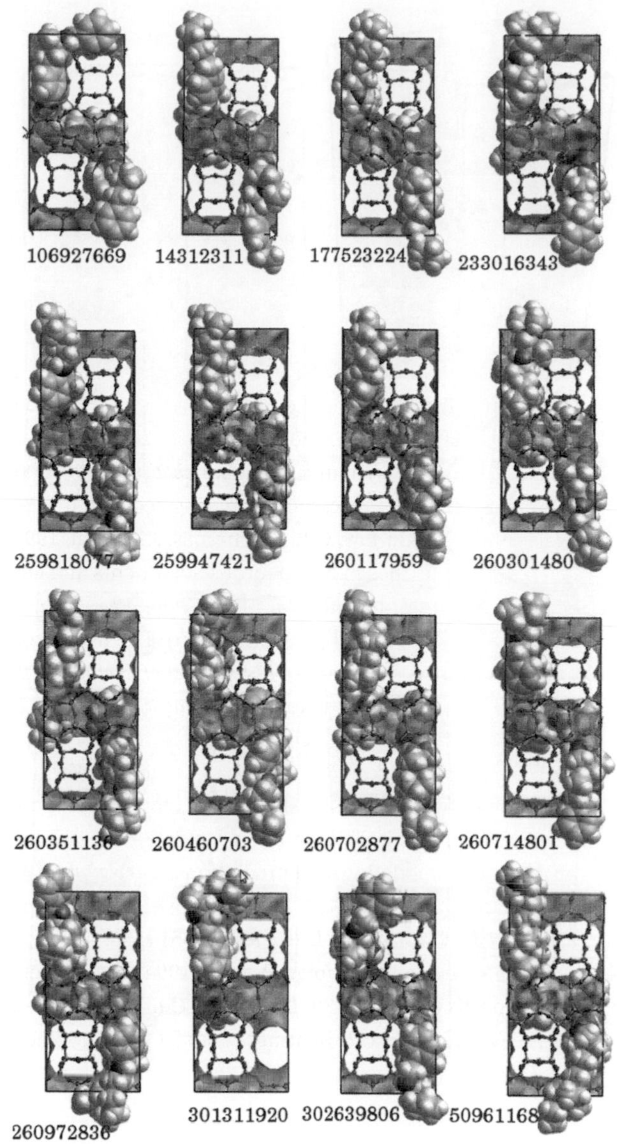

Figure 3.9 Projection along [010] of some OSDAs from Table 3.6 in BEA zeolite.

to the selection of OSDAs for the selective synthesis of zeolites is a promising field but there are issues that need to be implemented such as the treatment of counteranions (hydroxide or fluoride) in a first step, and the presence of aluminium in a second step, in order to have more realistic models whose comparison to experiments is more coherent. MC techniques are of limited accuracy in finding the optimum OSDA position and, although having been validated, need further refinement. Finally, 3D descriptors related to OSDA dimensions and zeolite pore size and shape must be added to the list of MT descriptors and, by doing so, optimized results can be obtained, shedding light on the complex field of rational design of zeolites.

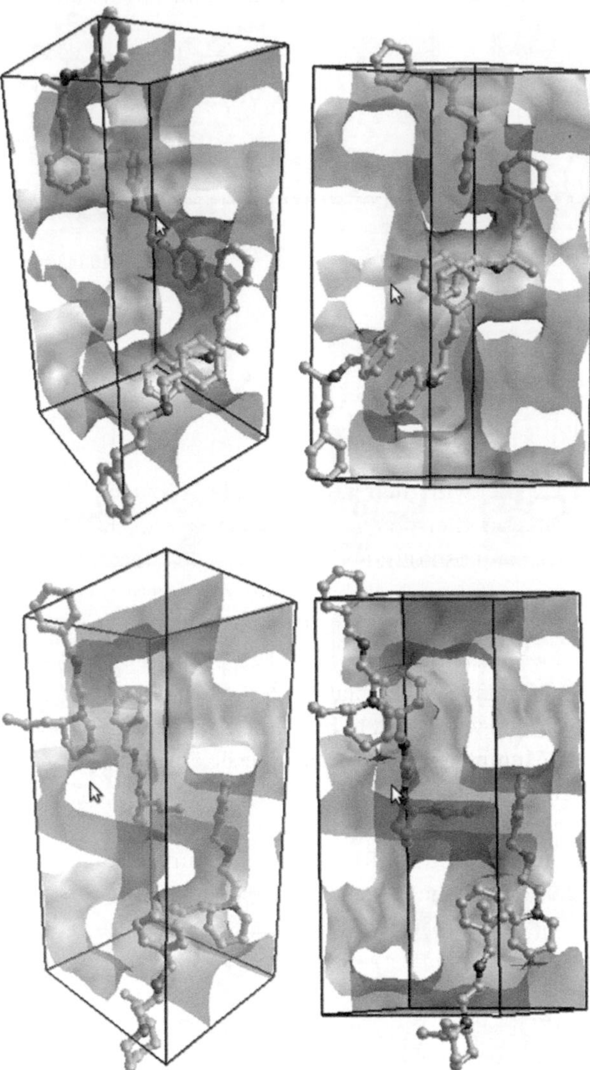

Figure 3.10 Two views of the fit of "L-shaped" 106927669 (top) and 2609672836 (bottom) OSDAs in BEA zeolite. These are two particularly good OSDA candidates for the selective synthesis of BEA.

Table 3.7 Zeo-OSDA van der Waals stabilization energies (kJ/mol Si) of selected OSDAs suitable candidates to synthesize BEA.

	MLR	ANN	Atomic model
EMol id	$E_{BEA\text{-}SDA}$	$E_{BEA\text{-}SDA}$	$E_{BEA\text{-}SDA}$
14312311	−13.3	−7.0	−5.6
50961168	-8.3	-3.8	-6.8
106927669	−12.6	−4.2	−7.0
177523224	−13.5	−7.2	−7.3

Table 3.7 (Continued)

	MLR	ANN	Atomic model
EMol id	$E_{BEA\text{-}SDA}$	$E_{BEA\text{-}SDA}$	$E_{BEA\text{-}SDA}$
179716557	−13.7	−6.8	−4.8
233016343	−14.3	−4.5	−6.5
259818077	−11.4	−5.2	−6.4
259947421	−11.8	−7.5	−6.4
260117959	−12.3	−5.3	−6.2
260301480	−11.4	−8.3	−4.9
260351136	−11.5	−8.5	−6.2
260460703	−11.3	−5.9	−6.7
260702877	−13.7	−7.6	−6.5
260714801	−11.7	−4.3	−6.8
260972836	−14.0	−5.8	−7.4
301308413	−13.4	−5.2	×
301311920	−11.4	−6.9	−3.7
301326260	−11.5	−6.3	×
302639806	−11.6	−6.2	−7.1

Loading in BEA is always four OSDA molecules per u.c., except in the case of 301311920 (2 OSDAs/u.c.).

Acknowledgments

German Sastre thanks the Science and Technology Commission of Generalitat Valenciana (GVA) for funding through project PROMETEO-2021-077. María Gálvez Llompart thanks GVA for a postdoctoral contract APOSTD/2019/055. We thank Prof. Fernando Rey for useful suggestions. We thank Centro de Cálculo Científico (SGAI-CSIC) for the use of their computational facilities.

References

1 Zhou, Z. (2021). *Journal of Materials Chemistry A* and materials advances editor's choice web collection: machine learning for materials innovation. *J. Mater. Chem. A* 9 (3): 1295–1296.
2 Campesato, O. (2020). *Artificial Intelligence, Machine Learning, and Deep Learning Stylus Publishing*. LLC.
3 Ceriotti, M., Clementi, C., and von Lilienfeld, O.A. (2021). Machine learning meets chemical physics. *J. Chem. Phys.* 154: 160401.
4 Srinivasan, S., Batra, R., Chan, H. et al. 2021). Artificial intelligence-guided *de novo* molecular design targeting COVID-19. *ACS Omega* 6 (19): 12557–12566.
5 Artrith, N., Butler, K.T., Coudert, F. et al. (2021). Best practices in machine learning for chemistry. *Nat. Chem.* 13 (6): 505–508.
6 Lemm, D., von Rudorff, G.F., and von Lilienfeld, O.A. (2021). Machine learning based energy-free structure predictions of molecules, transition states, and solids. *Nat. Commun.* 12 (1): 1–10.

7 Batra, R. (2021). Accurate machine learning in materials science facilitated by using diverse data sources. *Nature* 589: 524–525.

8 Sastre, G., Cantin, A., Diaz-Cabañas, M.J. et al. (2005). Searching organic structure directing agents for the synthesis of specific zeolite structures: an experimentally tested computational study. *Chem. Mater.* 2005 (17): 545.

9 Jensen, Z., Kwon, S., Schwalbe-Koda, D. et al. (2021). Discovering relationships between OSDAs and zeolites through data mining and generative neural networks. *ACS Cent. Sci.* 7 (5): 858–867.

10 Moliner, M., Rey, F., and Corma, A. (2013). Towards the rational design of efficient organic structure-directing agents for zeolite synthesis. *Angew. Chem., Int. Ed. Engl.* 52: 13880–13889.

11 Gálvez-Llompart, M., Gálvez, J., Rey, F. et al. (2020). Identification of new templates for the synthesis of BEA, BEC, and ISV zeolites using molecular topology and Monte Carlo techniques. *J. Chem. Inf. Model.* 60 (6): 2819–2829.

12 Galvez-Llompart, M., Cantín, A., Rey, F. et al. (2019). Computational screening of structure directing agents for the synthesis of zeolites. A simplified model. *Z. Kristallogr.* 234: 451–460.

13 Archer, R.H., Zones, S.I., and Davis, M.E. (2010). Imidazolium structure directing agents in zeolite synthesis: exploring guest/host relationships in the synthesis of SSZ-70. *Microporous and Mesoporous Mater.* 130 (1–3): 255–265.

14 León, S. and Sastre, G. (2020). Computational screening of structure-directing agents for the synthesis of pure silica ITE zeolite. *J. Phys. Chem. Lett.* 11: 6164–6167.

15 Jensen, Z., Kim, E., Kwon, S. et al. (2019). A machine learning approach to zeolite synthesis enabled by automatic literature data extraction. *ACS Cent. Sci.* 5 (5): 892–899.

16 Moliner, M., Román-Leshkov, Y., and Corma, A. (2019). Machine learning applied to zeolite synthesis: the missing link for realizing high-throughput discovery. *Acc. Chem. Res.* 52 (10): 2971–2980.

17 Kwak, S.J., Kim, H.S., Park, N. et al. (2021). Recent progress on a distribution over zeolite frameworks: linking theories and experiments. *Korean J. Chem. Eng.* 38: 1117–1128.

18 Daeyaert, F., Ye, F., and Deem, M.W. (2019). Machine-learning approach to the design of OSDAs for zeolite beta. *Proc. Natl. Acad. Sci.* 116 (9): 3413–3418.

19 Schwalbe-Koda, D. and Gómez-Bombarelli, R. (2021). Benchmarking binding energy calculations for organic structure-directing agents in pure-silica zeolites. *J. Chem. Phys.* 154 (17): 174109.

20 Muraoka, K., Chaikittisilp, W., and Okubo, T. (2020). Multi-objective *de novo* molecular design of organic structure-directing agents for zeolites using nature-inspired ant colony optimization. *Chem. Sci.* 11 (31): 8214–8223.

21 Statsoft, I. (2012). Statistica (data analysis software system) 10.0.

22 Nawi, N.M., Ransing, M.R., and Ransing, R.S. (2006). An improved learning algorithm based on the Broyden-Fletcher-Goldfarb-Shanno (BFGS) method for back propagation neural networks. *Sixth International Conference on Intelligent Systems Design and Applications.* 1: 152–157.

23 Johansson, E.M., Dowla, F.U., and Goodman, D.M. (1991). Backpropagation learning for multilayer feed-forward neural networks using the conjugate gradient method. *Int. J. Neural Syst.* 2 (4): 291–301.

24 Dumitru, C. and Maria, V. (2013). Advantages and disadvantages of using neural networks for predictions. *Ovidius Univ. Ann. Series Econom. Sci.* 13 (1): 444–449.

25 Mukherjee, K. and Colón, Y.J. (2021). Machine learning and descriptor selection for the computational discovery of metal–organic frameworks. *Mol. Simul.* 47 (10–11): 1–21.

26 Wanyonyi, F.S., Fidelis, T.T., Mutua, G.K. et al. (2021). Role of pore chemistry and topology in the heavy metal sorption by zeolites: from molecular simulation to machine learning. *Comput. Mater. Sci.* 195: 110519.

27 Zhang, Z. (2016). Variable selection with stepwise and best subset approaches. *Ann. Transl. Med.* 4 (7): 136.

28 Maitra, S. and Yan, J. (2008). Principle component analysis and partial least squares: two dimension reduction techniques for regression. *CAS Casualty Actuarial Society Discussion Paper Program*.

29 Zanni, R., Galvez-Llompart, M., Garcia-Domenech, R. et al. (2020). What place does molecular topology have in today's drug discovery. *Expert Opin. Drug Discov.* 15 (10): 1133–1144.

30 Galvez, J., Galvez-Llompart, M., Zanni, R. et al. (2013). Molecular topology: dissimilar similarities. *Drug Discov. Today Technol.* 10 (4): e475–e481.

31 Gálvez, J., Gálvez-Llompart, M., and García-Domenech, R. (2012). Molecular topology as a novel approach for drug discovery. *Expert Opin. Drug Discov.* 7 (2): 133–153.

32 Galvez, J.M., Villar, V., Galvez-Llompart, M. et al. (2011). Chemistry explained by topology: an alternative approach. *Comb. Chem. High Throughput Screen.* 14 (4): 279–283.

33 Zanni, R., Galvez-Llompart, M., Garcia-Pereira, I. et al. (2019). Molecular topology and QSAR multi-target analysis to boost the *in silico* research for fungicides in agricultural chemistry. *Mol. Divers.* 23 (2): 371–379.

34 Galvez-Llompart, M., Zanni, R., Galvez, J. et al. (2020). Molecular topology QSAR strategy for crop protection: new natural fungicides with chitin inhibitory activity. *ACS Omega* 5 (27): 16358–16365.

35 Gálvez, J., Gálvez-Llompart, M., and García-Domenech, R. (2010). Application of molecular topology for the prediction of the reaction times and yields under solvent-free conditions. *Green Chem.* 12 (6): 1056–1061.

36 Zanni, R., Galvez-Llompart, M., and Galvez, J. (2021). Computational analysis of macrolides as SARS-CoV-2 main protease inhibitors: a pattern recognition study based on molecular topology and validated by molecular docking. *New J. Chem.* 45 (19): 8654–8675.

37 Galvez, J., Zanni, R., Galvez-Llompart, M. et al. (2021). Macrolides may prevent severe acute respiratory syndrome coronavirus 2 entry into cells: a quantitative structure activity relationship study and experimental validation. *J. Chem. Inf. Model.* 61 (4): 2016–2025.

38 Galvez-Llompart, M., Ocello, R., Rullo, L. et al. (2021). Targeting the JAK/STAT pathway: a combined ligand-and target-based approach. *J. Chem. Inf. Model.* 61 (6): 3091–3108.

39 Pla-Franco, J., Gálvez-Llompart, M., Gálvez, J. et al. (2011). Application of molecular topology for the prediction of reaction yields and anti-inflammatory activity of heterocyclic amidine derivatives. *Int. J. Mol. Sci.* 12 (2): 1281–1292.

40 Srivastava, V., Selvaraj, C., and Singh, S.K. (2021). Chemoinformatics and QSAR. In: *Advances in Bioinformatics*, 183–212. Springer.

41 Costa, P., Evangelista, J.S., Leal, I. et al. (2021). Chemical graph theory for property modeling in QSAR and QSPR: charming QSAR & QSPR. *Mathematics* 9 (1): 60.

42 Cramer, R.D., Patterson, D.E., and Bunce, J.D. (1988). Comparative molecular field analysis (CoMFA). 1: Effect of shape on binding of steroids to carrier proteins. *J. Am. Chem. Soc.* 110 (18): 5959–5967.

43 Wadlinger, R.L., Rosinski, E.J., and Kerr, G.T. (1967) US Patent 3,308,069A.

44 Newsam, J.M., Treacy, M.M.J., Koetsier, W.T. et al. (1988). Structural characterization of zeolite-beta. *Proc. R. Soc. A* 420: 375–405.

45 Daeyaert, F. and Deem, M.W. (2019). Design of organic structure directing agents for polymorph A zeolite beta. *J. Mat. Chem. A* 7 (16): 9854–9866.

46 Coronas, J. (2010). Present and future synthesis challenges for zeolites. *Chem. Eng. J.* 156: 236–242.

47 Daeyaert, F. and Deem, M.W. (2019). Design of organic structure directing agents for chiral zeolite beta A. *Chem. Select* 4 (12): 3531–3537.

48 Mauri, A. (2020). alvaDesc: a tool to calculate and analyze molecular descriptors and fingerprints. In: *Ecotoxicological QSARs (ed.* K. Roy*)*, 801–820. Springer.

49 Chicco, D., Warrens, M.J., and Jurman, G. (2021). The coefficient of determination R-squared is more informative than SMAPE, MAE, MAPE, MSE and RMSE in regression analysis evaluation. *Peer J. Comput. Sci.* 7: e623.

50 Consonni, V. and Todeschini, R. (2008). New spectral indices for molecule description. *Matrix* 1: 2.

51 Dearden, J.C. (2017). The use of topological indices in QSAR and QSPR modeling. In: *Advances in QSAR Modeling (ed.* K. Roy*)*, 57–88. Springer.

52 Hemmer, M.C., Steinhauer, V., and Gasteiger, J. (1999). Deriving the 3D structure of organic molecules from their infrared spectra. *Vib. Spectrosc.* 19 (1): 151–164.

53 Consonni, V., Todeschini, R., Pavan, M. et al. (2002). Structure/response correlations and similarity/diversity analysis by GETAWAY descriptors. 2: Application of the novel 3D molecular descriptors to QSAR/QSPR studies. *J. Chem. Inf. Comput. Sci.* 42 (3): 693–705.

54 Consonni, V., Todeschini, R., and Pavan, M. (2002). Structure/response correlations and similarity/diversity analysis by GETAWAY descriptors. 1: Theory of the novel 3D molecular descriptors. *J. Chem. Inf. Comput. Sci.* 42 (3): 682–692.

55 Majumdar, S. and Basak, S.C. (2018). Beware of external validation! A comparative study of several validation techniques used in QSAR modelling. *Curr. Comput. Aided Drug Des.* 14 (4): 284–291.

56 Kearns, M. and Ron, D. (1999). Algorithmic stability and sanity-check bounds for leave-one-out cross-validation. *Neural Comput.* 11 (6): 1427–1453.

57 Veerasamy, R., Rajak, H., Jain, A. et al. (2011). Validation of QSAR models-strategies and importance. *Int. J. Drug Des. Discov.* 3: 511–519.

58 https://www.emolecules.com

4

Generating, Managing, and Mining Big Data in Zeolite Simulations

Daniel Schwalbe-Koda and Rafael Gómez-Bombarelli

Department of Materials Science and Engineering, Massachusetts Institute of Technology, Massachusetts, USA

4.1 Introduction

Accelerated materials innovation is imperative to address global challenges in energy, the environment, or healthcare [1,2]. Mitigating climate change, for example, requires a fast decarbonization of the economy, including hard-to-decarbonize sectors such as industry or transportation [3]. Enabling this transition requires discovering catalysts for sustainable chemical transformations [4]. However, catalyst discovery has long relied on trial-and-error. Finding stable and affordable materials that can catalyze given chemical reactions with high activity/selectivity requires exploring a combinatorial number of structures and compositions, often aggravated by the high-dimensional problem of synthesizing the materials themselves. As such, strategies to accelerate the design of novel catalysts are urgently required.

Zeolites are among the catalysts of interest for sustainable applications. Their mature use in the petrochemical industry makes them interesting candidates for a variety of reaction pathways involving thermocatalysis [5,6], especially because these materials are known to have a large topological diversity [7,8]. Unfortunately, zeolites are no exception to the trial-and-error discovery efforts typically seen in the field of catalysis. Due to strong phase competition effects between polymorphs, finding cost-effective synthesis routes for the synthesis of these materials requires labor-intensive experimental efforts to be overcome [9,10]. Choosing the optimal synthesis conditions is further complicated by the wide array of synthesis parameters, long crystallization times that slow down serial iterative research, and the associated costs of the reactions.

On the other hand, data-driven methods hold promise for avoiding such intensive experimentation by downselecting the most promising candidate structures/synthesis routes using computational methods, thus prior to experimentation [11–16]. In the last few years, the materials and chemistry communities have experienced a surge of works related to data-intensive methods, including high-throughput simulations or machine learning. In line with programs such as the Materials Genome Initiative in the United States [17], several efforts were made toward automating the generation, handling, curation, and publicizing of data related to materials discovery and synthesis. In the decade prior to the establishment of this program, several new materials had already been discovered through computation, for many applications such as superconductors [18], batteries [19], thermoelectrics [20], or electrocatalysts [12]. These successes planted the seeds for improving data-sharing efforts, which later drove a surge in research for materials informatics and applied machine learning in recent years [21–28]. Enabling researchers around the world to benefit from data curated for classes of materials, therefore, is a driving force in materials discovery.

AI-Guided Design and Property Prediction for Zeolites and Nanoporous Materials, First Edition. Edited by German Sastre and Frits Daeyaert.

Data has always been abundant in the field of zeolites. With more than 250 structures recognized by the International Zeolite Association (IZA), strong industrial interest, and over six decades of research in the field [29], large volumes of data have been created in the form of journal articles, patents, or simulations in the last few years. Nevertheless, the application of data science methods to zeolite discovery has so far been limited. The diversity of synthesis conditions and challenges in proposing reproducible synthesis routes may have hindered advances in the field. One of the major entrance barriers toward the application of data science methods for zeolite discovery is the absence of machine-readable experimental data [30]. Independent efforts, such as the databases of structures and verified syntheses by the IZA, are good examples of accessible and useful data that serves the community in an important way. Given the variety of possibilities in zeolite discovery, however, implementing similar concepts using purely computational data could help accelerate zeolite synthesis.

In this chapter, we review our efforts in creating and managing computational databases for supporting zeolite synthesis. In particular, we focus on our work combining simulations and data-driven methods to understand and predict the role of organic structure-directing agents (OSDAs) in templated zeolite synthesis. In the first part of this chapter, we briefly review advances in classical simulation techniques applied in a high-throughput setting to materials discovery, particularly those programs focused on curating databases for materials properties. Then, we discuss a few existing databases in the zeolite community. In the second part, we describe our methodology to create a database of OSDAs for zeolites, along with the reasoning behind design and implementation choices. We detail the algorithms, benchmarks, and design principles that support our computational approach to templated zeolite synthesis. Finally, we outline a few opportunities for zeolite discovery using data-driven methods. Advances in theoretical modeling, simulations, and machine learning may pave the way for faster, cost-effective zeolite discovery and their applications in sustainable processes.

4.1.1 Computational Materials Databases

The use of high-performance computing (HPC) to investigate a large number of materials systematically has become ubiquitous in the last decade, particularly due to a larger accessibility of computational resources. Nevertheless, computational databases of calculated materials and their properties started as early as 2002. One example is the Electronic Structure Project [31,32], where a variety of crystal structures from the Inorganic Crystal Structure Database (ICSD) were investigated in the search for scintillator materials. Although the candidates investigated in the first study did not exhibit the targeted performance, the work showed how systematic analyses of materials could guide experimentation, with potential for finding better candidates under extended criteria.

Leveraging this principle of screening materials prior to experimentation, several works leveraged the power of HPC to discover new materials [11–16], setting the foundation for major collaborative initiatives in the 2010s. Platforms such as the Harvard Clean Energy Project, for example, screened millions of organic photovoltaic compounds for solar cell applications using a combination of crowd sourced computation, database management, and quantum chemistry [33]. Others, such as CatApp, provided a user-friendly interface to aid the design of catalysts based on precomputed surface properties and scaling relations [34]. However, general-purpose databases of computed materials properties only started with platforms such as the Materials Project [35] or AFLOW [36,37]. Both initiatives developed an extensive computational infrastructure to automate and deploy simulations to HPC centers without human intervention. In the early stages of the Materials Project, more than 33,000 structures had been calculated, most of which came from the ICSD.

Similarly, the AFLOW project created a workflow to automatically calculate properties of inorganic alloys, intermetallics, and other compounds in a variety of structures. As at the time of writing this chapter, the AFLOW project has more than 3,562,000 compounds totaling over 705 million calculated properties, and the Materials Project provides almost 145,000 compounds and a variety of predicted properties.

The development of automation infrastructures by different groups has lead to the creation of different databases, including the Open Quantum Materials Database [38], the AiiDA project [39,40], the Computational 2D Materials Database [41], the Catalysis-Hub [42], the Open Catalyst project [43], and many others. While each database often provides different properties, simulation parameters, etc., the unifying theme among these projects and infrastructures is to make the data accessible by design. The leaders of these projects remain committed to disseminating data, aiding *in silico* materials design through reproducible simulation protocols, web interfaces, calculations, and database management. This centralized management enables calculations to be performed systematically and made available for researchers in a fast, reliable way.

Very often, different research groups can perform their own calculations independently and share their data in a variety of formats. While findable-accessible-interoperable-reusable (FAIR) data principles [44] should be encouraged, persistent storage of calculated structures can be a challenge to independent researchers. To address this demand, a series of platforms offer materials data storage in a public and accessible way. The Novel Materials Discovery (NOMAD) database [45], for example, currently houses more than 50 million calculations across different groups, softwares, and materials classes. Although the calculations are stored for up to 10 years, the data processing features from the NOMAD repository automatically extract the most important information about calculations performed with a variety of simulation packages. This is similar to the Materials Data Facility [46,47], Materials Cloud [48], or ioChem-bd [49], which provide storage, application programming interfaces (APIs), and integration tools for the published data. Repositories analogous to the ICSD, but focused mostly on predicted or experimental structures, also exist. The (Predicted) Crystallography Open Database (COD) [50,51], for instance, stores crystal structures of a variety of materials beyond inorganics as contributed by various groups. While the web services of these databases are not meant to be as comprehensive as those offered by the aforementioned projects, databases such as COD and PCOD centralize diverse materials structures in an open-access library. Finally, solutions such as Zenodo or FigShare are general-purpose repositories beyond the area of materials, but can help storing and sharing the data in the raw formats provided by researchers.

Thanks to the combination of high-throughput computation and data sharing, these and other independent projects have spurred the development and discovery of materials for several applications. Systematic generation, machine-readability, and reusability of the data are important features driving the success of these projects. From enabling researchers to interact with the data at a high level to providing raw results that incentivize reproducibility and further computational investigations, databases are helping to shift the discovery paradigm in materials science. These outcomes are guiding the generation of data for several new materials classes and applications, setting the standards for new computational developments around database creation and management in the field.

4.1.2 Zeolite Databases

Zeolites are amenable to enumeration and database management due to their diverse polymorphism. As an example, the number of distinct zeolite structures currently recognized by the IZA surpasses 250 [29]. Rigorously evaluated by the Structure Commission of this Association, new zeolites that are experimentally realized, characterized, and resolved from the crystallographic perspective are added to the Structural Database of the IZA. This is perhaps one of the most used databases in the field, as it categorizes and fully describes the structures representing each polymorph. As of 2021, the database has the form of a web interface with different entries containing properties, literature data, etc. for each framework. The IZA database curates structural data in the form of human-readable format, and is continuously updated when new frameworks are approved or when new functionalities are added. Examples of recent features in the website include addition of nuclear magnetic resonance (NMR) spectra for some zeolites or tools that help calculating pore properties.

The diversity of zeolite topologies is not limited to known structures. Computer-enumerated structures of zeolites not yet realized, so-called hypothetical zeolites, have been generated into large collections of data with varying frameworks and topologies that have been investigated by different researchers. Although this topic is treated in greater extent in other chapters of this book, a few examples are mentioned here to showcase the different forms of hypothetical zeolite databases. One of the major hypothetical zeolite databases was proposed by Treacy et al., who enumerated millions of four-connected graphs leading to zeolite structures, and later refined them using simulations to obtain tens of thousands of promising hypothetical structures [52]. The database is divided into categories to represent their feasibility according to energetic constraints, and provides a web interface to enable queries of structural and topological information. Another well-known database of hypothetical zeolites was developed by Deem et al. using a Monte Carlo strategy to propose and refine new structures, and later duplicating these frameworks using topological data. With this procedure, Deem et al. found over 2.7 million zeolite structures [53], about 330,000 of which had energy within 30 kJ/mol SiO_2 above quartz, as computed using the Sanders-Leslie-Catlow core-shell potential [54] widely used in silicates. Along with the calculated properties of each zeolite, this dataset was deposited at the PCOD. In contrast with previous approaches, Li et al. did not enumerate the space of all possible four-connected frameworks, but focused on a class of structures known as ABC-6 zeolites [55]. This family of zeolites corresponds to zeolites formed by intercalating three different layers A, B, and C, in a hexagonal unit cell. Several known frameworks are part of this family, such as CAN, GME, OFF, CHA, and others. The dataset was used to derive feasibility criteria for ABC-6 zeolites, and to realize new structures.

In addition to databases of zeolite structures, synthesis datasets are important to inform and reproduce the synthesis of zeolite frameworks. Along with its Structure Database, the IZA manages a database of verified syntheses of zeolites [56]. The collection provides a series of synthesis conditions for several known frameworks, all of which have been independently reproduced by other researchers. Given the challenges associated with the realization of these materials, this database works as a reliable "recipe book" for zeolite synthesis, although it is far from comprehensive due to the intense efforts in reproducing and curating synthesis routes in the literature. The online version of the database offers human-readable data in the form of text files, graphs, e.g., for X-ray diffraction patterns, and images.

As most synthesis recipes for zeolites are published independently by individual research groups, curating synthesis pathways to become machine-readable requires substantial effort. One early attempt in applying modern data science methods for synthesis prediction in zeolites was performed by Moliner et al. [57], who used a combination of high-throughput experimentation and a neural network to predict the phase outcomes from the reaction conditions. As machine learning is often data-intensive, several experimental results from the group's internal data were used to train the neural network model and further validate it with new experiments.

Although digitizing experimental data from internal sources such as laboratory notebooks is an important step toward the application of data science in the field of zeolites, unifying synthesis recipes published in the literature requires substantially more effort. Xu et al. created a database of 1585 reactions from articles, patents, and internal experimental results for AlPO zeolites [58]. This data enabled preliminary studies on classification models for inorganic recipes from different sources.

A comprehensive effort in curating datasets of zeolite syntheses has been performed by Jensen et al. in recent years [59,60]. The authors used automated literature extraction to process synthesis recipes in the zeolite literature. Tens of thousands of articles published in the last six decades related to zeolites were parsed from publishers' web pages and analyzed using natural language processing (NLP). The NLP models were trained to recognize descriptions of materials syntheses [61] after a subset of data points were labeled by researchers. Using this method, the authors analyzed the synthesis conditions leading to low- and high-density Ge-containing zeolites [59] and trained a generative model to produce OSDAs similar to the ones from the literature [60]. The same data extracted from the literature was used to validate representations of interzeolite transformations and intergrowths using graph theory [62]. While the NLP pipeline is not strictly necessary for extracting the data, as demonstrated by works where significant insights are derived from smaller, hand-curated datasets [63], the automated text analysis allows much larger datasets to be created and made available for the public. The datasets from Jensen et al. are available in machine-readable formats and can help the development of new data-driven approaches for zeolite synthesis. This combination of experimental data and insights, particularly for reactants as diverse as OSDAs, are invaluable for advancing zeolite discovery and improving the catalytic behavior of the materials.

4.2 Database of OSDAs for Zeolites

Developing a comprehensive database for OSDAs for zeolites has been a challenge for the zeolite community for years. As mentioned previously and in other chapters of this book, several works succeeded in using computational methods to design templates for targeted zeolites. The approaches vary from shape matching [64], genetic algorithms [65–67], multi-dimensional optimization [68], generative models [60], and others [69–72]. However, other than the summary metrics and textual description of the computational methods in the corresponding articles, machine-readable datasets arising from these OSDA generation procedures and simulations are rarely made available for the use of the general public. Although some works disclose author-specific formats for the data [67], reusing and reproducing the results requires expertise in computational tools. This requirement hinders experimental validation of predicted results, as it becomes hard for independent experimental groups to access the computational data beyond the results directly reported in articles.

A FAIR OSDA database, therefore, would allow users to investigate findings of existing and new templates for zeolite synthesis through a frictionless experience. Ideally, the database would also allow users to compare the performance of different templates for various frameworks with compatible metrics, thus being predictive of synthesis outcomes prior to the experiments. Finally, the database would integrate literature data with simulation results, which would have two purposes: i) calibrating theoretical results based on past experimental results; and ii) recording the successes of OSDA-based zeolite synthesis as a reference. Although collating synthesis results from decades of efforts from the zeolite community into a single database is labor-intensive, it could increase reproducibility in their synthesis and expedite progress in the discovery of new structures.

In this section, we discuss our attempts toward a database of OSDAs for zeolite synthesis. We start discussing our efforts in accelerating simulations of zeolite-OSDA pairs, which vary from developing algorithms to accelerate the generation of poses for calculations, as well as calibrating force field energies with higher levels of theory. Then, we explain the pipeline used to perform a high-throughput screening of OSDAs for zeolites, including the details regarding database management. Using this data, we explain the importance of defining new binding energy metrics to facilitate the analysis, and how geometric and electrostatic descriptors may aid the data analysis. Finally, we explain how we developed a database to makes this data accessible to the field, allowing users to interact with hundreds of thousands of zeolite-OSDA pairs.

4.2.1 Developing a Docking Algorithm

The first requirement for simulating the templating effects of OSDAs toward zeolites is generating initial conformations of the molecules inside the cavities of the nanoporous material. This so-called docking algorithm is analogous to those in ligand-protein binding communities, which are often interested in predicting drug affinities toward certain protein pockets [73]. Whereas there is no lack of software packages that enable researchers to attempt docking of ligands in proteins, the zeolite community has few strategies to computationally dock OSDAs in zeolites within computational pipelines to predict binding. Therefore, a simulation platform for predicting binding affinities between OSDAs and zeolites has to first generate algorithms and software packages to produce structures which will later be simulated.

To minimize the computational resources employed in the high-throughput virtual screening, a docking algorithm has to satisfy two major constraints: i) producing physically reasonable poses; and ii) performing operations in a computationally efficient fashion. So far, docking algorithms such as Monte Carlo docking [74,75], performing Fourier transforms of the structural space [66,67], avoiding steric hindrance [68], and directly growing an OSDA inside the zeolite [76,77], have been reported in the field. Despite the variety of methods, no tool had been made available for public use. Thus, we started by developing an algorithm and an open-source software package that could generate zeolite-OSDA poses for further simulation. We called our software Voronoi Organic-Inorganic Docker (VOID) [78], which was built upon the engines of Zeo++ [79,80] and pymatgen [81].

Given a few similarities between docking ligands on proteins or OSDAs in zeolites, we borrowed the jargon from the molecular recognition field to describe the problem of OSDA docking (Figure 4.1). Thus, zeolite structures are denominated hosts, OSDAs are guests, and each zeolite-molecule pair are referred to as complexes [73]. A complex denotes a unique zeolite-OSDA pair, while a pose is a particular arrangement of the atoms of the joint zeolite-OSDA system. Each complex can have multiple poses associated with it, since there are many possible ways of arranging OSDAs inside the cavities. The number of templates per unit cell is referred to as loading. The choice of the unit cell may influence the loading of the systems, especially for pore-based zeolites, and should be taken into consideration when generating a variety of poses.

Using this notation, the docking procedure corresponds to producing poses from complexes. The first step, therefore, is to calculate relaxed structures for the (unloaded) host and the guest in vacuum. Once the structures have been optimized, e.g., through force field (FF) or density functional theory (DFT) approaches, different systems are used as input for the rigid docking algorithm.

For example, one molecule can have several low-energy conformations, and one structure may be represented by different choices of unit cells. In many cases, a molecule could be a good fit for a given pore, yet not be commensurate with the unit cell. Therefore, creating a diverse set of poses is essential to better explore the configuration space of zeolite-OSDA pairs and obtain more accurate predictions of systems with good binding energies.

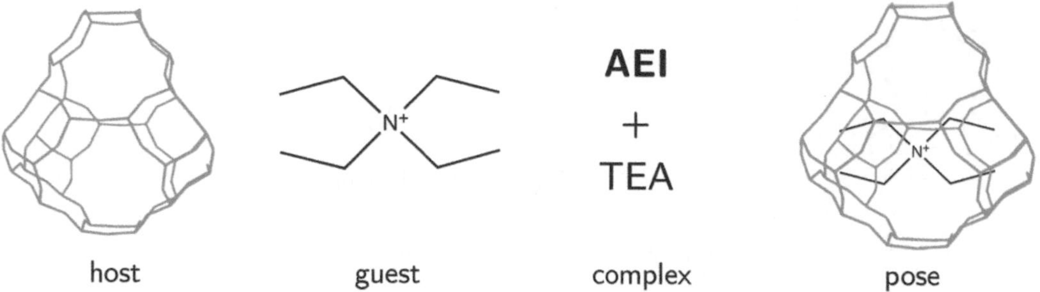

Figure 4.1 Naming convention we adopted to refer to zeolite-OSDA systems. The naming is exemplified with the AEI-tetraethylammonium (TEA) pair.

Nevertheless, exhaustively sampling the configuration space of several molecules and zeolites is an expensive task. Particularly for flexible molecules with a large number of conformers, sampling the molecular geometry space may be extremely expensive, even with force fields. Similarly, each host has an infinite number of unit cells with an arbitrarily large number of atoms. An useful approximation, therefore, is to minimize the number of initial structures to the primitive cells and low-energy conformers, except in the case of pore-based zeolites with short unit cell parameters. Then, both the host and the guest are treated as rigid bodies at docking time, thus considering their ground state geometries and unit cells as input coordinates for the docking algorithm. This first approximation then ignores the effects of conformational changes during docking and enables the dedicated algorithm to focus solely on the docking. Energy minimizations and calibrations are then addressed in a second step.

Even without energy constraints, finding poses leading to low-energy structures requires some computational metric of fitness or pose quality. One possible way of quantifying the goodness-of-fit between guests and hosts is to measure the distances between all atoms in the framework with respect to all atoms in the template, and penalize systems for which the molecules are overly close to the framework. The loss can then be simply a one-zero loss, or a more sophisticated function. This enables the docking algorithm to generate poses with reasonable configurations, even if not guided by total energy calculations.

Traditional Monte Carlo docking algorithms rely on the metrics and approximations described above to insert a template into a host. By performing a series of random actions such as rotating or translating the (rigid) molecule, the move is accepted according to some probability dependent on the improvement in the fitness function. After a series of Monte Carlo moves, the final host-guest arrangement is accepted as the final pose. When increasing the loading, thus docking the n-th molecule into the zeolite, a pose containing $(n - 1)$ molecules is used as an input to the procedure above. While the Monte Carlo docking algorithm is simple and does not require any previous information on the initial structures, it requires constant calls to the fitness function. Particularly when thousands of moves are required, as in zeolites with large unit cells, the computational cost required to create poses may become too expensive. Furthermore, since the Monte Carlo moves are sequential, it is difficult to leverage parallel computations except by running multiple independent simulations, which will produce more answers, but will not reduce wall time.

To bypass this problem and generate as many high-quality poses as possible with low computational cost, we proposed the use of Voronoi decomposition to sample the void space of the host prior to docking (Figure 4.2a) [78]. Then, instead of relying on random translations to prevent clashing of atoms between the zeolite framework and the template as in Monte Carlo docking, we fix the center of mass of the OSDA onto the Voronoi nodes of the framework with maximum distance toward the zeolite. This guarantees that the OSDA geometry will start in the points of

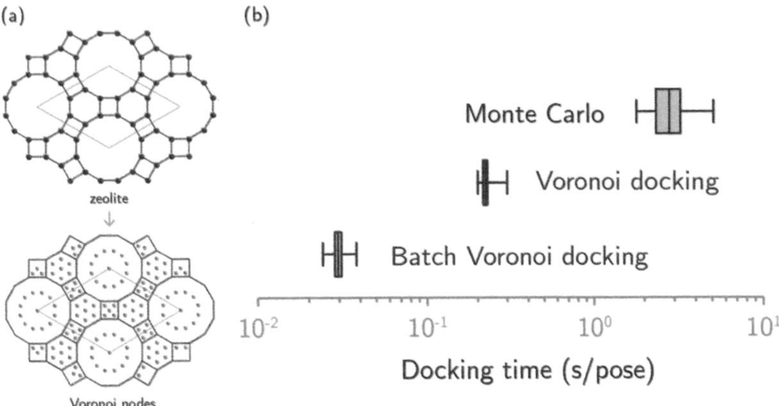

Figure 4.2 (a) Example of sampling Voronoi nodes for the AFI framework. Red circles are tetrahedral atoms, and green circles are Voronoi nodes. Oxygen atoms are omitted. (b) Pose generation rate for the trimethyladamantylammonium molecule docked on CHA zeolite, as benchmarked by the VOID package. When docking is parallelized with 50 poses per docking move, the batched Voronoi docking method is about two orders of magnitude faster than Monte Carlo docking.

maximum distance toward the framework, eliminating the need for random translations. Although rotations of the rigid body guest are still required, they can be easily parallelized by writing matrix multiplications using tensors. Therefore, for one single starting geometry, hundreds of random rotations can be attempted at once, instead of relying on the serial approach from Monte Carlo. This Voronoi docking algorithm is capable of generating as many poses for a given complex as the Monte Carlo approach, but at a much faster pace. Even without parallelization, the docking can be performed almost an order of magnitude faster than its Monte Carlo counterpart, as translation moves are bypassed. With parallelization, this value reaches almost a two orders of magnitude improvement (Figure 4.2b). Thus, the code is able to generate poses from complexes in an efficient manner, in a first step toward the high-throughput simulation of zeolites.

Importantly, this achievement is made possible by the availability of open-source packages such as Zeo++ and pymatgen. Using these packages as requirements, the Voronoi Organic-Inorganic Docker (VOID) software is also available as an open-source code. We believe that disclosing software for reproducing calculations is an important step in increasing the reproducibility of computational works, and should be available alongside publications in the field. This could enable other users to move the field forward without the need to reprogram all tools.

4.2.2 Calibrating Binding Energy Predictions

Once the pose has been created from a complex, the second step toward predicting the templating ability of an OSDA toward a zeolite is to calculate the binding affinity between the two. The gold-standard binding metric would be free energies of binding at high levels of theory such as DFT. However, these calculations are computationally very expensive. In particular, two different classes of approximations are at play: one is that free energies require extensive sampling compared with energies at zero temperature. The other is that energies from first-principles calculations are much more costly than empirical models such as FFs. A surrogate metric to quantify the templating ability of an OSDA is to obtain its binding energy toward the given pose from total energy calculations. Then, it is assumed that poses with the lowest binding energies represent the most favorable zeolite-OSDA pairs.

Over the years, computing host–guest interactions has relied on the use of a broad set of methods, varying from accurate quantum chemistry calculations [82–90] to FFs to plain shape matching [64,76,91,92]. Furthermore, simulations of OSDA–zeolite interactions at the FF level have been performed using a wide variety of functional forms and parametrizations, including CVFF [70, 93–96], Dreiding [66–68,97,98], COMPASS [99], UFF [100,101], and others [65,75,102–105]. In addition, binding energy calculations can be performed in multiple ways using different simulation parameters. This includes optimizing hosts and guests at constant pressure [75,105] or volume [68,93,96,100,101,106], performing molecular dynamics (MD) simulations on separate systems [66,68,97,98,102,104–106], or keeping the geometry of the host constant while optimizing the guest conformation inside the confined space [87,88,93,96,100,101]. From physical principles alone, the tradeoffs are between the choice of simulation parameters and the cost, accuracy, and reproducibility of binding energy calculations of zeolite-OSDA pairs. Before selecting the simulation parameters of production runs, therefore, it is important to benchmark the theory against experiments or reliable simulations to obtain the best possible accuracy-cost trade-off. Only a few experimental data points exist for the thermodynamics of OSDA binding in zeolites [107], which hinders an adequate calibration. Moreover, these experimental studies lack the chemical diversity that would be explored in screening undertakings. Thus, to rationalize the selection of simulation parameters in zeolite-OSDA simulations, we created a dataset of 272 poses comprised of neutral OSDAs docked in different zeolite frameworks to study correlations between binding energies from FF and higher-accuracy reference DFT calculations [108].

The definition of one possible binding energy is shown in Figure 4.3a. With this particular method, the pose is optimized at constant volume. Then, the binding energy is obtained by freezing both the host and the guest at the optimized coordinates, and simulating them separately. Figure 4.3b shows the correlations between the different binding energies. As the FF is not parametrized to reproduce DFT binding energies, the absolute values and ranges of energies vary substantially according to the method. However, we observed that DFT binding energies are usually well-correlated with one another, despite the usage of different simulation parameters. Capturing trends in binding energies, therefore, may be enough for initial screening purposes, where templates are ranked according to their goodness-of-fit. Nevertheless, even DFT energies do not show a perfect correlation among themselves. Despite the reasonable precision of DFT simulations, the choice of simulation parameters may influence not only the predicted energies but also the computational costs. For example, binding energies from expensive MD simulations are correlated with energies obtained from plain optimization approaches. Although this result could be ascribed to the short trajectory lengths of the MD simulations, similar results were found for FF calculations as long as 100 ps. This result has important implications, showing that within the Dreiding FF [109] employed in the benchmark, MD simulations do not bring substantially more information than simple optimizations followed by binding energies derived from the frozen pose method. Remarkably, binding energies from the Dreiding FF at constant volume correlate well with those from DFT approaches, suggesting that the FF is a reasonable approximation for ranking OSDAs based on their binding energies. Therefore, significant computational cost can be saved by computing binding energies without MD trajectories.

This analysis of accuracy of the binding energies goes beyond the computational efficiency of such calculations. Since binding energies are computed with respect to a pose, deriving the binding affinity for a complex may require considerable sampling of poses. One way to assign a binding energy for a complex is to follow a "variational principle," in which the pose that minimizes the binding energy between an OSDA and a zeolite is used as the reference value for a complex. Therefore, the larger the sampling of the configuration space, the closer the estimated binding energy should be to the ground state of the zeolite-OSDA pair. However, this approach requires

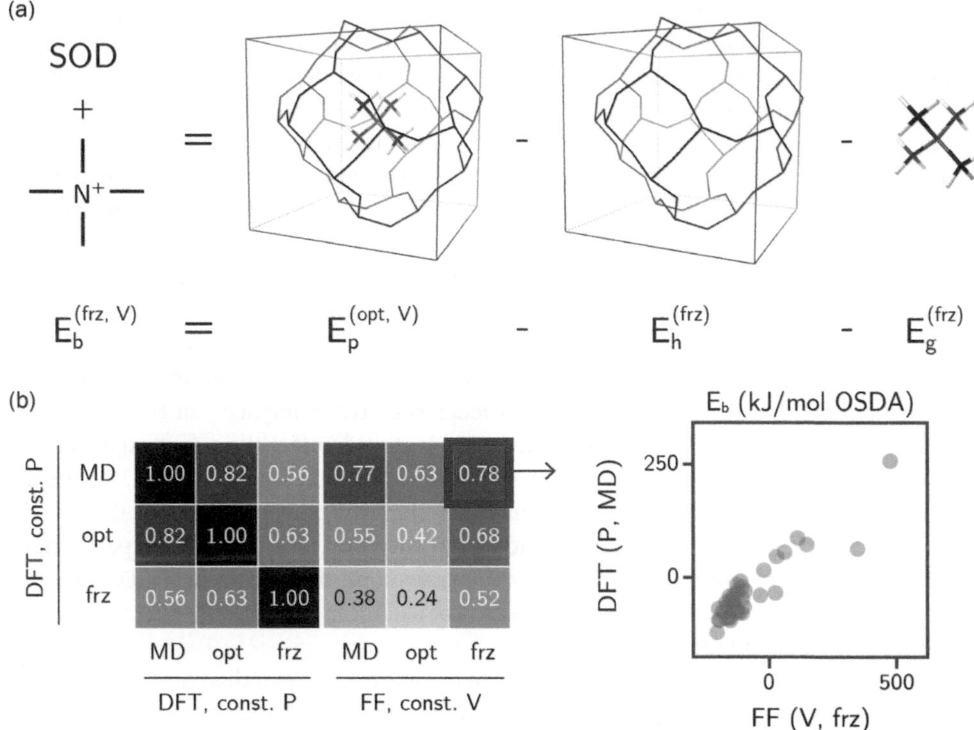

Figure 4.3 (a) Schematic on how to calculate the binding energy (E_b) according to the frozen pose method. Poses are first optimized at constant volume (energy $E_p^{(opt,V)}$). Then, the host and each guest are frozen in their optimized positions and calculated separately (energies $E_h^{(frz)}$ and $E_g^{(frz)}$). (b) Spearman's correlation coefficient between binding energies calculated with DFT and the Dreiding FF across different simulation parameters. The frozen pose method has high correlation with binding energies derived from *ab initio* MD simulations at constant pressure (right plot). *Source:* Panel (b) was adapted from [108].

ensuring that calculations are reproducible across different poses. For example, convergence in the energy of isolated and joint structures is expected to obtain accurate and reproducible binding energies. If the simulation method leads to a smooth potential energy surface, then optimization algorithms or MD integrators should be able to sample the configuration space in a similar manner. However, this cannot be assumed for a general-purpose FF applied to the specific case of zeolites. To understand the effects of different initialization parameters in the final energy of zeolite-OSDA pairs, we have performed several calculations of unloaded zeolites using a diverse set of optimizers. Then, the energy of each zeolite is compared with the lowest energy obtained for that same framework. We found that depending on the optimization pathway, the simulation converges to different energies, even for systems devoid of OSDAs [108]. This represents a challenge for binding energy calculations, as a good template for a zeolite may be labeled with an erroneously poor binding energy due to local minima in the potential energy surface. More importantly, several binding energies rely on reference energies to be computed, and differences in optimization can lead to systematic errors in the binding affinities computed for given zeolite-OSDA pairs. As a conclusion, therefore, avoiding the use of separate calculations to obtain binding energies is preferable in terms of reproducibility.

A similar effect can be seen when different initial structures are from the same host, i.e., geometries obtained using different simulation methods are used as an input for the binding energy calculations. As simulations at constant volume were observed to be more correlated with DFT than their constant pressure counterparts, it is important to understand how changing the initial unit cell volume affects the final binding energy. To perform this analysis, we started from the variational principle stated above for the binding energies of a complex. If the calculation method is robust to small changes in the unit cell, differences between binding energies from distinct complexes are expected to be smaller. For practical purposes, this robustness is desirable to increase the reproducibility of the calculations and ensure that sampling the configuration space of poses is effective. With this concept in mind, we performed calculations for different complexes, at an average of 12 poses per complex, using different methods. We found, once again, that binding energies obtained from optimizations and MD simulations using the Dreiding FF are outperformed by frozen pose calculations using this parametrization [108]. This suggests that some methods capture the sampling of the configurational space in a better way, both in terms of accuracy as well as in reproducibility. Understanding the role of each parameter prior to deploying simulations at large scales is paramount to ensure that the resulting data is clean, reproducible, and useful.

4.2.3 Performing and Analyzing High-Throughput Screening Calculations

Equipped with efficient algorithms for generating poses and calculating OSDA-zeolite binding energies, we proceeded to deploy calculations in a high-throughput manner [10]. As mentioned in the Introduction, several infrastructures already exist to perform first-principles simulations on a large scale. However, they are generally not applicable to the supramolecular recognition problems seen in the zeolite community. Thus, we developed software; e.g., the GULPy package [108], to act as an interface to simulation packages of interest, such as the General Utility Lattice Program. In addition, we worked with database management to assign relationships between structures in a physically meaningful way. Figure 4.4 illustrates how different objects interact with each other within a database. Using the notation introduced above, we show how a given OSDA (chemical

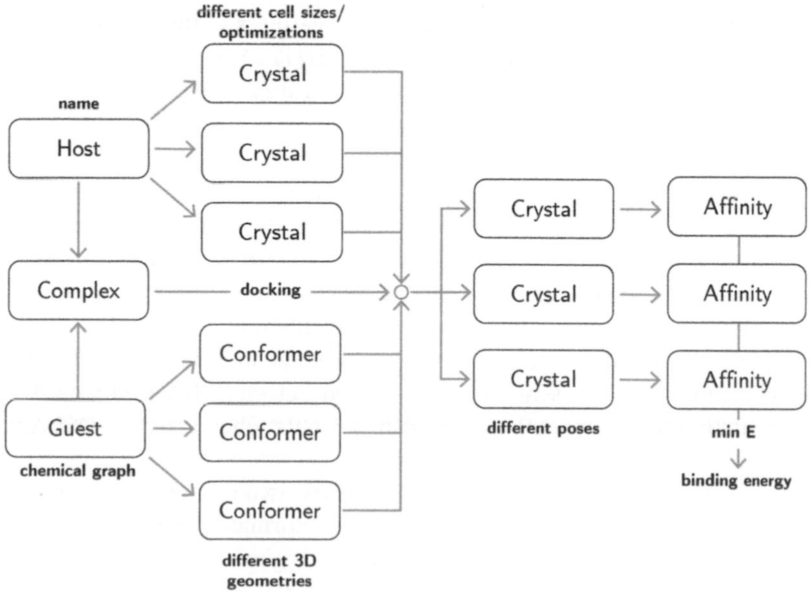

Figure 4.4 Simplified diagram of database schema created to manage hosts, guests, poses, and binding affinities.

graph of type Guest) and a zeolite topology (denoted as a Host) relate to each other through a Complex. A pose is generated from Conformers and Crystals, as well as their respective complexes. Finally, Affinities can be obtained for each pose. This so-called schema enables the simulation data to be scheduled, parsed, and stored in an organized manner, which is particularly important in big-data approaches. In this regard, custom-made databases provide enough flexibility to appropriately store the data for future analysis.

Deploying calculations in a high-throughput fashion requires orchestrating software to automate the communication between the database and the HPC resources. Each task, such as calculating conformers for a molecule, docking a molecule into a zeolite, or calculating their binding affinity, is assigned its own job, and submitted in parallel to a supercomputer cluster. The advantage of this parallelized approach is to batch jobs with similar expected runtimes into a single submission, which can be estimated using, for instance, the number of atoms of the substrate. Optimizations in the algorithms and in the job submission process can lead to substantial savings in computational cost. For example, using the batched Voronoi docking algorithm instead of the Monte Carlo docking method, and using the frozen pose binding energies instead of MD-derived ones could save orders of magnitude in computational cost. In our work, we simulated over 586,000 poses with these methods, leading to almost 1 million CPU-hours in computational runtime. If these methods had not been employed, we estimate, from the results used to calibrate the binding energies, that over 440 million CPU-h would have been used to calculate the same binding energies, with a poorer agreement with the theory likely due to reproducibility errors. The codebase also included automated detection and handling of errors, both arising from the simulation, and from the HPC outages.

After each job was run and parsed, the data was stored in the database for further analysis. In our case of zeolite-OSDA pairs, the target is to predict templating strengths from different values of binding energies. By simulating over 112,000 unique complexes, we obtained a binding matrix between all known zeolites and all known OSDAs. The simplest approach to analyze such data is to follow past works and select one target zeolite framework to be analyzed. When OSDAs are ranked for a given framework, selecting good candidates for that particular structure becomes a matter of minimizing the total binding energy of the complexes, regardless of the loadings. This method has demonstrated to be efficient toward a series of discoveries [110,111]. We call this ranking the "directivity" of an OSDA toward the target zeolite, which measures how good a candidate is across all the possible other templates.

The problem of downselecting OSDAs based solely on their binding energy toward a target framework is that templates selected by optimizing binding energies may be more selective toward a different topology rather than the desired one. This effect comes from the strong phase competition observed in zeolites, and is an important parameter to be controlled in the synthesis of these materials. Therefore, analyzing this competition effect requires selecting molecules which simultaneously minimize the binding energy toward the target framework while destabilizing all the other possible competing phases. It becomes useful, therefore, to rank zeolites for a given OSDA, thus attempting to predict the outcome of the synthesis results when a given OSDA is employed in the synthesis. We call this ranking the "competition energy" of a zeolite when a certain OSDA is used (Figure 4.5).

Finally, when ranking these binding energies, it is useful to define two extra parameters: the energy reference across rows/columns of the binding matrix, and the normalization of the energy. First, the energy reference is arbitrary given a ranking, but it enables a higher interpretability of the method when used correctly. To provide an useful parameter of directivity/competition rankings, we decided to define the energy reference as the second lowest binding energy across OSDAs/

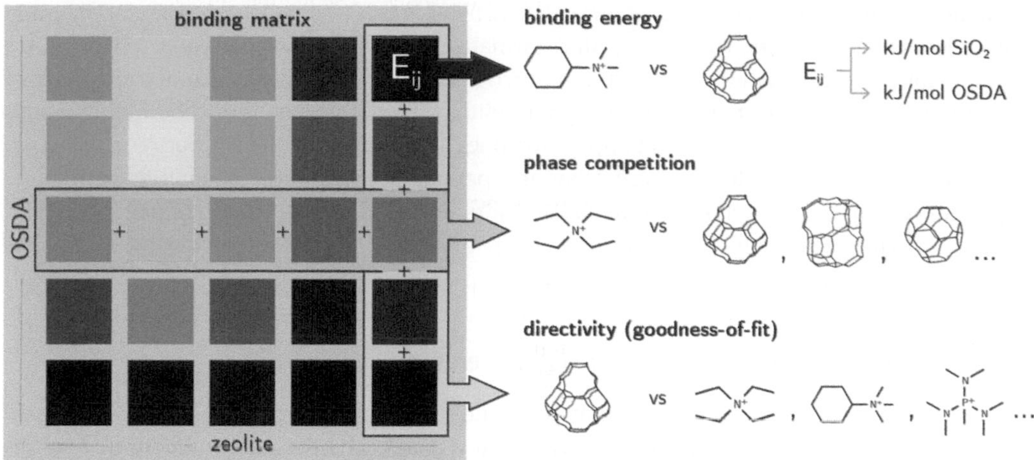

Figure 4.5 Schematic of binding energy matrix for zeolite-OSDA pairs. Each element of the binding matrix is a binding energy normalized either by SiO_2 or OSDA. Along columns (zeolites), OSDAs are compared to verify how directive they are toward a given framework. Along rows (OSDAs), zeolites are compared according to their binding energy to quantify the phase competition effect.

zeolites in the binding matrix. Using the second-best guest or host binder allows us to quantify the difference between the absolute best binder with respect to its strongest competing phase or alternative template. From the data analysis perspective, comparing phase competition effects in zeolites then becomes verifying how negative is the competition energy for a given framework, with the most negative systems better in terms of binding. The design principle, therefore, becomes obtaining OSDAs which are reasonably directive (within a few kJ/mol of the best binders), but greatly stabilize only the framework of interest (competition energy as negative as possible).

The second parameter is the normalization. Since each pose has a different loading of OSDAs, and each framework has its own number of tetrahedral sites, computing binding energies in a way that is invariant to the choice of the unit cell is important. To do so, we normalize the total binding energy by two metrics: i) the number of tetrahedral atoms in a zeolite, thus obtaining a binding energy per SiO_2, or ii) the number of OSDAs in a pose, leading to binding energies normalized per number of OSDAs. Although these two energies are basically related by a proportional factor for a given zeolite-OSDA pair, they play different roles in determining the binding energies of the system overall, and have important interpretations. For example, very small OSDAs may greatly stabilize a framework, but at the expense of high loadings that may not be achieved in practice. Computing the correct OSDA-OSDA interactions may lead to improvements of this metric, but this comes with its own challenges [108]. On the other hand, zeolites with different building units may be stabilized in different ways. Zeolites with building units large enough to contain an inorganic cation such as K+, but small enough to contain an OSDA, tend to have "empty regions," even after docking. This "unused space" in the simulated structure (but occupied in the actual synthesis procedure) ends up reflecting on different baselines of binding energy per SiO_2, which would make the analysis of phase competition harder. Therefore, analyzing binding energies according to different energy normalizations enables rationalizing the OSDA design for a target zeolite.

However, analyzing different metrics at once becomes challenging when several parameters are available. In addition, relying on different interpretations can be misleading in terms of overcoming trial-and-error in zeolite synthesis. To bypass this problem, we proposed a metric that unifies all the binding energies described above under a single parameter. Since ranking both the directivity and competition energies is challenging, we proposed that a good OSDA for a zeolite should simultaneously minimize all energies at once, thus achieving OSDAs that have optimal trade-off in terms of loading and affinity. To do so, we first normalize the ranking between each OSDA using a physics-inspired representation from partition functions. Then, we perform a geometric average of the binding energies to obtain a new ranking of binding, which is weighted by the binding matrix and a value of temperature. While the temperature does not change the ranking between the structures, it affects the magnitude of energies that are spanned in the templating energy, with lower temperatures favoring more stable zeolite-OSDA pairs across all metrics. This templating energy, therefore, provides a single metric of binding energy to analyze the interactions between zeolites and OSDAs, facilitating the analysis of the data.

4.2.4 Recalling Synthesis Outcomes from the Literature

Although several qualitative interpretations of binding energies in zeolite-OSDAs have been shown to be predictive of a few synthesis outcomes, it was important to verify whether the results are broadly predictive across several topologies and OSDAs. In particular, quantifying the predictive power of the theoretical approach may enable assessing the limitations of the theory and determine where they can improve. However, performing experimental validations on the theory across many frameworks is costly, and would hardly be comprehensive. To bypass this problem, we leveraged the literature data on OSDAs for zeolites curated by Jensen et al. [60], which contains thousands of synthesis routes across hundreds of distinct OSDAs. These OSDAs were calculated with the high-throughput approach, and a binding matrix constructed based on these results.

To verify whether the theory is capable of recalling the past literature results, we proposed the use of a recall curve, which ranks the OSDAs/zeolites according to a binding energy metric, then verifies the cumulative recall of the literature based on this ranking (Figure 4.6). While the machine learning community often uses the receiver operating characteristic (ROC) curve to perform such assessments, the ROC analysis requires the use of true negative data points. However, in the case of zeolites crystallized with OSDAs, there are few true negative data points. The fact that a certain zeolite has never been observed to be crystallized with one OSDA does not prove that such realization is impossible. On the contrary, given the large number of zeolite-template pairs and the high-dimensional synthesis conditions, it is impossible to state, with absolute certainty, that a given template will never crystallize a given framework. Exceptions to this observation may be cases where the template is substantially larger than any of the cavities of the zeolite framework. However, the theory is automatically able to capture this result, where high binding energies or the inability of perform a docking reflect that a template is extremely unlikely to synthesize a given zeolite.

Using the normalized area under the recall curve as a figure of merit, we quantified the agreement between the theory and all past synthesis results in the zeolite literature. However, to obtain statistically meaningful recall metrics, we restricted the analysis to zeolites with more than 10 different OSDAs, which can then be ranked and compared. In 34 out of these 40 most common zeolites in the literature the recall curve had area above 0.5 (Figure 4.6b), indicating it is predictive of the synthesis outcomes for most structures [10]. In contrast with typical machine learning approaches that use past synthesis results as input for a model, our simulations do not require any input from the literature, and thus are predictive purely from first principles. Despite several approximations in the simulations, including the lack of electrostatic interactions, the general purpose of the FF employed, and the pure silica framework, this agreement between simulations and past outcomes suggests that the zeolite topology is directed by dispersion interactions, even though heteroatom distributions play a major role in the nucleation and stability of the framework.

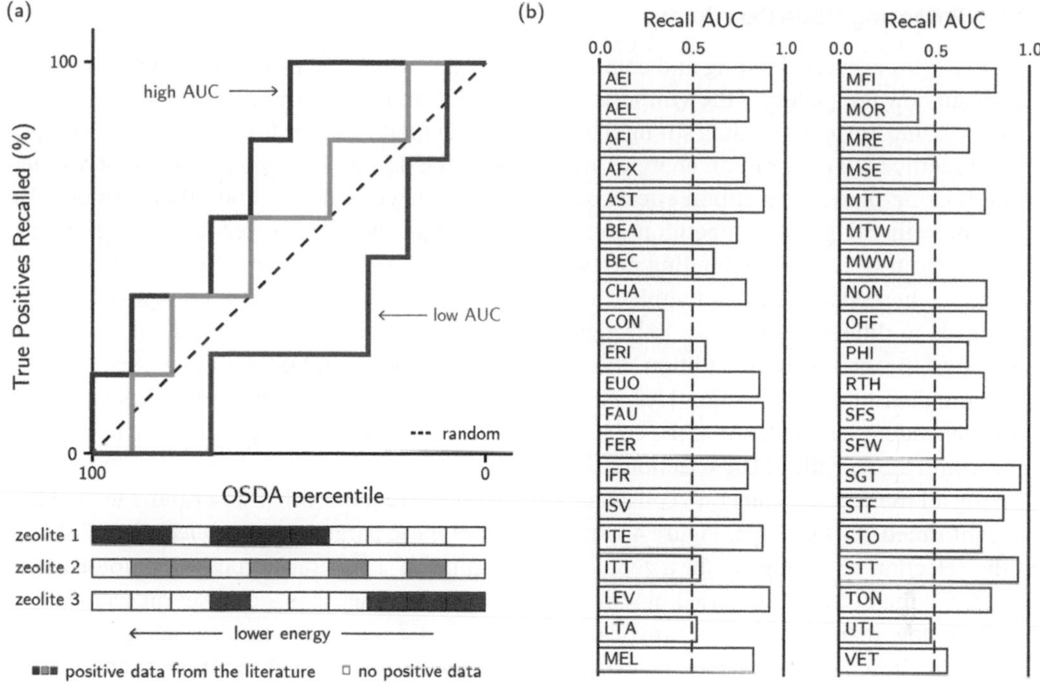

Figure 4.6 Schematic of literature recall. (a) Diagram on how recall curves are constructed. For a given zeolite, OSDAs are ranked according to a binding energy metric such as the templating energy. Then, the cumulative percentage of recalled zeolite-OSDA pairs is plotted on the *y*-axis. A high area under the curve indicates that most syntheses routes for that particular zeolite involve OSDAs with low binding metric. (b) Area under the recall curves for 40 different zeolites from the literature. *Source:* Diagrams adapted from [10].

The disagreement between the theory and the experiments for a few zeolites can also be interpreted from a different perspective. For the case of zeolites such as MOR or MTW, whose agreement between the theory and the literature is small (Figure 4.6b), it can be suggested that OSDAs are not the major drivers of the crystallization of these structures. In fact, these frameworks are often crystallized in OSDA-free conditions, and their occurrence in the literature is mostly due to the synthesis conditions other than the template. In the case of other frameworks such as CON [10], the formation of molecular aggregates [112] may contrast with the inability of the theory to predict OSDA–OSDA interactions appropriately, as well as the challenges associated to producing and docking molecular clusters instead of individual molecules. This is also the case of systems such as LTA [10], where supramolecular interactions [113] are not fully captured by the FF. Thus, a few challenges to the prediction of binding energies remain, and are detected by the literature metrics.

Overall, the agreement between the FF calculations and the literature is remarkable. Despite the lack of electrostatic interactions in the charge-neutral Dreiding FF, the templating effect is well-modeled and reproduces the experimental outcomes. In addition, problems with zeolite structural optimization, for which the FF is not optimized, are bypassed with frozen-pose binding energies. Finally, the agreement suggests that the docking procedure is able to sample the configuration space of molecules for zeolites. Altogether, these computational tools are predictive of synthesis outcomes in zeolites and can be used for the first-principles design of templates for these materials.

4.2.5 Proposing OSDA Descriptors

Equipped with accurate binding and competition energies which are proven to be predictive, we can design new templates for the synthesis of zeolites. The most straightforward way to design an OSDA is to maximize the binding strength between the template and the framework using computational methods. If competition energies are available, the design of templates can be performed by minimizing phase competition effects between the targeted structure and other frameworks, either through the phase competition metrics or the aggregated parameter templating energy. However, even with accurate binding metrics, several other parameters are often taken into consideration when selecting a new template for zeolite synthesis. For example, the shape of the molecule leads to differences in phase selectivity between two zeolites, and can influence the final outcome of the synthesis. In addition, the charge distribution of an OSDA can be used to tailor aluminum distributions of zeolites [10]. Finally, as the cost associated with the zeolite synthesis is highly dependent on the choice of the OSDA, factors such as synthetic complexity have to be taken into account along with OSDA selection.

Taking all these parameters into consideration, therefore, requires other descriptors to enable a better informed OSDA design. Figure 4.7 shows a few of these parameters we selected for informing the selection of a template for a zeolite. One example of a general-purpose descriptor in the literature is the SCScore, proposed by Coley et al. [114]. This metric is a data-learnt feature that quantifies how many reactions are involved in the synthesis of a given template using a reaction network database as training data. Instead of relying purely on chemical intuition, therefore, the SCScore quantifies a metric of synthesis complexity of a given template and can inform the selection of new OSDAs. Although obtaining the true cost of the synthesis involves determining the yield of the organic reactions, the difficulty in purifying the templates, the existence of stereoisomers, etc., the SCScore offers one metric to separate the complex from simple molecules. Often it is desirable to minimize the SCScore of a template while also maximizing its binding strength. Since the OSDA synthesis complexity and its binding energy toward a particular framework are independent parameters, both can be visualized separately.

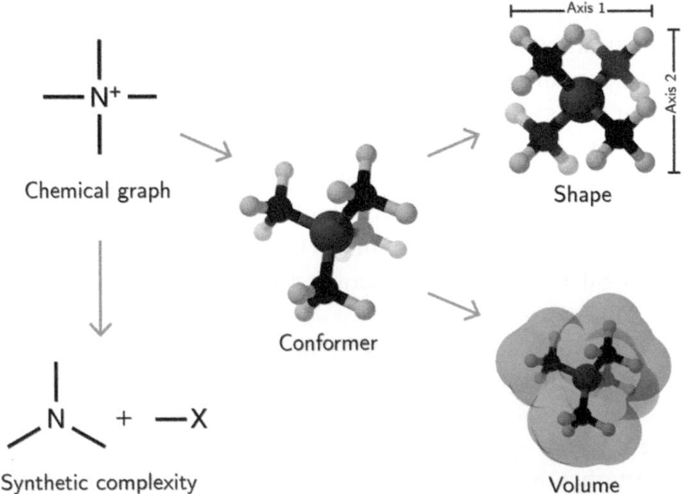

Figure 4.7 Different descriptors used to downselect OSDAs. The chemical graph enables the calculation of 3D conformers, from which geometric properties such as volume or shape are derived. The synthetic complexity can be calculated with retrosynthesis-aware metrics such as SCScore. *Source:* Adapted from [114].

Not all descriptors are independent in OSDA design. One of the simplest descriptors that can be attributed to an OSDA is its volume. Since the pore-filling ability of a template is closely related to its selectivity toward a given framework, quantifying the conformational volume of the molecule is paramount to design structures that maximize the occupancy of tailored void spaces in targeted frameworks. Thus, molecular volume and binding strength are often intertwined. However, computing the volume of a molecule is not uniquely defined. One convention we used to obtain a volume for OSDAs is to take one conformer for a given chemical graph; for example, the ground state conformer, as calculated using molecular mechanics force fields [115,116], and use a voxelized approach to measure how much space each conformer occupies. This method partitions the three-dimensional (3D) space into small cubes (voxels) and determines if a cube is occupied or not based on criteria of distances and atomic radii. In relatively rigid molecules, the volume does not change substantially with the choice of conformer. However, in molecules with more degrees of freedom, the conformer may determine if the molecule fits or not altogether inside a zeolite pore or cavity, with pronounced volume changes. This is the case, for example, of molecules whose shape may adopt a "folded" configuration versus an "open" conformer when docked in different zeolite cages [117]. This highlights again the importance of creating several poses starting from different conformers during the docking procedure.

One interesting outcome of quantifying the volume of OSDAs is that it is a positive, interpretable, and unidimensional descriptor that affects the maximum binding strength that can be achieved in a zeolite framework. When plotting the binding affinities of OSDAs for a given zeolite with respect to the volume, often a minimum of binding is found for a particular volume, leading to a V-shaped plot. This plot informs that molecules with ideal pore-filling ability share similar volumes, and can be extremely useful for distinguishing outliers in molecules with different sizes. While this is clear for the case of small-pore zeolites (Figure 4.8a), the advantage of using such a descriptor is further demonstrated in the case of pore-based frameworks such as MFI (Figure 4.8b). As the docking procedure is performed for all guests and hosts for a given database, molecules with different sizes may be packed inside a framework with different loadings. However, since free energies of binding were not computed and guest–guest interactions cannot be easily quantified within the approximations we adopted, the probabilities of forming molecular aggregates may not be directly translated into low binding energies. Often, therefore, it may be desirable to find the largest possible OSDA volumes that minimize the binding energy, while also bypassing potential artifacts from these theoretical results. In Figure 4.8b, the tetrapropylammonium OSDA stands out

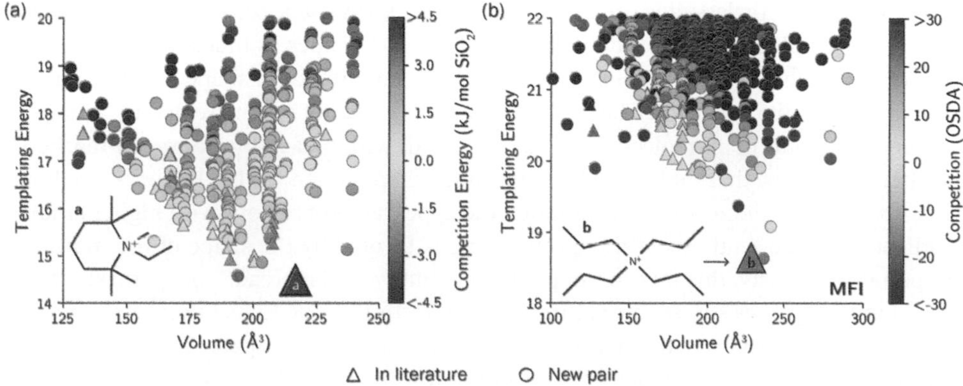

Figure 4.8 (a) Relationship between the volume of an OSDA and its templating energy toward an AEI; and (b) MFI frameworks. The best OSDAs for these zeolites are shown in the plot. *Source:* Adapted from [10].

as one of such candidates for the synthesis of MFI. Given that this molecule was the first to enable the synthesis of ZSM-5, the hypothesis captures an intuition that may be useful in the design of new OSDAs for zeolites.

The volume alone is not sufficient to predict good templates for given zeolites. Figure 4.8 shows that even molecules sharing similar volumes exhibit a wide range of binding energies, despite their similar occupancy of the space. These energy differences are mostly due to differences in molecular shape. Since zeolite pores and cavities have different shapes and topologies, it is expected that templates with different shapes lead to completely different binding energies, even when sharing identical volumes. However, quantifying molecular shape is often performed with high-dimensional descriptors. In the drug design community, the task of comparing molecules according to their shapes led to the creation of several representations for quantitative structure-activity predictions and models. In the zeolite community, these representations have been employed in many ways, including in supervised or unsupervised learning tasks [60,67,118]. While most machine learning works for OSDA design are recent, one of the first representations of phase selectivity based on shape was performed by Boyett et al. [92], where moments of inertia of molecules were used to distinguish between molecules with similar synthesis outcomes. One drawback of all these representations is that they often rely on dimensionality reduction algorithms to be visualized. Furthermore, the interpretability of the representation is often lost during the featurization [60], hindering tasks such as clustering molecules with similar shapes.

To create a low-dimensional, interpretable representation of the molecular shape, we proposed using two principal axes of molecules as a surrogate for the shape of the molecule. To obtain this shape representation, the Cartesian coordinates of the molecular conformer are projected into a two-dimensional (2D) plane fitted to these atomic coordinates. This operation is equivalent of performing a principal component analysis (PCA) of the distribution of atomic positions in the 3D space, then projecting it onto a 2D space. Note that while other works use a PCA to reduce the dimensionality of a set of representation vectors [60], thus losing interpretability in the process, our shape descriptor is merely a projection that preserves the metric of the Cartesian space, thus allowing a higher interpretability. After this procedure, the ranges of the projections in this plane are measured, and two axes are derived. In our notation, Axis 1 is the longer axis along this plane, while Axis 2 is the shorter one. When the Cartesian coordinates are given in Angstrom (Å), the units of these axes are also described in Å. Furthermore, the projection operation is dependent on the conformer, so we selected the ground-state conformer for each molecule as the reference.

The 2D representation of a molecular shape is unarguably limited, as shape is often thought of using 3D attributes. Nevertheless, it is useful for visualizing OSDA shape selectivity in zeolite synthesis by encoding binding energies as dependent variable and shapes as independent variables. Figure 4.9 shows how these visualizations shed light on the importance of shape in selecting adequate molecules for zeolite synthesis. The 2D histogram of Figure 4.9 is constructed by averaging the binding energy (color) of all molecules whose shape fall within each bin (hexagon markers). The color differences show that molecules having similar shapes often lead to better/worse binding energies, according to the zeolite. Furthermore, the appearance of regions with higher binding strength indicate that the synthesis of the given zeolite is favored by the shape of the molecule. Indeed, despite its simplicity, this metric can predict qualitatively the breadth of synthesis conditions that lead to the synthesis of zeolites [10]. More interestingly, due to its interpretability, this metric allows finding molecules with intermediate behavior in the shape space. We have used this property of the representation to design molecules for intergrowth zeolites, showing that

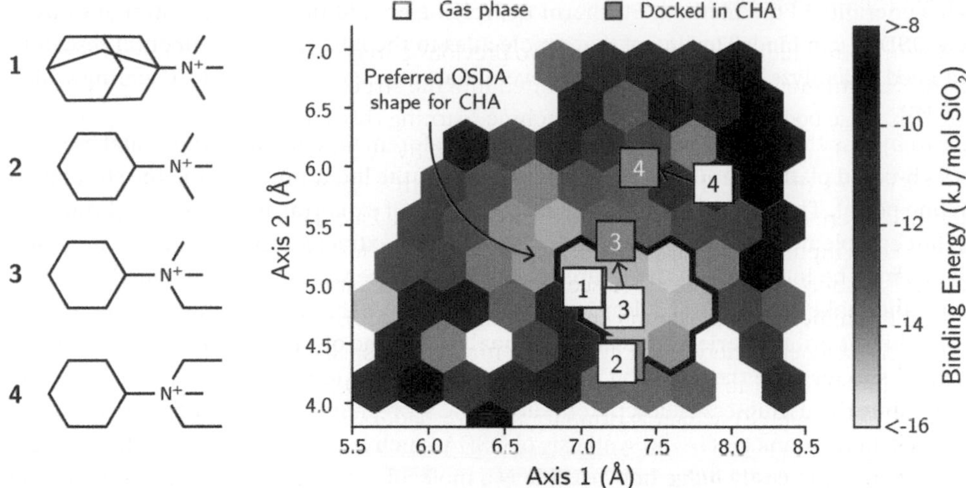

Figure 4.9 Dependence of an OSDA shape and its binding energy toward CHA zeolite. Each hexagon is colored with the average binding energy of all OSDAs that fall within that area. White squares indicate the shape of ground-state conformers of OSDAs 1–4 (left) optimized in the gas phase, while gray squares indicate their respective shapes after structural optimization in CHA zeolite. *Source:* Adapted from [10].

bi-selectivity of OSDAs can be achieved when molecules have similar binding metrics and intermediate shapes toward the end members of the frameworks of interest [10,117].

Finally, simple charge distributions of OSDAs can be quantified using geometric quantities. Similar to the shape metric, sophisticated representations can be proposed in order to capture the electrostatics of each OSDA, which could be useful for tailoring active site distributions in zeolites. However, instead of using complex representations, we proposed a one-dimensional (1D) descriptor for quantifying differences of aluminum distributions in CHA zeolites according to the OSDA employed in its synthesis [10]. The descriptor is created by taking the distance between the quaternary nitrogen atom and the center of mass of the molecule. In the case of the CHA zeolite, the longer the distance between them, the closer the positive charge is to the end of the elongated *cha* cavity. This has been shown to modulate the proportion of paired aluminum sites in *d6r* units in the CHA zeolite. Although this electrostatic descriptor cannot capture differences of shape nor treat molecules with more than one positive charge, it correlates linearly with the predicted fraction of paired aluminum sites in *d6r* units of zeolites. Extending this representation toward more complex molecules may help screening for OSDAs with different charge distributions, thus favoring different aluminum distributions in zeolite synthesis.

4.2.6 Designing with Interactivity

Together, the descriptors enable screening for OSDAs with good binding metrics, shape selectivity, low synthetic complexity, and tailored charge distributions. While computational methods can enable the optimization of all these parameters at once, determining the weights and cost functions for the optimization is a non-trivial task. For example, quantifying the trade-offs between shape matching and SCScore could require system-specific cost functions, reducing the immediate generality of the method. Furthermore, relying on additional computational steps to select new OSDAs can hinder testing of these molecules in the lab, since experimental researchers would need to analyze the molecules with several data-intensive steps before deciding which systems to try.

In order to bypass these problems and inspired by the aforementioned materials databases, we created a web-based platform unifying the design principles, the literature data, and the simulation results in one portal. The database offers only the possibility of exploring template-based routes at this time, but complementary inorganic synthesis parameters extracted from the literature or predicted by ML may be integrated in the future. The web interface was designed to enable users to filter, select, and make data-informed decisions on the molecules that would be desired for a given framework without prior experience in computational work. One of the most practical utilities of this approach is to leverage the chemical intuition of researchers to select new OSDAs in zeolite synthesis. Despite the usefulness of descriptors such as SCScore, it is unclear how to map chemical graphs to important parameters in the synthesis of OSDAs such as yield. On the other hand, experienced researchers can easily judge how feasible is a molecular structure, what is its intellectual novelty, and so on. Therefore, this human–computer partnership enables users to select templates for zeolite synthesis using the parameters described above and make better decisions based on theoretical results.

Since the key human input is how to assign relative weights to the computational descriptors, we facilitated this selection using interactive sliders that enable users to filter molecules based on binding, shape, and complexity metrics (Figure 4.10). Several plots representing the different binding metrics are displayed along with sliders that add/remove templates to the graphs based on their descriptors. Although all the trade-offs between descriptors cannot be fully quantified through data mining of computational metrics alone, the accessibility of the user interface may enable a better selection of OSDAs, even if qualitative.

From the software engineering and user experience perspective, implementing these computational tools requires accessing the database of simulations, plotting them in an interactive way, and updating the filters as the users perform their selections. Our web platform, named Organic Structure-directing agents DataBase (OSDB), was created using Django interfacing a PostgreSQL database through psycopg. The structured schema of these databases allows for an organization of the data and easy deployment to the cloud. Once the web interface requests information from the database, interactive plots are created using Bokeh. This tool has several plotting and interaction functionalities, which were extended using JavaScript to implement tooltips showing molecules, sliders, and so on. It also interfaces well with Python and HTML webpages.

With a comprehensive, unified platform for OSDA design for zeolites, there are numerous exciting possibilities for designing zeolite synthesis. For example, researchers can use OSDB as a reference database for template design without having to recalculate hundreds of thousands of zeolite-OSDA pairs. This data accessibility may help expedite zeolite research worldwide, using FAIR data principles, and is inspired by the existence of curated databases such as the exemplary one from the IZA. In addition, the literature data can help increase the reproducibility of the syntheses, which is essential in the field. Finally, the modular computational approach can be extended to provide different analyses in the field, similar to what already exists in the materials community in platforms such as Materials Project or AFLOW. Overall, design of materials and synthesis using interactivity can aid scientists in making data-driven decisions without compromising the possibilities for discovery starting from exploratory concepts.

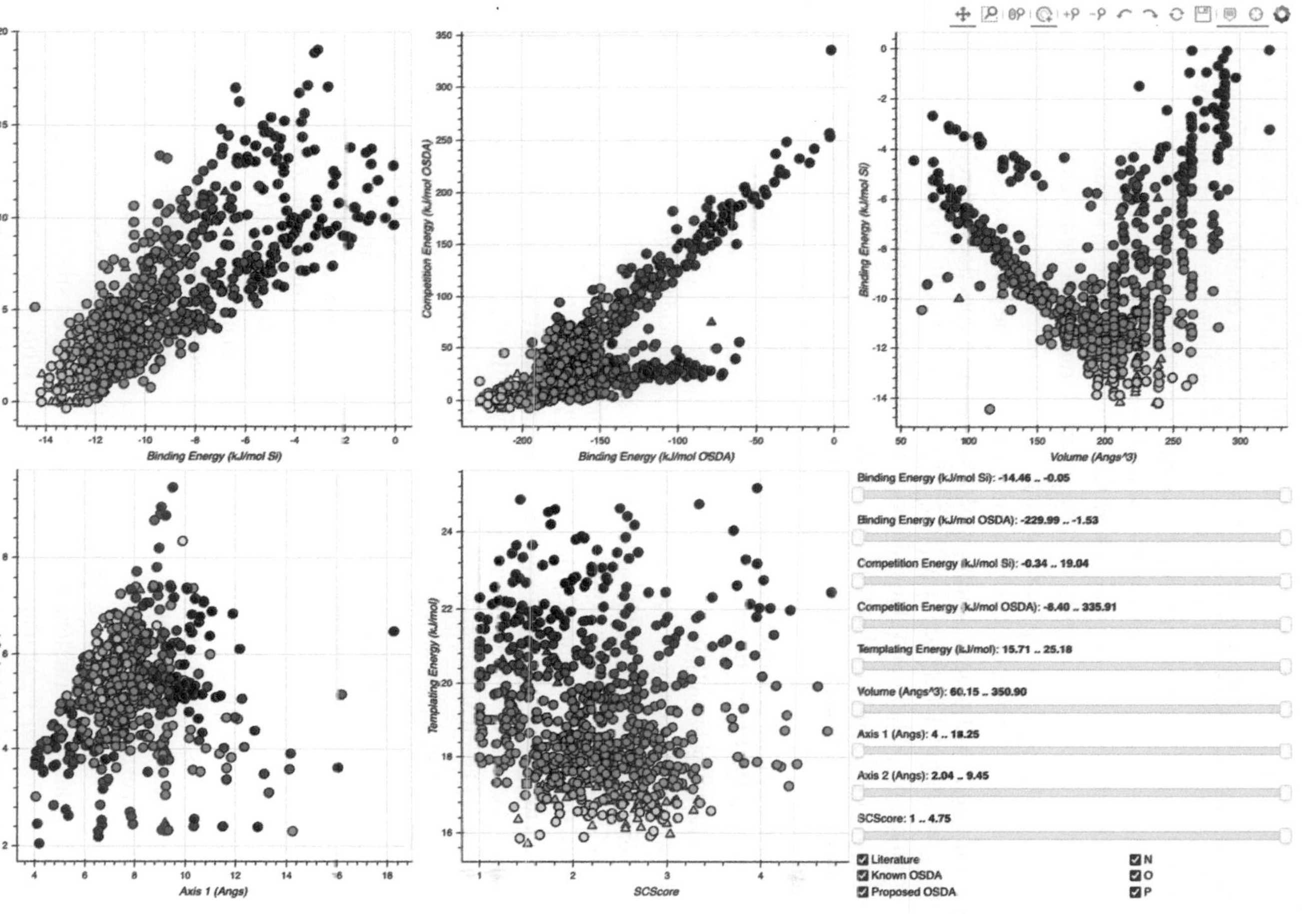

Figure 4.10 Graphical interface of interactive filters for downselecting OSDAs for a given zeolite as implemented in OSDB. Each data point represents an OSDA, which can be visualized upon hovering the cursor over the dots. The sliders are used to include/exclude points within certain ranges of synthesis descriptors.

4.3 Outlook

In this chapter, we analyzed the importance of data curation for zeolite discovery, focusing particularly on databases of OSDAs for zeolites. As data-driven approaches to the design of new systems are highly dependent on data generation methods, we explored how high-throughput computation can aid the discovery of new templates for zeolite synthesis. Using insights learned from the materials community and extending them to synthesis, we devised algorithms, benchmarked simulations, gathered literature, and analyzed data to expedite the synthesis of zeolites.

Despite the advances in these computational methods, several challenges exist for future discovery and applications. For example, despite the comprehensive approach to simulating known zeolite frameworks, the first-principles synthesis of hypothetical frameworks remains elusive. Although recent works have attempted to propose novel templates for their synthesis [98,119], overcoming phase competition with known frameworks is one major issue. The pipeline described in our work may help finding templates for novel zeolite structures. As the theory does not rely on past experimental data to propose new templates, simulation of hypothetical zeolites is feasible. In particular, the combination of competition and shape descriptors can guide the synthesis of new structures. However, selecting the hypothetical candidates among the millions of structures predicted to be thermodynamically accessible remains a problem. Traditional "structural feasibility" descriptors have been studied in zeolites for more than a decade [53,120–122], but the community has not yet realized a completely new framework entirely from first principles. While high-throughput host-guest simulations may one day help in realizing one of such structures, selecting which frameworks to simulate remains a problem [123].

Recent data-driven methods have been proposed to try to elucidate such issues [124,125], but more work needs to be done in this field. Increasing the size of the OSDA chemical space under exploration could be beneficial when proposing new molecules. While the pipeline described here was demonstrated mostly for literature OSDAs when studying templates, it is not restricted to these data points. Novel molecules proposed to synthesize the known frameworks have been successful in achieving the desired results [10], given the generality of the simulations. However, creating a database of OSDAs to simulate requires careful deliberation. Genetic algorithms can be used for proposing new OSDAs to be simulated [66,67], but the outcomes do not always lie within the chemical space of typical OSDAs and suffer from unfeasible constraints, such as low carbon-to-nitrogen ratios, presence of undesired substructures, etc. Modern generative models improve over this constrained molecular data generation [126,127], but molecules are often unfeasible despite their constrained diversity [60]. Finally, restricted generation algorithms may be a solution for proposing novel OSDAs [68], but a strategy to accelerate the exploration of the chemical space is required. Overall, downselecting templates from the space of plausible OSDAs to the subspace of feasible, commercially relevant, and selective OSDAs may require a combination of data-driven methods, simulations, and experiments.

Another frontier in expediting computational zeolite research lies in accelerating first-principles simulations. In many applications such as adsorption, diffusion, or catalysis, few force fields have been proposed to quantify interactions between zeolites and guest species. As such, simulations often rely on expensive DFT calculations, often coupled with MD, nudged elastic band, and enhanced sampling strategies [128]. Creating methods capable of simulating structures with accuracy similar to DFT, but computational cost closer to that from force fields, is essential to improve the understanding of new structure-synthesis-properties relationships in zeolites. One alternative to this is simulating zeolites using neural network (NN) force fields. These techniques have been successfully employed for various molecular and solid systems over the last few decades [129–138].

However, the chemical and structural diversity of zeolites hinders the application of NNs toward these materials. In particular, NN struggle to extrapolate beyond the training set. On the other hand, simulations of catalytic reactions or enhanced sampling require sampling high-energy (i.e., low probability) states unlikely to be found in training sets derived from MD simulations. To bypass this problem, we have proposed an algorithm to train NN force fields in a physics-informed, data-efficient manner [139]. Our algorithm uses the concept of adversarial attacks from the machine learning literature to propose new geometries to sample using high-cost methods such as DFT. These geometries correspond to points that balance thermodynamic likelihood with uncertainty in the machine learning predictions, thus likely being the most informative for the NN force fields being trained. In addition to several examples of molecules, we demonstrated the usefulness of this algorithm for training an NN force field for zeolite-OSDA pairs [139]. Despite the chemical and topological diversity of these systems, the final NNs were able to generate MD trajectories with better physical constraints than the ones trained with traditional methods. If extended to simulate systems where the chemical diversity is not so severe, the approach could enable creating faster and more accurate simulations of zeolites for several systems.

Finally, the data-informed decisions for zeolite synthesis can be extended in many ways. The web-based platform can unify even more synthesis and literature data for zeolites, particularly as the field continues to move forward. Improving the modularization of the pipeline can improve the development of new computational methods for zeolites, offer better APIs for machine readability, and may interface existing software for decentralized prediction of properties. Another interesting possibility is the selection of inorganic synthesis conditions. While domain expertise can often inform synthesis conditions, very few works have analyzed the ideal conditions to produce certain zeolite frameworks. The combination of literature data and machine learning techniques may help exploration of the high-dimensional synthesis conditions in the zeolite world [63]. Finally, computational discovery of template-free synthesis routes for zeolites is still underdeveloped [62,140–146]. We have rationalized classes of solid-state transformations in these materials [62], but predicting outcomes of zeolite recrystallizations has not yet been performed. Given the industrial relevance of template-free synthesis, predictive methods to understand crystallization can better inform synthesis conditions of the materials. Overall, advances in computation can guide experimentation in zeolites by avoiding trial-and-error, making data accessible, and aiding decision-making with demonstrated theoretical principles.

References

1 Tabor, D.P., Roch, L.M., Saikin, S.K. et al. (2018). Accelerating the discovery of materials for clean energy in the era of smart automation. *Nat. Rev. Mater.* 3 (5): 5–20. doi:10.1038/s41578-018-0005-z.

2 Chen, H., Engkvist, O.,Wang, Y. et al. (2018). The rise of deep learning in drug discovery. *Drug Discovery Today* 23 (6): 1241–1250. doi:10.1016/j.drudis.2018.01.039.

3 IPCC (2018). Summary for policymakers. In: *Global Warming of 1.5°C* (ed.V. Masson-Delmotte, P. Zhai, and H.O. Pörtner, et al.). An IPCC Special Report. https://archive.ipcc.ch/report/sr15/pdf/sr15_citation.pdf and https://www.ipcc.ch/2018/10/08/summary-for-policymakers-of-ipcc-special-report-on-global-warming-of-1-5c-approved-by-governments/

4 Davis, S.J., Lewis, N.S., Shaner, M. et al.(2018). Net-zero emissions energy systems. *Science* 36 (6396): eaas9793. doi:10.1126/science.aas9793.

5 Vermeiren, W. and Gilson, J.-P. (2009). Impact of zeolites on the petroleum and petrochemical industry. *Top. Catal.* 52 (9): 1131–1161. doi:10.1007/s11244-009-9271-8.

6 Yi, L., Lin, L., and Jihong, Y. (2017). Applications of zeolites in sustainable chemistry. *Chem* 3 (6): 928–949. doi:10.1016/j.chempr.2017.

7 Davis, M.E. (2002). Ordered porous materials for emerging applications. *Nature* 417 (6891): 813–821. doi:10.1038/nature00785.

8 Jiyang, L., Corma, A., and Jihong, Y. (2015). Synthesis of new zeolite structures. *Chem. Soc. Rev.* 44 (20): 7112–7127. doi:10.1039/c5cs00023h.

9 Gallego, E.M., Teresa Portilla, M., Paris, C. et al. (2017). *Ab initio* synthesis of zeolites for pre-established catalytic reactions. *Science* 355 (6329): 1051–1054. doi:10.1126/science.aal0121.

10 Schwalbe-Koda, D., Kwon, S., Paris, C. et al.(2021). A priori control of zeolite phase competition and intergrowth with high-throughput simulations. *Science* 374 (6565): 308–315. doi:10.1126/science.abh3350.

11 Shoichet, B.K. (2004). Virtual screening of chemical libraries. *Nature* 432 (7019): 862–865. doi:10.1038/nature03197.

12 Greeley, J., Jaramillo, T.F., Bonde, J. et al. (2006). Computational high-throughput screening of electrocatalytic materials for hydrogen evolution. *Nat. Mater.* 5 (11): 909–913. doi:10.1038/nmat1752.

13 Alapati, S.V., Karl Johnson, J., and Sholl, D.S. (2006). Identification of destabilized metal hydrides for hydrogen storage using first principles calculations. *J. Phys. Chem. B* 110 (17): 8769–8776. doi:10.1021/jp060482m.

14 Subramaniam, S., Mehrotra, M., and Gupta, D. (2008). Virtual high throughput screening (vHTS): a perspective. *Bioinformation* 3 (1): 14–17.

15 Setyawan, W., Gaume, R.M., Lam, S. et al. (2011). High-throughput combinatorial database of electronic band structures for inorganic scintillator materials. *ACS Comb. Sci.* 13 (4): 382–390. doi:10.1021/co200012w.

16 Armiento, R., Kozinsky, B., Fornari, M. et al. (2011). Screening for high-performance piezoelectrics using high-throughput density functional theory. *Phys. Rev. B* 84 (1): 014103.

17 Holdren, J.P. and National Science and Technology Council (US) (2011). *Materials Genome Initiative for Global Competitiveness*. Executive Office of the President. 6: National Science and Technology Council. https://www.mgi.gov.

18 Xiang, X.-D., Xiaodong, S., Briceno, G. et al. (1995). A combinatorial approach to materials discovery. *Science* 268 (5218): 1738–1740. doi:10.1126/science.268.5218.1738.

19 Ceder, G., Chiang, Y.-M., Sadoway, D.R. et al. (1998). Identification of cathode materials for lithium batteries guided by first-principles calculations. *Nature* 392 (6677): 694–696. doi:10.1038/33647.

20 Madsen, G.K.H. (2006). Automated search for new thermoelectric materials: the case of LiZnSb. *J. Am. Chem. Soc.* 128 (37): 12140–12146. doi:10.1021/ja062526a.

21 Butler, K.T., Davies, D.W., Cartwright, H. et al. (2018). Machine learning for molecular and materials science. *Nature* 559 (7715): 547–555. doi:10.1038/s41586-018-0337-2.

22 Sanchez-Lengeling, B. and Aspuru-Guzik, A. (2018). Inverse molecular design using machine learning: generative models for matter engineering. *Science* 361 (6400): 360–365. doi:10.1126/science.aat2663.

23 Jonathan Schmidt, M.R., Marques, G., Botti, S. (2019). Recent advances and applications of machine learning in solid-state materials science. *NPJ Comput. Mater.* 5 (1): 83. doi:10.1038/s41524-019-0221-0.

24 Chen, C., Zuo, Y., Weike, Y. (2020). A critical review of machine learning of energy materials. *Adv. Energy Mater.* 10 (8): 1903242. doi:10.1002/aenm.201903242.

25 Jablonka, K.M., Ongari, D., Moosavi, S.M. (2020). Big-data science in porous materials: materials genomics and machine learning. *Chem. Rev.* 120 (16): 8066–8129. doi:10.1021/acs.chemrev.0c00004.

26 Musil, F., Grisafi, A., Bartók, A.P. (2021). Physics-inspired structural representations for molecules and materials. *Chem. Rev.* 121 (16): 9759–9815. doi:10.1021/acs.chemrev.1c00021.

27 Nandy, A., Duan, C., Taylor, M.G. (2021). Computational discovery of transition-metal complexes: from high-throughput screening to machine learning. *Chem. Rev.* 121 (16): 9927–10000. doi:10.1021/acs.chemrev.1c00347.

28 Huang, B. and Anatole von Lilienfeld, O. (2021). *Ab initio* machine learning in chemical compound space. *Chem. Rev.* 121 (16): 10001–10036. doi:10.1021/acs.chemrev.0c01303.

29 Baerlocher, C. and McCusker, L.B. (2021). Database of zeolite structures. http://www.iza-structure.org/databases.

30 Moliner, M., Román-Leshkov, Y., and Corma, A. (2019). Machine learning applied to zeolite synthesis: the missing link for realizing high-throughput discovery. *Acc. Chem. Res.* 52 (10): 2971–2980. doi:10.1021/acs.accounts.9b00399.

31 Klintenberg, M., Derenzo, S.E., and Weber, M.J. (2002). Potential scintillators identified by electronic structure calculations. *Nucl. Instrum. Methods Phys. Res. Section A Accelerat. Spectromet. Detect. Assoc. Equip.* 486 (1–2): 298–302. doi:10.1016/S0168-9002(02)00723-4.

32 Ortiz, C., Eriksson, O., and Klintenberg, M. (2009). Data mining and accelerated electronic structure theory as a tool in the search for new functional materials. *Comput. Mater. Sci.* 44 (4): 1042–1049. doi:10.1016/J.COMMATSCI.2008.07.016.

33 Hachmann, J., Olivares-Amaya, R., Atahan-Evrenk, S. et al. (2011). The Harvard clean energy project: large-scale computational screening and design of organic photovoltaics on the world community grid. *J. Phys. Chem. Lett.* 2 (17): 2241–2251. doi:10.1021/jz200866s.

34 Hummelshøj, J.S., Abild-Pedersen, F., Studt, F. et al. (2012). CatApp: a web application for surface chemistry and heterogeneous catalysis. *Angew. Chem. Int. Ed.* 51 (1): 272–274.

35 Jain, A., Ong, S.P., Hautier, G. et al. (2013). Commentary. The materials project: a materials genome approach to accelerating materials innovation. *APL Mater.* 1 (1): 11002. doi:10.1063/1.4812323.

36 Curtarolo, S., Setyawan, W., Gus, L.W. et al. (2012). AFLOW: an automatic framework for high-throughput materials discovery. *Comput. Mater. Sci.* 58: 218–226.

37 Calderon, C.E., Plata, J.J., Toher, C. et al. (2015). The AFLOW standard for high-throughput materials science calculations. *Comput. Mater. Sci.* 108: 233–238. doi:10.1016/J.COMMATSCI.2015.07.019.

38 Saal, J.E., Kirklin, S., Aykol, M. et al. (2013). Materials design and discovery with high-throughput density functional theory: the Open Quantum Materials Database (OQMD). *JOM* 65 (11): 1501–1509.

39 Pizzi, G., Cepellotti, A., Sabatini, R. et al. (2016). AiiDA: automated interactive infrastructure and database for computational science. *Comput. Mater. Sci.* 111: 218–230. doi:10.1016/J.COMMATSCI.2015.09.013.

40 Merkys, A., Mounet, N., Cepellotti, A. et al. (2017). *A posteriori* metadata from automated provenance tracking: integration of AiiDA and TCOD. *J. Cheminform.* 9 (1): 56. doi:10.1186/s13321-017-0242-y.

41 Haastrup, S., Strange, M., Pandey, M. et al. (2018). The computational 2D materials database: high-throughput modeling and discovery of atomically thin crystals. *2D Materials* 5 (4): 042002. doi:10.1088/2053-1583/aacfc1.

42 Winther, K.T., Hoffmann, M.J., Boes, J.R. et al. (2019). Catalysis-Hub.org: an open electronic structure database for surface reactions. *Scientific Data* 6 (1): 1–10. doi:10.1038/s41597-019-0081-y.

43 Chanussot, L., Das, A., Goyal, S. et al. (2021). Open CATALYST 2020 (OC20) dataset and community challenges. *ACS Catal.* 11 (10): 6059–6072. doi:10.1021/acscatal.0c04525.

44 Wilkinson, M.D., Dumontier, M., Aalbersberg, I.J. et al. (2016). The FAIR guiding principles for scientific data management and stewardship. *Scientific Data* 3 (1): 160018. doi:10.1038/sdata.2016.18.

45 Draxl, C. and Scheffler, M. (2018). NOMAD: the FAIR concept for big data-driven materials science. *MRS Bulletin* 43 (9): 676–682. doi:10.1557/mrs.2018.208.

46 Blaiszik, B., Chard, K., Pruyne, J. et al. (2016). The materials data facility: data services to advance materials science research. *JOM* 68 (8): 2045–2052. doi:10.1007/s11837-016-2001-3.

47 Blaiszik, B., Ward, L., Schwarting, M. et al. (2019). A data ecosystem to support machine learning in materials science. *MRS Commun.* 9 (4): 1125–1133. doi:10.1557/mrc.2019.118.

48 Talirz, L., Kumbhar, S., Passaro, E. et al. (2020). Materials Cloud, a platform for open computational science. *Scientific Data* 7 (1): 299. doi:10.1038/s41597-020-00637-5.

49 Álvarez-Moreno, M., de Graaf, C., López, N. et al. (2015). Managing the computational chemistry big data problem: the ioChem-BD platform. *J. Chem. Inf. Model.* 55 (1): 95–103. doi:10.1021/ci500593j.

50 Gražulis, S., Chateigner, D., Downs, R.T. et al. (2009). Crystallography open database: an open-access collection of crystal structures. *J. Appl. Crystallogr.* 42 (4): 726–729. doi:10.1107/S0021889809016690.

51 Armel, L.B. (2005). Inorganic structure prediction with GRINSP. *J. Appl. Crystallogr.* 38 (2): 389–395. doi:10.1107/S0021889805002384.

52 Treacy, M.M.J., Rivin, I., Balkovsky, E. et al. (2004). Enumeration of periodic tetrahedral frameworks. II: Polynodal graphs. *Microporous Mesoporous Mater.* 74 (1–3): 121–132. doi:10.1016/j.micromeso.2004.06.013.

53 Deem, M.W., Pophale, R., Cheeseman, P.A. et al. (2009). Computational discovery of new zeolite-like materials. *J. Phys. Chem. C* 113 (51): 21353–21360. doi:10.1021/jp906984z.

54 Sanders, M.J., Leslie, M., and Catlow, C.R.A. (1984). Interatomic potentials for SiO_2. *J. Chem. Soc. Chemcomm.* (19): 1271–1273. doi:10.1039/c39840001271.

55 Yi, L., Xu, L., Liu, J., Duan, F. et al. (2015). *In silico* prediction and screening of modular crystal structures via a high-throughput genomic approach. *Nat. Commun.* 6 (1): 8328. doi:10.1038/ncomms9328.

56 Synthesis Commission of the International Zeolite Association (2001). *Verified Synthesis of Zeolitic Materials*, 2e. Amsterdam: Elsevier Science.

57 Moliner, M., Serra, J.M., Corma, A. et al. (2005). Application of artificial neural networks to high-throughput synthesis of zeolites. *Microporous Mesoporous Mater.* 78 (1): 73–81. doi:10.1016/j.micromeso.2004.09.018.

58 Yu, J. and Xu, R. (2010). Rational approaches toward the design and synthesis of zeolitic inorganic open-framework materials. *Acc. Chem. Res.* 43 (9): 1195–1204. doi:10.1021/ar900293m.

59 Jensen, Z., Kim, E., Kwon, S. et al. (2019). A machine learning approach to zeolite synthesis enabled by automatic literature data extraction. *ACS Cent. Sci.* 5 (5): 892–899. doi:10.1021/acscentsci.9b00193.

60 Jensen, Z., Kwon, S., Schwalbe-Koda, D. et al. (2021). Discovering relationships between OSDAs and zeolites through data mining and generative neural networks. *ACS Cent. Sci.* 7 (5): 858–867. doi:10.1021/acscentsci.1c00024.

61 Kim, E., Huang, K., Tomala, A. et al. (2017). Machine-learned and codified synthesis parameters of oxide materials. *Scientific Data* 4 (1): 170127. doi:10.1038/sdata.2017.127.

62 Schwalbe-Koda, D., Jensen, Z., Olivetti, E. et al. (2019). Graph similarity drives zeolite diffusionless transformations and intergrowth. *Nat. Mater.* 18 (11): 1177–1181. doi:10.1038/s41563-019-0486-1.

63 Muraoka, K., Sada, Y., Miyazaki, D. et al. (2019). Linking synthesis and structure descriptors from a large collection of synthetic records of zeolite materials. *Nat. Commun.* 1 (1): 4459. doi:10.1038/s41467-019-12394-0.

64 Gies, H. and Marker, B. (1992). The structure-controlling role of organic templates for the synthesis of porosils in the systems SiO_2/template/H_2O. *Zeolites* 12 (1): 42–49. doi:10.1016/0144-2449(92)90008-D.

65 Lewis, D.W., Freeman, C.M., and Catlow, C.R.A. (1995). Predicting the templating ability of organic additives for the synthesis of microporous materials. *J. Phys. Chem.* 99 (28): 11194–11202. doi:10.1021/j100028a022.

66 Pophale, R., Daeyaert, F., and Deem, M.W. (2013). Computational prediction of chemically synthesizable organic structure directing agents for zeolites. *J. Mater. Chem. A* 1 (23): 6750–6760. doi:10.1039/c3ta10626h.

67 Daeyaert, F., Fengdan, Y., and Deem, M.W. (2019). Machine-learning approach to the design of OSDAs for zeolite beta. *Proc. Natl. Acad. Sci. USA* 116 (9): 3413–3418. doi:10.1073/pnas.1818763116.

68 Muraoka, K., Chaikittisilp, W., and Okubo, T. (2020). Multiobjective *de novo* molecular design of organic structure-directing agents for zeolites using nature-inspired ant colony optimization. *Chem. Sci.* 11 (31): 8214–8223. doi:10.1039/D0SC03075A.

69 Catlow, C.R.A., Coombes, D.S., Lewis, D.W. et al. (1998). Computer modeling of nucleation, growth, and templating in hydrothermal synthesis. *Chem. Mater.* 10 (11): 3249–3265. doi:10.1021/cm980302o.

70 Nakagawa, Y., Lee, G.S., Harris, T.V. et al. (1998). Guest/host relationships in zeolite synthesis: ring-substituted piperidines and the remarkable adamantane mimicry by 1-azonio spiro [5.5] undecanes. *Microporous Mesoporous Mater.* 22 (1–3): 69–85. doi:10.1016/S1387-1811(98)00106-1.

71 Wagner, P., Nakagawa, Y., Lee, G.S. et al. (2000). Guest/host relationships in the synthesis of the novel cage-based zeolites SSZ-35, SSZ-36, and SSZ-39. *J. Am. Chem. Soc.* 122 (2): 263–273. doi:10.1021/ja990722u.

72 Burton, A.W., Zones, S.I., and Elomari, S. (2005). The chemistry of phase selectivity in the synthesis of high silica zeolites. *Curr. Opin. Colloid Interface Sci.* 10 (5–6): 211–219. doi:10.1016/j.cocis.2005.08.005.

73 Sliwoski, G., Kothiwale, S., Meiler, J. et al. (2014). Computational methods in drug discovery. *Pharmacol. Res.* 66 (1): 334–395. doi:10.1124/pr.112.007336.

74 Freeman, C.M., Catlow, C.R.A., Thomas, J.M. et al. (1991). Computing the location and energetics of organic molecules in microporous adsorbents and catalysts: a hybrid approach applied to isometric butenes in a model zeolite. *Chem. Phys. Lett.* 186 (2–3): 137–142, 11. doi:10.1016/S0009-2614(91)85119-H.

75 Gálvez-Llompart, M., Cantín, A., Rey, F. et al. (2019). Computational screening of structure directing agents for the synthesis of zeolites: a simplified model. *Z. Kristallogr. Cryst. Mater.* 234 (7–8): 451–460. doi:10.1515/zkri-2018-2132.

76 Lewis, D.W., Willock, D.J., Catlow, C.R.A. et al. (1996). *De novo* design of structure-directing agents for the synthesis of microporous solids. *Nature* 382 (6592): 604–606. doi:10.1038/382604a0.

77 Willock, D.J., Lewis, D.W., Catlow, C.R.A. et al. (1997). Designing templates for the synthesis of microporous solids using *de novo* molecular design methods. *J. Mol. Catal. A Chem.* 119 (1–3): 415–424. doi:10.1016/S1381-1169(96)00505-5.

78 Schwalbe-Koda, D. and Gómez-Bombarelli, R. (2021). Supramolecular recognition in crystalline nanocavities through Monte Carlo and Voronoi network algorithms. *J. Phys. Chem. C* 12 (5): 3009–3017. doi:10.1021/acs.jpcc.0c10108.

79 Willems, T.F., Rycroft, C.H., Kazi, M. et al. (2012). Algorithms and tools for high-throughput geometry-based analysis of crystalline porous materials. *Microporous Mesoporous Mater.* 149 (1): 134–141. doi:10.1016/J.MICROMESO.2011.08.020.

80 Pinheiro, M., Martin, R.L., Rycroft, C.H. et al. (2013). Characterization and comparison of pore landscapes in crystalline porous materials. *J. Mol. Graph. Model.* 44: 208–219. doi:10.1016/j.jmgm.2013.05.007.

81 Ong, S.P., Richards, W.D., Jain, A. et al. (2013). Python materials genomics (pymatgen): a robust, open-source python library for materials analysis. *Comput. Mater. Sci.* 68: 314–319. doi:10.1016/J.COMMATSCI.2012.10.028.

82 Svelle, S., Tuma, C., Rozanska, X. et al. (2009). Quantum chemical modeling of zeolite-catalyzed methylation reactions: toward chemical accuracy for barriers. *J. Am. Chem. Soc.* 131 (2): 816–825. doi:10.1021/ja807695p.

83 Göltl, F., Gruneis, A., Bučko, T. et al. (2012). van der Waals interactions between hydrocarbon molecules and zeolites: periodic calculations at different levels of theory, from density functional theory to the random phase approximation and Møller-Plesset perturbation theory. *J. Chem. Phys.* 137 (11): 114111. doi:10.1063/1.4750979.

84 Shang, J., Gang, L., Singh, R. et al. (2014). Adsorption of CO_2, N_2, and CH_4 in Cs-exchanged chabazite: a combination of van der Waals density functional theory calculations and experiment study. *J. Chem. Phys.* 140 (8): 84705. doi:10.1063/1.4866455.

85 Piccini, G., Alessio, M., Sauer, J. et al. (2015). Accurate adsorption thermodynamics of small alkanes in zeolites. *Ab initio* theory and experiment for H-Chabazite. *J. Phys. Chem. C* 119 (11): 6128–6137. doi:10.1021/acs.jpcc.5b01739.

86 Zhang, Y., Yu, J., Yeh, Y.-H. et al. (2015). An adsorption study of CH_4 on ZSM-5, MOR, and ZSM-12 zeolites. *J. Phys. Chem. C* 119 (52): 28970–28978. doi:10.1021/acs.jpcc.5b09571.

87 Fang, H., Kamakoti, P., Zang, J. et al. (2012). Prediction of CO_2 adsorption properties in zeolites using force fields derived from periodic dispersion-corrected DFT calculations. *J. Phys. Chem. C* 116 (19): 10692–10701. doi:10.1021/jp302433b.

88 Fang, H., Awati, R., Boulfelfel, S.E. et al. (2018). First-principles-derived force fields for CH_4 adsorption and diffusion in siliceous zeolites. *J. Phys. Chem. C* 122 (24): 12880–12891. doi:10.1021/acs.jpcc.8b03267.

89 Sauer, J. (2019). *Ab initio* calculations for molecule-surface interactions with chemical accuracy. *Acc. Chem. Res.* 52 (12): 3502–3510. doi:10.1021/acs.accounts.9b00506.

90 Rehak, F.R., Piccini, G., Alessio, M. et al. (2020). Including dispersion in density functional theory for adsorption on flat oxide surfaces, in metal-organic frameworks and in acidic zeolites. *Phys. Chem. Chem. Phys.* 22 (14): 7577–7585. doi:10.1039/d0cp00394h.

91 Lok, B.M., Cannan, T.R., and Messina, C.A. (1983). The role of organic molecules in molecular sieve synthesis. *Zeolites* 3 (4): 282–291. doi:10.1016/0144-2449(83)90169-0.

92 Boyett, R.E., Stevens, A.P., Ford, M.G. et al. (1996). A quantitative shape analysis of organic templates employed in zeolite synthesis. *Zeolites* 17 (5–6): 508–512. doi:10.1016/S0144-2449(96)00073-5.

93 Zones, S.I., Nakagawa, Y., Yuen, L.T. et al. (1996). Guest/host interactions in high silica zeolite synthesis: [5.2.1.02.6]tricyclodecanes as template molecule. *J. Am. Chem. Soc.* 118 (32): 7558–7567. doi:10.1021/ja954246t.

94 Millini, R., Frigerio, F., Bellussi, G. et al. (2003). *A priori* selection of shape-selective zeolite catalysts for the synthesis of 2,6-dimethylnaphthalene. *J. Catal.* 217 (2): 298–309. doi:10.1016/S0021-9517(03)00071-X.

95 Hong, S.B., Lecar, E.G., Wright, P.A. et al. (2004). Synthesis, structure solution, characterization, and catalytic properties of TNU-10: a high-silica zeolite with the STI topology. *J. Am. Chem. Soc.* 126 (18): 5817–5826. doi:10.1021/ja031981t.

96 Gómez-Hortigüela, L., Furio Corà, C., Catlow, R.A. et al. (2004). Computational study of the structure-directing effect of benzylpyrrolidine and its fluorinated derivatives in the synthesis of the aluminophosphate AlPO-5. *J. Am. Chem. Soc.* 126 (38): 12097–12102. doi:10.1021/ja0481023.

97 Shi, L., Jiyang, L., Duan, F. et al. (2005). $[C_3N_2H_{12}]$-$[MnAl_3P\,4O_{17}]$ • $[H_3O]$: a manganese (II)-substituted aluminophosphate with zeotype AFN topology. *Microporous Meso-Porous Mater.* 85 (3): 252–259. doi:10.1016/j.micromeso.2005.06.023.

98 Donghui, J. and Hong, S.B. (2019). Targeted synthesis of a zeolite with pre-established framework topology. *Angew. Chem. Int. Ed.* 58 (39): 13845–13848. doi:10.1002/anie.201909336.

99 Millini, R., Carluccio, L.C., Carati, A. et al. (2004). ERS-12: a new layered tetramethylammonium silicate composed by ferrierite layers. *Microporous Mesoporous Mater.* 74 (1–3): 59–71. doi:10.1016/j.micromeso.2004.06.007.

100 Burton, A.W., Lee, G.S., and Zones, S.I. (2006). Phase selectivity in the syntheses of cage-based zeolite structures: an investigation of thermodynamic interactions between zeolite hosts and structure directing agents by molecular modeling. *Micropor. Mesopor. Mater.* 90 (1–3): 129–144. doi:10.1016/j.micromeso.2005.11.022.

101 Zones, S.I., Burton, A.W., Lee, G.S. et al. (2007). A study of piperidinium structure-directing agents in the synthesis of silica molecular sieves under fluoride-based conditions. *J. Am. Chem. Soc.* 129 (29): 9066–9079. doi:10.1021/ja0709122.

102 Jaramillo, E., Grey, C.P., and Auerbach, S.M. (2001). Molecular dynamics studies of hydrofluorocarbons in faujasite-type zeolites: modeling guest-induced cation migration in dry zeolites. *J. Phys. Chem. B* 105 (49): 12319–12329. doi:10.1021/jp011997a.

103 Sastre, G., Sandra Leiva, M.J., Sabater, I.G. et al. (2003). Computational and experimental approach to the role of structure-directing agents in the synthesis of zeolites: the case of Cyclohexyl Alkyl Pyrrolidinium salts in the synthesis of β, EU-1, ZSM-11, and ZSM-12 zeolites. *J. Phys. Chem. B* 107 (23): 5432–5440. doi:10.1021/jp027506j.

104 Sastre, G., Cantin, A., Diaz-Cabañnas, M.J. et al. (2005). Searching organic structure directing agents for the synthesis of specific zeolitic structures: an experimentally tested computational study. *Chem. Mater.* 17 (3): 545–552. doi:10.1021/cm049912g.

105 Chawla, A., Li, R., Jain, R. et al. (2018). Cooperative effects of inorganic and organic structure-directing agents in ZSM-5 crystallization. *Mol. Syst. Des. Eng.* 3 (1): 159–170. doi:10.1039/c7me00097a.

106 Gómez-Hortigüela, L., Hamad, S., López-Arbeloa, F. et al. (2009). Molecular insights into the self-aggregation of aromatic molecules in the synthesis of nanoporous aluminophosphates: a multilevel approach. *J. Am. Chem. Soc.* 131 (45): 16509–16524. doi:10.1021/ja906105x.

107 Zones, S.I., Jayanthi, K., Pascual, J. et al. (2021). Energetics of the local environment of structure-directing agents influence zeolite synthesis. *Chem. Mater.* 33 (6): 2126–2138. doi:10.1021/acs.chemmater.0c04796.

108 Schwalbe-Koda, D. and Gomez-Bombarelli, R. (2021). Benchmarking binding energy calculations for organic structure-directing agents in pure-silica zeolites. *J. Chem. Phys.* 154 (17): 174109. doi:10.1063/5.0044927.

109 Mayo, S.L., Olafson, B.D., and Goddard, W.A. (1990). DREIDING: a generic force field for molecular simulations. *J. Phys. Chem.* 94 (26): 8897–8909. doi:10.1021/j100389a010.

110 Schmidt, J.E., Deem, M.W., and Davis, M.E. (2014). Synthesis of a specified, silica molecular sieve by using computationally predicted organic structure-directing agents. *Angew. Chem. Int. Ed.* 53 (32): 8372–8374. doi:10.1002/anie.201404076.

111 Brand, S.K., Schmidt, J.E., Deem, M.W. et al. (2017). Enantiomerically enriched, polycrystalline molecular sieves. *Proc. Natl. Acad. Sci. USA* 114 (20): 5101–5106. doi:10.1073/pnas.1704638114.

112 Toby, B.H., Khosrovani, N., Dartt, C.B. et al. (2000). Structure-directing agents and stacking faults in the CON system: a combined crystallographic and computer simulation study. *Microporous Mesoporous Mater.* 39 (1): 77–89. https://doi.org/10.1016/S1387-1811(00)00178-5.

113 Corma, A., Rey, F., Rius, J. et al. (2004). Supramolecular self-assembled molecules as organic directing agent for synthesis of zeolites. *Nature* 431 (7006): 287–290. doi:10.1038/nature02909.

114 Coley, C.W., Rogers, L., Green, W.H. et al. (2018). SCScore: synthetic complexity learned from a reaction corpus. *J. Chem. Inf. Model.* 58 (2): 252–261. doi:10.1021/acs.jcim.7b00622.

115 Halgren, T.A. (4 1996). Merck molecular force field. I: Basis, form, scope, parameterization, and performance of MMFF94. *J. Comput. Chem.* 17 (5–6): 490–519. doi:10.1002/(SICI)1096-987X (199604)17:5/6<490::AID-JCC1>3.0.CO;2-P.

116 Tosco, P., Stiefl, N., and Landrum, G. (2014). Bringing the MMFF force field to the RDKit: implementation and validation. *J. Cheminform.* 6 (1): 37. doi:10.1186/s13321-014-0037-3.

117 Schwalbe-Koda, D., Corma, A., Román-Leshkov, Y. et al. (2021). Data-driven design of biselective templates for intergrowth zeolites. *J. Phys. Chem. Lett.* 12 (43): 10689–10694. doi:10.1021/acs.jpclett.1c03132.

118 León, S. and Sastre, G. (2020). Computational screening of structure-directing agents for the synthesis of pure silica ITE zeolite. *J. Phys. Chem. Lett.* 1 (15): 6164–6167. doi:10.1021/acs.jpclett.0c01734.

119 Boruntea, C.-R., Sastre, G., Lundegaard, L.F. et al. (2019). Synthesis of high-silica erionite driven by computational screening of hypothetical zeolites. *Chem. Mater.* 31 (22): 9268–9276. doi:10.1021/acs.chemmater.9b01229.

120 Sastre, G. and Corma, A. (2010). Predicting structural feasibility of silica and germania zeolites. *J. Phys. Chem. C* 114 (3): 1667–1673. doi:10.1021/jp909348s.

121 Yi, L., Jihong, Y., and Ruren, X. (2013). Criteria for zeolite frameworks realizable for target synthesis. *Angew. Chem. Int. Ed.* 52 (6): 1673–1677. doi:10.1002/anie.201206340.

122 Salcedo Perez, J.L., Haranczyk, M., and Zim-Mermann, N.E.R. (2019). High-throughput assessment of hypothetical zeolite materials for their synthesizeability and industrial deployability. *Z. Kristallogr. Cryst. Mater.* 234 (7–8): 437–450. doi:10.1515/zkri-2018-2155.

123 Lin, L.-C., Berger, A.H., Martin, R.L. et al. (2012). *In silico* screening of carbon-capture materials. *Nat. Mater.* 11 (7): 633–641. doi:10.1038/nmat3336.

124 Helfrecht, B.A., Semino, R., Pireddu, G. et al. (2019). A new kind of atlas of zeolite building blocks. *J. Chem. Phys.* 151 (15): 154112. doi:10.1063/1.5119751.

125 Helfrecht, B.A., Pireddu, G., Semino, R. et al. (2012). Ranking the synthesizability of hypothetical zeolites with the sorting hat. *ArXiv* 211: 13764,.

126 Gómez-Bombarelli, R., Wei, J.N., Duvenaud, D.K. et al. (2018). Automatic chemical design using a data-driven continuous representation of molecules. *ACS Cent. Sci.* 4 (2): 268–276. doi:10.1021/acscentsci.7b00572.

127 Schwalbe-Koda, D. and Gomez-Bombarelli, R. (2020). Generative models for automatic chemical design. (eds. K. T. Schütt and Stefan Chmiela et al.). *Lecture Notes in Physics* 968: 445–467. doi:10.1007/978-3-030-40245-7_21.

128 Speybroeck, V.V., Hemelsoet, K., Joos, L. et al. (2015). Advances in theory and their application within the field of zeolite chemistry. *Chem. Soc. Rev.* 44 (20): 7044–7111. doi:10.1039/C5CS00029G.

129 Behler, J. and Parrinello, M. (2007). Generalized neural-network representation of high-dimensional potential-energy surfaces. *Phys. Rev. Lett.* 98 (14): 146401. doi:10.1103/PhysRevLett.98.146401.

130 Behler, J. (2011). Neural network potential-energy surfaces in chemistry: a tool for large-scale simulations. *Phys. Chem. Chem. Phys.* 13 (40): 17930. doi:10.1039/c1cp21668f.

131 Gastegger, M. and Marquetand, P. (2015). High-dimensional neural network potentials for organic reactions and an improved training algorithm. *J. Chem. Theory Comput.* 11 (5): 2187–2198. doi:10.1021/acs.jctc.5b00211.

132 Morawietz, T., Singraber, A., Dellago, C. et al. (2016). How van der Waals interactions determine the unique properties of water. *Proc. Natl. Acad. Sci.* 113 (30): 8368 LP – 8373. doi:10.1073/pnas.1602375113.

133 Artrith, N. and Urban, A. (2016). An implementation of artificial neural-network potentials for atomistic materials simulations: performance for TiO_2. *Comput. Mater. Sci.* 114: 135–150. https://doi.org/10.1016/j.commatsci.2015.11.047.

134 Botu, V., Batra, R., Chapman, J. et al. (2017). Machine learning force fields: construction, validation, and outlook. *J. Phys. Chem. C* 121 (1): 511–522. doi:10.1021/acs.jpcc.6b10908.

135 Schütt, K.T., Sauceda, H.E., Kindermans, P.-J.-J. et al. (2018). SchNet: a deep learning architecture for molecules and materials. *J. Chem. Phys.* 148 (24): 241722. doi:10.1063/1.5019779.

136 Ang, S.J., Wang, W., Schwalbe-Koda, D. et al. (2021). Active learning accelerates *ab initio* molecular dynamics on reactive energy surfaces. *Chem.* 7 (3): 1–32. doi:10.1016/j.chempr.2020.12.009.

137 Mueller, T., Hernandez, A., and Wang, C. (2020). Machine learning for interatomic potential models. *J. Chem. Phys.* 152 (5): 50902. doi:10.1063/1.5126336.

138 Cheng, B., Mazzola, G., Pickard, C.J. et al. (2020). Evidence for supercritical behaviour of high-pressure liquid hydrogen. *Nature* 585 (7824): 217–220. doi:10.1038/s41586-020-2677-y.

139 Schwalbe-Koda, D., Tan, A.R., and Gómez-Bombarelli, R. (2021). Differentiable sampling of molecular geometries with uncertainty-based adversarial attacks. *Nat. Commun.* 12 (1): 5104. doi:10.1038/s41467-021-25342-8.

140 Xie, B., Song, J., Ren, L., Yanyan, J. et al. (2008). Organotemplate-free and fast route for synthesizing beta zeolite. *Chem. Mater.* 20 (14): 4533–4535. doi:10.1021/cm801167e.

141 Marler, B. and Gies, H. (2012). Hydrous layer silicates as precursors for zeolites obtained through topotactic condensation: a review. *Eur. J. Mineral.* 24 (3): 405–428. doi:10.1127/0935-1221/2012/0024-2187.

142 Maldonado, M., Oleksiak, M.D., Chinta, S. et al. (2013). Controlling crystal polymorphism in organic-free synthesis of Na-zeolites. *J. Am. Chem. Soc.* 135 (7): 2641–2652. doi:10.1021/ja3105939.

143 Eliášová, P., Opanasenko, M., Wheatley, P.S. et al. (2015). The ADOR mechanism for the synthesis of new zeolites. *Chem. Soc. Rev.* 44 (20): 7177–7206. doi:10.1039/C5CS00045A.

144 Goel, S., Zones, S.I., and Iglesia, E. (2015). Synthesis of zeolites via interzeolite transformations without organic structure-directing agents. *Chem. Mater.* 27 (6): 2056–2066. doi:10.1021/cm504510f.

145 Itabashi, K., Kamimura, Y., Iyoki, K. et al. (2012). A working hypothesis for broadening framework types of zeolites in seed-assisted synthesis without organic structure-directing agent. *J. Am. Chem. Soc.* 134 (28): 11542–11549. doi:10.1021/ja3022335.

146 Tendeloo, L.V., Gobechiya, E., Breynaert, E. et al. (2013). Alkaline cations directing the transformation of FAU zeolites into five different framework types. *ChemComm* 49 (100): 11737–11739. doi:10.1039/c3cc47292b.

5

Co-templating in the Designed Synthesis of Small-pore Zeolite Catalysts

Ruxandra G. Chitac[1], Mervyn D. Shannon[2,3], Paul A. Cox[3], James Mattock[3], Paul A. Wright[1], and Alessandro Turrina[2]

[1] *EaStCHEM School of Chemistry, University of St Andrews, St Andrews, United Kingdom*
[2] *Johnson Matthey Technology Centre, Billingham, United Kingdom*
[3] *School of Pharmacy and Biomedical Sciences, University of Portsmouth, Portsmouth, United Kingdom*

5.1 Introduction

The importance of small-pore zeolites and zeotypes in catalytic reactions such as methanol-to-olefins (MTO) catalysis and the selective catalytic reduction (SCR) of NO_x from diesel exhausts using ammonia, and in selective adsorption of small molecules such as carbon dioxide, has led to intense study of their synthesis and structure-property relationships. This is well-covered in a number of excellent recent reviews [1,2]. Small-pore zeolites are defined as having a pore size, or window diameter of 3–4 Å, which usually results from their porosity being accessible only via 8-membered rings (bounded by 8 tetrahedrally-coordinated framework cations and 8 oxygen atoms, 8MRs). As a class of zeolites or zeotypes, they make up 41% of the total number of known framework structures (Figure 5.1) and a third of them have three-dimensionally-connected porosity that makes them highly desirable for applications [3]. Thus, it is useful to consider what methodology is the most efficient for synthesizing these structures in a controlled manner. An in-depth review of the work carried out so far in 8MR zeolite syntheses has been published by Dusselier and Davis [1].

Compared with medium or large pore frameworks (which have access to pores typically constrained by 10Rs or 12Rs, respectively), small pore zeolites often have pore space comprising of well-defined cavities rather than channels, although there are some exceptions. Small pore cavity structures with low framework Si/Al ratios are typically formed in the presence of inorganic cations as structure directing agents, often with low specificity. Examples include zeolites A (LTA, prepared with Na), chabazite, merlinoite, ZK-5 (K), and Rho (Na, Cs) [4]. However, for higher Si/Al frameworks, charge balance is most effectively achieved with bulky organic structure directing agents (OSDAs), often alkylammonium templates, where the OSDAs achieve favorable interaction energies within the cavities, acting as templates for them. By corollary, this enables improved structure direction of the crystallization of cavity materials to be an effective synthetic approach. For example, perhaps the most catalytically important small pore zeolite, SSZ-13 (a high silica

AI-Guided Design and Property Prediction for Zeolites and Nanoporous Materials, First Edition. Edited by German Sastre and Frits Daeyaert.

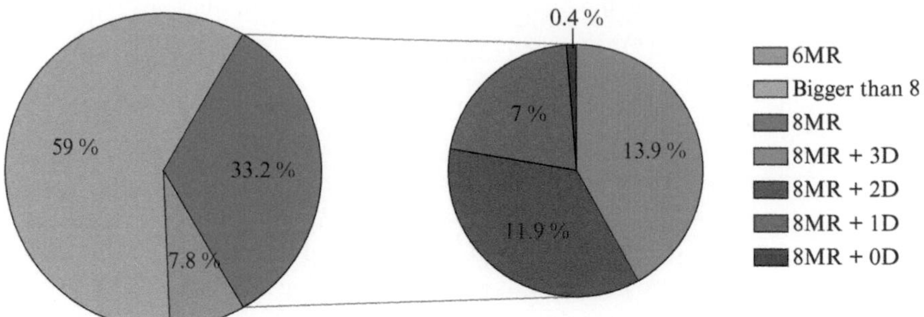

Figure 5.1 Number of frameworks with maximum pore size of 6MR, 8MR, and larger than 8MR. The number of 8MR frameworks are split into percentages based on connectivity of pores. *Source:* Data from the IZA database in September 2021 [3].

chabazite used in NH_3 SCR) is crystallized with high specificity by the trimethylammonium adamantane ion [5]; other examples include high Si/Al or pure SiO_2 forms of LTA, MER, etc [6–8].

Careful inspection of small pore framework structures, either observed or hypothetical, shows that many contain more than one type of cavity [9], so that under the right synthetic conditions improved selectivity to these materials might be achieved by the use of more than one template, designed specifically for the different cavity types. The concept of co-templating was first mentioned in the literature by Castro et al. [10] in 2007, in the synthesis of small-pore SAPOs STA-7 (SAV) and STA-14 (KFI), two three-dimensional (3D) connected small pore frameworks. At that point, using templates designed for specific cavities was a relatively new concept and tackling the synthesis of frameworks with multiple types of cavities was an even more challenging task. Since that report, the scope and practice of adding more than one organic template has expanded in different directions, where the organic additives can perform functions other than the more classical templating. We will therefore start this chapter by defining what we mean by co-templating, among these different literature approaches.

In particular, we will present examples of syntheses where co-templating has led to the discovery of new topologies or new compositions for small-pore zeolites. The control that co-templating can bring over the composition of a material will also be discussed. This can be in terms of tuning the chemical composition and elemental distribution of the framework or from the point of view of controlling the structure of intergrowths in predictable and quantifiable ways, so they can be considered a distinct family of novel materials.

Most of the new materials presented here have resulted from continual and close interaction with computational methods, and the role of computation will be highlighted to emphasize the added value this brings to targeted synthesis, not least by the optimal use of laboratory resources. We will therefore make a case here for the high potential of co-templating in the synthesis of small-pore zeolites, particularly when framed in the rapid developments occurring in computational chemistry.

5.1.1 Definitions: Templates and Structure Directing Agents; Co-templating; Dual Templating; Mixed Templating

Most literature in co-templating addresses the use of organic molecules to direct zeolite crystallization and some distinction can be made between the terms templates and structure directing agents (SDAs) based on the degree of structural fit and even rotational freedom of the species involved in their zeolite host, where a "true" template is considered to exhibit the closer fit.

We also consider that in certain syntheses, inorganic cations (and even anions such as fluoride) have an important structure directing role, where they have an unambiguous relationship to siting in a small cavity, for example, in contrast with the ion being situated in a "window," pore or an ion-exchange site in the structure. Examples of inorganic "templates" include K^+ cations for the *can* cavities alongside various OSDAs to synthesize ABC-6 type materials (e.g., UZM-12 (ERI) or ZSM-34 (ERI-OFF intergrowth) [11,12]) or F^- anions in *d4r* cavities along with quinuclidine in the *sod* cavities for the synthesis of AlPO-16 (AST) [13,14]. A summary of these definitions and relevant examples can be found in Table 5.1.

Co-templating in the context of this chapter, therefore, refers to the case where different cavities in the same material are templated by different species, either more than one organic and/or organic and inorganic. There are also rare examples of OSDAs that have acted together within the same cavity, as well as in different cavities. This means that one molecule alone does not have to be the "best fit" if there are other molecules that can "co-operate" to lead to the formation of a specific cavity [15].

Looking beyond new framework synthesis, the use of more than one template for the same cavity has been used to control the composition and stability of the final product [16–19]. Here SDAs are no longer required only to template the cavities but also need to fulfil other requirements based on the target composition, including incorporation of catalytic species, controlling Al distribution and stability, or even to reduce cost. For examples where two molecules compete for the same cavity (with a view to improving properties) we prefer the term dual templating.

Table 5.1 Definitions of commonly used terms in publications on zeolite preparation and examples to illustrate their meaning.

Term	Definition	Examples
Structure-directing agent (SDA)	Inorganic or organic additive in the synthesis gel that enables the crystallization of a specific phase	Organic: quaternary ammonium cations (aluminosilicates) or amines (SAPOs and AlPOs), phosphonium cations, polycyclic molecules, etc. inorganic: K^+, Na^+, Cs^+, Li^+, Ca^{2+}, Sr^{2+}, etc.
Template	A molecule (or ion) that is an SDA and a good fit in terms of shape and size for a cavity in a framework	Organic: tetramethltetraazacyclotetradecane (tetramethylcyclam) for the *sav* cavity [10] inorganic: K^+ in the *can* cavity [21]
Co-templating	Synthetic approach where two or more molecules (or ions) that are good fits for different cavities of a framework are used as additives and together lead to crystallization of a specific topology	Organic–organic: STA-7 (tetramethylcyclam in the larger cavity of the SAV framework and tetraethylammonium (TEA^+) in the smaller cavity) [10] Organic–inorganic: STA-30 (K^+ in the *can* cavity and 1,8- (diazabicyclo[2.2.2] octane) octyl (diDABCO-$C8^{2+}$) bromide in the *swy* cavity) [21]
Mixed-templating/ dual-templating	Synthetic approach where multiple SDAs are used	Property/composition control using two SDAs for same cage for cost-efficiency [5]; one-pot syntheses including transition metals or P-modifying agents [16] Charge-density mismatch (CDM) approach – various SDAs with different charge densities used, but some act as "void fillers" and others act as templates [22] Addition of additives to modify texture, introduce hierarchical porosity [20]

Finally, there are references to dual or mixed template systems where one of the organic additives acts as a true template while another acts to change the microstructure, for example surfactant cations in the preparation of hierarchical zeolites, which may or may not be incorporated [20]. This is outside the scope of this chapter.

5.2 SAPO Zeotypes: "Model" Systems for Co-templating

Some of the clearest examples of co-templating of nanoporous solids are observed in the synthesis of aluminophosphate zeotypes and their substituted variants. Although not strictly zeolites, they possess very similar structural features and are themselves important catalysts. Aluminophosphate zeotypes are usually prepared using organic structure directing agents, which may be amines, protonated at the pH values of synthesis, or alkylammonium cations (although recent work by Seo et al. [23] shows that inorganic cations can be used together with OSDAs in the synthesis of certain aluminophosphate structures, including small pore materials). When preparing substituted $AlPO_4$ frameworks, either via metal dications, $M^{2+} \leftrightarrow Al^{3+}$, or silicon, $Si^{4+} \leftrightarrow P^{5+}$, positively charged OSDAs balance the negative charge imparted to the framework by the substitutions. In the absence of alkali metal cations, the interaction between the OSDA and the framework strongly directs the crystallization of metalloaluminophosphates (MAPOs) and silicoaluminphosphates (SAPOs) [24]. As a result, SAPOs that contain cavities in their framework structures provide good examples of co-templating with two different OSDAs.

5.2.1 The CHA-AEI-SAV-KFI System

SAPO-34 is the most important of the aluminophosphate materials, because of its catalytic performance in the MTO reaction and, when copper-loaded, in the SCR of NO_x [1]. SAPO-34 has the CHA topology, with a single cavity type (*cha*) that is templated by many simple OSDAs, including morpholine or TEA^+ cations. The structure comprises *d6r* units, all in the same orientation, linked via 4MRs to give *cha* cavities (Figure 5.2). There is then 3D connectivity of these *cha* cavities via 8MR windows. Three other closely-related frameworks can be envisaged as built up from *d6r*

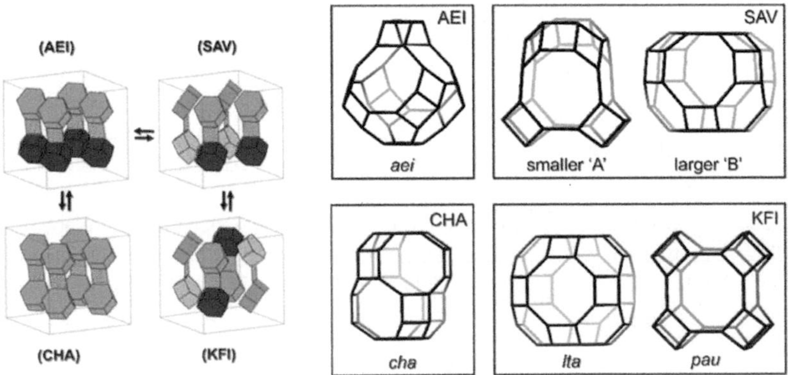

Figure 5.2 Left: the orientation of the *d6rs* in CHA, AEI, SAV, and KFI. Right: the other cavities present in the frameworks. All figures portrayed with Si-Si connectivities and structures obtained from the IZA database [3].

units linked by 4MRs (Figure 5.2), each with 3D connectivity via 8MR windows. All have now been prepared as materials and so have 3-letter codes (AEI, SAV and KFI). While the CHA and AEI frameworks contain a single type of cavity, the SAV and KFI framework types have two types of cavities, and this proved the key to their synthesis in hitherto unobserved SAPO forms by a co-templating approach.

The SAV framework was first observed in studies using the tetraazamacrocycle, tetramethylcyclam (TMTACT), as an effective OSDA [25]. Whereas the MgAPO and SAPO compositions crystallized as STA-6 (SAS), with a one-dimensional (1D) small pore channel system, CoAPO and ZnAPO compositions gave the SAV structure, containing larger and smaller types of cavities. The more catalytically interesting SAPO form of SAV therefore became an important target, and an empirical co-templating approach was adopted toward its synthesis. Remarkably, a mixture of either TMTACT or the simpler tetraazacyclotetradecane (cyclam) with TEA^+ cations was highly effective in giving pure SAPO SAV (STA-7), with the cyclam templating the larger cavity and TEA^+, in the *tg.tg* configuration, the smaller cavity (Figure 5.3). This was confirmed experimentally by single crystal X-ray diffraction and solid-state NMR. Co-templates such as di-*iso*-propylamine and di-*n*-propylamine resulted in mixtures of SAPO STA-6 with minority STA-7 phase.

Extending this approach toward a SAPO with the KFI topology type as a target, it was known that the *lta* cavity found in this structure had been observed to be templated by the azaoxacryptand Kryptofix K222, in $AlPO_4$-42 (LTA). The problem then reduced to choosing a suitable template for the *pau* cavity of the KFI framework. Modeling was used to screen ten readily available amines and alkylammonium cations inside an AlPO model of the KFI structure type, in order to assess their suitability to stabilize the *pau* cavity. The most favorable binding energy ($-177.8\ kJ\ mol^{-1}$) was again obtained for TEA^+, although in the *tt.tt* configuration in this cavity [10]. TEA^+ was therefore selected as the second OSDA in the experimental synthesis mixture for the target material, along with K222.

Experimentally, the MgAPO and SAPO forms of KFI, STA-14, were then successfully prepared using a co-templating approach with K222 and TEA^+, in which both OSDAs were measured to be present by NMR, and single crystal diffraction confirmed the predicted TEA^+ configuration. Attempts to synthesize the KFI structure type using K222 with the other candidate co-templates all proved to be unsuccessful, demonstrating the ability of the modeling approach to correctly identify suitable OSDAs for the co-templating approach. The co-templating syntheses were developed further to prepare a range of SAPO compositions of STA-7 and STA-14, and their catalytic properties and coking behavior were measured in the MTO reaction, where they showed similar

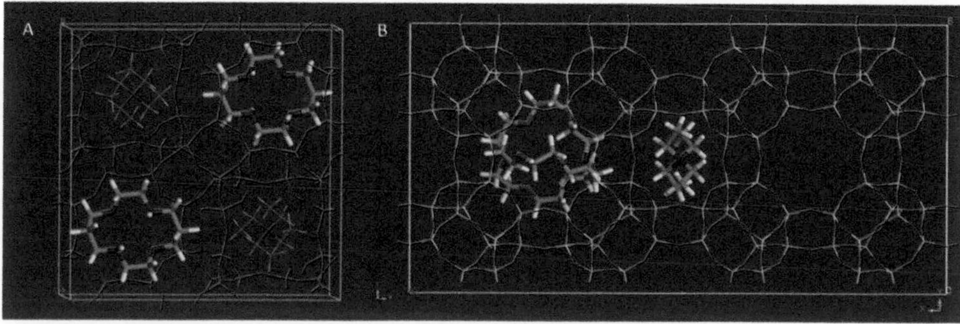

Figure 5.3 Modeled positions of co-templates within frameworks. A: TEA^+ in the *tg.tg* conformation in the smaller cavity "A" and cyclam in the larger cavity "B" of AlPO SAV. B: TEA^+ in the *tt.tt* conformation in the *pau* cavity and K222 in the *lta* cavity of AlPO KFI. *Source:* Composed of figures previously published by M. Castro [10].

conversion performance to SAPO-34, although with different hydrocarbon pools and characteristic small olefin selectivities [26].

One of the initial reasons for using azamacrocycles as OSDAs was their ability to complex cations such as Ni^{2+} and Cu^{2+}, the latter of which is of importance in SCR of NO_x. Ni-cyclam, for example, is an effective SDA for SAPO STA-6 [27]. It was therefore straightforward to modify the co-templating method to prepare Cu-SAPO-STA-7, where Cu-cyclam occupies the larger cavities in the as-prepared material and TEA^+ the smaller cavities [28]. Calcination gives Cu-SAPO-STA-7 with Cu^{2+} cations evenly distributed throughout the structure. This was found to have good performance in both CH_4- and NH_3-mediated SCR of NO_x [28,29]. Similar one-pot synthesis approaches to the inclusion of the Cu^{2+} approach have since been used to prepare Cu-SAPO-34 and also Cu-SAPO-18 by dual-templating approaches, in which copper-polyamine complexes have been used in conjunction with amines or alkylammonium cations as templates for either the *cha* or *aei* cavities [29,30]. This enables control over the level of Cu^{2+} included in the final catalyst.

5.2.2 Development of a Retrosynthetic Co-templating Approach for ABC-6 Structure Types

A more recent example of the combination of target structure choice, computational simulations, and prior synthetic knowledge in designed synthesis of SAPOs can be found in the work by Turrina et al., who used a "retrosynthetic co-templating method" to prepare two small-pore SAPOs, SAPO-56 (AFX) and STA-18 (SFW), as well as a large-pore material, STA-19 (GME), in the ABC-6 family [31]. In this family, the frameworks are built from the stacking of layers of 6-membered rings (6MRs) parallel to the *ab*-plane, connected by 4MRs along the *c*-direction. There are three different positions for these 6MRs, related through translation in the hexagonal *ab*-plane, centered at *x*, *y* coordinates of (0,0), (⅓,⅔), and (⅔,⅓) and denoted A, B, and C. Each stacking sequence within this family can then be described using these letters. The CHA topology type, for example, has the AABBCC stacking sequence.

The three ABC-6 topologies AFX, SFW, and GME share a common feature, the *gme* cavity, but also have their own distinctive larger cavities (or channels for GME) due to the different stacking of the 6MRs. Initial computational modeling focused on finding the most appropriate template for the *gme* cavity. Various amines and ammonium cations were screened (Figure 5.4) and it was found that trimethylamine, TrMA (or trimethylammonium in its protonated version) gave the best fit for the cavity based on its size and shape, quantified as the non-bonding interaction with the cavity (Table 5.2).

In the case of SAPO-56, the starting point for the larger cavity (*aft*) template was chosen based on the synthesis of the aluminosilicate AFX–SSZ-16, streamlining the OSDA screening process both computationally and in the lab [32]. Indeed, the use of trimethylamine and 1,4-(diazabicyclo[2.2.2]octane)butyl (diDABCO-$C4^{2+}$) was shown to be the optimum co-template pair for SAPO-56 (Figure 5.5).

For the SFW topology, the OSDA that had been used to prepare its aluminosilicate counterpart, SSZ-52, was much shorter than the *sfw* cavity, so that two molecules are required per cavity with the consequence that there is considerable intergrowth in that material. Other computational work supported the observation that a longer template would be more appropriate [33]. This piece of information, coupled with the fact that the *sfw* cavity is ~5 Å longer than an *aft* cavity, but otherwise similar in shape, led to the computational prediction that diDABCO-$C6^{2+}$ would be the optimum co-template for a SAPO with the SFW topology (Figure 5.5). Indeed, when trimethylamine was used as a co-template with diDABCO-$C6^{2+}$, it successfully led to the preparation of fault-free SAPO (SFW), STA-18.

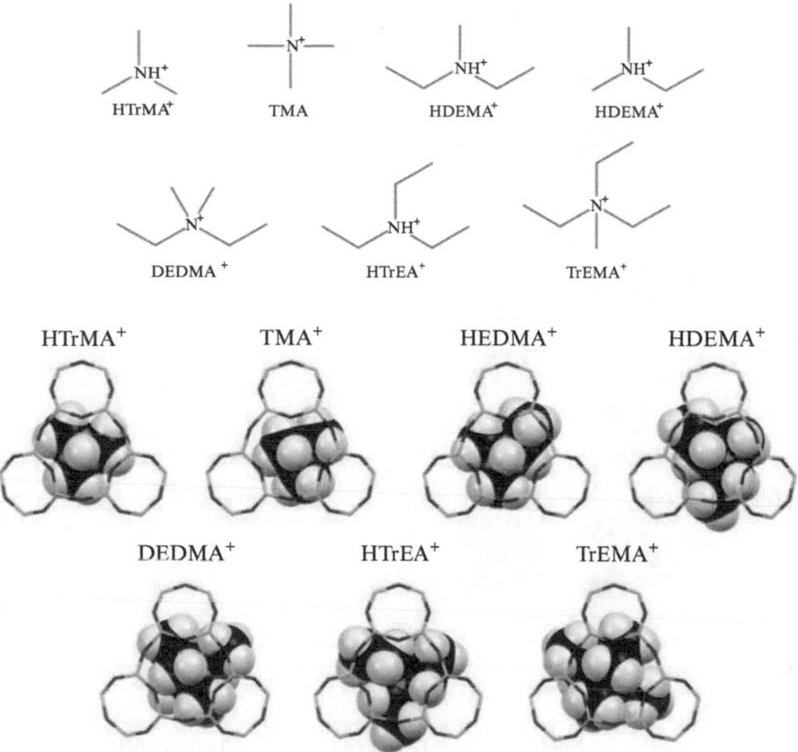

Figure 5.4 Sketches of the amines screened as templates for the *gme* cavity and their computationally calculated fits in the AlPO *gme* cavity. *Source:* Figure reprinted with permission from Turrina et al. [31]. Copyright 2016 American Chemical Society.

Table 5.2 Binding energies of the amines screened as templates for the *gme* cavity.

SDA	Binding energy/kcal mol^{-1}
$HTrMA^+$	−136
TMA^+	−116
$HEDMA^+$	−133
$HDEMA^+$	−128
$DEDMA^+$	−74
$HTrEA^+$	−117
$TrEMA^+$	−63

Binding energy calculated as the difference in energy between the total energy of the system and the isolated framework and isolated template.
Source: From Turrina et al. [31].

These successes demonstrated the feasibility of designed co-templated synthesis in the SAPO system. However, during the retrosynthetic synthesis of SAPO (SFW), a second phase, STA-20, was observed to form via serendipitous co-templating under specific conditions. STA-20 also has a structure within the ABC-6 family, albeit with double and single 6MRs, which was solved by X-ray powder diffraction and electron diffraction, with the help of a hypothetical zeolite database [15]. Its SWY framework is composed of columns of alternating *can* and *d6r* cavities, adjacent to *gme* and *swy*

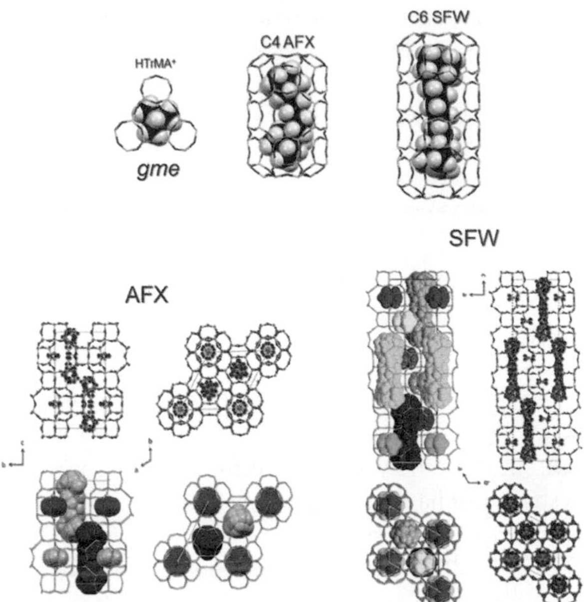

Figure 5.5 (Above) OSDAs selected for each cavity in the target structures in SAPO-56 and SAPO STA-18 (SFW); and (below) modeled locations of all possible template positions in them. *Source:* Figure reprinted with permission from Turrina et al. [31]. Copyright 2016 American Chemical Society.

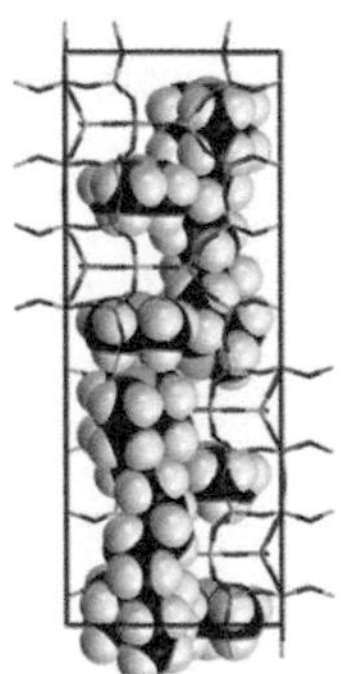

Figure 5.6 The lowest-energy positions of diDABCO-C6^{2+} and trimethylamine in the AlPO SWY framework calculated computationally. *Source:* Figure reprinted with permission from Turrina et al. [15]. Copyright 2017 American Chemical Society.

cavities (the latter being the longest ABC-6 cavity observed). Solid state NMR showed both templates were present, and modeling suggested that trimethylamine and diDABCO-C6^{2+} would template the larger *gme* and *swy* cavities of the framework, respectively. However, the diDABCO-C6^{2+} template is appreciably shorter than the *swy* cavity, and modeling suggests that trimethylamine also occupies the *swy* cavity, in which it is found alongside diDABCO-C6^{2+}. The positions found during modeling were also used as a starting point for structure refinement which, together with elemental analysis, supported the hypothesis that diDABCO-C6^{2+} and trimethylamine co-template in the *swy* cavity (Figure 5.6).

The examples of SAPOs with the SAV, KFI, AFX, SFW, and SWY topologies all show the potential to use the co-templating approach with OSDAs to prepare materials with two or more cavity types in this compositional regime. For high selectivity to target phases over competing phases that crystallize from only one of the templates (typically GIS or SOD for trimethylamine, and AlPO-5 for the longer templates), both OSDAs must be similarly effective in their structure direction.

5.3 Co-templating Aluminosilicate Zeolites

Attempts to synthesize aluminosilicate zeolites via the use of OSDAs, while extremely effective, is complicated by the effects of alkali metal cations typically present in the synthesis, usually as hydroxides, to ensure alkaline pH suitable for crystallization. The interaction of these cations with the crystallizing frameworks and also with the OSDAs themselves is incompletely understood, although the effects are important and, in some cases, clearly demonstrated (i.e., K^+ in the *can* cavity of zeolites such as L, erionite, and offretite). A recent perspective by Shin et al. gives examples of this [34]. Therefore, we have broadened the definition of co-templating to include cases where cations show a clear structure relationship to the cavities they occupy in the as-prepared zeolites.

5.3.1 Inorganic/Organic Co-templates

5.3.1.1 Targeting new phases in the RHO family using divalent cations

One of the most spectacular examples of the targeted synthesis of zeolites containing multiple cavity types is to be found in the studies of Hong et al., in which they have prepared a family of embedded isoreticular zeolites of increasing complexity, ranging from zeolite Rho through PST-29, paulingite, ZSM-25, PST-20, PST-25, and PST-26/28 (Figure 5.7) [35,36]. These structures, starting from RHO, are expanded to higher generations by inserting additional *d8r* and 18-hedral $[4^{12}8^6]$ *pau* cavities between the *lta* cavities along the unit cell edges. Space in between the scaffolds is filled with embedded cavities to form fully tetrahedrally connected frameworks.

While state-of-the-art electron microscopy enabled the structure solution of the framework structure of the fourth member of the family, ZSM-25 (Figure 5.7), computational modeling made an important contribution to understanding the most likely positions of the ~40 TEA^+ OSDA cations in the unit cell, present along with Na^+ cations and water molecules. The structure of ZSM-25 is very complex, with five different types of cavities present and over 4500 atoms in the unit cell. With large numbers of guest molecules/cations present in the structure, it is important to have a good starting model for the NaTEA-ZSM-25 structure prior to undertaking Rietveld refinement of the X-ray data. The binding energies for the TEA^+ cation in each of the five different cavities in the ZSM-25 structure were determined using a force field approach. The results showed that the most likely positions for the TEA^+ cations were inside the *pau*, $[4^66^28^6]$ and $[4^{12}6^88^6]$ (*lta*) cavities, each of which is big enough for the TEA^+ cation to be included without any unfavorable close contacts. The modeled positions of the TEA^+ inside the different cavities are shown in Figure 5.8, and the calculated interaction energies are shown in Table 5.3. For the cavities showing favorable energies, the interaction is significantly more favorable for the slightly smaller *pau* and $[4^66^28^6]$ cavities than with the larger $[4^{12}6^88^6]$ *lta* cavity. The interaction energies with the smaller $[4^78^5]$ and $[4^68^4]$ cavities are much less favorable, i.e., more positive (Table 5.3), because they are smaller, and this leads to close contacts between the cavity and the TEA^+. The conclusion from the modeling data was that the TEA^+ cations would adopt sites in the $[4^66^28^6]$, $[4^{12}8^6]$ *pau*,

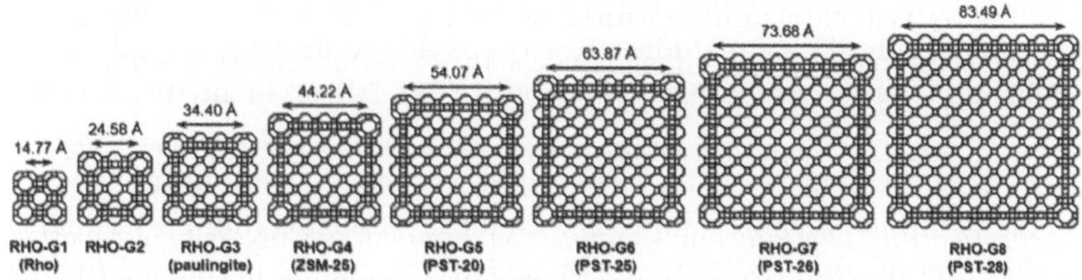

Figure 5.7 Framework structures of RHO, PST-29 (RHO-G2), PAU (RHO-G3), and ZSM-25 (RHO-G4) and other members of the RHO family of embedded isoreticular zeolites. *Source:* Figure reprinted with permission from Shin et al. [37]. Copyright 2016 Wiley.

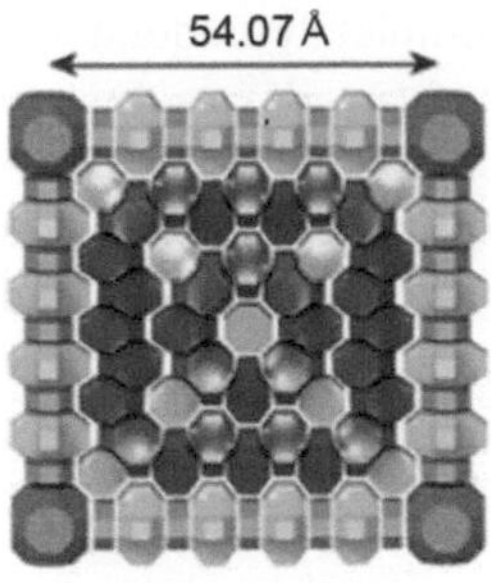

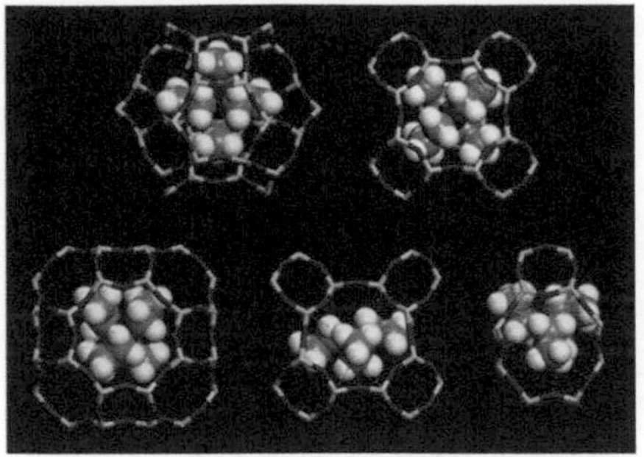

Figure 5.8 Energy-minimized location of TEA^+ cations in (clockwise, from top left) $[4^6 6^2 8^6]$, $[4^{12} 8^6]$ *pau*, $[4^6 8^4]$, $[4^7 8^5]$, and $[4^{12} 6^8 8^6]$ *lta* cavities in the ZSM-25 framework. Dashed lines indicate energetically unfavorable close contacts. *Source:* Figure reprinted with permission from Guo et al. [35]. Copyright 2015 Nature.

Table 5.3 Calculated interaction energies of TEA^+ cations in different cavities of the ZSM-25 framework.

Cavity type	Interaction energy of TEA^+/kcal mol^{-1}
$[4^6 6^2 8^6]$	−52.7
$[4^{12} 8^6]$ (*pau*)	−51.8
$[4^{12} 6^8 8^6]$ (*lta*)	−39.6
$[4^7 8^5]$	−3.7
$[4^7 8^5]$	+8.6

and $[4^{12} 6^8 8^6]$ *lta* cavities. Since there are two *lta*, 18 *pau* and 24 $[4^6 6^2 8^6]$ cavities and *ca.* 40 TEA^+ cations per unit cell, it was concluded that most of the *pau* and $[4^6 6^2 8^6]$ cavities are occupied by TEA^+ cations. Having established that it was likely that TEA^+ cations occupy the $[4^{12} 8^6]$ *pau* and $[4^6 6^2 8^6]$ cavities, a single TEA^+ cation was included in each of these two types of cavities for the starting structural model for ZSM-25. This model was used as the starting point for the X-ray refinement of the ZSM-25 structure, which yielded a result that was in excellent agreement with experiment, confirming the usefulness of modeling in successfully locating OSDAs within such a complex host structure.

Having determined the structure of ZSM-25, and identified the likely templating role of TEA^+, it was then possible to suggest structures for any number of hypothetical higher-generation frameworks using the same rules of expansion. Hong et al. then had the insight that in zeolite minerals, some of the extra 8MR features observed in abundance in the higher-generation hypothetical structures are observed naturally in zeolites containing alkaline earth metal cations, such as Sr^{2+}. Remarkably, by careful synthetic mixed cation studies, it was possible to direct the synthesis of some of these target higher-generation structures, either pure (PST-20) or in mixtures (PST-25/26/28) [37].

Although this example of co-templating has some very specific features, it does outline the fine balance of interactions of inorganic and organic templates with the crystallizing framework that is responsible for the crystallization of complex frameworks, and which can be harnessed to give novel multi-cavity materials.

5.3.1.2 Designed synthesis of the aluminosilicate SWY, STA-30

Databases of hypothetical zeolite structures, together with their lattice energies, provide viable targets for aluminosilicate zeolite synthesis [9,38]. Another practical approach is to examine those materials already prepared but with alternative compositions, and devise routes to their synthesis as aluminosilicates. Several ABC-6 systems were known to have been prepared first as SAPOs and later as zeolites, including AFX (SAPO-56 and SSZ-16) [39,40]. The discovery of the SAPO STA-20 (SWY), with *can*, *gme* and *swy* cavities, provided such a target for cross-compositional synthesis, which was subsequently achieved by inorganic/organic co-templating.

In the studies on the SAPO SWY, trimethylamine was found to be the best template for the *gme* cavity [31], but amines are not usually effective templates in zeolite synthesis. Furthermore, modeling revealed that the diDABCO-C6^{2+} is significantly shorter than the *swy* cavity and trimethylamine is also included there. As a result, significant adjustments had to be made in the synthesis design. First, it was recognized that SWY is related to the ERI structure (Figure 5.9) and since erionite can be prepared from K containing gels, where the K^+ cations template the *can* cavities, an alternative inorganic SDA/organic SDA templating approach was adopted. In this, the *can* and *swy* cavities of the structure were targeted, and diDABCO molecules were chosen as likely templates for the long *swy* cavities [21]. Indeed, computational modeling led to the choice of diDABCO-C8^{2+} as the most suitable OSDA for the *swy* cavity (Figure 5.9), as discussed below.

The resulting planned co-templated synthesis, suitably optimized, gave highly crystalline SWY over a range of Si/Al values. The material was stable to calcination and activation, and subsequently modified by copper cation inclusion to give an active catalyst for the NH_3 SCR of NO_x, which was the original target.

Synthesis of the SWY topology in aluminosilicate form is an excellent example of how the co-templating approach, coupled with computational modeling, can lead to efficient synthesis of new topologies with a variety of compositions. This is a valuable pathway for producing new materials that might be of interest for various industrial applications as it makes the initial discovery more time and resource efficient, making the trial-and-error process in the lab redundant or shortening it significantly.

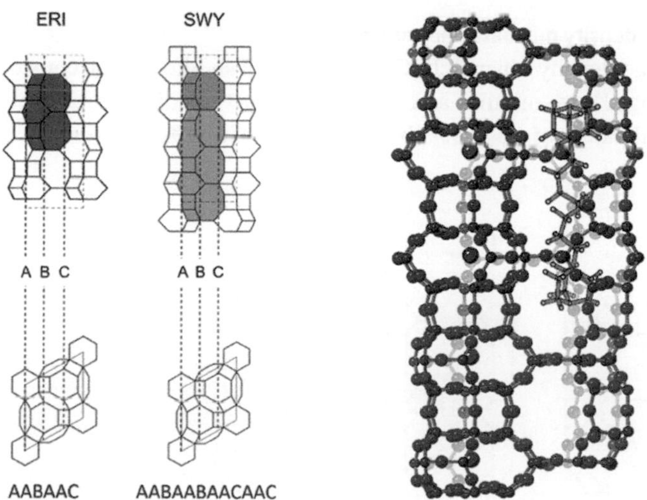

Figure 5.9 (Left) The relationship between the ERI and SWY frameworks in terms of ABC layer stacking sequences. (Right) Co-templates in aluminosilicate STA-30 (K in *can* cavity – purple spheres and diDABCO-C8 in the *swy* cavity). *Source:* Figure reprinted with permission from Chitac et al. [21]. Copyright 2021 American Chemical Society.

(i) Modeling Templates in Elongated Cavities

The selection of diDABCO-type OSDAs for SAPO-56 (AFX), SAPO STA-18 (SFW), and zeolite STA-30 (SWY) were all based on computational modeling of host–guest interactions between the organic cations and elongated cavities in the frameworks [21,31]. The calculations predicted the optimum length for the polymethylene chain connecting the two DABCO end-groups based on the length of the cavity, as confirmed by the synthetic results. Thus, diDABCO-C4^{2+} was a good fit for the *aft* cavity in SAPO-56, while being too short for the *sfw* cavity in STA-18, which required the longer diDABCO-C6^{2+}. Similarly, in the design of the zeolite form of SWY, it was predicted that diDABCO-C8^{2+} was the best fit for the *swy* cavity, with longer templates twisting to fit in the cavity (Figure 5.10). The synthetic results for STA-18 and STA-30 synthesized with OSDAs predicted to require twisting to fit within their respective large cavity, as reflected quantitatively by higher torsion energies, show that these templates give less crystalline materials.

(ii) Other Effects of Inorganic Cations *in* Zeolite Syntheses

The SWY example is a well-defined example of inorganic and organic cations acting as co-templates. More generally, the inclusion of inorganic cations can have other more subtle effects. For example, cooperation between the inorganic and organic cations in the gel for preparing CHA zeolites has achieved control of Al distribution in the framework as well as reduction in the amount of N,N,N-trimethyl-1-admantylammonium ($TMAda^+$). The work presented by Di Iorio et al. [18] shows how choosing Na^+ or K^+, along with $TMAda^+$, can influence whether Al sites are paired or isolated. Similarly, due to the size of the two inorganic cations, they preferentially occupy different sites in the framework, which leads to the possibility of using less OSDA when K^+ is present, as it can effectively co-template the *cha* cavity (by competing with the $TMAda^+$ cation). The complete understanding of the mechanism at play in the study mentioned above was possible due to computational calculations that allowed the simulation of various cationic positions and the energy advantage offered by one arrangement over the other. This is then another example of the possible depth of knowledge that can be achieved when theory and experiment come together.

5.3.1.3 Co-templating and the charge density mismatch approach

Multiple template systems in the charge-density mismatch (CDM) approach should also be considered in a discussion of co-templating. Here, organic and inorganic structure directing agents (SDAs) cooperate to enable the crystallization of a desired phase [22]. One of the “templates” has a

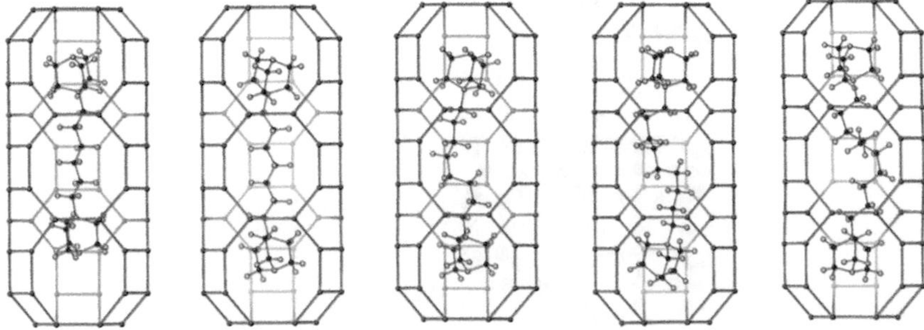

Figure 5.10 The fit of diDABCO-C6^{2+}, -C7^{2+}, -C8^{2+}, -C9^{2+}, and -C10^{2+} (from left to right) in the *swy* cavity of silicate SWY. *Source:* Figure reprinted with permission from Chitac et al. [21]. Copyright 2021 American Chemical Society.

double role, because it acts as a mineralizing agent in the first step of gel synthesis, where it fulfils the role of the low-charge density SDA (e.g., tetrapropylammonium -TPA^+) that does not lead to crystallization of the aluminosilicate species formed in solution. The crystallization (or high-charge density) SDAs (alkylammonium and alkali metal cations) added after aging of the aluminosilicate solution induce the formation of the desired phase and are included in its pores. For UZM-12 (ERI) prepared by the CDM approach, for example, the crystallization SDAs are incorporated into the framework (K^+ in *can* and a diammonium cation in *eri*), while the low-charge density molecule (TEA^+ or TPA^+) is not part of the final product, so this case can be viewed as a co-templating approach from the perspective of the crystallization SDAs [11].

The incorporation of both the low- and high-charge density SDAs only occurs under very specific conditions; this can be considered a co-templating approach if the SDAs direct the formation of different cavities. UZM-5 (UFI) crystallizes from a gel containing TEA^+ as the low-charge density SDA and tetramethylammonium (TMA^+), along with Na^+ as high charge density SDAs. All of these cations are incorporated into the final solid but act as both templates (TMA^+ in *wbc*) and "void fillers" by occupying the same cavity with other cations without specifically directing its formation (TEA^+, TMA^+, and/or Na^+ in the *lta* cavities) [41]. Hence, this remains simply a mixed-template approach.

5.3.2 Two Organic Templates in Zeolite Synthesis

Up until now, the only example where co-templating a small pore zeolite topology type with two different OSDAs has been reported is the preparation of aluminosilicate AFT (named SSZ-112) [42,43]. The AFT topology belongs to the subset of ABC-6 structures in which the repeated stacking sequence can be described using only *d6rs*. Apart from *d6rs*, the AFT structure possesses three different cavity types: *gme*, *cha*, and *aft*. The AFT structure was discovered back in 1989 as an aluminophosphate (AlPO-52) and it was synthesized using a combination of TEAOH and tripropylamine (Pr_3N) as OSDAs [44]. While the structure-directing affinity of TEAOH for the *cha* cavity is well-known [45], there is no evidence in the literature that Pr_3N plays the role of template for the *aft* cavity.

The aluminosilicate version of AFT was obtained by combining hexamethonium with 1-propyl-1-methylpiperidinium (1P1MPi) or 1-butyl-1-methylpyrrolidinium (1B1MPy) via an inter-zeolite conversion synthesis approach using zeolite Y as the reactant. The van der Waals (vdW) interaction energies reported in the paper by Xie [43], calculated via molecular mechanics calculation using CVFF force field within the Material Studios software, show a more favorable energy of 0.7 kJ (mol $Si)^{-1}$ for 1B1MPy within a *cha* cavity (one molecule loaded per cavity), while the energy difference is 0.2 kJ (mol $Si)^{-1}$ in favor of 1P1MPi for the *aft* cavity (with two molecules loaded per cavity). The vdW energies for hexamethonium were not reported. No further information on the location of the two OSDAs in SSZ-112 was given.

Even in this SSZ-112 case, the diffraction pattern shows some peak broadening resulting from an intergrowth structure. Indeed, co-templating with OSDAs, particularly in the ABC-6 system, often leads to intergrowths, and these are considered below in the final section as a separate category of novel co-templated zeolite materials.

5.3.2.1 Applications of Dual/Mixed Organic Templating

Most work related to more general use of more than one organic SDA to control small pore zeolite properties is found in CHA zeolites, due to their importance in catalysis. There are various examples where the partial replacement of the effective but expensive $TMAda^+$ has been attempted. Zones achieved this by using an N,N,N-trialkylbenzyl quaternary ammonium cation, which acted in concert with the $TMAda^+$ to template the *cha* cavity and reduce the amount of SDA necessary to crystallize pure CHA [5].

The dual template method can also be used to include catalytically active copper cations in the pore structure of zeolites SSZ-13 (CHA) and SSZ-39 (AEI) [46,47], when one of the templates is strongly structure directing for the zeolite, and the second is a copper polyamine complex, stable under the conditions of synthesis, as described above for Cu-SAPO-18. It is then possible to adjust the copper content of the final catalyst.

Another example of compositional modification is the use of $TMAda^+$ along with a tetraalkylphosphonium cation (P-modifying agent: P-MA) to control the P content introduced into the CHA zeolite [16]. This is desirable because zeolites such as P-ZSM-5 display enhanced stability over unmodified ZSM-5, which is used as a fluid catalytic cracking (FCC) additive [48]. While medium-pore ZSM-5 can be modified post-synthetically by treatment with phosphoric acid, this is not possible for small-pore zeolites due to their window size and *in situ* inclusion of P-species is an attractive alternative. The phosphonium cation acts as a pore-filler initially, but when the zeolite undergoes calcination at 600°C, it leads to the formation of various phosphate species that interact with the aluminosilicate framework. This P-modified form has been shown to have increased thermal stability and a better catalyst performance, effectively improving the properties of the final product by stabilizing the Al [16,49]. The same improvement has been shown for a P-modified AEI. By using a mixture of N,N-diethyl-2,6-dimethylpiperidinium and tetraethylphosphonium (TEP^+) cations, Kakiuchi et al. have synthesized a material that has similar SCR activity, but significantly improved thermal stability compared to the non-modified version [17].

The AFX framework topology has also been the subject of mixed templating for Si/Al control and P modification. Mitani et al. used diDABCO-$C4^{2+}$ in conjunction with TEP^+ to synthesize a P-modified AFX zeolite that maintained crystallinity up to a higher temperature compared to the aluminosilicate counterpart [50], and Tsunoji et al. have studied the effect of templates and SDAs on Si/Al and the properties of various P-MAs in AFX. They synthesized a series of materials with N,N,N',N'-tetraethylbicyclo[2.2.2]oct-7-ene-2,3:5,6-dipyrrolidinium ($TEBOP^{2+}$), Na^+, TMA^+, tetramethylphosphonium (TMP^+), and TEP^+ in various combinations (Figure 5.11) [19]. They achieved the increase of Si/Al ratio when Na^+ was replaced with TMA^+, which they correlated with the difference in charge density between the two SDAs that templated the *gme* cavity. The P-MA size influenced the cavity that it templated in the final product and led to different P species in the calcined zeolite.

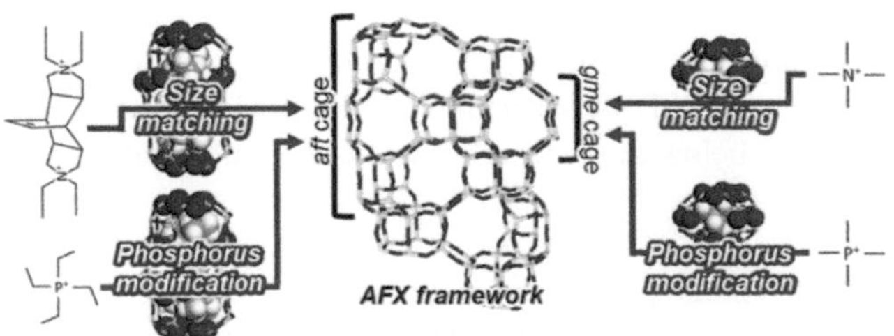

Figure 5.11 The fit of the OSDAs and P-MAs in the AFX framework. *Source:* Figure reprinted with permission from Tsunoji et al. [19]. Copyright 2021 Elsevier.

5.4 Intergrowth Zeolite Structures as Co-templated Materials

The co-templating approach has also been successfully exploited in the synthesis and structural control of disordered (intergrowth) zeolite and zeotype structures, which can usefully be considered together, as a distinctive category of novel materials. In this section we will discuss some examples where co-templating was used to prepare disordered materials.

Unlike ordered structures which exhibit 3D framework periodicity, disordered structures lack periodicity in at least one crystallographic direction. This phenomenon is also called stacking disorder of structurally invariant Periodic Building Units (PerBUs), where the stacking sequence of the PerBU deviates from periodic ordering over isolated stacking "errors" all the way to random stacking sequences. Chemical disorder (i.e., different cations on a particular site), dynamic disorder (i.e., rotational disorder of template molecules), and structural disorder (i.e., disordered molecules in the cavities of zeolite frameworks) are excluded from this definition. The physicochemical properties and thc catalytic behavior of a disordered structure might be different to those of the ordered counterpart. Therefore, synthesizing these materials and gaining control over their structural ratios are of high importance for the design of zeolites with specific sorption and catalytic properties. The determination of the types of stacking defects, the proportions in which they occur, and their exact location are very important to fully understand the materials' properties. The combination of Transmission Electron Microscopy (TEM) with the simulation of PXRD patterns of disordered models has successfully been employed to characterize most of the disordered zeolite structures reported in the literature.

The two main groups of small pore zeolite structures prone to structural disorder are CHA/AEI and different zeolites in the ABC-6 family. CHA and AEI are structurally related by the different arrangement of the *d6rs* (Figure 5.2), while the ABC-6 structures are built up from 6MRs linked by 4MRs with different stacking arrangements along one axis (Figure 5.12) [3]. Examples of

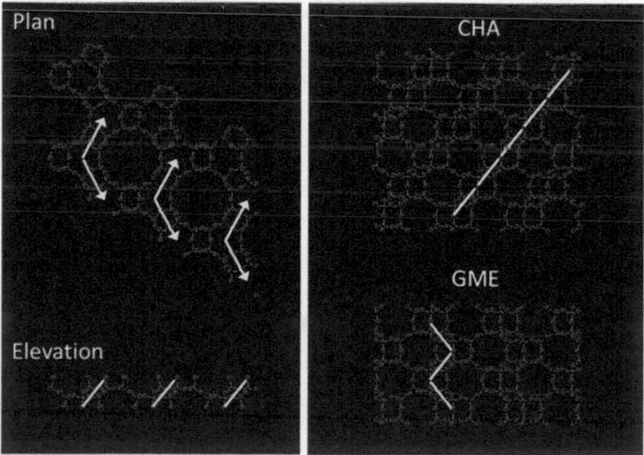

Figure 5.12 The left panel shows the atoms in the 0.5 nm layers in plan (001) and elevation (100) views that stack to form ABC-6 *d6r* structures, whilst the right panel shows the stacking in CHA and GME structures. The arrows in the plan view indicate the higher 6MR as can be seen in the elevation. The yellow lines in the right panel run from the centre of a *d6r* to the nearest 0.5 nm higher or lower in the *c*-direction and highlight the stacking by translation (CHA) or mirror (GME) of the individual layers. These lines are also used schematically in the DIFFaX model in Table 5.5.

disordered materials belonging to the latter families are: CHA/GME [52,53], Babelite [54], ZTS-1 and ZTS-2 (CHA/AFX) [55], Zeolite T and ZSM-34 (ERI/OFF) [56,57], and JMZ-11 (*cha* – *aft* – "*sfw*-GME tail") [58].

The structures of both SAPO-34 [45] and SAPO-18 [59] are composed entirely of *d6r* building blocks, linked by 4MRs. The difference lies in the mode of stacking: in CHA all *d6r*s have the same orientation (inversion center), whereas in AEI the orientation of the *d6r*s is alternated (mirror plane). This creates two different cavity types: *cha* and *aei*. Both consist of six 8MRs connected via 4MRs and enclose the same volume. However, their shape is different: the *cha* cavity is shaped like a cylinder while the *aei* cavity is shaped like a pear. Thanks to their 8MR window size (3.8 × 3.8 Å), pore shape, and strengths of the acid sites created by silicon substitution, both SAPOs have found industrial use in the methanol-to-olefin (MTO) conversion process [60,61]. The small pore opening provides the selectivity for light olefins, while the internal large cavities allow aromatic intermediates to form via the most accepted "hydrocarbon pool" mechanism [62]. However, larger aromatics like phenanthrene and pyrene can generate coke with consequent deactivation of the catalyst. It has been reported that the Si content and lower strength of the Brønsted acid sites in SAPO-18 can benefit their lifetime [63–65]. Moreover, SAPO-18 has shown higher yield for propene and C4 while SAPO-34 has the higher ethylene yield [66].

These factors, combined with the structure similarity of these two topologies, have compelled many researchers to study SAPO AEI/CHA intergrowths for MTO. The synthesis of this intergrowth material was reported for the first time by ExxonMobil [51]. Tetraethylammonium was used as only OSDA, most likely because both SAPO-18 and SAPO-34 can be prepared using this OSDA. The control over the ratio of AEI/CHA was achieved by careful control of the Si/Al ratio of the starting gel and the heating rate of the hydrothermal crystallization. To determine the AEI/CHA ratio, DIFFaX [67] was used in the patent assuming two separate intergrowth phases were present, while dissolution AFM studies carried out by Smith et al. have shown a change in disorder between the core and surface regions [68]. Following this, several papers have been published where the ratio of AEI/CHA was controlled via a co-templating approach. The approach has also been explored as a method of controlling this type of intergrowth [69,70]. For example, Guo et al. have synthesized SAPO-34/SAPO-18 by using a mixture of diethylamine (CHA-directing), N,N′-diisopropylethylamine (AEI-directing), and triethylamine (both CHA- and AEI-directing) as OSDAs. Aside from the ratio of AEI/CHA, the Si incorporation, morphology, and crystal sizes could also be controlled by the types, amounts, and combinations of the OSDAs [70]. In terms of catalytic activity, a higher content of CHA led to a higher selectivity of light olefins, while higher content of AEI exhibited longer life time and lower C2/C3 ratio.

The co-template approach has also been used to synthesize many intergrowth materials belonging to the ABC-6 family. Two examples will be discussed: ZTS-1/ZTS-2 [55] and JMZ-11 [58]. ZTS-1 and ZTS-2 are novel CHA/AFX aluminosilicate intergrowths. The two end-member topologies belong to a sub-class of ABC-6 framework structures in which the repeated stacking sequence can be described using only *6MR*: AABBCC(A) for CHA and AABBAACC(A) for AFX. While CHA only contains *cha* cavities, AFX contains two different cavity types, *aft* and *gme*, which are approximately 5 Å shorter and longer, respectively, than a *cha* cavity. For the synthesis of ZTS-1, OSDAs were selected to template separately the *cha* cavity of the CHA portion (N,N,N-trimethyl-1-adamantylammonium) and the *aft* cavity of the AFX portion (1,10-(1,4-butanediyl)bis(1-azonia-4-azabicyclo[2,2,2]octane). The ratio of CHA/AFX in the two materials was determined by matching the experimental XRD patterns collected on the calcined hydrated samples with the DIFFaX simulation and by the amount of *aft*-directing SDA measured by TG/DTA analysis. ZTS-1 was described as approximately 20/80 CHA/AFX. Due to the limited synthesis conditions adopted to prepare ZTS-1, it was not possible for the authors to control the ratio of CHA/AFX by simply changing the ratio of the two OSDAs. In fact, ZTS-2, which shows an inverted ratio of CHA/AFX (80/20), was synthesized by fully replacing the *cha* directing agent, N,N,N-trimethyl-1-adamantylammonium, with K^+ cations, known to provide a structure-directing effect for CHA, especially at low silica to alumina ratios (SAR). Overall, this work has shown how the co-templating approach can be used to prepare a new intergrowth material with a partial control over the ratio of the two structural components. Notably, Xie very recently reported the synthesis of a CHA/AFX intergrowth aluminosilicate using two different OSDAs selected by molecular mechanics calculation [43].

Two of the authors of this book have recently reported a new family of intergrowth materials known as JMZ-11, where the relative proportions of cavities of *cha*, *aft* and those in an "*sfw*-GME" tail are controlled by co-templating. JMZ-11 was discovered during an attempt to reduce the SAR of AEI below 10. Two members of this family, JMZ-11A and JMZ-11B, were prepared using N,N-dimethyl-3,5-dimethylpiperidinium (3,5-DMP) or 1,1-diethyl-2,6-dimethylpiperidinium (2,6-DMP) as a single OSDA, while two other members were synthesized using a co-templating approach in order to control the proportion of cavities. 3,5-DMP and 1,3-bis(1-adamantyl)imidazolium (BAI) or 1-adamantyl trimethylammonium (1-ATMAH) were used for JMZ-11C and JMZ-11D, respectively (Table 5.4).

JMZ-11 matcrials arc ABC-6 *d6r* structures. The known ordered crystalline phases in the ABC-6 *d6r* system are: GME, CHA, AFT, AFX, and SFW [3]. Babelite is a well-known ABC-6 *d6r* disordered intergrowth, which is considered to be a stochastic intergrowth of CHA and GME with a 50% probability that CHA stacking is followed by another CHA layer and a 50% probability that it faults to a GME layer [54]. This single fault probability, p, (equal to 0.5 in babelite) describes the stacking and, if we consider the distribution of cavities larger than *gme*, there is a monotonic decrease in the fraction of cavities of size *n* in the material such that $f_{n+1} = (1-p)f_n$ for $n \geq 2$. Here *n* is the number of *d6r* layers along the sidewall of a cavity: for *gme* it is *1*, *cha 2*, *aft 3*, *sfw 4*, whilst larger cavities are just referred to by this number *n*. The fraction of *gme* cavities can be ignored since it is completely determined by the fractions of larger cavities: the number of *gme* cavities associated with a cavity of size *n* (≥*2*) is easily shown to be *n–2*. We can think of babelite as a material composed entirely of a "*cha-gme*" tail and very small proportions of very large cavities are expected.

Table 5.4 OSDAs used in JMZ-11 materials and the resulting cavity distributions.

			Cavities distribution by matched XRD			
			(% By Number)			(% By Volume)
Sample	**SDA-1**	**SDA-2**	***cha***	***aft***	**"*sfw*-GME" Tail**	**"*sfw*-GME" Tail**
JMZ-11A	3,5-DMP	–	54	23	23	68
JMZ-11B	2,6-DMP	–	57	10	33	65
JMZ-11C	3,5-DMP	BAI	37	56	7	20
JMZ-11D	3,5-DMP	1-ATMAH	50	15	35	74

In JMZ-11 materials, the tail begins with *sfw* cavities and the relation $f_{n+1} = (1-r)f_n$ now applies only for $n \geq 4$. The proportions of *cha* and *aft* cavities are controlled independently of such a tail in JMZ-11 materials by the co-templating approach.

The structure of the JMZ-11 materials was initially characterized by powder X-ray diffraction. It was immediately clear that the materials were ABC-6 *d6r* structures from the value of the *a*-lattice parameter, that they were disordered or intergrowths from the broadened peaks in the patterns, and that they did not correspond to the simulated diffraction patterns of any of the sequence of CHA-GME intergrowths, including babelite, described in the ABC-6 section of Intergrowth Families in the IZA Structure Commission Database of Zeolite Structures. Those simulations were made using DIFFaX [67] software and were based on the stacking of sheets 5 Å thick composed of a 6-ring on the A sites linked to a 6-ring on the B site. The stacking of these sheets is then either by mirrors to create *d6r* GME stacking or translation (or inversion) to create *d6r* CHA stacking. In the model, the probability that the next layer is a CHA stacking is (1−p) or that it is a GME stacking is p. It does not matter what went before; the stacking of these 5 Å layers is stochastic. The stacking can also be described as Reichweite 0, meaning there is no memory effect of previous layers. Several years earlier we had realized that, to generate all of the known ordered ABC-6 *d6r* structures based on this 5 Å layer, memory of the two previous layers was needed (a Reichweite 2 model [67]) to generate ordered SFW and hence, for example, intergrowths of CHA and SFW. Four probabilities (p, q, r, s) are needed to describe these situations (Table 5.5). More generally, they can also describe the intergrowths in JMZ-11, which display

Table 5.5 Probability matrix for Reichweite 2 model.

	1	2	3	4	5	6	7	8
	/	/	/	/	\	\	\	\
	(/)	(/)	(\)	(\)	(/)	(/)	(\)	(\)
	(/)	(\)	(/)	(\)	(/)	(\)	(/)	(\)
1	1−p	–	–	–	p	–	–	–
2	q	–	–	–	1−q	–	–	–
3	–	r	–	–	–	1−r	–	–
4	–	1−s	–	–	–	s	–	–
5	–	–	s	–	–	–	1−s	–
6	–	–	1−r	–	–	–	r	–
7	–	–	–	1-q	–	–	–	q
8	–	–	–	p	–	–	–	1−p

Pure CHA = ... 111111 ....

Or .... 888888 ....

Pure GME = ... 363636 ....

Pure AFX = ... 257425742574 ....

Pure AFT = ... 2157842157842157 ....

Pure SFW = ... 253253253 ....

Or .... 746746746 ....

At the top, the 8 possible sequences of 3 layers are shown indicating either translation from one layer to the next or mirror from one layer to the next. The sequence of stacking for each of the known ABC-6 ordered phases is also given from which it is trivial to determine the values of p, q, r, and s.

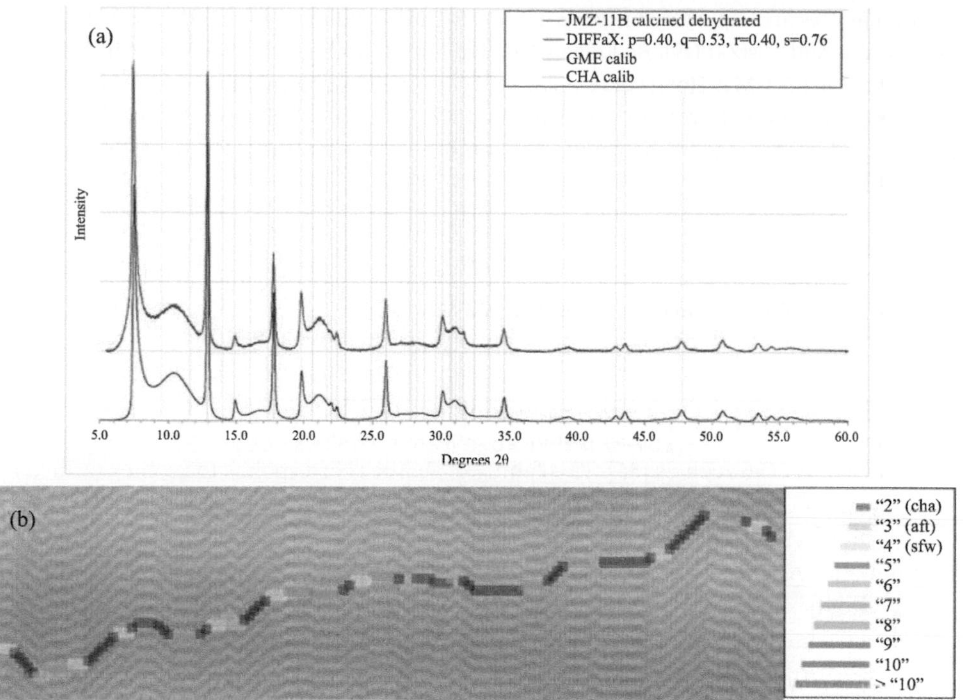

Figure 5.13 Example of JMZ-11B showing: (a) the XRD data and DIFFaX simulation; and (b) and example ADF STEM image annotated with the cavity sequence along the *c*-direction.

"*sfw*-GME" tails with *cha* and *aft* populations that differ widely from the iterative equation for stochastic intergrowths of CHA-GME. For example, JMZ-11B has very few *aft* cavities relative to *cha* and *sfw* (Figure 5.13).

The most striking feature in the powder XRD pattern is the broad peak in the region of 9.5°2θ (Cu K_α). In stochastic CHA-GME intergrowths, this does not shift from the 9.5°2θ position up to 50% GME content (after which it shifts to the region of 11.5° – a GME peak), but in JMZ-11 materials it generally shifts to either lower or higher values.

Initial simulations using the Reichweite 2 model were carried out somewhat blind, but annular dark field STEM imaging was key to getting a good starting set of probabilities (p, q, r, s) for the DIFFaX modeling work. This provided irrefutable evidence for the presence of a "*sfw*-GME" tail in all the samples of JMZ-11 that was not simply part of the "*cha*-GME" tail in stochastic CHA-GME intergrowths. The information needed from the microscopy to obtain approximate values of p, q, r, and s is the distribution of cavity sizes and the average number of *cha* cavities in blocks of *cha* cavities. This requires identification of 200–400 cavities (*cha* or larger) in sequence along the *c*-direction (Figure 5.14). This cannot be realistically achieved in a single image; 10 to 20 images are more likely. Even with this number, the sampling of the largest cavities will be poor.

Molecular dynamics calculations were carried out to determine the different distribution of *cha*, *aft*, and consequently "sfw-GME" tail between the four JMZ-11 materials. The cavities and OSDAs studied are shown in Figure 5.15. As shown in Table 5.6, the 2,6-DMP binds well in *cha* and *sfw* cavities with two molecules fitting in *sfw*, whereas it does not bind well in the *aft* cavity.

Therefore, *aft* cavities are suppressed in JMZ-11-B. 3,5-DMP binds best in the *aft* cavity and tends to enhance *aft* content (JMZ-11A). BAI also binds well in *aft* and boosts *aft* content in JMZ-11C in conjunction with 3,5-DMP. However, 1-ATMAH binds well in *cha* cavities and boosts *cha* cavities when used in conjunction with 3,5 DMP in JMZ-11D. In conclusion, the co-templating approach has enabled to control the cavity distribution of this complex family of intergrowth ABC-6 materials.

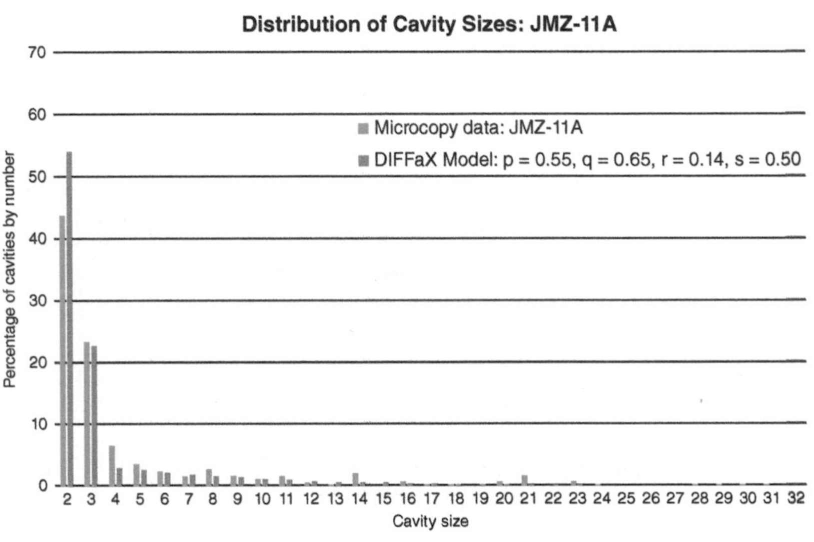

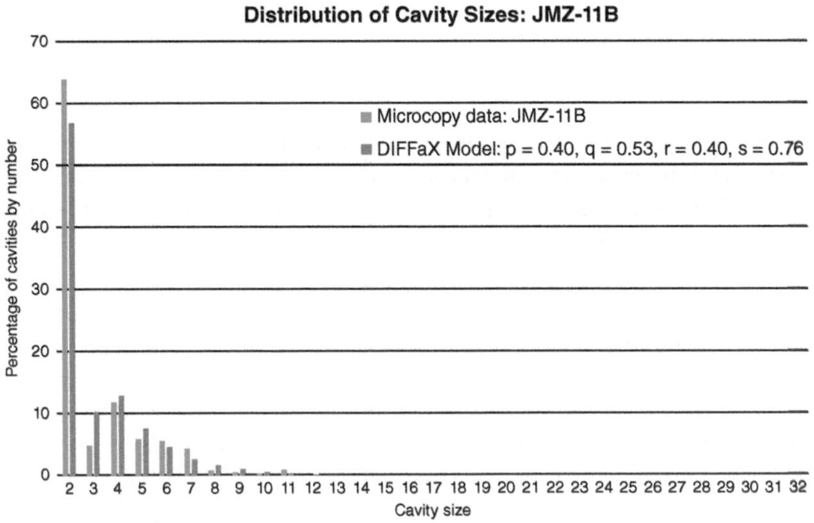

Figure 5.14 Distribution of cavity sizes from ADF STEM images (blue) and the those corresponding to the best DIFFaX match to the powder XRD pattern from calcined evacuated samples (orange) for the corresponding samples of JMZ-11. The mean value of the size of blocks of *cha* cavities is also significant. For the models deduced from DIFFaX matching to XRD this is 1.8 for JMZ-11A, 2.5 for JMZ-11B, 1.1 for JMZ-11C, and 4.5 for JMZ-11D.

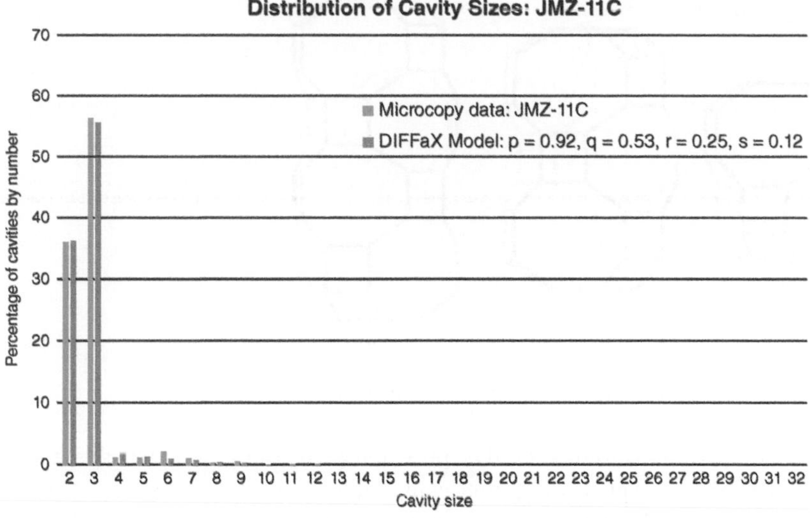

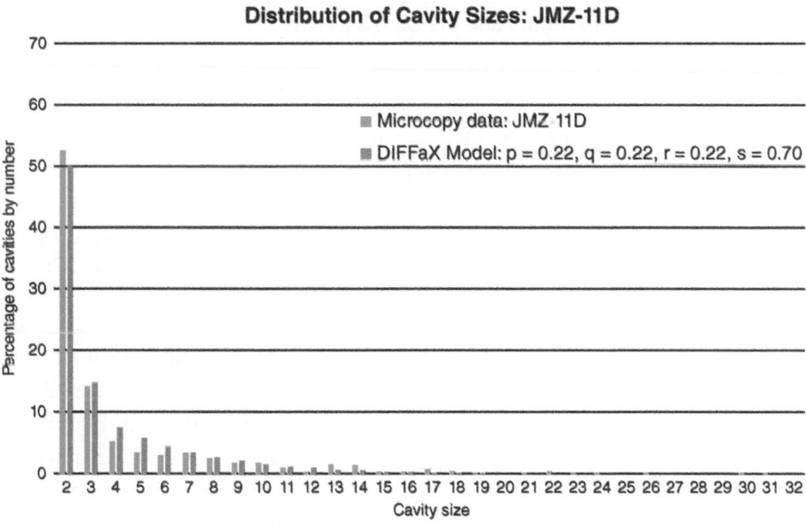

Figure 5.14 (Continued)

Table 5.6 Binding energies calculated using the COMPASS force field [71] within the Forcite module in the program Materials Studio [72].

	Binding energy (kJ mol^{-1} SiO_2)			
OSDA	***cha***	***aft***		***sfw***
	1 OSDA x cavity	1 OSDA x cavity	2 OSDA x cavity	2 OSDA x cavity
3,5-DMP	−26.4	−13.9	**−27.2**	−37.5
2,6-DMP	**−29.3**	−14.0	−24.9	**−41.2**
1-ATMAH	−30.0	–	–	–
BAI	–	−18.0	–	–

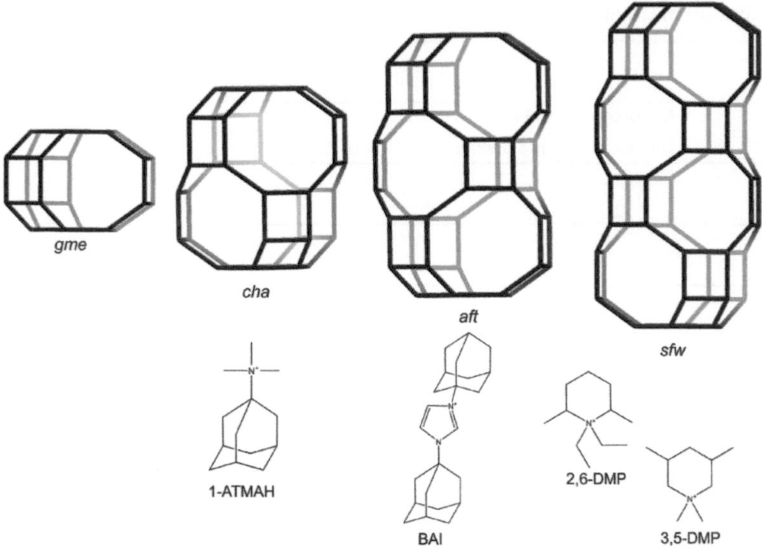

Figure 5.15 Types of cavities in JMZ-11 materials (top) and the OSDAs used in their preparation (bottom).

A summary of all co-templating examples discussed so far can be found in Table 5.7.

5.5 Discussion

Combining potential structure directing agents, including the use of two organic species, has considerable scope for the synthesis of materials with novel zeolite structures and improved compositions and textures. It can be a useful approach in both exploratory preparations as well as targeted syntheses. The trial-and-error approach has the potential to lead to the crystallization of new phases that contain multiple types of cavities that could be valuable for new applications, if the templates act cooperatively. However, a route to targeted co-templated materials would be much more efficient. An ideal designed synthesis would involve identification of an energetically feasible framework with more than one cavity type, assessment of its similarity to known structures, and computational insight, specifically into potential templates and more widely into the mechanisms of nucleation and crystallization.

Computational methods can contribute to improve the success rate of designed synthesis in several ways: enumeration and energetic feasibility of hypothetical frameworks; modeling template-framework interactions; understanding fundamental steps in crystal nucleation and growth; and in parallel, to apply machine learning (ML) methods to make predictive use of experimental results.

There are already extensive libraries of potential zeolite structures [9,38] for which the energetics have been evaluated using molecular mechanics programs such as GULP [75]. Subsets that contain more than one type of cage can readily be extracted: the ABC-6 system is a particularly fertile area for this approach [9].

Understanding the fit of individual SDAs in zeolite frameworks has been made routine by computational modeling, as described above, and a synthetic approach is most effective when it is informed by such calculations. There are undoubtedly cases where the consideration of multiple OSDAs together is essential in order to avoid misleading outcomes. A good example is provided by

Table 5.7 Materials synthesized through co-templating, listed with the SDAs that were used for the specific cages.

Zeolite topology	Material name and composition	SDA-Cage pair	Literature references
AST	AlPO-16	quinuclidine – *sod* F^- – *d4r*	Bennett et al. 1991 [13]
LTA	LTA-type aluminophosphate	K222 – *lta* F^- – *d4r*	Schreyeck et al. 1997 [73]
SAV	STA-7 (SAPO)	*tg.tg* TEA^+ – small cavity cyclam – large cavity	Castro et al. 2007 [10]
	Cu-SAPO STA-7	*tg.tg* TEA^+ – small cavity Cu-cyclam – large cavity	Picone et al. 2011 [28]
ERI	UZM-12 (aluminosilicate)	$[Me_3N(CH_2)_4NMe_3]^{2+}$ or $[Me_3N(CH_2)_6NMe_3]^{2+}$ – *eri* K^+ – *can*	Miller et al. 2007 [74]
		$[Me_3N(CH_2)_xNMe_3]^{2+}$ (x = 4–6), $[Me_2EtN(CH_2)_yNMe_2Et]^{2+}$ (y = 4–6), $[MeEt_2N(CH_2)_zNMeEt_2]^{2+}$ (z = 4–5), diMePyrrolidinium-Cn^{2+} (n = 4–6) and diMePiperidinium-Cm^{2+} (m = 4–5) – *eri* K^+ – *can*	Lee et al. 2010 [11]
KFI	STA-14 (MgAPO and SAPO)	K222 – *lta* TEA^+ – *pau*	Castro et al. 2009 [26]
AFX	SAPO-56	diDABCO-$C4^{2+}$ – *aft* TrMA – *gme*	Turrina et al. 2016 [31]
	Aluminosilicate, P-modified zeolites	$TEBOP^{2+}$, Et_4P^+ – *aft* Na^+, Me_4N^+, Me_4P^+ – *gme*	Tsunoji et al. 2021 [19]
SFW	STA-18 (SAPO)	diDABCO-$C6^{2+}$ – *sfw* TrMA – *gme*	Turrina et al. 2016 [31]
SWY	STA-20 (SAPO)	diDABCO-$C6^{2+}$ – *swy* TrMA – *gme* and *swy*	Turrina et al. 2017 [15]
	STA-30 (aluminosilicate)	diDABCO-Cn^{2+} (n = 7–10) – *sfw* K^+ – *can*	Chitac et al. 2021 [21]
CHA/AFX intergrowth	ZTS-1/ZTS-2 (aluminosilicate)	N,N,N-trimethyl-1-adamantylammonium – *cha* 1,10-(1,4-butanediyl)bis(1-azonia-4-azabicyclo[2,2]octane – *aft*	Naraki et al. 2017 [55]
ABC-6 *d6r* structure	JMZ-11	1-ATMAH – *cha* BAI – *aft* 2,6-DMP^+ – *sfw* 3,5-DMP^+ – *sfw*	Shannon et al. 2020 [58]

the early work of Lewis et al. [76]. They used force field calculations to show the importance of considering more than one template to rationalize why TPA^+ synthesizes the MFI structure, while tetrabutylammonium (TBA^+) synthesizes MEL. A single TBA^+ cation has a good interaction energy with the MFI framework, much better than TPA^+ (Table 5.8). However, this result does not consider how the OSDAs might pack inside the host structure. As soon as additional OSDAs are included in the calculation, it becomes clear that a second TBA^+ cation destabilizes the first one due to steric considerations and they cannot pack efficiently in the MFI structure without incurring a very significant energetic penalty (Table 5.9).

Where packing of the OSDAs may give rise to close contact, therefore, multiple OSDAs must be taken into consideration in the models used. This is likely to be much more of an issue where the framework has interconnecting channels rather than, for example, large cavities that contain a single OSDA. However, even when modeling OSDAs within cavities, it is important that the molecule remains within the confines of the cavity. In the co-templating of STA-7 highlighted earlier in this chapter, one of the reasons that DIPA is not a good OSDA to pair with cyclam for the synthesis of the target, is that the DIPA molecule protrudes outside of its target *pau* cavity, destabilizing the cyclam in the adjacent cavity.

Most force field-based calculations employ a short-range cut-off, typically at around 8–10 Å, so that any short-range interactions between atoms, such as van der Waals forces, are considered to be zero for atoms that exceed the specified cut-off value. However, it should be borne in mind that charged OSDAs will exert electrostatic interactions that have long-range effects. This is good justification for looking at multiple templates for charged OSDAs, although the whole question of charge distributions is a difficult one because we often do not know the exact distribution of Al ions in a zeolite structure, Again, this may have an important effect on the interaction between the framework and the OSDA.

The work discussed in this chapter has concentrated on the design of templates for specific targets. Whilst it can be seen that there have been significant successes resulting from this approach, it does have disadvantages. In particular, the use of computationally designed OSDAs for a specific

Table 5.8 Non-bonded energies for single molecules of TPA and TBA in the MFI and MEL frameworks.

Template/framework	Stabilization energy/kJ mol^{-1}
TPA^+/MFI	−133.9
TBA^+/MFI	−165.5
TBA^+/MEL	−159.5
TPA^+/MEL	−119.9

Table 5.9 Stabilization energy of template/framework combinations when two OSDA molecules are included at adjacent intersection sites.

Template/framework	Stabilization energy/kJ mol^{-1}
TPA^+/MFI	−29.7
TBA^+/MFI	+14.9
TBA^+/MEL	−18.3
TPA^+/MEL	−8.5

target can often result in the synthesis of competing frameworks. One way of combatting this is to use high performance computing to run high-throughput screening [77]. This is a "brute force" approach where large numbers of calculations are run to determine the binding energy of candidate OSDAs within a large number of structures in order to improve the likelihood of finding a suitable OSDA to synthesize the target. Whilst this does potentially improve the accuracy of template design, the emphasis is still on the calculation of binding energies, which omits a large number of additional variables that may affect the suitability of the OSDA for the desired structure. These other variables, such as Si/Al ratio or inorganic cations, may have an impact on the final product or the necessary synthesis conditions [78].

Including additional components into calculations, such as water or inorganic ions, has been demonstrated in several studies and often yields a valuable insight into their role in the synthesis process. A good example is provided by the work of Gómez-Hortigüela et al. who have developed a protocol to give an insight into the competition between the OSDA and water during synthesis [79]. One interesting finding from this study is that in the case of using triethylamine (TrEA) in the synthesis of the AFI structure, large amounts of water are incorporated inside the framework. This is attributed to the relatively low interaction between TrEA and the framework, leading to higher incorporation of water molecules to provide the necessary additional thermodynamic stabilization needed for the structure to crystallize. In effect, water is acting co-operatively with the OSDA to direct the formation of the product, thus highlighting the importance of the inclusion of water in this particular system.

Whilst it is possible to generate models with increasing numbers of additional components, some judgement needs to be made as to whether or not the additional time and effort required to increase the complexity of the model is necessary. For example, in designing a new OSDA for a target material, it seems reasonable to try to maximize the interaction between the OSDA and the framework without relying on additional interactions via occluded water molecules. However, such approximations could lead to incorrect conclusions where subtle effects from water or inorganic ions may have a strong influence on the experimental outcome. Comparisons between experiment and theory will continue to inform theoreticians about the limitations of the approximations that are made. As time evolves, more sophisticated models will become more routine as we seek to understand computationally and predict more subtle effects caused by the various components in the synthesis process. This could relate to the initial zeolite oligomerization steps in silicate and aluminosilicate systems [80] and the influence of alkylammonium cations to determine which oligomeric species are present [81].

In parallel, it is important to make use of the massive amounts of experimental data available from some 70 years of zeolite synthesis. Not only will this provide a knowledge of competing phases, but could also provide a starting point for suitable synthetic conditions (gel composition, temperature, time) which remain difficult to predict through general principles. One way to extract this data can be via the developing field of machine learning. Despite the relatively recent inception of artificial intelligence (AI), or machine-learning, in zeolite synthesis, there are already a number of highly influential contributions which will be discussed in the subsequent chapters of this book. The promising progress in this direction is demonstrated by the analysis of the synthesis feasibility of hypothetical structures, published by Li et al., the extraction of information and trends in zeolite synthesis by Jensen et al., or the work by Muraoka et al., in which they extract links between synthesis and structure descriptors based on previously published experimental work [82–84].

Large databases of hypothetical structures are now available to provide a wealth of targets for novel synthesis [38,85]. Advances in the use of software, such as that written by Jensen et al., show that existing literature can be analyzed in order to find zeolite structures and reaction conditions to form accessible databases to aid in material design [83]. Similarly, a database has been created for MOFs by Li et al. that scans the Cambridge Structural Database to identify desired targets [86]. Tools such as the Python library STK created by the Jelfs group and the RDkit have made the creation of machine learning software increasingly accessible [87,88].

The review from 2019 by Moliner et al. describes the current state of machine learning and its potential in solving the complex problem of zeolite synthesis [78]. The future advancements possible through machine learning could prove to be a useful tool coupled with the concept of co-templating that might allow researchers in the future to understand how the synthetic conditions affect and interact with the SDAs to lead to the formation of any desired structure. The realm of machine learning is beyond the scope of this chapter and will be discussed later in the book.

5.6 Conclusions

Co-templating of zeolites, in the strict sense defined here, involves the structure direction of at least two cavity types in the same material by well-fitting organic or inorganic species. This has been achieved for a relatively small number of crystallographically well-ordered small-pore zeotypes (aluminophosphates LTA, SAV, KFI, SWY) and zeolites (ERI, SWY, and possibly PST-20 and AFT). Nevertheless, together with advances in structure prediction and computational modeling, it does offer potential routes to materials with multi-cage topology types either not yet observed or with novel framework compositions and so offers an extension to the highly effective single OSDA approach described elsewhere in this book.

It may be, however, that the most widespread application of the co-templating method can in future be in the synthesis of intergrowth structures containing more than one cage type, readily achieved in zeolites as the results of stacking disorder of structural layers. Such structures are expected to have catalytic properties that are intimately related to the distribution of the cages throughout individual crystallites. Now that advances in structural techniques and analysis enable such materials to be accurately characterized and consequently readily distinguished, this opens up a wide compositional space for modeling synthetic and catalytic studies

Acknowledgments

Ruxandra G. Chitac thanks the University of St Andrews and Johnson Matthey for funding. The authors gratefully acknowledge the EPSRC (Designed Synthesis of Zeolites: EP/S016201/1, EP/S016147/1). Paul A. Wright acknowledges the Royal Society (Industry Fellowship: INF\R2\192052) for funding.

References

1 Dusselier, M. and Davis, M.E. (2018). Small-pore zeolites: synthesis and catalysis. *Chem. Rev.* 118: 5265–5329.

2 Kemp, K.C., Min, J.G., Choi, H.J. et al. (2020). Small gas adsorption and separation in small-pore zeolites BT: new developments. In: *Adsorption/Separation of Small Molecules by Zeolites* (ed.S. Valencia and F. Rey), 1–30. Cham: Springer International Publishing.

3 IZA database http://www.iza-structure.org/databases (accessed 29 September 2021).

4 Robson, H. and Lillerud, K.P. (ed.) (2001). *Verified Syntheses of Zeolitic Materials*. Elsevier.

5 Zones, S.I. (2007). Preparation of molecular sieves using a structure directing agent and an N,N,N-Trialkyl benzyl quaternary ammonium cation. US 8,007,763 B2.

6 Corma, A., Rey, F., Rius, J. et al. (2004). Supramolecular self-assembled molecules as organic directing agent for synthesis of zeolites. *Nature* 431: 287–290.

7 Park, M.B., Jo, D., Jeon, H.C. et al. (2014). Zeolite synthesis from a charge density perspective: the charge density mismatch synthesis of UZM-5 and UZM-9. *Chem. Mater.* 26: 6684–6694.

8 Barrett, P.A., Valencia, S., and Camblor, M.A. (1998). Synthesis of a Merlinoite-type zeolite with an enhanced Si/Al ratioviapore filling with tetraethylammonium cations. *J. Mater. Chem.* 8: 2263–2268.

9 Li, Y., Li, X., Liu, J., Duan, F. et al. (2015). *In silico* prediction and screening of modular crystal structures via a high-throughput genomic approach. *Nat. Commun.* 6: 8328.

10 Castro, M., Garcia, R., Warrender, S.J. et al. (2007). Co-templating and modelling in the rational synthesis of zeolitic solids. *Chem. Commun.* 33: 3470.

11 Lee, J.H., Park, M.B., Lee, J.K. et al. (2010). Synthesis and characterization of ERI-Type UZM-12 zeolites and their Methanol-to-Olefin performance. *J. Am. Chem. Soc.* 132: 12971–12982.

12 Zhou, F., Tian, P., Liu, Z. et al. (2007). Synthesis of ZSM-34 and its catalytic properties in Methanol-to-Olefins reaction. *Chinese J. Catal.* 28: 817–822.

13 Michael Bennett, J. and Kirchner, R.M. (1991). The structure of As-synthesized AIPO4-16 determined by a new framework modeling method and rietveld refinement of synchrotron powder diffraction data. *Zeolites* 11: 502–506.

14 Patarin, J., Schott-Darle, C., Le Goff, P.Y. et al. (1995). Rietveld refinement of the tetragonal variant of AlPO4-16 prepared in fluoride medium. *Stud. Surf. Sci. Catal.* 98: 256–257.

15 Turrina, A., Garcia, R., Watts, A.E. et al. (2017). STA-20: an ABC-6 zeotype structure prepared by co-templating and solved via a hypothetical structure database and STEM-ADF imaging. *Chem. Mater.* 29: 2180–2190.

16 Yamasaki, Y., Tsunoji, N., Takamitsu, Y. et al. (2016). Synthesis of phosphorus-modified small-pore zeolites utilizing tetraalkyl phosphonium cations as both structure-directing and phosphorous modification agents. *Microporous Mesoporous Mater.* 223: 129–139.

17 Kakiuchi, Y., Yamasaki, Y., Tsunoji, N. et al. (2016). Synthesis of phosphorus-modified AEI zeolites derived by the dual-template method as a durable catalyst with enhanced thermal/hydrothermal stability for selective catalytic reduction of NO_x by NH_3. *Chem. Lett.* 45: 122–124.

18 Di Iorio, J.R., Li, S., Jones, C.B. et al. (2020). Cooperative and competitive occlusion of organic and inorganic structure-directing agents within chabazite zeolites influences their aluminum arrangement. *J. Am. Chem. Soc.* 142: 4807–4819.

19 Tsunoji, N., Tsuchiya, K., Nakazawa, N. et al. (2021). Multiple templating strategy for the control of aluminum and phosphorus distributions in AFX Zeolite. *Microporous Mesoporous Mater* 321: 111124.

20 Wu, L., Degirmenci, V., Magusin, P.C.M.M. et al. (2012). Dual template synthesis of a highly mesoporous SSZ-13 zeolite with improved stability in the Methanol-to-Olefins reaction. *Chem. Commun.* 48: 9492.

21 Chitac, R.G., Bradley, J., McNamara, N.D. et al. (2021). Designed synthesis of STA-30: a small-pore zeolite catalyst with topology type SWY. *Chem. Mater.* 33: 5242–5256.

22 Lewis, G.J., Miller, M.A., Moscoso, J.G. et al. (2004). Experimental charge density matching approach to zeolite synthesis. *Stud. Surf. Sci. Catal.* 154 (A): 364–372.

23 Seo, S., Ahn, N.H., Lee, J.H. et al. (2019). Combined alkali-organoammonium structure direction of high-charge-density heteroatom-containing aluminophosphate molecular sieves. *Angew. Chemie Int. Ed.* 58: 9032–9037.

24 Wright, P.A. (2008). *Microporous Framework Solids*, Cambridge, 443. UK: Royal Society of Chemistry.

25 Wright, P.A., Maple, M.J., Slawin, A.M.Z. et al. (2000). Cation-directed syntheses of novel zeolite-like metalloaluminophosphates STA-6 and STA-7 in the presence of azamacrocycle templates. *J. Chem. Soc. Dalt. Trans.* 8: 1243–1248.

26 Castro, M., Warrender, S.J., Wright, P.A. et al. (2009). Silicoaluminophosphate molecular sieves STA-7 and STA-14 and their structure-dependent catalytic performance in the conversion of methanol to olefins. *J. Phys. Chem. C* 113: 15731–15741.

27 Garcia, R., Philp, E.F., Slawin, A.M.Z. et al. (2001). Nickel complexed within an Azamacrocycle as a structure directing agent in the crystallization of the framework metalloaluminophosphates STA-6 and STA-7. *J. Mater. Chem.* 11: 1421–1427.

28 Picone, A.L., Warrender, S.J., Slawin, A.M.Z. et al. (2011). A co-templating route to the synthesis of Cu SAPO STA-7, giving an active catalyst for the selective catalytic reduction of NO. *Microporous Mesoporous Mater.* 146: 36–47.

29 Turrina, A., Eschenroeder, E.C.V., Bode, B.E. et al. (2015). Understanding the structure directing action of copper–polyamine complexes in the direct synthesis of Cu-SAPO-34 and Cu-SAPO-18 catalysts for the selective catalytic reduction of NO with NH_3. *Microporous Mesoporous Mater.* 215: 154–167.

30 Martínez-Franco, R., Moliner, M. et al. (2014). Direct synthesis design of Cu-SAPO-18, a very efficient catalyst for the SCR of NO_x. *J. Catal.* 319: 36–43.

31 Turrina, A., Garcia, R., Cox, P.A. et al. (2016). Retrosynthetic co-templating method for the preparation of silicoaluminophosphate molecular sieves. *Chem. Mater.* 28: 4998–5012.

32 Lobo, R.F., Zones, S.I., and Medrud, R.C. (1996). Synthesis and Rietveld refinement of the small-pore zeolite SSZ-16. *Chem. Mater.* 8: 2409–2411.

33 Davis, T.M., Liu, A.T., Lew, C.M. et al. (2016). Computationally guided synthesis of SSZ-52: a zeolite for engine exhaust clean-up. *Chem. Mater.* 28: 708–711.

34 Shin, J., Jo, D., and Hong, S.B. (2019). Rediscovery of the importance of inorganic synthesis parameters in the search for new zeolites. *Acc. Chem. Res.* 52: 1419–1427.

35 Guo, P., Shin, J., Greenaway, A.G. et al. (2015). Family with expanding structural complexity and embedded isoreticular structures. *Nature* 524: 74–78.

36 Lee, H., Shin, J., Choi, W. et al. (2018). PST-29: a missing member of the RHO family of embedded isoreticular zeolites. *Chem. Mater.* 30: 6619–6623.

37 Shin, J., Xu, H., Seo, S. et al. (2016). Targeted synthesis of two super-complex zeolites with embedded isoreticular structures. *Angew. Chemie Int. Ed.* 55: 4928–4932.

38 Pophale, R., Cheeseman, P.A., and Deem, M.W. (2011). A database of new zeolite-like materials. *Phys. Chem. Chem. Phys.* 13: 12407–12412.

39 Wilson, S.T., Broach, R.W., Blackwell, C.S. et al. (1999). Synthesis, characterization and structure of SAPO-56, a member of the ABC double-six-ring family of materials with stacking sequence AABBCCBB. *Microporous Mesoporous Mater.* 28: 125–137.

40 Zones, S.I. (1985). Zeolite SSZ-16. US patent 4508837A.

41 Park, M.B., Ahn, N.H., Broach, R.W. et al. (2015). Crystallization mechanism of zeolite UZM-5. *Chem. Mater.* 27: 1574–1582.

42 Xie, D. and Jensen, K.O. (2019). Molecular sieve SSZ-112, its synthesis and use. US patent 10,343,927 B2.

43 Xie, D. (2021). Rational design and targeted synthesis of small-pore zeolites with the assistance of molecular modeling, structural analysis, and synthetic chemistry. *Ind. Eng. Chem. Res.* 60: 15403–15415.

44 Bennett, J.M., Kirchner, R.M., and Wilson, S.T. (1989). Synthesis and idealized topology of AIPO4-52, a new member of the ABC Six-ring family. In: *Zeolites: Facts, Figures, Future Part A: Proceedings of the 8th International Zeolite Conference* (ed.P.A. Jacobs and R.A. van Santen), vol. 49, 731–739. Elsevier.

45 Lok, B.M., Messina, C.A., Patton, R.L. et al. (1984). Silicoaluminophosphate molecular sieves: another new class of microporous crystalline inorganic solids. *J. Am. Chem. Soc.* 106: 6092–6093.

46 Martínez-Franco, R., Moliner, M., Thogersen, J.R. et al. (2013). Efficient one-pot preparation of Cu-SSZ-13 materials using cooperative OSDAs for their catalytic application in the SCR of NO_X. *ChemCatChem* 5: 3316–3323.

47 Martín, N., Boruntea, C.R., Moliner, M. et al. (2015). Efficient synthesis of the Cu-SSZ-39 catalyst for $DeNO_x$ applications. *Chem. Commun.* 51: 11030–11033.

48 Han, L., Ouyang, Y., Xing, E. et al. (2020). Enhancing hydrothermal stability of framework Al in ZSM-5: from the view on the transformation between P and Al species by solid-state NMR spectroscopy. *Chinese J. Chem. Eng.* 28: 3052–3060.

49 Tsunoji, N., Osuga, R., Yasumoto, M. et al. (2021). Controlling hydrocarbon oligomerization in phosphorus-modified CHA zeolite for a long-lived Methanol-to-Olefin catalyst. *Appl. Catal. A Gen.* 620: 118176.

50 Mitani, E., Yamasaki, Y., Tsunoji, N. et al. (2018). Synthesis of phosphorus-modified AFX zeolite using a dual-template method with tetraethylphosphonium hydroxide as phosphorus modification agent. *Microporous Mesoporous Mater.* 267: 192–197.

51 Mertens, M.M., Verberckmoes, A., Janssen, M.J. et al. (2010). Crystalline intergrowth material, its synthesis and its use in the conversion of oxygenates to olefins. US patent 2010/0028679 A1.

52 Skeels, G.W., Sears, M., Bateman, C.A. et al. (1999). Synthesis and characterization of Phi-Type zeolites LZ-276 and LZ-277: faulted members of the ABC-D6R family of zeolites. *Microporous Mesoporous Mater.* 30: 335–346.

53 Vennestrøm, P.N.R., Lundegaard, L.F., and Jensen, B. (2018). Novel disordered Abc-6 molecular sieve. US Patent WO 2018/086975 Al.

54 Szostak, R. and Lillerud, K.P. (1994). Babelite: the random member of the ABC-D6R family of zeolites. *J. Chem. Soc. Chem. Commun.* 20: 2357.

55 Naraki, Y., Ariga, K., Nakamura, K. et al. (2017). ZTS-1 and ZTS-2: novel intergrowth zeolites with AFX/CHA structure. *Microporous Mesoporous Mater.* 254: 160–169.

56 Breck, D.W. and Acara, N.A. (1960). Crystalline zeolite T. US patent 2,950,952.

57 Rubin, M.K., Rosinski, E.J., and Plank, C.J. (1978). Crystalline zeolite ZSM-34 and method of preparing the same. US patent 4,086,186.

58 Shannon, M., Turrina, A., and Yang, S. (2020). Molecular sieve intergrowths of Cha and Aft having an "Sfw-GME Tail," methods of preparation and use. US patent 2020/0316572 A1.

59 Chen, J., Thomas, J.M., Wright, P.A. et al. (1994). (SAPO-18): a new microporous solid acid catalyst. *Catal. Letters* 28: 241–248.

60 Mertens, M.M. (2009). Synthesis and use of AEI structure-type molecular sieves. US Patent WO 2009/117186 Al, 2009.

61 Li, J., Wei, Y., Liu, G. et al. (2011). Comparative study of MTO conversion over SAPO-34, H-ZSM-5 and H-ZSM-22: correlating catalytic performance and reaction mechanism to zeolite topology. *Catal. Today* 171: 221–228.

62 Dahl, I.M. and Kolboe, S. (1993). On the reaction mechanism for propene formation in the MTO reaction over SAPO-34. *Catal. Letters* 20: 329–336.

63 Marcus, D.M., Song, W., Ng, L.L. et al. (2002). Aromatic hydrocarbon formation in HSAPO-18 catalysts: cage topology and acid site density. *Langmuir* 18: 8386–8391.

64 Aguayo, A.T., Gayubo, A.G., Vivanco, R. et al. (2005). Role of acidity and microporous structure in alternative catalysts for the transformation of methanol into olefins. *Appl. Catal. A Gen.* 283: 197–207.

65 Wragg, D.S., Akporiaye, D., and Fjellvåg, H. (2011). Direct observation of catalyst behaviour under real working conditions with X-Ray diffraction: comparing SAPO-18 and SAPO-34 methanol to olefin catalysts. *J. Catal.* 279: 397–402.

66 Wang, Y., Chen, S.-L., Jiang, Y.-J. et al. (2016). Influence of template content on selective synthesis of SAPO-18, SAPO-18/34 Intergrowth and SAPO-34 molecular sieves used for methanol-to-olefins process. *RSC Adv.* 6: 104985–104994.

67 Treacy, M.M.J., Newsam, J.M., and Deem, M.W. (1991). A general recursion method for calculating diffracted intensities from crystals containing planar faults. *Proc. R. Soc. London. Ser. A Math. Phys. Sci.* 433: 499–520.

68 Smith, R.L., Sławiński, W.A., Lind, A. et al. (2015). Nanoporous intergrowths: how crystal growth dictates phase composition and hierarchical structure in the CHA/AEI system. *Chem. Mater.* 27: 4205–4215.

69 Zhao, D., Zhang, Y., Li, Z., Wang, Y. et al. (2017). Synthesis of SAPO-18/34 intergrowth zeolites and their enhanced stability for dimethyl ether to olefins. *RSC Adv.* 7: 939–946.

70 Guo, L., Zhu, W., Miao, P. et al. (2018). Intergrowth silicoaluminophosphate molecular sieves synthesized and their catalytic performances for methanol to olefins reaction. *Ind. Eng. Chem. Res.* 57: 10398–10402.

71 Sun, H., Jin, Z., Yang, C. et al. (2016). COMPASS II: extended coverage for polymer and drug-like molecule databases. *J. Mol. Model.* 22: 1–10.

72 *Dassault Systèmes BIOVIA Materials Studio* (2020). San Diego: Dassault Systèmes.

73 Schreyeck, L., D'Agosto, F., Stumbe, J. et al. (1997). Synthesis of the LTA-Type AlPO4 in the presence of the Diazapolyoxa Macrocycle "Kryptofix 222." *Chem. Commun.* 13: 1241–1242.

74 Miller, M.A., Lewis, G.J., Moscoso, J.G. et al. (2007). Synthesis and catalytic activity of UZM-12. *Stud. Surf. Sci. Catal.* 170: 487–492.

75 Gale, J.D. (1997). GULP: a computer program for the symmetry-adapted simulation of solids. *J. Chem. Soc. Faraday Trans.* 93: 629–637.

76 Lewis, D.W., Freeman, C.M., and Catlow, C.R.A. (1995). Predicting the templating ability of organic additives for the synthesis of microporous materials. *J. Phys. Chem.* 99: 11194–11202.

77 Lee, Y., Barthel, S.D., Dłotko, P. et al. (2018). High-throughput screening approach for nanoporous materials genome using topological data analysis: application to zeolites. *J. Chem. Theory Comput.* 14: 4427–4437.

78 Moliner, M., Román-Leshkov, Y., and Corma, A. (2019). Machine learning applied to zeolite synthesis: the missing link for realizing high-throughput discovery. *Acc. Chem. Res.* 52: 2971–2980.

79 Gómez-Hortigüela, L., Pérez-Pariente, J. et al. (2009). Insights into structure direction of microporous aluminophosphates: competition between organic molecules and water. *Chem. – A Eur. J.* 15: 1478–1490.

80 Freeman, E.E., Neeway, J.J., Motkuri, R.K. et al. (2020). Understanding initial zeolite oligomerization steps with first principles calculations. *AIChE J.* 66.

81 Ciantar, M., Trinh, T.T., Michel, C. et al. (2021). Impact of organic templates on the selective formation of zeolite oligomers. *Angew. Chemie* 133: 7187–7192.

82 Li, Y., Yu, J., and Xu, R. (2013). Criteria for zeolite frameworks realizable for target synthesis. *Angew. Chemie Int. Ed.* 52: 1673–1677.

83 Jensen, Z., Kim, E., Kwon, S. et al. (2019). A machine learning approach to zeolite synthesis enabled by automatic literature data extraction. *ACS Cent. Sci.* 5: 892–899.

84 Muraoka, K., Sada, Y., Miyazaki, D. et al. (2019). Linking synthesis and structure descriptors from a large collection of synthetic records of zeolite materials. *Nat. Commun.* 10: 4459.

85 Majda, D., Paz, F.A.A., Friedrichs, O.D. et al. (2008). Hypothetical zeolitic frameworks: in search of potential heterogeneous catalysts. *J. Phys. Chem. C* 112: 1040–1047.

86 Li, A., Bueno-Perez, R., Wiggin, S. et al. (2020). Enabling efficient exploration of metal–organic frameworks in the Cambridge structural database. *CrystEngComm* 22: 7152–7161.

87 Turcani, L., Berardo, E., and Jelfs, K.E. (2018). STK: a python toolkit for supramolecular assembly. *J. Comput. Chem.* 39: 1931–1942.

88 Landrum, G.A. (2022). RDKit: Open-source cheminformatics software. http://rdkit.org (accessed 1 March 2022).

6

Computer Generation of Hypothetical Zeolites

Estefania Argente[1], Soledad Valero[1], Alechania Misturini[2], Michael MJ. Treacy[3], Laurent Baumes[4], and German Sastre[2]

[1] *Valencian Research Institute for Artificial Intelligence (VRAIN); Universitat Politècnica de València, Valencia, Spain*
[2] *Instituto de Tecnología Química UPV-CSIC; Universitat Politècnica de València, Valencia, Spain*
[3] *Department of Physics, Arizona State University, Tempe, Arizona, USA*
[4] *ExxonMobil, Clinton, New Jercey, USA.*

6.1 Introduction

Zeolites account for 250 years of history. However, it was only from the 1930s that the science of zeolites really took off [1]. Experimentally speaking, the synthesis of a new zeolite framework that was designed computationally still represents a formidable challenge. For decades, synthesis of zeolites was considered to be essentially an unpredictable task, although it was obvious from the increasing number of new zeolites being synthesized, that the simple rule of tetrahedral coordination and microporosity allowed a very large family of materials. Crystallographic principles were the only guiding rule for a while to generate zeolites not yet synthesized, called hypothetical. In this sense, a large effort has been devoted to predict new zeolitic crystal structures. Started by the pioneering work of Wells [2], the topology-based approaches have been further applied by Smith [3] and O'Keeffe [4,5]. It was not until 1996 that two breakthroughs came onto the scene. Methods to represent zeolites by unique mathematical symbols were developed. Treacy et al. [6–8] represented zeolites first as directed graphs, with symbols determined from tables of operators, which were then embedded as a topology by simulated annealing. Delgado-Friedrichs et al. [9] represented structures via a unique key that contained all of the essential graph and topological information. This led to the development of the program *Systre* [10], which impressively solved the long-standing mathematical problem of determining graph uniqueness for three-dimensional (3D) periodic structures. Soon after, the first database of hypothetical zeolites was made public [8], with considerable functionalities to search according to structural and topological parameters.

This chapter focuses on the computational generation of inorganic crystal structures, specifically zeolites. Zeolites are crystalline, microporous materials, typically tectosilicates or aluminosilicates, commonly used as commercial adsorbents and catalysts, making a large family of, currently, 255 synthesized topologies [11].

The growing interest in the synthesis and characterization of new materials with specific physico-chemical functionalities and economic profitability has spanned the computational high-throughput (HT) search of hypothetically feasible materials. Therefore, in the increasing effort that the scientific community is making for the design and understanding of materials, the role of modeling methods is decisive. Synthesis techniques have improved in the last decade with the introduction of HT methods [12], making it possible to systematically vary some synthesis variables to cover a whole search area, increasing the probability of finding a new material. This has given rise to new structures, some of which are certainly complex, and whose structural resolution raises fresh challenges.

AI-Guided Design and Property Prediction for Zeolites and Nanoporous Materials, First Edition. Edited by German Sastre and Frits Daeyaert.

Sometimes the experimental difficulty lies not in the synthesis of a new zeolite but in its structural resolution. Zeolite Structure Determination (ZSD) [13] focuses on determining the structures of existing compounds using the zeolite information, such as space group, density, and cell parameters, obtained by the current experimental characterization techniques (X-ray diffraction, electron microscopy, etc.). As compared to other inorganic materials, structure determination of zeolites is more complicated because they are usually difficult to prepare as large single crystals. Without knowing the structure, new zeolites cannot be properly studied.

Several computational model-building methods have been used for ZSD, aiming to find the atom positions in the unit cell (UC). Algorithms usually work only with T-atoms, since they allow the reconstruction of the whole framework, moving atom positions until they respect all constraints defined, such as framework connectivity, bonding geometry, framework density, and/or certain agreement with experimental observations [13]. During the past three decades, two global optimization techniques have been developed: i) techniques based on the Monte Carlo method, such as ZEFSAII [14], FraGen [15], and SCIBS [6–8]; and ii) techniques based on genetic algorithms (GA), such as EZs [16], P-GHAZ [17], and zeoGAsolver [18,19]. GAs allow solving complex combinatorial problems, on which exploring all possibilities becomes a daunting problem.

In the following section, we will describe briefly how GAs work. Using zeoGAsolver as an example, we will also explain how they can be used for zeolite structure determination and prediction.

6.2 Genetic Algorithms

Genetic algorithms (GA) represent a subdomain of artificial intelligence techniques, and they employ adaptive techniques for the solution of search and optimization problems [20,21]. GAs can find specific sub-spaces on which solutions are more probably located. The basic principles of GAs were established by Holland [20], and have since been used in many application areas [22,23].

Unlike traditional optimization or search methods, GAs simultaneously explore different regions of probable solutions, not just one [24]. To do this, GAs work with a population of individuals, and in each step or generation of the GA, recombination (crossover and mutation) and selection operators are applied to this population, allowing to perform exploration and exploitation in the search space [25]. Exploration is the process of visiting completely new regions of a search space, while exploitation is the process of visiting those regions of a search space within the vicinity of previously visited points. Moreover, an evaluation or fitness function needs to be defined for the GA and applied to the individuals of the GA. This fitness function denotes the optimality of each individual. Figure 6.1 shows a basic evolutionary flow of GAs that maintain a fixed population size. Starting from an initial generation, the fitness function determines whether each individual can be selected to form new individuals in the selection process. The individuals of the next generation have a large probability of performing better and have a higher fitness than the previous one. Depending on the desired strategy, new individuals may replace the parents in the next generation, or compete with them in a selection process, in order to maintain a stable population. The individuals of the new generation are created using the combination of crossover and mutation operators. The crossover operator exchanges genetic information between the selected parents in order to give a better and promising offspring. The mutation operator randomly alters certain genes on the chromosomes (i.e., the codification of the individual) to maintain the genetic diversity of the population and avoid premature convergence of the GA to suboptimal solutions (i.e., convergence to local maxima).

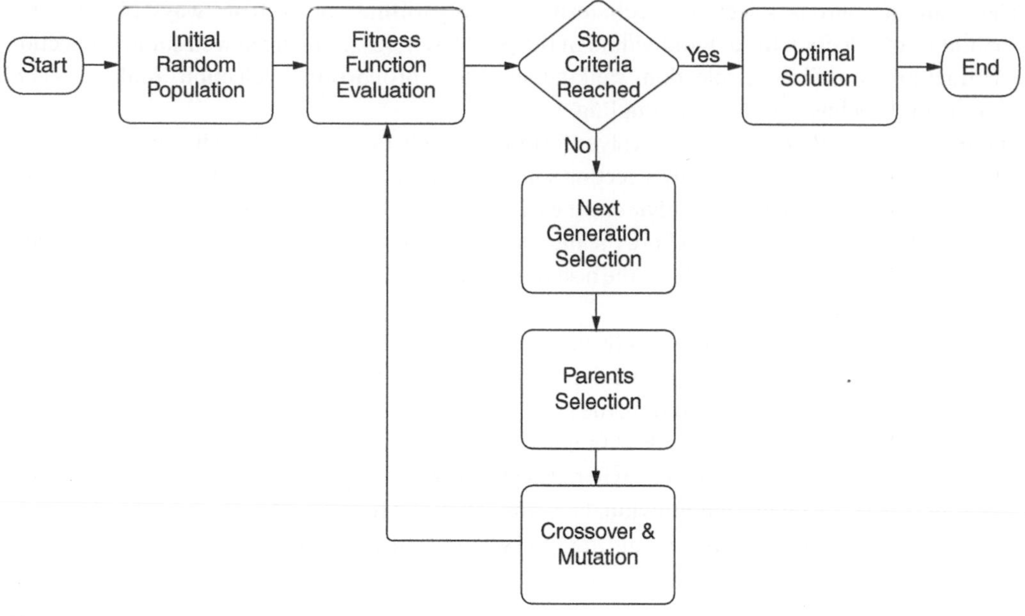

Figure 6.1 Evolutionary flow of genetic algorithms.

6.2.1 Codification of Genetic Algorithms

Genetic algorithms (GA) do not work directly with the parameters or variables to optimize, but rather with a codification of them. The way in which such coding is performed largely determines the choice of the operators to be used, as well as the final performance of the GA. Binary coding, based on fixed strings of ones and zeros, has been widely used since the beginning of GAs, as it normally leads to a very simple and efficient implementation of the mutation and crossover operators. The current trend is to use those non-binary representations that are more suitable for the problem to be addressed. In this sense, one of the most convenient possibilities for solving problems whose variables belong to continuous search spaces is the use of *real coding*, where each individual or chromosome is formed by a vector of real numbers, and each gene represents one of the variables to be optimized. GAs employing this type of coding have provided excellent results in solving various optimization problems in continuous domains [21,26].

6.2.2 Selection Operators for Genetic Algorithms

Throughout the process followed by a GA, it is often necessary to select individuals from those proposed by the algorithm, either to be recombined or to form part of the next generation. This selection can be made in different ways and according to different criteria, depending on the optimization strategy followed by the GA. Normally, the selection operator is employed to improve the average quality of a population. Thus, it determines the individuals of better quality (in terms of the fitness function), that have a greater chance of being selected for applying the recombination operators (crossover, mutation), and lead to the next generation. Therefore, this operator is focused on the exploitation of promising regions of the search space [25]. Moreover, when descendants do not replace parents and both are added to the next generation, the selection operator can be applied to adjust the size of the population obtained after the recombination operators, selecting those individuals who have a good balance between their aptitude and diversity.

There are numerous selection mechanisms and algorithms, as well as ways of assigning probabilities of selection to each individual in the population [27]. The most well-known selection algorithms are tournament selection, ranking selection, proportional selection, roulette wheel selection, and stochastic universal sampling.

The *tournament selection* [27,28] involves forming subsets of individuals (with fixed size) from the current generation, by randomly selecting some of them. For example, individuals can be taken five at a time, selecting only one individual at each time. To perform this selection, a "tournament" is performed within each subset, where the individuals are compared with each other, considering as winners those individuals that have the best fitness (which will form part of the group of selected individuals).

The *ranking selection* [29] allows effective control of the probability with which the best individuals are selected, compared to the average probability of selecting the rest of the individuals (selection pressure). In each generation, the individuals of the population are ordered according to their fitness, assigning a score to each of them. Thus, the worst individual is assigned a score of 1, while the best is assigned a score of N, where N is the size of the population. From this ranking, a probability p_i is assigned to each individual i by the following function, where $max + min = 2$ and $1 \leq max \leq 2$, with max and min being the lowest and highest values of the fitness among all individuals.

$$p_i = \frac{1}{N}\left(min + \frac{(max - min)(i-1)}{N-1}\right)$$

A variant of this type of selection is *ranking selection*, the rank-space method [30], which seeks to maintain a balance between the quality of the individuals selected and their diversity, to avoid excessive loss of genetic material. In this case, the score an individual receives is the result of the combination of the score received according to its fitness and diversity.

The *proportional selection* [20,22] method assigns selection probabilities according to the relative objective fitness of individuals. The probability p_s of an individual C_i being selected is calculated as:

$$P_s(C_i) = \frac{f(C_i)}{\sum_{j=1}^{N} f(C_j)}$$

where f is the fitness function used to determine the quality of each individual in the population under study. In this way, individuals with above-average fitness tend to be copied more times than those below average. Based on this proportional selection, there are two principal probability assignment ways: the roulette wheel selection and stochastic universal sampling.

The *roulette wheel selection* [22] is like playing a roulette wheel in a casino. A portion of the wheel is assigned to each individual of the population according to its fitness value, thus the individual with higher fitness occupies more space of the wheel, and vice versa. Each individual can be selected only once. Then, all individuals can be selected by randomly spinning the wheel N times.

The *stochastic universal sampling* (SUS) [31] is a sampling algorithm of a variation of roulette wheel selection with minimum spread and zero bias. In contrast with a single selection pointer employed in a roulette wheel mechanism, SUS uses M evenly spaced pointers to sample all the solutions with a single random spin, where M is the number of individuals required to be selected. The single random pointer 1 is produced in the interval [0, F/M], where F is the summation of all individual fitnesses. Then, M individuals are selected by M pointers, starting with point 1 and spaced by F/M. Therefore, the weaker members with low fitness in the population have probability to be chosen and the unfair feature of the fitness proportional selection is reduced.

Figure 6.2 shows the application of the roulette wheel selection and stochastic universal sampling to the same set of individuals, for selecting 4 individuals (M = 4). In the roulette wheel selection, the roulette will be spun 4 times, getting 4 selection pointers. In this stochastic universal sampling example, all the individuals are selected in a unique spin of the roulette. Four pointers are placed over the roulette, spaced "F/M" between them, to be selected. For this case, M = 4 and F = 100, thus the position of the first pointer is selected randomly in the interval [0, 100/4] (15 in this example).

6.2.3 Crossover Operators for Genetic Algorithms

Crossover operators have an exploiting nature, since their purpose is to exchange genetic material between selected individuals [32]. The philosophy followed tries to improve the population from genetic exchange among promising candidates, with the idea that their descendents will yield better results since they have combinations of genes from good individuals. Normally not all the individuals of a generation are crossed, but a crossover probability (P_{cross}) is applied to select, by some of the mechanisms presented above, the individuals that would act as progenitors. The way in which this operator is applied depends upon the representation or coding of the problem. Examples of crossover operators in real coding are: Wright's heuristic crossing [33], linear crossing [33], simple crossing [33,34], arithmetic crossing [34], discrete crossing [35], BLX-α crossover [21], line-extended crossing [35], intermediate extended crossing [35], FCB crossover [36], and linear BGA crossing [37].

BLX-α (in particular α = 0.5), logical FCB, and linear crossovers have become the best crossover operators for real coding GAs [38]. All these crossover operators consider the exploration intervals for obtaining offspring genes. In particular, BLX-α crosses two parents in order to obtain new

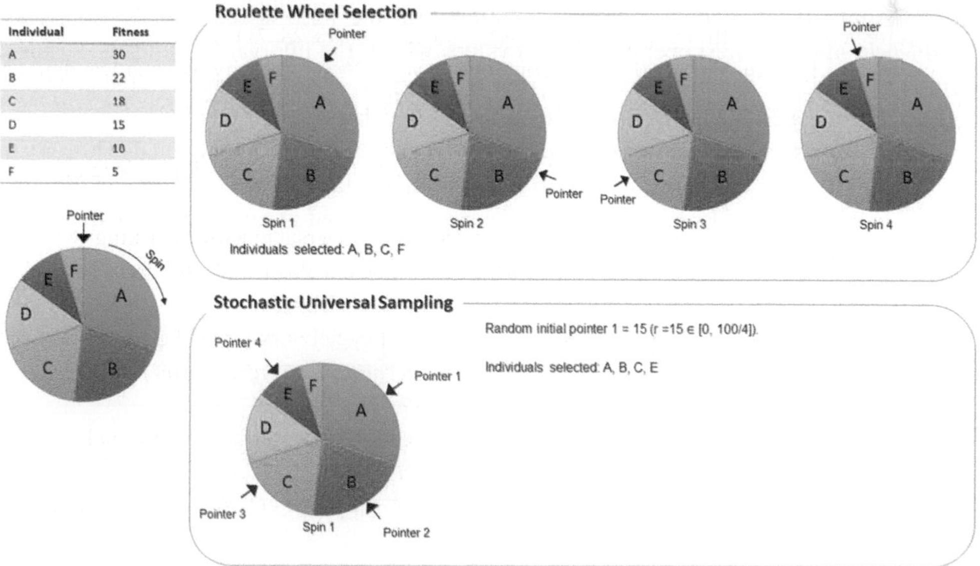

Figure 6.2 Application of the roulette wheel selection and stochastic universal sampling operators.

offspring and the degree of diversity introduced by this operator is determined by the α value (the higher the α value, the higher diversity). More specifically, from two parents $C_1 = (c^1_1 ... c^1_n)$ and $C_2 = (c^2_1 ... c^2_n)$, where c^i_x represents a gene x of the chromosome C_i, we obtain a descendant $H = (h_1,...,h_i,...,h_n)$ where each random h_i gene is assigned a value within the interval $[c_{min} - I \cdot \alpha, c_{max} + I \cdot \alpha]$, where $c_{max} = \max(c^1_i, c^2_i)$, $c_{min} = \min(c^1_i, c^2_i)$, $I = c_{max} - c_{min}$.

6.2.4 Mutation Operators for Genetic Algorithms

Mutation operators have an exploratory nature, which is intended to preserve the genetic diversity of the population, since they allow the appearance of new genetic material. Through the mutation, it is possible to explore new areas of the search space that have not appeared before. Basically, a mutation operator alters the genetic content of an individual at random. The way in which this alteration is carried out depends largely on the type of coding used, as well as its degree of incidence. Generally, the operator is applied following a given mutation probability (P_{mut}), further indicating the number of genes to be altered when it is applied. A high P_{mut} introduces too much noise into the search system, preventing the convergence of a GA, whereas a low P_{mut} keeps the balance between maintaining diversity within the population and the degree of convergence achieved by the GA.

In the case of binary coding, the gene or genes to be modified are randomly selected and their value is swapped, i.e., if the gene currently has a value of "0," it is assigned a "1" and vice versa [20,22]. A comparison of this type of operator and the crossover operator on binary coding can be seen in Figure 6.3.

When using real coding there are numerous methods for applying the mutation operator, such as aleatory mutation, non-uniform mutation, Mühlebein mutation, and Real Number Creep.

In *aleatory mutation* [34], the selected gene to mutate is assigned a uniform random r value within the gene domain. In *non-uniform* mutation [34], a random gene is selected and assigned a non-uniform random value, that depends on the current generation number, the maximum number of generations, and a shape parameter, which determines the degree of dependence on the number of iterations. This operator performs a uniform search within the initial space when the current generation number is small, transforming into a more localized search as generations pass.

The *Mühlebein mutation* [35] operator obtains values within the interval $(c_i - rang_i, c_i + rang_i)$, with a high probability of generating values close to c_i, where $c_i \subset (a_i, b_i)$ is the selected gene and $rang_i$ defines the mutation range, normally established as $0.1*(b_i - a_i)$.

The *Real Number Creep* [39] mutation operator allows generating new individuals close to an individual located at a good local maximum. To accomplish this task, the value of the genes in the chromosome is increased or decreased by a small amount obtained at random. The user should determine the maximum shift that is allowed.

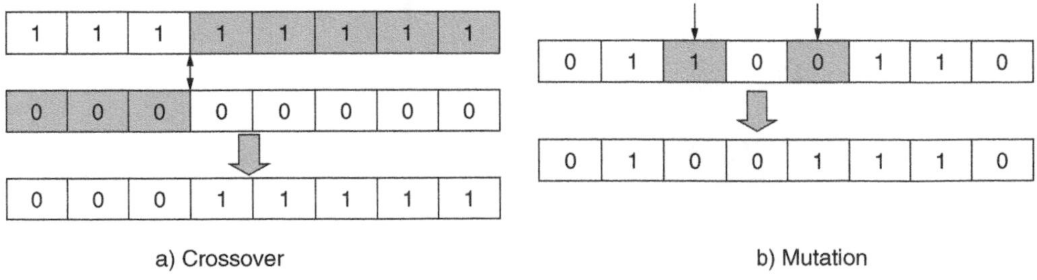

Figure 6.3 Example of basic crossover and mutation operators on binary chromosomes.

As we have seen, there are different selection, mutation, and crossover operators that can be applied in a genetic algorithm. In the next section, we will detail how GAs have been used in the field of zeolite structure determination and prediction.

6.3 Algorithms for Zeolite Structure Determination and Prediction

Like other solid-state crystalline materials, zeolite structures can be determined by techniques such as X-ray diffraction, electron diffraction, neutron diffraction, etc., as well as by computer-aided structure construction, or by a combination of these techniques [13]. In general, the zeolite structure model contains two pieces of information that will be used in a GA: i) the space group and cell parameters, which can be obtained from indexing the diffraction patterns; and ii) the atomic coordinates of the T (tetrahedral) atoms.

In comparison to X-ray crystallography and electron microscopy methods, model building methods aim to find the arrangement of the atoms of the structure directly in real space. During a model building procedure, individual atoms are moved together to form a series of feasible zeolite structures under the constraints of structure, connectivity, bond geometry, structure density, and/or agreement with experimental observations. The goal of these methods is to find the structure solution that is the most feasible or in agreement with experimental observations, among all the models constructed.

The so-called Zeolite Structure Prediction (ZSP) consists of the theoretical generation of a previously unknown structure. If the corresponding compound already exists, it is equivalent to Zeolite Structure Determination (ZSD), also known as a posteriori structure prediction. If it does not yet exist, ZSP can predict new zeolite structures with interesting microporous features, thus providing a huge pool of potential candidates for future-oriented synthesis. Figure 6.4 depicts the ZSD process.

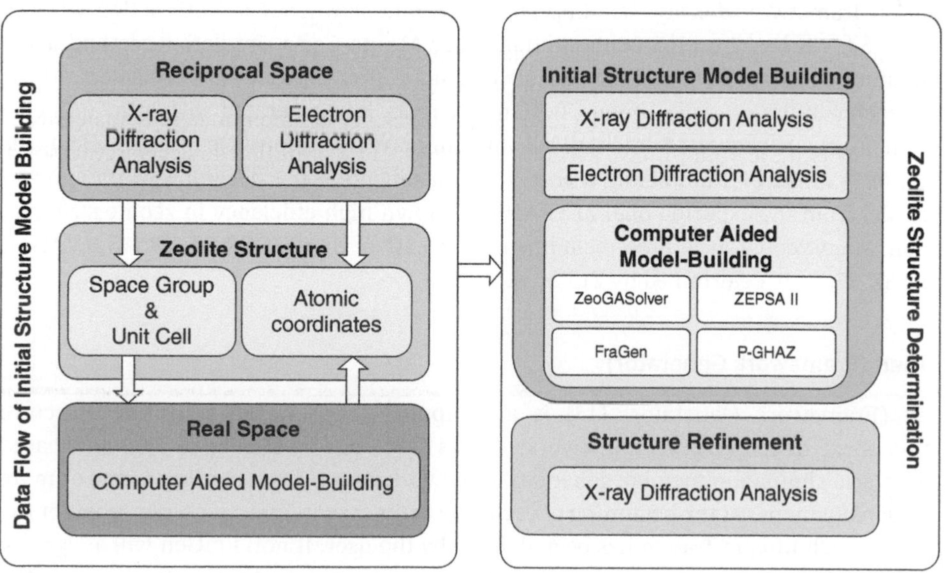

Figure 6.4 Zeolite Structure Determination process.

Table 6.1 General comparison of computer-aided approaches.

Software	Aim	Method	Results
ZEFSAII	ZSP and ZSD	MC, SA	Deem's Database Hypo. Structures (PCOD) ECR-9, UiO-6, UiO-7, ERS-7, MCM-47, SSZ-55, SSZ-77
FraGen	ZSP and ZSD	MC, PT	Hypothetical Zeolite Database ZSM-18, SSZ-51
SCIBS	ZSP	MC, SA	Atlas of Prospective Zeolite Structures ZSM-10
GRINSP	ZSP	MC	Predicted Crystallography Open Database
EZs	ZSP	GA	JBW, BIK, ABW, SOD, CHA
P-GHAZ	ZSP and ZSD	GA	ITE, MER, CHA, HZM#1, HZM#2, HZM#3
zeoGAsolver	ZSD	GA	ITE, ITW, AEI, ATS, CDO, 7 hypothetical zeolites

ZSP (Zeolite Structure Prediction), ZSD (Zeolite Structure Determination). Methods: MC (Monte Carlo); SA (Simulated Annealing); PT (Parallel Tempering); and GA (Genetic Algorithms).

During the last three decades, two different global optimization techniques have been developed: i) techniques based on different versions of the Monte Carlo (MC) method, such as Simulated Annealing (SA) [14,40] and Parallel Tempering (PT) [15]; and ii) techniques based on evolutionary theory, such as GAs [41] and differential evolution [42]. Computer-aided model building methods relevant to both zeolite structure determination (ZSD) and zeolite structure prediction (ZSP) are listed in Table 6.1.

6.3.1 ZEFSAII

ZEFSAII [14] is a free computer program developed by Deem et al., based on the Monte Carlo simulation method. It aims to generate the most feasible zeolite structures in real space under a set of restraints on framework density, bonding geometry, and Powder X-ray diffraction (PXRD) intensities. ZEFSAII is based on the configuration-based Monte Carlo simulation method using space group, unit cell parameters, and the number of unique T-atoms as input parameters. At the beginning of each simulation cycle, unique T-atoms are randomly generated within the unit cell, and equivalent atoms are generated automatically by symmetry operations. Next, ZEFSAII adjusts the locations of T-atoms by minimizing a cost function reflecting the deviation of the current structural model from the expected one. ZEFSAII has shown high efficiency in zeolite structure determination. Many zeolite structures, including ECR-9, UiO-6, UiO-7, ECR-34, ERS-7, MCM-47, SSZ-55, and SSZ-77, were determined by ZEFSAII.

6.3.2 FraGen (Framework Generator)

The FraGen (Framework Generator) [15] is a computer program for real-space structure determination of extended inorganic frameworks. It was developed by Li et al. in 2012 and, based on the Monte Carlo simulation method, generates framework structures according to one or more user-defined density maps. At the beginning of each simulation cycle, FraGen checks whether the site symmetry of each unique T-atom has been defined by the user. If not, FraGen will assign the site symmetry for each atom according to various forms of constraints. FraGen then adjusts the location of each atom by minimizing a cost function through parallel tempering.

ZEFSAII and FraGen have also been used for a priori ZSP working without any experimental input, thus being capable of generating millions of hypothetical zeolite frameworks.

6.3.3 SCIBS (Symmetry-Constrained Intersite Bonding Search)

SCIBS (Symmetry-Constrained Intersite Bonding Search) [6–8] enumerates all possible four-valent networks within each space group given the number of unique T-atoms. Using this program, a database hosted by the http://www.hypotheticalzeolites.net website [43] has been created and has been freely available since April 2004. The database was constructed using a symmetry-constrained inter-site link search method. This method lists all possible 4-connected networks within each type of space group given the number of unique tetrahedral vertices, n_T. With this technique, graphs of up to $n_T = 7$ have been achieved for some highly symmetric space groups. For this, a sophisticated simulated annealing strategy was implemented to embed graphs in real space as SiO_2 composition. In addition to simulated annealing, an adaptation of the GASP program by Wells [44] is applied in the early stages. This efficiently forces the 4-valent T-atoms into as regular a tetrahedron as possible. Also, a simple genetic algorithm is used in conjunction with the 12 parallel anneals of each structure (parallel tempering). In addition, a simple cost function favoring a regular tetrahedral arrangement of the neighboring silicon atoms, based on the function given by Boisen et al. [45], is employed to emulate the final zeolitic geometry. Low-energy candidates are then optimized with the General Utility Lattice Program (GULP) [46,47], which provides a more standardized framework energy.

The SCIBS method is guaranteed to find all possible 4-valent graphs, given the space group and the number of unique T-atoms. This inevitably gives rise to a combinatorial explosion of graphs as the number of unique T-atoms increases. Consequently, the results in different space groups grow at different rates (Figure 6.5). The number of low-energy topologies, i.e., topologies with regular SiO_4 tetrahedra, grows much more slowly. Nevertheless, over 10 million reasonably low-energy structures are listed in the database.

Building of the database is an inherently parallel process, with each space group and number of unique T-atoms being assigned to separate processors. The early database was built using 64 processors, running non-stop for over 2 years.

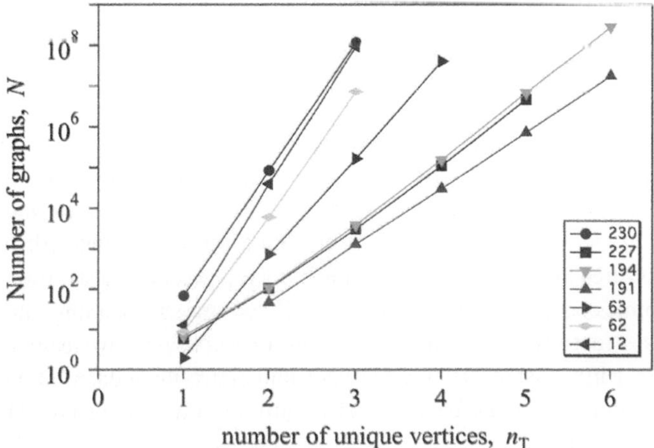

Figure 6.5 Combinatorial explosion of possible graphs with increasing number of unique tetrahedral atoms, n_T. *Source:* Delgado-Friedrichs et al. [48].

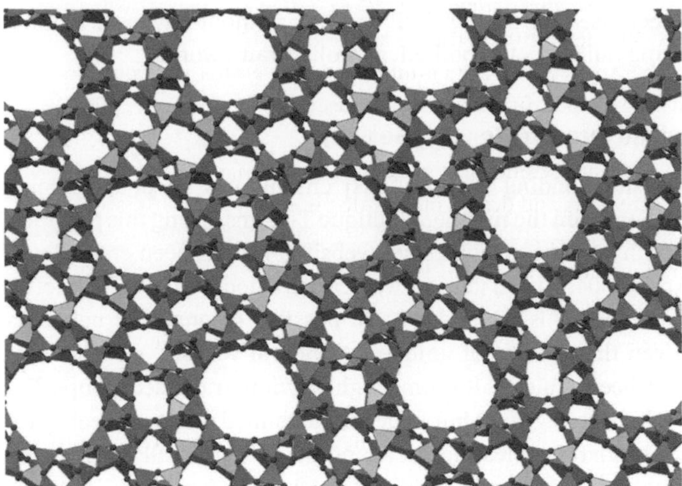

Figure 6.6 Layered zeolite structure 191_3_239. This has $n_T = 3$ unique tetrahedra in the layer group *p6/mmm*. This structure is 4-tetrahedra thick, and all tetrahedra are fully 4-valent.

The SCIBS method lends itself easily to the problem of finding layered, tubular, and ball zeolite structures. These are fully-connected 4-valent structures. Figure 6.6 shows, as an example, a layered zeolite that is 4-tetrahedra thick. All tetrahedra are fully 4-connected with no −OH hydroxy group terminations molecules needed at the layer surfaces. There are 80 layer groups (as opposed to 230 space groups: there are no cubic layer groups). A database of layered structures is under construction [43].

All of the simpler zeolites proposed to the Structure Commission of the International Zeolite Association ($n_T \leq 3$) since 2005 were found to be already in the hypothetical zeolite structure database. Of course, the ultimate goal is to find ways to guide the synthesis chemists to synthesize the attractive candidates listed in the database.

It is worth emphasizing that a graph can have more than one topology. Usually, the various topologies are not all of equal low energy, and only the lowest-energy topology is kept in the database.

6.3.4 TTL GRINSP (Geometrically Restrained Inorganic Structure Prediction)

GRINSP (Geometrically Restrained Inorganic Structure Prediction) [49] allows the exploration of the possibilities of occurrence of 3-, 4-, 5-, and 6-connected 3D networks, thus facilitating the generation of all possible framework structures constructed exclusively from corner-sharing polyhedra. The cost function employed by the algorithm is based on matching the interatomic distances of the model with the ideal distances provided by the user. The Monte Carlo algorithm first finds the selected candidate structures after verification of the expected geometry, and then optimizes the cell parameters and atomic coordinates. Using these predicted models, the software produces the characteristics of the isostructural compounds that would be obtained by cationic substitutions. Thanks to this algorithm, a huge list of CIF files of hypothetical polymorphs of boron oxide, zeolites, aluminum fluorides, among others, has been compiled and is freely available in the PCOD (Predicted Crystallography Open Database).

Despite the importance of Zeolite Structure Determination in the field of chemical industry, and GAs being a technique that seems suitable for this domain, we have only found three relevant applications of GAs for ZSD: EZs, P-GHAZ, and zeoGAsolver.

6.3.5 EZs (Exclusive Zones)

EZs (Exclusive Zones) [16] is based on a previous work by Woodley et al. [41], also based on GAs, but focused on the prediction of dense inorganic oxides and fluorides. EZs was presented in 2004, when Woodley et al. applied it in zeolite structure prediction [16]. EZs is a tool that imposes into the crystalline space some zones in which atoms are forbidden to enter, this leading to simulate the micropore and small voids in zeolites. EZs divide the unit cell into $2n$ small spaces by n Cartesian Grids. The lattice points representing the small boxes nearby are set as genes of the structure. Thus, EZs simplifies the spatial relationship of atoms of the zeolite into $23n$ independent lattice points to facilitate gene swapping. The fitness function (or cost function) consists of a weighted series of additive terms, comprising: i) the discrepancy between the ionic charge and the sum of the bond valences between each ion and its nearest neighbors; ii) a term based on the Coulombic interaction energy between similar ions; iii) the discrepancy between the expected and actual first coordination numbers of the candidate; iv) a term based on the Coulombic interaction energy between anions and cations; v) a bond valence term between similarly charged ions; and vi) the discrepancy between the expected and actual second coordination numbers of the candidate.

Initially, several random candidate structures are created using the known unit cell dimensions and constituent ions. Each candidate is then evaluated using the cost function. The selection of 2 M candidates is statistically weighted using the cost function values of the candidates, such that the better candidates are exponentially more likely to be chosen. These 2 M parent candidates are then paired up for crossing with a probability of P_{cross} and mutating with a probability P_{mut}, to create two new candidate structures referred to as children. Since a binary coding is used in EZs, two points are randomly chosen along a parent's chromosome and the middle section is swapped when crossing to produce children's chromosomes.

6.3.6 P-GHAZ (Parallel Genetic Hybrid Algorithm for Zeolites)

P-GHAZ (Parallel Genetic Hybrid Algorithm for Zeolites) [17] improves Baumes et al. [50] approach, using a modified modelization of the objective function to find hypothetical zeolite structures close to the thermodynamic feasibility criterion. A population made of random atoms is initialized. Each individual from this population represents the initial atoms coordinates (x,y,z) belonging to the asymmetric unit (AU). The selection operator uses a tournament between the actual and a random individual. The P-GHAZ algorithm manages only T-atom positions. The oxygen atoms are added at the end of the algorithm at the center of two T-T connected atoms. At each generation, a crossover operator and a heuristic mutation are applied. P-GHAZ uses two types of crossover operators: i) one-point crossover, that performs a crossover between the atoms of two parents at a random position; and ii) a uniform crossover, that performs a crossover between the atoms of two parents with a ratio of 50%. Moreover, it applies a mutation operator that allows an atom randomly selected to be moved. The atom is moved along the axes (x,y,z) and the algorithm retains the best movement. The individuals are selected for the mutation with a defined probability. The fitness function considers the following aspects: i) the connection between all atoms in the unit cell, as all generated T-atoms must be properly connected to each other through an O atom to create a viable zeolite; ii) the formation of tetrahedra, as each T-atom needs 4 bonds with the other T-atoms to be stable; iii) the creation of rings formed by X-tetrahedra or X-rings, as zeolites are characterized by these X-rings, which are the origin of the porosity and therefore of their ability to adsorb molecules, so these X-rings have to be large enough to be interesting; iv) the existence of 3-rings, which consists of three successive tetrahedral rings that are very unlikely to appear in pure

Table 6.2 Comparison of GA-based ZSP approaches.

#T in AU	zeoGAsolver	EZs	P-GHAZ
1 T	LTA NPO (*) SOD (*)	SOD CHA ABW FAU	CHA (*) DFT (*)
2 T	NPT BIK AHT	BIK JBW	
3 T	AEI NSI ATS ITW		~~ITW~~
4 T	CDO ITE RTH AFN		ITE ~~RTH~~
5 T	~~STF~~ ~~CFI~~		CFI ~~STF~~ ~~ISV~~

The (*) symbol indicates a zeolite found as a side product when searching for another target. A crossed out target indicates that the tool has not been able to find it.

silica or aluminosilicate systems, thus being penalized if detected; v) the distances between atoms, as two T-atoms must be at the right distance (between 2.4 Å and 3.6 Å); and vi) the TTT angles between T-atoms. These last two penalties define the energetic stability of the potential zeolite.

6.3.7 zeoGAsolver

zeoGAsolver [18,19], based on genetic algorithms, uses domain-dependent crossover and selection operators for zeolite structure determination. Using the density, cell parameters, and symmetry of a zeolite sample whose resolution cannot be achieved by analysis of X-ray diffraction (XRD) data, the software attempts to locate the coordinates of the T-atoms of the zeolite unit cell using a fitness function, defined through different penalties that consider space group restrictions, tetra-coordination restrictions, and tetrahedron planarity, among others. By testing the software for known zeolites such as LTA (Zeolite A), AEI (SSZ-39), ITW (ITQ-12), and others, the algorithm has found not only most of the target zeolites, but also new hypothetical zeolites whose viability is confirmed by energetic and structural criteria.

Table 6.2 compares results of the three GA-based softwares available. ZeoGAsolver can determine zeolite structures from 1 to 4 T-atoms in the asymmetric unit; EZs can determine zeolites with 1 or 2 T-atoms in the asymmetric unit; whereas P-GHAZ has been able to find zeolites with up to 4 or 5 T-atoms in the asymmetric unit. However, P-GHAZ has not been able to find some of the target zeolites (e.g., RTH, ITW, ISV, STF), whereas zeoGAsolver can find practically all of its target zeolites so far (up to 4 T-atoms in the asymmetric unit), and new hypothetical zeolites as interesting side products.

6.4 zeoGAsolver: A Specific Example of Genetic Algorithm for ZSD

As an example of genetic algorithms for ZSD, here we describe zeoGAsolver in detail. After setting up and initializing steps, the iterative process starts, in which the algorithm repeatedly: (a) evaluates the fitness of each possible zeolite; (b) applies the crossover operators onto the population for generating new individuals; and (c) applies a selection operator that maintains a suitable population size and enables good individuals to survive. Next, we briefly describe these steps.

6.4.1 Setting Up and Coding Scheme

Real coding codification [51] is used in zeoGAsolver to represent zeolites, so each chromosome is composed of the T-atoms (T1, T2, ...,Tn) of the zeolite asymmetric unit; and each Ti atom is represented by its coordinates (*xi,yi,zi*). Every coordinate is a gene, so each Ti atom is a "triple gene." For example, the representation of the chromosome for the zeolite AEI (with three T-atoms in its asymmetric unit) is: AEI = (TAEI1, TAEI2, TAEI3) = ((0.2733, 0.4042, 0.0521), (0.1128, 0.2289, 0.0606), (0.1126, 0.0369, 0.1664)). All the atoms that compose a zeolite unit cell can be obtained by applying the symmetry operations (defined by the space group of the zeolite) to the T-atoms of the asymmetric unit.

As input parameters, zeoGAsolver employs the zeolite structure parameters, i.e., space group, density, and unit cell parameters ($a,b,c,\alpha,\beta,\gamma$), which can be obtained by XRD, electron microscopy, etc. In fact, from the analysis of the XRD data, one or several space groups can be specified, as well as the cell parameters. Moreover, the chemical analysis gives the density of the material, from which a T-atom range of atoms in the UC is obtained, as well as its corresponding T-atom range of atoms in the AU.

The parameters for controlling the GA lifecycle are established at this stage: population size, GA termination criterion, crossover probability, selection probability, crossover operators to be used, such as BLX-α, and size of the pool of chromosomes for the selection operator. The tunability of these parameters makes the tool configurable and flexible enough to carry out zeolite structure searching for different types of zeolites.

6.4.2 Initialization

In the initialization step, an initial population of individuals (represented by their chromosomes) is obtained. In zeoGAsolver, the user can select between generating a typical random initial generation, where (x,y,z) positions of T-atoms are randomly set, or generating a more sophisticated random initial generation, named *Initial Generation with Wyckoff* positions (IGW). Although random, the initialization process takes into account: i) the T-atom range of the AU; ii) the coordinate constraints specified in the space group considered, which imply restrictions for (x,y,z) coordinates of each T-atom; iii) the T-atom range of the UC; and iv) the Wyckoff positions specified in the space group. In this last case, the algorithm first obtains the multiplicity values related to the Wyckoff positions of the space group. Next, the possible combination of operations are calculated, according to the T-atom range of the UC and the T-atom range of the AU, so then a combination list of Wyckoff multiplicities is obtained. Finally, the individuals of the initial population are randomly created, obtaining random values that obey the combination list of Wyckoff multiplicities. Using Wyckoff positions, more promising individuals satisfy the target density (number of T-atoms in the UC) in the starting generation.

6.4.3 Fitness Evaluation

In the fitness evaluation step, zeoGAsolver evaluates each individual, using the fitness function, to determine whether an individual complies with the restrictions to be qualified as a zeolite. A correct definition of the fitness function is crucial for any GA algorithm to evolve well. Although the complete set of conditions that makes a zeolite feasible is still under research [13], the structural analysis of existing zeolites allows extracting the guiding principles that define zeolite feasibility. Some of these principles have been part of the common background in zeolite science since early times, and thus they have been considered in the design of the fitness function of other published works. For example, tetra-coordination of atoms, T–T bond distances (that need to be always very close to 3.05 Å) or T-T-T angles. The zeoGAsolver fitness function is defined in Equations (6.1) and (6.2).

$$\text{Fitness} = 1/(1 + \text{Penalties}) \tag{6.1}$$

$$\begin{aligned} \text{Penalties} = {} & w_{TC} \bullet P_{TC} + w_{GC} \bullet P_{GC} + w_{SG} \bullet P_{SG} + w_D \bullet P_D \\ & + w_A \bullet P_A + w_{AA} \bullet P_{AA} + w_{TP} \bullet P_{TP} + w_M \bullet P_M \end{aligned} \tag{6.2}$$

The *Penalties* function is a non-negative real number that represents the total penalization of a possible zeolite structure which does not comply with the restrictions to be a zeolite. Each of these restrictions is represented by a specific sub-penalty P_i, which is a real number ($P_i \geq 0$) and its weight w_i, that reflects the importance of each sub-penalty for the whole fitness evaluation function. Thus, the range of *Fitness* is inside the real interval [0,1]. An ideal structure of a zeolite gets *Fitness* = 1 and *Penalties* = 0, but values near to 1 are a good approximation, and can also be a zeolite or be easily transformed into a zeolite.

The cited sub-penalties are detailed as follows:

TetraCoordinationPenalty (P_{TC}) checks whether each T-atom of the UC has 4 connections (bonds) with other T-atoms. Specifically, T-T atoms are connected if their distance is within the interval *Bond_dist* = [Dist_{min}, Dist_{max}] (this interval can be specified at the setting-up step; recommended values are [2.8, 3.4] Å). Thus, when a T-atom does not have 4 bonds, the structure is penalized.

GlobalCoordinationPenalty (P_{GC}) determines whether the global connection number of all the T-atoms of the individual in the UC can be compensated. For example, an individual resulting in 16 T-atoms in its UC, where one T-atom is tri-coordinated, another is penta-coordinated, and the other 14 atoms are tetra-coordinated, is not penalized here, because the tri-coordinated and penta-coordinated atoms compensate each other.

SpaceGroupPenalty (P_{SG}) penalizes an individual when it does not comply with the T-atom *coordinate constraint rules* specified in its Space Group.

DensityPenalty (P_D) penalizes an individual if the number of T-atoms of its UC does not satisfy the input density range (i.e., T-atom range of the UC).

AnglesPenalty (P_A) verifies whether the angles (T-T-T) formed by the bonds between T-atoms of the UC are within the desired interval ranges [52]. Figure 6.7a shows the angle penalty function, where the maxima and minima are taken from reference [52], depending on the abundance of ring sizes maximizing certain angles.

AnglesAveragePenalty (P_{AA}) checks whether the average of all 6 T-T-T angles formed using each T-atom as central is within the desired interval range [52]. Figure 6.7b shows this penalty function.

TetrahedronPlanarityPenalty (P_{TP}) checks how "planar" is the tetrahedron around a central atom, through the dihedral angle formed by three pairs of adjacent planes T_i-T_c-T_j, with T_c being the central atom and all combinations i,j ($i,j = 1 - 4, i \neq j$). If these dihedral angles are not within *Dihedral_angles = [Dih_{min}, Dih_{max}]* then the structure is penalized. This interval can be configured at the setting-up step. Recommended values are those in the interval [84, 96] degrees.

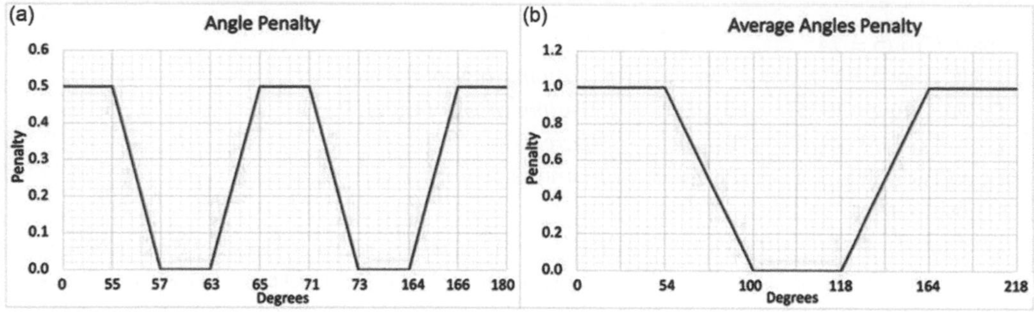

Figure 6.7 (a) Angle and (b) Average Angles Penalties functions.

MolecularityPenalty (P_M) penalizes an individual when its unit cell does not create a continued material, but it forms a molecule. For each T-atom in the UC, it verifies whether there are bonds between this unit cell and its neighboring cells (i.e., atoms within *Bond_dist*). If there is no bond, the structure represents only a molecular structure, so it is penalized.

6.4.4 Crossover

When the crossover process is applied, all individuals are selected in pairs (*parent1*, *parent2*) to be crossed as parents using the *fast tournament selection* [28]. This method selects each parent as the individual with the best *Fitness* value from a pool of randomly selected ones. ZeoGAsolver applies a crossover probability before selecting it as a parent. Depending on this probability, if the individual should not be crossed, then it is not finally selected, but it is marked as crossed in order to avoid its future selection. Then, several crossover operators are applied: the multiple crossover *BLX-α* (with several values for the alpha parameter) and a domain-dependent hybrid crossover named *SCW*.

The multiple crossover *BLX-α* operator is a variant of Sanchez et al. [53]. This operator is applied to cross the genes of each two parents and obtain more than two offsprings. For these offspring, only the two most promising ones are selected, i.e., those with the highest fitness value. In this way, the *BLX-α* operator is applied over several possible zeolites and with different α values. In zeoGAsolver, a mutation operator is not used since the behavior of the BLX-α operator acts as a mutation function, because in order to obtain the final gene values it always uses a random component. In fact, it explores useful spaces that have not been reached to guarantee the population diversity, and it also prevents the fast search space reduction, i.e., the premature convergence.

The *Smart Crossover using Wyckoff* (SCW) information, proposed Liu et al. [18], and previously present in FraGen, is a domain-dependent crossover, which exchanges some T-atoms from both parents with the same multiplicity of Wyckoff positions. The SCW crossover first selects the same *k_positions* of T-atoms from both parents and then crosses those with the same multiplicity from the selected ones, obtaining two new offspring in each crossing. If parents have no T-atoms with the same final multiplicity, this crossover operator has no effect.

For example, given two individuals T_A and T_B (Figure 6.8), where each of them has 4 T-atoms in their AU, such that $T_A = (T_{A1\_16}, T_{A2\_8}, T_{A3\_4}, T_{A4\_4})$ and $T_B = (T_{B1\_16}, T_{B2\_4}, T_{B3\_8}, T_{B4\_8})$, where T_{A1_16} represents the first T-atom (A1), with multiplicity value 16 (which means that 16 T-atoms can be generated in the unit cell by symmetry operations based on this T-atom), and so on. Assuming that SCW crossover

Space Group = 74
T_UC_range = {32, 36} *T-atom range of the UC*
T_AU_range = {3, 5} *T-atom range of the AU*
$T_A = (T_{A1\_16}, T_{A2\_8}, T_{A3\_4}, T_{A4\_4})$ *parent1*
$T_B = (T_{B1\_16}, T_{B2\_4}, T_{B3\_8}, T_{B4\_8})$ *parent2*
k_positions = 2
$T_A \otimes_{SCW} T_B = (T_{A1\_16}, \mathbf{T_{A2\_8}}, T_{A3\_4}, \mathbf{T_{A4\_4}}) \otimes_{SCW} (T_{B1\_16}, \mathbf{T_{B2\_4}}, T_{B3\_8}, \mathbf{T_{B4\_8}}) \rightarrow \{(T_{A1\_16}, T_{B4\_8}, T_{A3\_4}, T_{A4\_4}), (T_{B1\_16}, T_{B2\_4}, T_{B3\_8}, T_{A2\_8}), (T_{A1\_16}, T_{A2\_8}, T_{A3\_4}, T_{B2\_4}), (T_{B1\_16}, T_{A4\_4}, T_{B3\_8}, T_{B4\_8})$

Figure 6.8 Example of application of SCW crossover, on parents T_A and T_B. Selected genes (k_positions = 2) to be considered by the operator are in bold.

selects *k_positions* = 2 T-atoms in each parent (in this example, positions 2 and 4), then it will exchange T_{A2} with T_{B4}, and T_{A4} with T_{B2}, since these two pairs of T-atoms have the same multiplicity.

Note that the BLX-α crossover is applied on the genes directly (i.e., over "x," "y," or "z" coordinates), whereas SCW is applied on the T-atoms formed by the triple-gene (*x,y,z*), thus it maintains the (*x,y,z*) relationship of each T-atom, but it crosses T-atoms between parents. Therefore, BLX-α generates new genetic material at gene level, whereas SCW generates new combinations of the genetic material. Moreover, SCW crossover is used in combination with BLX-α to profit from its exploratory behavior. When using SCW, the current individual with good T-atoms complying with Wyckoff positions has more probabilities to be found as target.

6.4.5 Population Reduction and Termination Criterion

ZeoGAsolver applies an elitist stochastic universal sampling process to maintain the desired predefined population size. Thus, the best individuals are first selected, to ensure its survival, and then the stochastic universal sampling [31] is applied, which is a kind of proportional selection method where individuals with better fitness have more surviving probabilities. Finally, the GA stops when it has done the maximum number of generations allowed or when the last generations do not have any significant fitness improvement.

6.5 Graphics Processing Units in Zeolite Structure Determination and Prediction

In previous sections, a survey of genetic algorithms (GAs) has been provided. Here, we review the use of graphic processing units (GPUs) to accelerate Evolutionary Algorithms (EAs).

Baumes et al. have developed a methodology based on GA, after integration in the EASEA library [54], using a GPU implementation [50,55]. Speed up on GPU was the main objective of these initial works. Here, we will review the different contributions where GPUs have been used to either solve the structure of synthesized unknown zeolites or to propose new theoretical frameworks.

6.5.1 Quick Presentation of GPU Cards

In 1965, Moore [56] predicted that the evolution of technology would double the number of transistors per square millimeter of *silicon* every year, and later, every two years. At first, most people mistook this law for the doubling of computer power every two years. This was not a big mistake because even though being able to put twice as many transistors on the same surface did not imply that computers would run twice as fast, it happened so that clock frequency (and more generally single core computer chips) more or less did. But what was true in the past may not hold forever, and we must observe that it is now several years that the clock frequency of personal computers (PCs) seems to be stuck below 4 GHz (independently of the verification of Moore's law). Therefore, even if clock speed is not everything, and certainly not the best performance indicator, the speed of a single core central processing unit (CPU) does not increase as much as before, even though (thanks to Moore's law) the number of transistors that chips can hold still increases at the same rate. At first, this extra space was used to increase cache size, until it became possible to host several processing units on the same chip, eventually leading to multi-core CPUs. Even though a dual-core chip does not make the computer run twice as fast, manufacturers can claim that their power is multiplied by two, which is only correct if the running application is parallel enough so that it can use both cores at the same time. In 2006, NVIDIA launched CUDA (Compute Unified Device Architecture) that provides a Software Development Kit (SDK) and Application Programming Interface (API), allowing a programmer to use a variation of the C language to code algorithms for GeForce 8 GPUs series. Then, NVIDIA came out with TESLA calculators, which are graphic cards without a graphic output, totally dedicated to scientific processing, that boast 4.3 TeraFlops for around $8.000. Such enormous advertised computing power is very attractive for users of inherently parallel, CPU-greedy, Monte Carlo, and evolutionary algorithms.

6.5.2 Efficient Parallelization of Evolutionary Algorithms on GPUs

In 2009, Maitre et al. [57] presented a straightforward implementation of a standard EA that evaluates its population in parallel on a GPU card. Tests done on a benchmark and a real-world problem using an old NVIDIA 8800GTX card and a newer but not top of the range GTX260 card, show a roughly 30× (resp. 100×) speedup for the whole algorithm compared to the same algorithm running on a standard 3.6 GHz PC. Knowing that much faster hardware is now already available, this opens new horizons to Evolutionary Computation (EC). Since these cards remain very difficult to program, the knowhow has been integrated into the EASEA language so that code for GPU (*-cuda* option) can be output effortlessly.

GPU cards are in fact very powerful. Massively parallel computers have (among others) one main drawback: all the elementary processors on the card are organized into larger multi-processors, that must all execute the same program, e.g., SPMD model for Single Program Multiple Data. Inside each multi-processor, all elementary processors must execute the same instruction at the same time but possibly on different data, e.g., SIMD model for Single Instruction Multiple Data. Standard EAs need to run an identical evaluation function on different individuals (that can be considered as different data), meaning that this is exactly what GPUs have been designed to deal with. The most basic idea that comes to mind when we want to parallelize an EA is to run the evolution engine in a sequential way on some kind of master CPU (potentially the host computer CPU), and when a new generation of children has been created, get all children to evaluate in parallel on a massively parallel computer.

This may, however, sound like a bad idea because at each generation it is necessary to transfer the whole population to the parallel computer and get the results back. In our implementation of the GPU algorithm, we first try to minimize the Weierstrass test function

$$W_{b,h}(x) = \sum_{i=1}^{\infty} b^{-ih} \sin\left(b^i x\right)$$

with $b > 1$ and $0 < h < 1$. The use of this test function allows us to define problems and GA parameters that are the best fit for the GPU architecture. As shown in Figure 6.9, GPUs are very efficient at evaluating very large populations. In addition, we can observe that transferring the individuals is quickly overcome using large populations, meaning that GPU appears to be suitable for problems with a reasonably high number of parameters, e.g., individual size.

After testing the system with the Weierstrass function, a simple fitness function based on the connectivity of T-atoms in zeolites is used. Because recent zeolite synthesis in discovery programs involves T-atoms that differ from the original formulation of aluminosilicates, such as Ge, P, or B (considering ITQ and SSZ materials, for example), there will not be any implicit chemical composition. The GA is employed in order to find the "correct" locations of T-atoms. As the T-T distance for bonded atoms lies in a fixed range ($Dist_{min}$, $Dist_{max}$) previously mentioned, the connectivity of each new configuration of T-atoms can be evaluated. The fitness function corresponds to the number of defects in the structure, and Fitness = f1 + f2, is defined as it follows:

1) All T-atoms should be linked to 4 and only 4 neighboring T-atoms, thus f1 = Abs(4-Number of Neighbors);
2) No T-atom should be too close, e.g., T-T < $Dist_{min}$, hence f2 = Number of Too Close T-atoms.

This very terse fitness function is used to quickly find candidates, which are then submitted to a second optimization/evaluation algorithm to minimize the energy of the system. The function took 7.66 s to evaluate 20,000 individuals, meaning that all in all, the speedup offered by the GPU card is nearly ×60 on the new GTX260 (Figure 6.10), which is not so bad, even if it is less than the ×105 observed on the Weierstrass function. In reference [58] the author obtained 70×–120× speed up using Rosenbrock's test function.

At these speeds, most algorithms reach premature convergence. Therefore, much larger populations need to be used, and this is exactly what GPUs are good at. Orders of magnitude increase in search speed allow exploring much wider and complicated search spaces. In [59], an asynchronous island model running on clusters of machines equipped with GPU cards is presented (Figure 6.11). This technique is used to further prevent premature convergence while facing extremely complex problems. Island model GAs are based on independent GAs which evolve separately and intermittently exchange genetic material. However, individuals should not be exchanged among islands too often. Local exploitation should take place so that a new spot is well-explored before the island migrates to another place toward the direction of another good immigrant, indicating that infrequent communication is actually an advantage over frequent communication. Note that communication is usually what prevents parallel machines from yielding a linear speedup with the number of machines. If the main process stops and waits for all evaluations to complete before moving to the next generation, the algorithm is synchronous, or asynchronous otherwise. Authors demonstrated that supra-linear speedup over one machine and linear speedup considering clusters of different sizes are obtained. This architecture has been employed to tackle the extremely difficult case of ITQ-43. Details about the architecture are given in the next section, dedicated to the zeolite structure problem.

6.5.3 Genetic Algorithms on GPUs for Zeolite Structures Problem

Here, we focus on both the prediction and determination of zeolites frameworks using GAs with GPU. In [50], the authors demonstrate the potential of GPU and GAs for the atomistic prediction/determination community. Solutions show new stable hypothetical structures with 6 to 14 T-atoms in the asymmetric units, taking into account only a restriction on the density. Unlike most predictive techniques, a simple fitness function is used to assess the viability of a structure, rather than evaluating an energy function. This should provide a good framework assessment, flexible

(a)

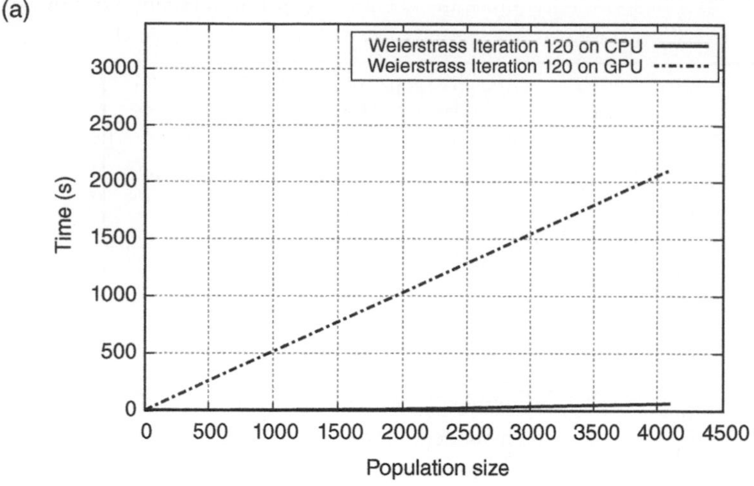

(b)

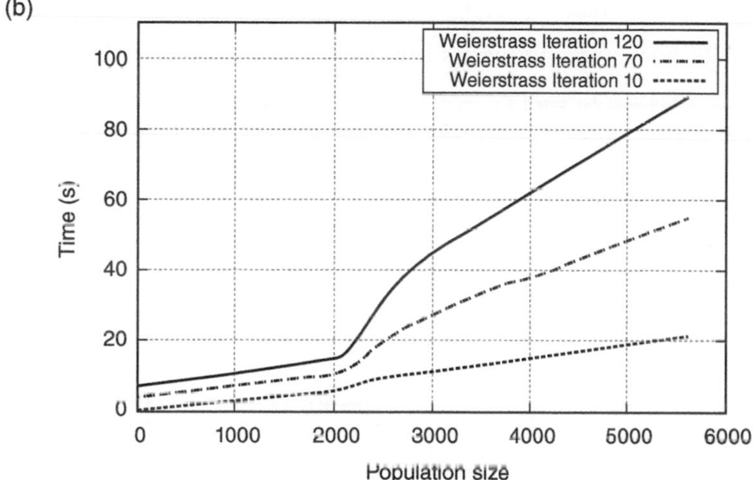

(c)

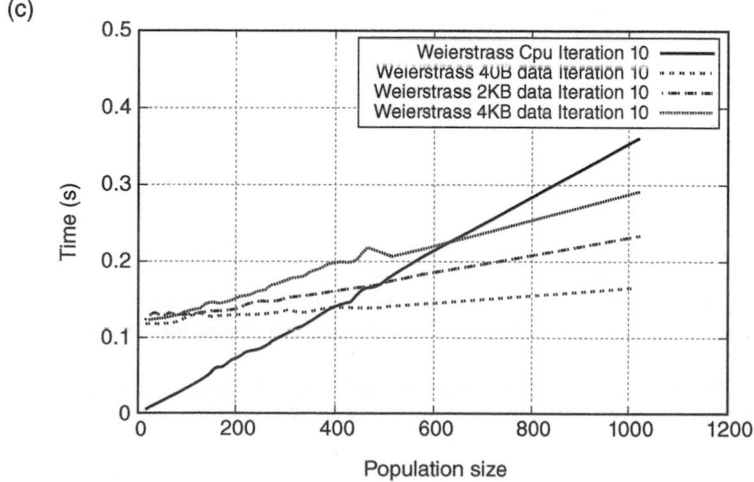

Figure 6.9 (a) CPU (top) and CPU + GPU (bottom) Evaluation times for increasing population sizes; (b) Influence of the computational cost of the fitness function on 8800GTX for increasing population sizes. Evaluation time is virtually constant up to 2048 individuals and linear afterwards; (c) Influence of individual size (40 B → 4 KB) on a very short evaluation function (10 iterations): overhead is overcome for a population between 400 and 650 individuals. *Source:* Figures taken from [57] with permission.

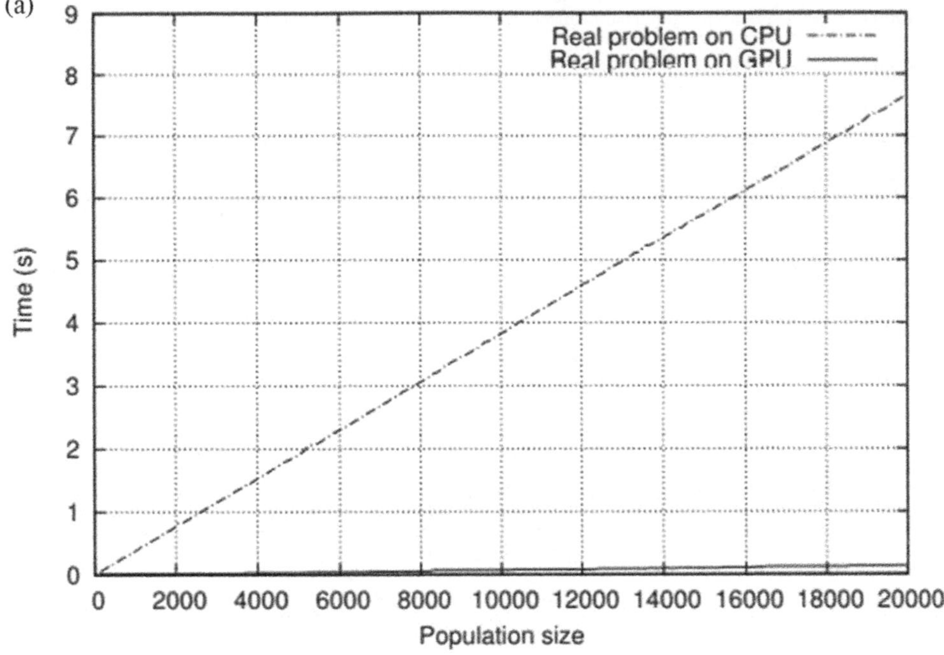

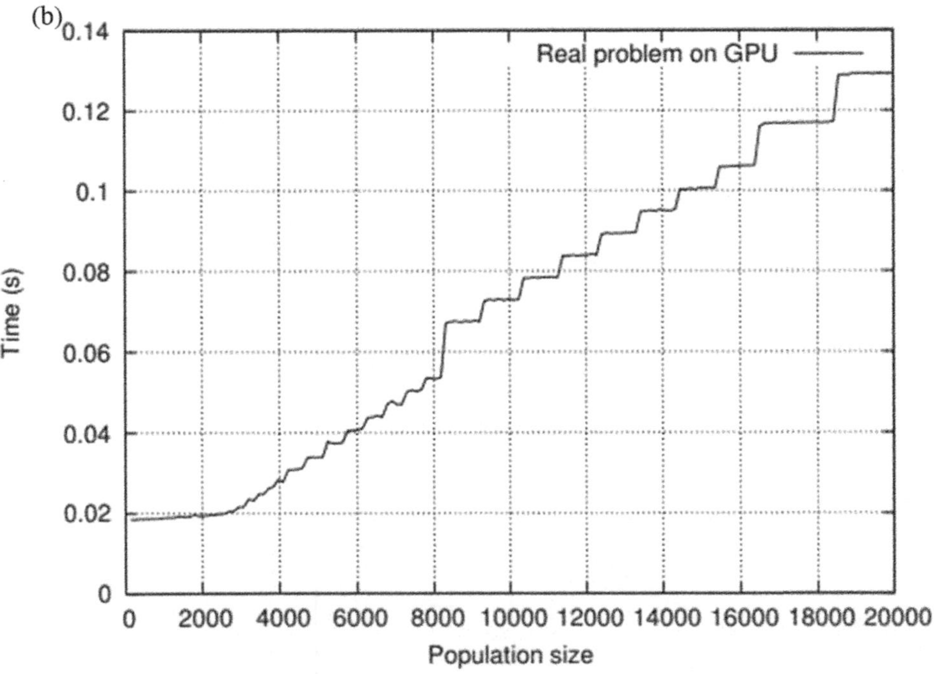

Figure 6.10 (a) CPU (top) and GPU (bottom) evaluation time for increasing population size; (b) Total time for GPU. Constant up to 2048 individuals and linear when the card is loaded. *Source:* Figures taken from [57] with permission.

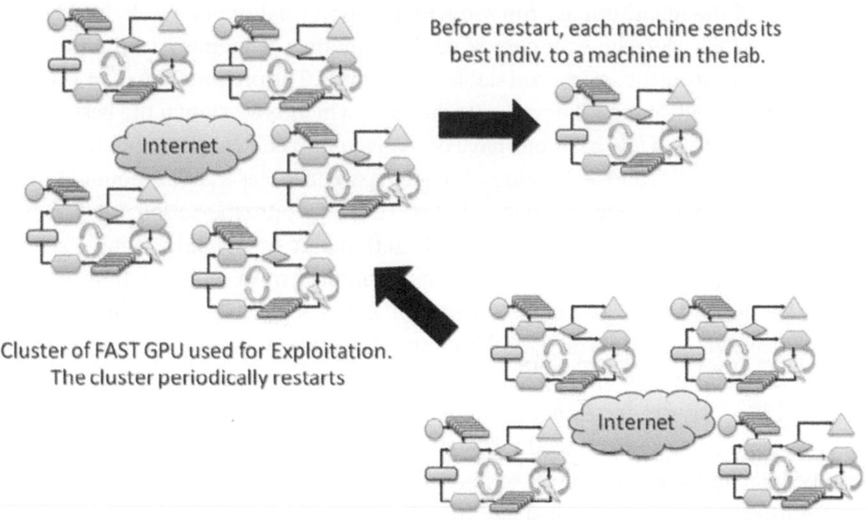
Before restart, each machine sends its best indiv. to a machine in the lab.
Internet
Cluster of FAST GPU used for Exploitation. The cluster periodically restarts
Internet
Cluster of SLOW machines used for exploration

Figure 6.11 Heterogeneity of machine and Island configuration for efficient optimization. *Source:* Figure taken from [59] with permission.

enough to not discard correct solutions while retaining only a reasonable number of candidate solutions that can easily be refined by routine procedures. We note that for the greatest part of the search space, the distribution of atoms does not refer to any realistic solution, for which a simple and fast assessment remains sufficient. Therefore, the algorithm purpose is to provide a practical quantity of different potential solutions that are stored in a database (DB) and refined on a second computer by a routine based on an interatomic potential technique using GULP, and among them, we expect the structure to be discovered. Only unspecified T-atoms are considered, and oxygen is omitted. The fitness function is reduced to geometric descriptors based on T-T-T angles, T-T distances, and a connectivity term. Best possible fitness is zero when all T-atoms are tetra-coordinated, and distances and average angles exactly correspond to optimized values, i.e., 3.07 Å and 109.51°, while deviations from this optimum sum penalties. Required inputs are the unit cell lattice parameters, dimensions and angles, and density range. Preliminary tests using this strategy have shown that each zeolite from the IZA website with 5 T-atoms can be solved in minutes (32,000 individuals were employed). In order to show that such a strategy allows finding stable structures, even containing a relatively high number of T-atoms (up to 14), the authors used three different configurations of UC, listed in Table 6.3, and a density of 96 T-atoms per UC ±8.

Table 6.3 Configurations of unit cells (V = 5550 Å^3). Note that these unit cell configurations do not correspond to frameworks previously investigated, and the resulting structures presented here are new.

Unit cell	*a*	*b*	*c*	α	β	γ	Space group
A	17.868	13.858	22.410	90	90	90	Imma (74)
B	13.858	17.868	22.410	90	90	90	Imma (74)
C	17.868	22.410	13.858	90	90	90	Ima2 (46)

Distances are expressed in Å and angles in degrees.

Three different instances of the algorithm are run sequentially for a total of one week. Nearly 400 structures are found, from which 88% show an energy per T-atom after refinement in the range of (−128.2,−129.8) eV/T-atom, indicating the selection of stable frameworks. The presentation of 14 T-atoms structures situates our methodology among the pool of techniques which have shown the best results in terms of T-atoms number, as only a few solutions have been given for zeolites with more than 10 to 12 T-atoms. The employment of GPUs allowed testing a broader space in a reasonable amount of time. In addition, the use of an island model better optimizes the ratio exploration/exploitation.

In 2017, Abdelkafi et al. [17]. provided a new objective function able to evaluate a viable zeolite structure and a genetic algorithm hybridized with a heuristic algorithm to more efficiently solve the ZSP. The so-called Parallel Genetic Hybrid Algorithm for Zeolites pursues a double objective: first, to solve more effectively zeolite structures for which the structure elucidation, in the absence of large enough single crystals, needs techniques such as modern electron diffraction that are efficient but heavier and expensive to implement. The second objective consists in the generation of thermodynamically-stable hypothetical zeolite structures with original topologies. The main idea of the test protocol is to use a particular set of parameters of known zeolites, to confirm that P-GHAZ is able to recover their experimental structure when initialized with a random position set of T-atoms in the AU. If the target zeolites are found, it can confirm that the objective function is working well. The approach demonstrates a good capacity to generate stable zeolitic frameworks using three real zeolites benchmarks (Table 6.4).

The structures with an optimal evaluation are validated. For that, the best candidates are energy minimized in order to properly place the oxygen and silicon atoms in an acceptable configuration, without taking into account electrostatic interactions from molecular modeling, and using the Universal force field (UFF) [60] as implemented in the Cerius2 software [61]. From the three zeolites benchmark, they validate six possible framework topologies: three are known and three are new (Figure 6.12). For the last three, a subsequent optimization is performed by using the non-polarizable Bushuev-Sastre (BS) force field, able to reproduce with high accuracy the structure and particularly the energetics of the pure silica zeolites [62]. The geometry optimization of the

Table 6.4 Target zeolites evaluated by P-GHAZ method.

Topologies	SpGrpNb	NbSymOp	NbAuAtoms	NbUcAtoms
ITE	63	16	4	64
RTH	12	8	4	32
ITW	12	8	3	24

Space group number (SpGrpNb), number of symmetry operations (NbSymOp), number of atoms in the AU (NbAuAtoms), number of atoms in the UC (NbUcAtoms).

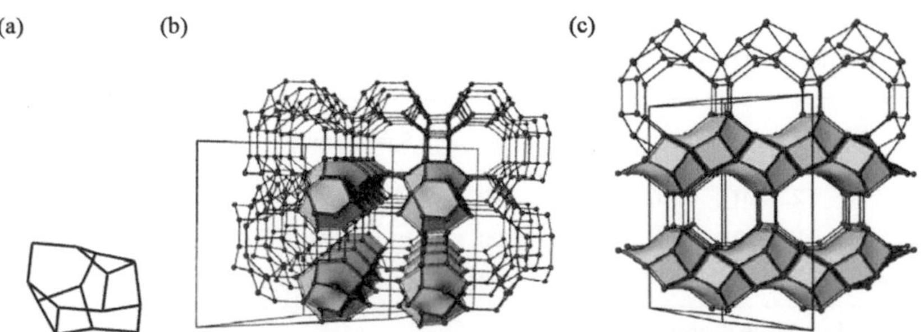

Figure 6.12 (a) New $[4^6 6^4]$ CBU found by P-GHAZ; (b) nets and tiling along [001]; and (c) along [101].

considered hypothetical zeolite structures within the purely siliceous form has been achieved using GULP. All the considered hypothetical structures have been first transformed within the P1 space group. The energy minimizations were carried under constant pressure, allowing to fully relax both the atomic positions and the unit cell parameters, applying the pseudo Newton–Raphson minimizing algorithm with the BFGS Hessian matrix updating scheme. The approach shows a capacity to recover unexpected framework topologies such as MER from the ITE parameters and CHA from the ITW parameters, while discovering novel zeolitic structures which could be synthesized by zeolite chemists.

6.5.4 GPUs in Island Model for Interrupted Zeolitic Frameworks

The island model mentioned in the previous section was developed in order to tackle the extremely complex case presented by the meso-microporous silico-germanate ITQ-43 [63], (Figure 6.13a). The model was first tested using the MFI zeolite, which contains 96 T-atoms in the unit cell (Figure 6.13b). Note that in the version used for this test, after all atoms have been randomly placed in the

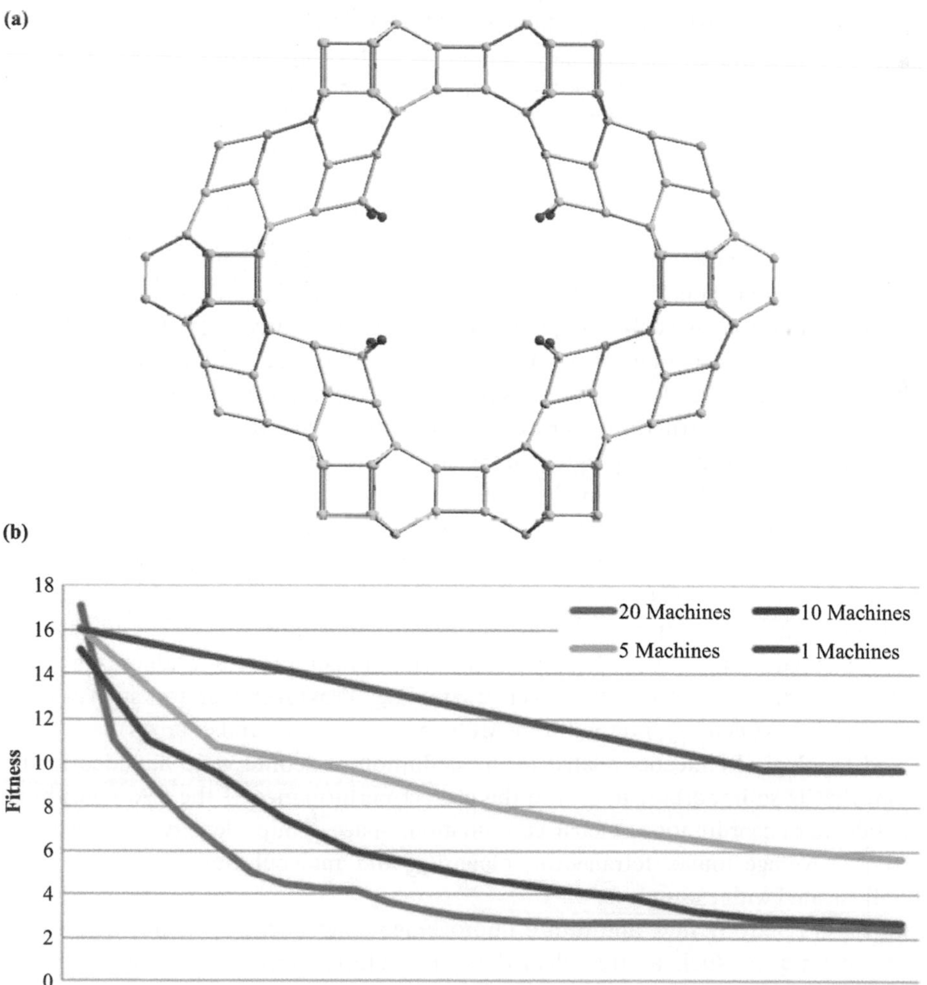

Figure 6.13 (a) Interrupted framework of ITQ-43; (b) Best fitness evolution for MFI and different number of machines in the island model. *Source:* Figure 6.13b taken from [59] with permission.

AU during the initialization procedure, each atom is iteratively and randomly selected and its position is translated toward the closest atom already present which is not tetra-coordinated. The final position is the optimal distance defined at 3.07 Å. The procedure stands for an improved initialization, made once at generation 0. Wickoff positions have also been integrated, allowing atoms at a distance inferior to 0.8 Å to be automatically placed at the corresponding special position onto the symmetry element defined by the space group, before the assessment of the solution. Note that the new position is virtual, i.e., calculated, in order to assess the structure viability, as $\{x,y,z\}$ coordinates are not modified in the genome code. The fitness function, that computes an approximated value of the real energy in the system, is minimized and based on geometric terms.

Note that a simple fully connected island model was not efficient enough to find interesting results for the ITQ-43. Two different clusters were created out of the 20 machines: one cluster of 16 machines that would do some exploration, and one cluster of 4 machines with an algorithm fit enough to exploit the best spots. The machines of the first cluster would periodically send their best individuals to the 4 other machines, while the 4 machines would exclusively exchange individuals between themselves. Their main goal was to find the best local value around the individuals sent by the cluster of 16 machines. If they did not manage to find any improvements, the 4 machines would simply restart simultaneously and wait for new suggestions coming from the 16 other machines. Periodically, the 4 machines would send the best individuals they found to an independent, slow machine in our lab that was not part of the optimization process. It would simply run to serve as an archive and collect the best individuals found to date.

6.6 Conclusions

Zeolites are widely used as catalysts, adsorbents, and ion-exchangers in the petrochemical and fine chemical industry. Although being constructed only from vertex-linked TO_4 tetrahedra, these tetrahedra can be connected in such a high number of different ways that the number of combinations exceeds, by several orders of magnitude, the current number of synthesized zeolites, hence the creation of the so-called field of hypothetical zeolites. Computer generation of hypothetical zeolites, as well as structure determination of synthesized but not solved zeolites, is a field that has experienced considerable progress in the last two decades.

Classic techniques such as Monte Carlo, Simulated Annealing, and Parallel Tempering have been used for these tasks, but also genetic algorithms have been used in the field of zeolite structure determination and prediction. Algorithms such as ZefsaII, FraGen, SCIBS, GRINSP, EZ, and P-GHAZ are briefly described, and our GA-based zeoGAsolver is described in detail.

A large number of different selection, mutation, and crossover operators that can be applied in a GA have been explained from scratch. This includes selection operators (tournament, ranking, proportional, roulette wheel, and stochastic universal sampling), crossover operators such as BLX-α, and mutation operators (aleatory, non-uniform, Mühlebein, and real number creep).

Structural and topological concepts related to the definition of zeolite were included in the fitness function, that is so important to ensure the correct performance of the GA algorithms. Properties include tetra-coordination, global coordination, space group, density, T-T distance, T-T-T angle, T-T-T average angle, tetrahedron planarity, and molecularity. They have been explained and illustrated with examples.

Still, the subject is relatively new and future improvements are needed to make algorithms more efficient. Techniques, such as tree algorithms, for refining topologies, have yet to be explored fully. Furthermore, automatic differentiation methods, which are powerful accelerators for conjugate-gradient strategies, seem to be under-utilized. The advent of low-cost GPUs presents many exciting opportunities. The important goal of guiding chemists in the synthesis of target structures in databases is still largely unsolved; that would be an important turning point in this field, and its fulfilment still seems to lie in the future. The present survey is written to facilitate an overall introductory insight, as well as to give the advanced reader ideas that hopefully will crystallize into more sophisticated and accurate software.

Acknowledgments

German Sastre thanks the Science and Technology Commission of Generalitat Valenciana (GVA) for funding through Prometeo 2021/077 and SGAI-CSIC for computational facilities. Alechania Misturini thanks GVA for predoctoral fellowship GRISOLIAP/2019/084.

References

1 Masters, A.F. and Maschmeyer, T. (2011). Zeolites: from curiosity to cornerstone. *Microporous Mesoporous Mater.* 142 (2–3): 423–438.

2 Wells, A.F. (1954). The geometrical basis of crystal chemistry. Part 1: *Acta Crystallogr.* 7 (8): 535–544.

3 Smith, J.V. (1980). Review of new crystal structures and mineralogy of zeolites and related materials. *Pure Appl. Chem.* 52 (9): 2131–2141.

4 O'Keeffe, M. and Hyde, B.G. (1978). On Si–O–Si configurations in silicates. *Acta Crystallogr. Sect. B* 34 (1): 27–32.

5 O'Keeffe, M. and Hyde, S.T. (1996). The asymptotic behavior of coordination sequences for the 4-connected nets of zeolites and related structures. *Zeitschrift fur Krist.* 211 (2): 73–78.

6 Treacy, M.M.J., Rao, S., and Rivin, I. (1992, 1993). A combinatorial method for generating new zeolite frameworks. In: *Proceedings from the Ninth International Zeolite Conference*, (ed.R. Von Ballmoos, J.B. Higgins, and M.M.J. Treacy), 381–388. Stoneham, MA: Butterworth-Heinemann.

7 Treacy, M.M.J., Randall, K.H., Rao, S. et al. (1997). Enumeration of periodic tetrahedral frameworks. *Zeitschrift für Krist. – Cryst. Mater.* 212 (11): 768–791.

8 Treacy, M.M.J., Rivin, I., Balkovsky, E. et al. (2004). Enumeration of periodic tetrahedral frameworks. II: Polynodal graphs. *Microporous Mesoporous Mater.* 74 (1–3): 121–132.

9 Friedrichs, O.D., Dress, A.W.M., Huson, D.H. et al. (1999). Systematic enumeration of crystalline networks. *Nature* 400 (6745): 644–647.

10 Delgado-Friedrichs, O. and O'Keeffe, M. (2003). Identification of and symmetry computation for crystal nets. *Acta Crystallogr. Sect. A Found. Crystallogr.* 5 (4): 351–360.

11 Baerlocher, C., McCusker, L.B., and Olson, D.H. (2007). *Atlas of Zeolite Frameworks Types*, 6th rev. ed. Amsterdam: Elsevier. http://www.iza-structure.org/databases.

12 Hagemeyer, A., Strasser, P., and Volpe, A.F. (2006). *High-Throughput Screening in Heterogeneous Catalysis*. Wiley-VCH.

13 Li, Y. and Yu, J. (2014). New stories of zeolite structures: their descriptions, determinations, predictions, and evaluations. *Chem. Rev.* 114: 7268–7316.

14 Falcioni, M. and Deem, M.W. (1999). A biased Monte Carlo scheme for zeolite structure solution. *J. Chem. Phys.* 110 (3): 1754–1766.

15 Li, Y., Yu, J., and Xu, R. (2012). FraGen: a computer program for real-space structure solution of extended inorganic frameworks. *J. Appl. Crystallogr.* 45 (4): 855–861.

16 Woodley, S.M., Catlow, C.R.A., Battle, P.D. et al. (2004). The prediction of inorganic crystal framework structures using excluded regions within a genetic algorithm approach. *Chem. Commun.* 4 (1): 22–23.

17 Abdelkafi, O., Idoumghar, L., Lepagnot, J. et al. (2017). Using a novel parallel genetic hybrid algorithm to generate and determine new zeolite frameworks. *Comput. Chem. Eng.* 98: 50–60.

18 Liu, X., Valero, S., Argenti, E. et al. (2017). Determining zeolite structures with a domain-dependent genetic algorithm. In: *Iberian Conference on Information Systems and Technologies, CISTI*, IEEE Computer Society.

19 Liu, X., Argente, E., Valero, S. et al. (2017). Applying genetic algorithms in chemical engineering for determining zeolite structures. In: *Advances in Intelligent Systems and Computing* (ed.H. Pérez García), J. Alfonso-Cendón, L. Sánchez González, et al.), 649: 34–43.

20 Holland, J.H. (1975). *Adaptation in Natural and Artificial System: An Introductory Analysis with Applications to Biology, Control, and Artificial Intelligence.* MIT Press.

21 Eshelman, L.J. and Schaffer, J.D. (1993). *Real-Coded Genetic Algorithms and Interval-Schemata*, 2e. 187–202. Elsevier.

22 Goldberg, D.E. (1989). Genetic algorithms in search, optimization, and machine learning. *Choice Rev. Online* 27 (2): 27-0936.

23 Siddique, N. and Adeli, H. (2013). *Computational Intelligence: Synergies of Fuzzy Logic, Neural Networks and Evolutionary Computing.* John Wiley & Sons.

24 Lim, T.Y. (2014). Structured population genetic algorithms: a literature survey. *Artif. Intell. Rev.* 41 (3): 385–399.

25 Crepinsek, M., Liu, S.H., and Mernik, M. (2013). Exploration and exploitation in evolutionary algorithms: a survey. *ACM Comput. Sur.* 45 (3): 35

26 Boyer, D.O. and Herv, C. (2001). Genetic algorithm with crossover based on confidence intervals as an alternative to least squares estimation for nonlinear models. In: *9th European Symposium on Artificial Neural Networks*, 193–198.

27 Goldberg, D.E. and Deb, K. (1991). *A Comparative Analysis of Selection Schemes Used in Genetic Algorithms*, 1e, Elsevier69–93.

28 Brindle, A. (1981). *Genetic Algorithms for Function Optimization.* Edmonton: University of Alberta.

29 Baker, J.E. (1985). Adaptive selection methods for genetic algorithms. In: *Proceedings of the 1st International Conference on Genetic Algorithms* (ed. J.J. Grefenstette), 101–111. Pittsburgh.

30 Winston, P.H. (1992). *Artificial Intelligence*, 3rd e. Addison-Wesley.

31 Baker, J.E. (1987). Reducing bias and inefficiency in the selection algorithm. In: *Proceedings of the Second International Conference on Genetic Algorithms on Genetic Algorithms and Their Applications*. 14–21.

32 Umbarkar, A.J. and Sheth, P.D. (2015). Crossover operators in generic algorithims: a review. *ICTACT J. Soft Comput.* 6 (1): 1083–1092.

33 Wright, A.H. (1991). Genetic algorithms for real parameter optimization. 1: 205–218.

34 Michalewicz, Z. (1992). *Genetic Algorithms + Data Structures = Evolution Programs.* Berlin, Heidelberg: Artificial Intelligence: Springer.

35 Mühlenbein, H. and Schlierkamp-Voosen, D. (1993). Predictive models for the breeder genetic algorithm. I: Continuous parameter optimization. *Evol. Comput.* 1 (1): 25–49.

36 Herrera, F., Herrera, F., Herrera-Viedma, E. et al. (1994). Fuzzy tools to improve genetic algorithms. *Proc. Second Eur. Congr. Intell. Tech. SOFT Comput.* 3: 1532–1539.

37 Schlierkamp-Voosen, D. and Mühlenbein, H. (1994). Strategy adaptation by competing subpopulations. In: *International Conference on Parallel Problem Solving from Nature*, (ed.Y. Davidor, H.P. Schwefel, and R. Männer), 866, LNCS, 199–208. Berlin, Heidelberg: Springer.

38 Herrera, F., Lozano, M., and Sánchez, A.M. (2003). A taxonomy for the crossover operator for real-coded genetic algorithms: an experimental study. *Int. J. Intell. Syst.* 18 (3): 309–338.

39 Davis, L. (1991). *Handbook of Genetic Algorithms* (ed.V.N. Reinhold). New York.

40 Deem, M.W. and Newsam, J.M. (1992). Framework crystal structure solution by simulated annealing: test application to known zeolite structures. *J. Am. Chem. Soc.* 114 (18): 7189–7198.

41 Woodley, S.M., Battle, P.D., Gale, J.D. et al. (1999). The prediction of inorganic crystal structures using a genetic algorithm and energy minimisation. *Phys. Chem. Chem. Phys.* 1 (10): 2535–2542.

42 Tremayne, M., Seaton, C.C., and Glidewell, C. (2002). Structures of three substituted Arenesulfonamides from X-ray powder diffraction data using the differential evolution technique. *Acta Crystallogr. Sect. B Struct. Sci.* 58 (5): 823–834.

43 Foster, M.D. and Treacy, M.M.J. (2004). *Atlas of Prospective Zeolite Structures* http://www.hypotheticalzeolites.net.

44 Wells, S.A. and Sartbaeva, A. (2015). GASP: software for geometric simulations of flexibility in polyhedral and molecular framework structures. *Mol. Simul.* 41 (16–17): 1409–1421.

45 Boisen, M.B., Gibbs, G.V., and Bukowinski, M.S.T. (1994). Framework silica structures generated using simulated annealing with a potential energy function based on an $H_6Si_2O_7$ molecule. *Phys. Chem. Miner.* 21 (5): 269–284.

46 Gale, J.D. (1997). GULP: a computer program for the symmetry-adapted simulation of solids. *J. Chem. Soc. – Faraday Trans.* 93 (4): 629–637.

47 Gale, J.D. and Rohl, A.L. (2003). The General Utility Lattice Program (GULP). *Mol. Simul.* 29 (5): 291–341.

48 Delgado-Friedrichs, O., Foster, M.D., O'Keeffe, M. et al. (2005). What do we know about three-periodic nets? *J. Solid State Chem.* 178 (8): 2533–2554.

49 Le Bail, A. (2005). Inorganic structure prediction with GRINSP. *J. Appl. Crystallogr.* 38 (2): 389–393.

50 Baumes, L.A., Kruger, F., Jimenez, S. et al. (2011). Boosting theoretical zeolitic framework generation for the determination of new materials structures using GPU programming. *Phys. Chem. Chem. Phys.* 13 (10): 4674–4678.

51 Herrera, F., Lozano, M., and Verdegay, J.L. (1998). Tackling real-coded genetic algorithms: operators and tools for behavioural analysis. *Artif. Intell. Rev.* 12 (4): 265–319.

52 Liu, X., Valero, S., Argente, E., Botti, V. et al. (2015). The importance of T···T···T angles in the feasibility of zeolites. *Zeitschrift fur Krist. – Cryst. Mater.* 230 (5): 291–299.

53 Sánchez, A.M., Lozano, M., García-Martínez, C. et al. (2008). Real-parameter crossover operators with multiple descendents: an experimental study. *Int. J. Intell. Syst.* 23 (2): 246–268.

54 Collet, P., Lutton, E., Schoenauer, M. et al. (2000). Take it EASEA. In: *In Parallel Problem Solving from Nature (PPSN VI)* (ed.M. Schoenauer, K. Deb, G. Rudolph et al.), 1917, 891–901. Springer Verlag.

55 Krüger, F., Maitre, O., Jiménez, S. et al. (2013). Generic local search (memetic) algorithm on a single GPGPU chip. *Nat. Comput. Ser.* 46: 63–81.

56 Moore, G.E. (2009). Cramming more components onto integrated circuits. Reprint. *Electronics* 38 (8): 114; Ff. *IEEE Solid-State Circuits Soc. Newsl.* 11 (3): 33–35.

57 Maitre, O., Baumes, L.A., Lachiche, N. et al. (2009). Coarse grain parallelization of evolutionary algorithms on GPGPU cards with EASEA. In: *Proceedings of the 11th Annual Conference on Genetic and Evolutionary Computation GECCO-2009*. 1403–1410.

58 Krüger, F., Maitre, O., Jiménez, S. et al. (2010). Speedups between ×70 and ×120 for a generic local search (Memetic) algorithm on a single GPGPU chip. In: *EvoApplications 2010. Lecture Notes in Computer Science*, (ed. C. Di Chio. S. Cagnoni, C. Cotta, et al.), 6024: 501–511.
59 Baumes, L.A., Kruger, F., and Collet, P. (2011). EASEA: a generic optimization tool for GPU machines in asynchronous island model. *Comput. Methods Mater. Sci.* 11 (3): 489–499.
60 Rappe, A.K., Casewit, C.J., Colwell, K.S. et al. (1992). UFF, a full periodic table force field for molecular mechanics and molecular dynamics simulations. *J. Am. Chem. Soc.* 114: 25.
61 Austin, N. (2000). Cerius2 program molecular simulations incorporated. San Diego, CA.
62 Bushuev, Y.G. and Sastre, G. (2009). Atomistic simulations of structural defects and water occluded in SSZ-74 zeolite. *J. Phys. Chem. C* 113: 25.
63 Jiang, J., Jorda, J.L., Yu, J. et al. (2011). Synthesis and structure determination of the hierarchical meso-microporous zeolite ITQ-43. *Science* 333 (6046): 1131–1134.

7

Numerical Representations of Chemical Data for Structure-Based Machine Learning

Gyoung S. Na

Korea Research Institute of Chemical Technology (KRICT), Daejeon, 141 Gajeong-ro, Republic of Korea

7.1 Machine Readable Data Formats

Chemical data is usually represented as chemical formulas, molecular structures, and other composite formats of feature vectors and chemical compounds. For machine learning, unstructured chemical data should be represented as numerical formats because the machine learning algorithms are basically mathematical functions mapping the numerical input values to the numerical target values. For example, fully-connected neural networks (FCNNs) [1] require the feature vectors as their inputs, and convolutional neural networks (CNNs) [2] require the input data of feature matrices or tensors. Thus, our first step of machine learning for chemical applications is to represent the chemical data as the numerical formats, such as feature vectors, feature matrices, and mathematical graphs.

7.1.1 Feature Vectors

Feature vectors is an most essential data format in machine learning. Formally, the feature vectors of the data are defined on a numerical space $\chi \in \mathbb{R}^d$, where d is the number of features in the feature vectors. Figure 7.1 presents three examples of the feature vectors for data representations. Numerically, we can describe an animal with its species, height, weight, and the other features. Also, we can present weather conditions using temperature, humidity, and the other features. Moreover, chemical experiments also can be represented as a feature vector of pressure, temperature, and employed chemical compounds. The main constraint of the feature vectors is that each feature in the feature vectors should be defined on $\mathbb{R}$. Although decision trees [3] allow categorical features, most machine learning algorithms are designed on the numerical inputs. Therefore, we need to convert the raw chemical data (e.g., atomic symbols) into the feature vectors on the numerical values or one-shot encodings.

The feature vectors are a natural and efficient way to numerically represent collected data. These feature vectors were popular in the initial stages of machine learning for chemical applications [4–6] because advanced machine learning methods that can handle more complex data representation of the chemical data had not yet been developed. However, the vector representations of the chemical data are still popular in chemical machine learning to integrate domain knowledge of chemistry and physics with machine learning methods [7, 8], because the feature vectors are the most flexible way to represent the chemical data. For example, most machine learning algorithms allow the vector-shaped inputs, and we can enter numerically represented domain knowledge into the prediction models simply by concatenating the knowledge to the vector-shaped inputs. Python Mendeleev[1] and RDKit[2]

AI-Guided Design and Property Prediction for Zeolites and Nanoporous Materials, First Edition. Edited by German Sastre and Frits Daeyaert.

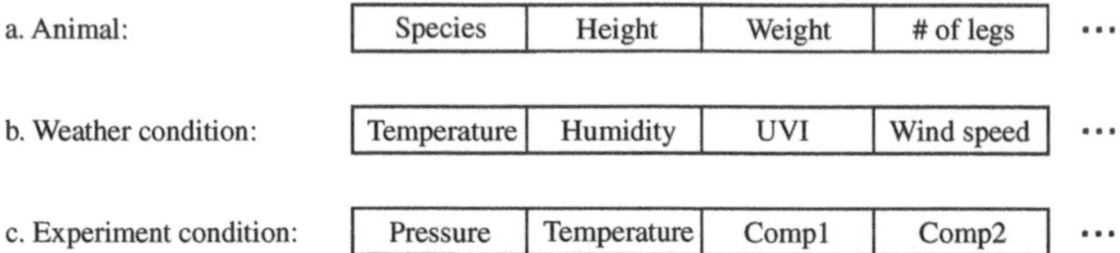

Figure 7.1 Examples of the feature vectors to represent the input data as the numerical formats. Each cell in the feature vectors is a numeric value in the real-valued space $\mathbb{R}$.

packages provide various modules and descriptors to convert the chemical data into the feature vectors.

Although the feature vectors are a useful way to efficiently construct a machine learning model for chemical applications, the feature vectors are not preferred in recent chemical machine learning due to information loss in representing the molecular and crystal structures. For example, relative positions and bonding information of the atoms in the molecular structures are inevitably distorted in converting the molecular structures into the feature vectors. In computer vision, numerous successes have been achieved by employing the matrix representation that can preserve the structural information of the image data [2], rather than using the primitive feature vectors. Therefore, we need to explore more advanced data representation that can preserve the geometric information of the molecular structure, if we want to fully exploit the structural information of the chemical compounds in machine learning.

7.1.2 Matrices

Numerical matrices are a rectangular array or table of numerical values. The matrices consist of their row or column vectors. For a matrix M, we denote its ith row and j^{th} column vectors as $M_{i,:}$ and $M_{:,j}$, respectively. In data science, the matrices are used to represent geometric information and structural characteristics of the data. For example, a geometry of particles in the n-body problem [9] on a 3D (three-dimensional) space can be represented by the matrices $M \in \mathbb{R}^{n\times3}$, where each row vector indicates XYZ coordinates of the particles. In addition to the particle's geometry, the matrix representations are useful to describe chemical systems and network structures [10].

The most popular research field of the matrix representation is computer vision to predict target values from the image data. The image data can be interpreted as a set of data points (pixels) in which each data point contains the numerical features encoding the colors [11] and the data points are organized in the lattice structure. Figure 7.2 illustrates the matrix representation of a black and white image of a handwritten digit. In the matrix representation, the pixels of the image are regarded as the elements of the matrix, and the relative positions between the pixels are preserved in the matrix. The colors of the pixels are represented as a binary value {0, 1} for black and white, respectively. In machine learning of computer vision, the image data is mainly handled with CNNs [2], which is the most popular neural network to predict target values from the image data. In addition to

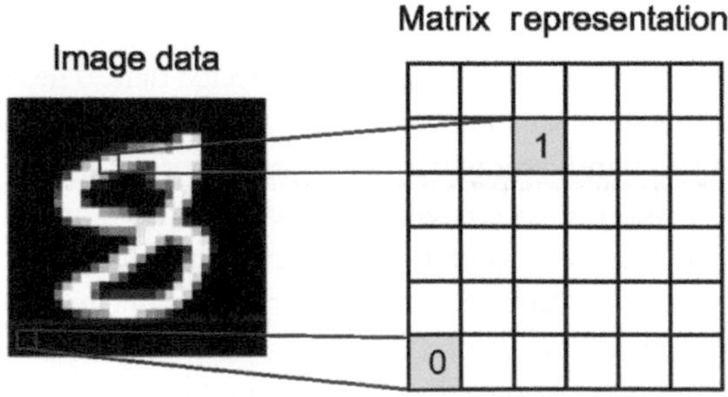

Figure 7.2 Image data of a handwritten digit and its matrix representation.

computer vision, the matrix representation of the structured data and pattern images has been widely applied to describe the molecular structures. The matrix representation of physical patterns to describe the molecular structures is commonly referred to as the molecular fingerprint [12]. With the image-based molecular fingerprints, various machine learning applications were proposed to predict molecular properties [12–14].

7.1.3 Mathematical Graphs

Mathematical graphs $G = (\mathcal{V}, \mathcal{U}, \mathbf{X}, \mathbf{E})$ are numerical representation to present structured data in non-Euclidean space, where $\mathcal{V}$ is a set of nodes; $\mathcal{U}$ is a set of edges; $X \in \mathbb{R}^{|\nu| \times d}$ is a d-dimensional node-feature matrix; and $E \in \mathbb{R}^{|u| \times m}$ is an m-dimensional edge-feature matrix. The mathematical graphs are popular to represent structured data in data science and machine learning, such as supply chain networks [15] and user relationships [16].

Also, the mathematical graphs can be used to numerically describe the chemical and geometric characteristics of the molecular structures. In recent deep learning, the mathematical graph has been widely used to represent the molecular and crystal structures [17–21]. In the graph-based representations, the atoms and their chemical bondings in the molecular and crystal structures are presented as the nodes and the edges, respectively. After presenting the atoms and the chemical bondings, chemical and physical attributes of the atoms and their bondings are respectively assigned to the edges in the graph. As a result, an attributed graph $G = (\mathcal{V}, \mathcal{U}, \mathbf{X}, \mathbf{E})$ is constructed from a molecular or crystal structure, where $\mathcal{V}$ is a set of atoms; $\mathcal{U}$ is a set of chemical bondings; $\mathbf{X}$ is a atom-feature matrix; and ε is a bonding-feature matrix. In Section 7.2, We will explore these graph-based representations for the molecular and crystal structures in machine learning.

7.2 Graph-based Molecular Representations

7.2.1 Chemical Representations of Molecular Structures

XYZ file format and simplified molecular-input line-entry system (SMILES) [22] are popular data formats to describe the molecular structures in data-driven chemistry. In the XYZ file, a molecular structure is described as a file containing atomic symbols and XYZ coordinates of the atoms in each row (Figure 7.3). In data-driven chemistry, various chemical datasets have been published with the XYZ files of extensive molecular structures [23, 24]. The XYZ representation of the molecular structures are useful when creating the fingerprints and graph representations of the molecules because we can derive most chemical and geometric attributes of the molecules from the atomic symbols and XYZ coordinates in the XYZ files. RDKit[3], which is a well-known cheminformatics library, provides various modules and functions to generate the molecular fingerprints and the molecular graphs.

Another popular representation method to represent the molecular structures is a string-based line notation called SMILES. SMILES was designed to encode the chemical and structural information of the molecules into the strings. For example, the molecular structure in Figure 7.3 can be represented by NC(=O)N in SMILES. Various molecular datasets have been published with the SMILES representations of the molecular structures, because SMILES is a lightweight and efficient method to construct molecular datasets [25–28]. However, although SMILES can store various chemical information such as bonds and rings of the molecules, SMILES sometimes causes information loss of the 3D geometry of the atoms because the geometric information can be distorted in converting the molecular structures into SMILES.

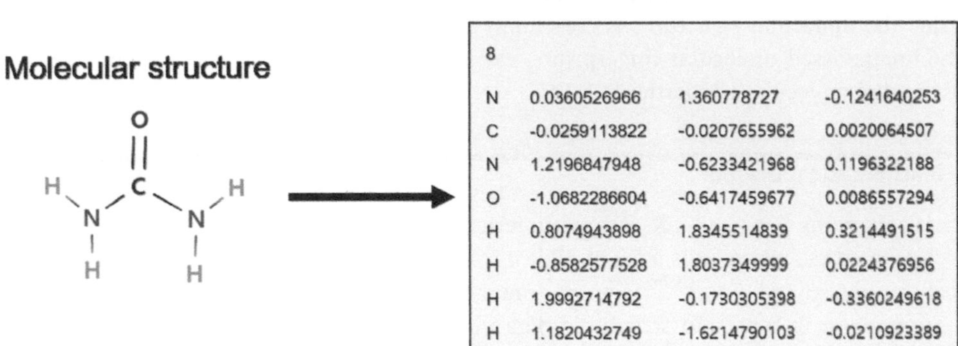

XYZ file format
Molecular structure
8
N 0.0360526966 1.360778727 -0.1241640253
C -0.0259113822 -0.0207655962 0.0020064507
N 1.2196847948 -0.6233421968 0.1196322188
O -1.0682286604 -0.6417459677 0.0086557294
H 0.8074943898 1.8345514839 0.3214491515
H -0.8582577528 1.8037349999 0.0224376956
H 1.9992714792 -0.1730305398 -0.3360249618
H 1.1820432749 -1.6214790103 -0.0210923389

Figure 7.3 XYZ file format to represent the geometric information of the example molecule.

7.2.2 Molecular Graphs

Recently, mathematical graphs have been widely applied to deep learning algorithms to numerically describe the molecular structures provided in XYZ files and SMILES. With the graph-based representations of the molecular structures, various machine learning algorithms achieved state-of-the-art accuracies in predicting molecular properties and discovering novel molecules [29, 30]. The goal of this chapter is to explain several methods to convert the molecular structures into the mathematical graphs to develop chemical applications based on state-of-the-art graph neural networks [18, 29, 31, 32].

In the graph-based representations, a molecular structure is described as a mathematical graph $G = (\mathcal{V}, \mathcal{U}, \mathbf{X}, \mathbf{E})$, where $\mathcal{V}$ is a set of atoms (nodes); $\mathcal{U}$ is a set of chemical bondings of the atoms (edges); $\mathbf{X} \in \mathbb{R}^{|\nu| \times d}$ is a d-dimensional atom-feature matrix; and $\mathbf{E} \in \mathbb{R}^{|u| \times q}$ is a q-dimensional bond-feature matrix. The atom and bond features are defined based on domain knowledge of target applications. For example, electrical attributes of the elements can be used to the atom features for the chemical applications that need to predict the electrical properties of the molecules.

The molecular graphs $G = (\mathcal{V}, \mathcal{U}, \mathbf{X}, \mathbf{E})$ are generated from the XYZ file or SMILES through the following three steps, as shown in Figure 7.4:

- **Step 1**: Identify the atoms and their chemical bondings from the XYZ file or SMILES of the molecules. In this step, chemical rules of domain experts or external cheminformatics libraries can be used to identify the atoms and their bondings.
- **Step 2**: Make an index for each atom, and assign a tuple of indices for each atomic bonding. Then, define a set of atoms and a set of atomic bondings using the indices.
- **Step 3**: Generate the molecular graph $G = (\mathcal{V}, \mathcal{U}, \mathbf{X}, \mathbf{E})$ from the indexed molecular structure by assigning the numerical features to the atoms and the atomic bondings. In the graph, the atoms and the atomic bondings are represented as the nodes and the edges, respectively. Note that the order of the rows in the node and edge feature matrices must match the indices of the atoms and the atomic bondings.

Although the graph conversion process for a single molecule was presented, the benchmark molecular datasets [24, 25, 27, 33] contain multiple molecular structures with XYZ file or SMILES for each molecule. That is, we should perform the graph conversion process in Steps 1–3 for each molecule to take the molecular datasets for machine learning. In the following sections, we will describe the processes to generate the molecular graphs from XYZ file and SMILES.

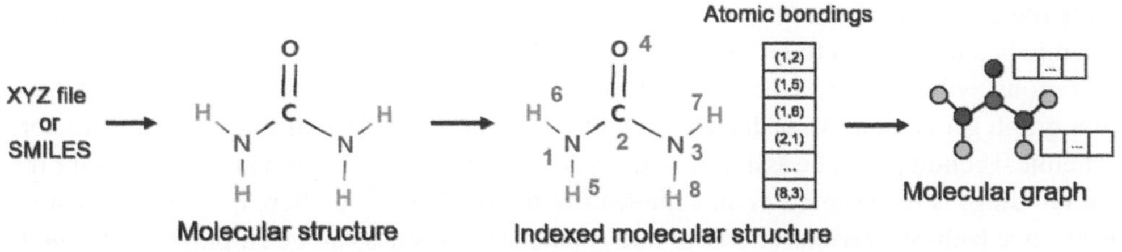

Figure 7.4 The overall process to generate the molecular graph from the XYZ file or SMILES. In the first step, the XYZ file or SMILES are converted into the molecular structure by the chemical rules of domain experts or external cheminformatics libraries. In the next step, indices are assigned to the atoms, and the atomic bondings are identified as the edges in the graphs based on the indices of the atoms. In the final step, the molecular graph is generated by assigning numerical features to the atoms (nodes) and the atomic bondings (edges), replaced as the nodes and the edges, respectively. Note that the order of the rows in the node and edge feature matrices must match the indices of the atoms and the atomic bondings.

7.2.3 XYZ File to Molecular Graph

To generate the molecular graphs, first, we need to extract the components of the attributed graphs $G = (\mathcal{V}, \mathcal{U}, \mathbf{X}, \mathbf{E})$ from XYZ files. However, it can be easily performed by reading XYZ files line by line. As shown in Figure 7.3, we can identify each atom and its XYZ coordinates by reading a line of the XYZ file. After identifying the atoms, elemental attributes from external libraries are assigned to the atoms. Therefore, a set of atoms V and a node-feature matrix X are generated for the molecular graph G. In this process, various elemental attributes can be employed to define the node features, such as atomic number, hybridization type, and formal charge. Furthermore, one-shot encoding of the categorical features of the atoms can be used for generating the node-feature matrix. In addition to the real-valued chemical and physical attributes of the atoms, content-based features that just indicate the existence of the element have been widely used in machine learning for chemical applications [29, 34], even though they do not contain any chemical and physical attributes of the atoms.

After reading the node information of the molecular graph, edge information including a set of edges $\mathcal{U}$ and an edge-feature matrix E should be calculated to complete the molecular graph. To determine the chemical bonding (edges) of the atoms from XYZ files, it is necessary to compute pairwise distance between the atoms in the molecules. From the XYZ file of the molecule $\mathcal{M}$, we can compute the pairwise distances between the atoms as:

$$d_{i,j} = \left\|\mathbf{x}_i - \mathbf{x}_j\right\|_2, \tag{7.1}$$

where $\mathbf{x}_i \in \mathbb{R}^3$ is a coordination vector of the ith atom in the XYZ file, and $\left\|\mathbf{x}\right\|_2$ is the Euclidean norm. In computing the atomic distances, angstrom (Å) is usually used to denote a unit of the atomic distance. The distance calculation in Equation 7.1 should be performed for all pairs of the atoms in the molecule to determine the existence of the chemical bondings between the atoms, i.e., a distance matrix $\mathbf{P} \in \mathbb{R}^{|\mathcal{V}| \times |\mathcal{V}|}$ should be calculated for all pairs of the atoms in $\mathcal{V}$.

After calculating the distance matrix $\mathbf{P}$, we need to determine the existence of the chemical bonding between the atoms to generate the edges of the molecular graph. For a threshold function $\theta(e_1, e_2)$, a set of edges $\mathcal{U}$ is generated by:

$$u = \left\{(i, j) P_{i,j} < \theta(e_i, e_j)\right\}, \tag{7.2}$$

where e_i is the atomic symbol of the ith atom. Note that the distance thresholds for each pair of atomic symbols should be defined by $\theta(e_1, e_2)$ based on domain knowledge in chemistry and physics. Various Python libraries for chem-informatics provide pre-defined $\theta(e_1, e_2)$ to support the molecular graph generation. After determining the edges of the molecular graph, feature vectors for the chemical bondings can be assigned to the generated edges, and the assigned features for the chemical bondings is represented as an edge-feature matrix $\mathbf{E} \in \mathbb{R}^{|\mathcal{U}| \times q}$, where q is the dimensionality of the edge features. Various chemical and geometric attributes can be employed to the edge features, and bonding types and bonding distances are also common for edge features in molecular graphs. Although the edge-feature matrix $\mathbf{E}$ is optional in graph neural networks [31, 32], some graph neural networks were designed to take the edge-feature matrix into their graph embedding process for generating more informative graph embeddings [18, 29]. Finally, the prediction problem to predict the target values $\mathbf{y}$ from the molecular structure G is defined as:

$$y = f(G), \tag{7.3}$$

where $f : \mathcal{G} \rightarrow \mathbb{R}^l$ is a neural network to project the non-Euclidean space $\mathcal{G}$ into the 1D (one-dimensional) target space.

7.2.4 SMILES to Molecular Graph

To generate the molecular graphs, we should extract and decode the chemical and structural information of the molecules from SMILES strings. Various Python packages to convert the SMILES representations into programmable molecular structures are publicly available, such as RDKit. Although a number of equally valid SMILES strings can be exist for a molecule (e.g., CCO and OOC for the structure of ethanol), they can be converted into the same representation in the molecular graph.

Wtih the SMILES decoding libraries, we can extract the atomic symbols and the molecular structures from SMILES strings. The extracted atomic symbols and the molecular structures are typically provided as a list of the categorical symbols and a list of the indexed edges, respectively. From the extracted information, we can generate a set of atoms $\mathcal{V}$ by assigning the extracted atomic symbols. Then, the node-feature matrix $\mathbf{X}$ is generated by assigning atomic attributes for each atomic symbols in $\mathcal{V}$. Similarly, the set of edges $\mathcal{U}$ and the edge-feature matrix $\mathbf{E}$ are generated by employing the indexed edges and assigning chemical attributes to the edges, respectively. One of the benefits of SMILES is that the public libraries automatically recognize the chemical bondings from the given SMILES based on domain knowledge in chemistry and physics. That is, we do not need to identify the chemical bondings from the geometry of the molecule.

7.2.5 Multiple Molecular Graph

In many chemical applications, molecule-to-molecule interactions play an important role in determining the synthesis processes and the molecular properties of the chemical products [35–37]. The datasets containing multiple molecular structures are typically provided as a collection of data points including multiple XYZ files or SMILES strings [34]. In the graph-based machine learning, these datasets containing multiple molecular structures can be easily abstracted as a list of the molecular graphs. For example, the data containing the chromophore and solvent molecules with their optical properties can be represented as a list $\mathcal{L} = [G_{chr}, G_{svt}, \mathrm{y}]$, where G_{chr} is the molecular graph of chromophore, G_{svt} is the molecular graph of solvent, and **y** is a target vector containing measured optical properties. Therefore, the prediction problems to predict target values **y** from K molecular structures are formally defined as:

$$y = f(G_1,\ldots,G_K). \tag{7.4}$$

To predict target values from the molecular interactions, graph neural networks should be designed to process the list of the molecular graphs, as shown in Equation 7.4. The most intuitive architecture to handle the multiple graphs is to employ K individual graph embedding layers for each kth molecular graph. In this architecture, each graph is converted into the latent feature vector by graph neural networks, and the feature vectors are concatenated [34]. Formally, graph neural networks of K individual graph embedding layers for K graph inputs can be defined as:

$$f(G_1,\ldots,G_K) = h(\phi(g_1(G_1)\oplus\ldots\oplus g_k(G_K))), \tag{7.5}$$

where $g_k : \mathcal{G} \to \mathbb{R}^{|\mathcal{V}|\times m}$ is the node aggregation layers to project the graph in $\mathcal{G}$ to m-dimensional latent space; ϕ is a readout function to generate the graph embedding from the latent embeddings of the nodes; and $h : \mathbb{R}^m \to \mathbb{R}^l$ is the dense layers to predict the lD target values from the graph embeddings. As another example, we can define graph neural networks for K-graph inputs based on the mean-based readout operation as:

$$f(G_1,\ldots,G_K) = h\left(\phi\left(\frac{1}{K}\sum_{k=1}^{K} g_k(G_k)\right)\right), \tag{7.6}$$

where gk have the same embedding dimensionality for all k. In addition to the network architectures in Equations 7.5 and 7.6, various approaches can be applied to handle the inputs of the multiple graphs [37].

7.3 Machine Learning with Molecular Graphs

So far, we have explored the numerical representations of the chemical data. Especially, we have studied the graph-based representations of the molecular structures, which can preserve both their chemical and structural characteristics. Graph neural networks consist of the graph embedding and dense layers. The graph embedding layers are designed to extract latent embeddings of the input graph, and the dense layers are implemented to predict the target values from the generated latent embeddings. To extract meaningful latent embeddings of the molecular graphs, various schemes of graph embedding have been studied in chemical science and machine learning. In this section, we will explore popular deep neural networks that were designed to process the graph-structured input data.

7.3.1 General Architecture of Graph Neural Networks

Various deep neural networks to handle the graph-structured data have been proposed in chemical science and machine learning [18, 29, 31, 32, 38]. The purpose of graph neural networks is to predict target values from input graph-structured data. Hence, graph neural networks can be defined as a function $f : \mathcal{G} \rightarrow \mathbb{R}^l$, where $\mathcal{G}$ is a non-Euclidean space of the mathematical graphs, and l is a dimensionality of the target variables. Most graph neural networks share a common architecture containing node aggregation (graph embedding) and prediction layers. Thus, graph neural networks can be generally described as a composite function $f = h \circ g$ of node aggregation layers g and dense layers h. The sub-functions g and h are constructed by stacking the node aggregation layers and the dense layers, respectively. In graph neural networks, the representation capabilities for the input graphs are fundamentally determined by the node aggregation scheme of g.

In graph-level prediction, an additional function called "readout" is typically inserted between the node aggregation and dense layers to generate graph-level embeddings from the node and the edge embeddings, i.e., graph neural networks are defined by $f = h \circ \phi \circ g$. The readout function can be defined from various approaches, such as mean-based operation [39], attention mechanism [39], and recurrent neural networks [40]. Like the node aggregation scheme, the readout function should also be carefully selected for accurate prediction of graph neural networks [39]. In the following sections, we will formally define each layer of graph neural networks.

The purpose of the node aggregation layer is to calculate latent representations of the nodes and the edges in the input molecular graph [31]. For a graph-structured data $G = (\mathcal{V}, \mathcal{U}, \mathbf{X}, \mathbf{E})$, the rth node aggregation layer $g^{(r)}$ can be presented as a general definition as:

$$\mathbf{Z}^{(r)}, \mathbf{H}^{(r)} = g^{(r)}(\mathcal{U}, \mathbf{Z}^{(r-1)}, \mathbf{H}^{(r-1)}), \tag{7.7}$$

where $\mathbf{Z}^{(r)}$ and $\mathbf{H}^{(r)}$ are the latent node-feature and edge-feature matrices of the rth node aggregation layer, respectively. Note that $r = 0$ indicates the input features, i.e., $\mathbf{Z}^{(0)}$ and $\mathbf{H}^{(0)}$ are the input node-feature matrix $\mathbf{X}$ and the edge-feature matrix $\mathbf{E}$, respectively. In the graph neural network, the inputs and the outputs of the node aggregation layer can be different for each node aggregation scheme. For example, the node aggregation layer can take only the input node feature matrix [31, 32, 38]. In the other method, the node aggregation layer takes both the node- and edge-feature matrices for the forward process [18]. Therefore, we need to feed the appropriate inputs into the node aggregation layers according to their definitions in the implementation.

In the dense layer, the target values are predicted from the latent node- and edge-feature matrices. Generally, fully-connected layers are used for the dense layers in graph neural networks [18, 31]. The rth dense layer $h^{(r)}$ is defined by:

$$\mathbf{s}^{(r)} = \delta^{(r)}(\mathbf{W}^{(r)}\mathbf{s}^{(r-1)} + \mathbf{b}^{(r)}), \tag{7.8}$$

where $\mathbf{s}^{(r)}$ is output latent features of the rth dense layer, $\mathbf{W}^{(r)}$ is a weight matrix of the layer, $\mathbf{b}^{(r)}$ is a bias vector of the layer, and $\delta^{(r)}$ is the activation function for generating nonlinear latent feature vectors. Note that $\mathbf{s}^{(0)}$ is the latent node, edge, or their combinational features from the last node aggregation layer $g^{(R)}$, where R is the number of node aggregation layers. In graph neural networks, the output of the last dense layer becomes the predicted target value of the network for the given graph-structured data G.

For the graph-level prediction, the readout function to generate a graph-level representation from the latent node- and edge-feature matrices is implemented between the node aggregation and dense layers in graph neural networks [18, 31]. The readout can be implemented from various methods [41–43]. Among the existing readout methods, the mean-based readout is the most popular in graph neural networks because it is simple to implement and efficient to inference. For a node aggregation function handling only the node features, the mean-based readout is defined by:

$$\mathbf{s}^{(0)} = \frac{1}{|\mathcal{V}|}\sum_{i=1}^{|\mathcal{V}|}\mathbf{Z}_i^{(R)}, \tag{7.9}$$

where $\mathbf{Z}_i^{(R)}$ is a row vector of the matrix $\mathbf{Z}^{(R)}$ and R is the number of the node aggregation layers in h.

Figure 7.5 presents the general architecture of graph neural networks with the readout function to predict the molecular properties $\mathbf{y}$ from the molecular graph G. For the input molecular graph, the node aggregation layers generate the latent node- and edge-feature matrices $\mathbf{Z}^{(R)}$ and $\mathbf{H}^{(R)}$, respectively. After the node aggregation, a molecule-level representation $\mathbf{s}^{(0)}$ is calculated by the readout function for the latent node- and edge-feature matrices. Finally, the molecular property $\mathbf{y}$ is predicted by the dense layers h for an input of the molecule-level representation $\mathbf{s}^{(0)}$. In graph neural networks, most algorithms follow this process to predict the graph-level target values from the input graph-structured data [18, 29, 31, 32, 38]. In the next section, we will briefly review popular state-of-the-art graph neural networks to build a practical application based on graph neural networks.

7.3.2 Graph Convolutional Network

Graph convolutional network (GCN) [31] is a widely used graph neural network to handle the graph-structured data in various applications [31, 34]. The node aggregation layer of GCN follows a mathematical scheme of the generalized convolution. The node aggregation layer in GCNs is called the graph convolution layer. For an input graph $G = (\mathcal{V}, \mathcal{U}, \mathbf{X}, \mathbf{E})$, the rth graph convolution layer is defined by:

$$\mathbf{Z}^{(r)} = \delta^{(r)}(\mathbf{A}\mathbf{Z}^{(r-1)}\mathbf{W}^{(r)}), \tag{7.10}$$

where $A \in \mathbb{R}^{|\mathcal{V}|\times|\mathcal{V}|}$ is an adjacency matrix of the graph, $W^{(r)} \in \mathbb{R}^{d_{(r-1)}\times d_r}$ is a trainable weight matrix to transform the $d_{(r\text{-}1)}$-dimensional features into the d_r-dimensional features, and $\delta^{(r)}$ is the activation function of the rth graph convolution layer.

The main concept of the graph convolution is to generate latent features of the nodes in the graph by collecting the node features of the neighborhood nodes connected by the edges. This operation is formally defined by the operation $\mathbf{AZ}^{(r\text{-}1)}$ in the graph convolution. In chemical applications, it can be interpreted that the latent features of the atoms are calculated based on the chemical and physical properties of the atoms connected by chemical bondings. Hence, the

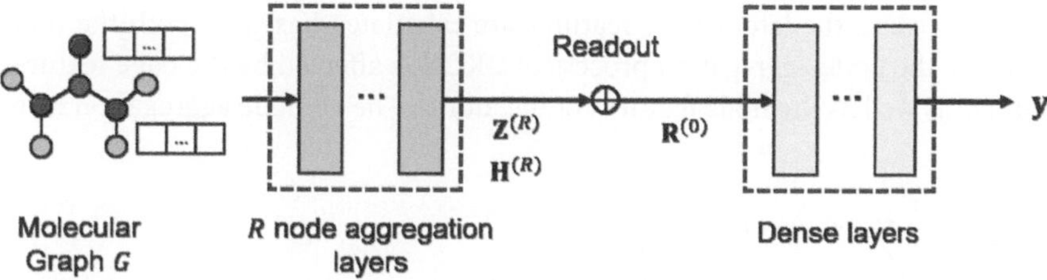

Figure 7.5 The general architecture of graph neural networks with the readout function to predict the graph-level molecular properties **y** from the input molecular graph G.

chemical and physical attributes of the atoms should be provided in the input node-feature matrix $\mathbf{X} = \mathbf{Z}^{(0)}$ in its rows. Since this concept of the graph convolution is fairly consistent with the chemical principles, GCNs have been widely applied to chemical applications.

7.3.3 Graph Attention Network

Graph attention network (GAT) [32] is a graph neural network that exploits an attention mechanism [44] for node aggregation. The attention mechanism is one of the most successful methods in machine learning, and it achieved state-of-the-art prediction performances in many machine learning applications [45, 46]. Among the various attention methods, GAT employs the self-attention mechanism that automatically calculates the importance of the objects within the input data. Formally, the latent features of the nodes in the graph are generated in the node aggregation layer of GATs as:

$$\mathbf{Z}_i^{(r)} = \delta^{(r)}\left(\alpha_{i,i}^{(r)}\mathbf{W}^{(r)}\mathbf{Z}_i^{(r-1)} + \sum_{j\in\mathcal{N}_i}\alpha_{i,j}^{(r)}\mathbf{W}^{(r)}\mathbf{Z}_j^{(r-1)}\right), \tag{7.11}$$

where $\mathcal{N}_i$ is a set of neighborhood indices of the ith node, and the attention coefficients $\alpha_{i,j}$ are computed by feeding a concatenated vector of $\mathbf{W}^{(r)}\mathbf{Z}_i^{(r-1)}$and $\mathbf{W}^{(r)}\mathbf{Z}_j^{(r-1)}$ to a trainable attention network [32]. In the training of GAT, the attention score is optimized to minimize the prediction error.

By employing the self-attention mechanism, GAT can generate informative latent features of the nodes by capturing important information from the edges of the neighborhood nodes. This concept is consistent with the atomic interactions in the molecular structures because the molecular properties are based on the atomic interactions. In the prediction, the trained attention scores play an important role to determine the molecular properties, since the attention mechanism captures the important chemical bondings by increasing the attention scores between the nodes. Although the self-attention mechanism imposes additions computational costs in generating the latent features of the nodes, it is empirically demonstrated that the self-attention mechanism can improve the prediction accuracies of the graph neural networks [32].

7.3.4 Continuous Kernel-based Convolutional Network

The continuous kernel-based convolution [17] was introduced to process the molecular and crystal graphs to predict molecular and crystalline properties. This operator is also known as the edge-conditioned convolution [47]. We refer to the graph neural networks with the continuous kernel-based convolutional operator as continuous kernel-based convolutional network (CKCN). Unlike GCNs and GATs, CKCN requires the edge-feature matrix as well as the node-feature matrix of the molecular and crystal graphs. The node aggregation scheme of CKCN is also known as the edge-conditioned convolution because the latent node features are calculated based on both the node and edge features. That is, the node aggregation process of CKCN is affected by the edge features of the chemical bondings as well as the node features of the atoms. The rth node aggregation layer of CKCN is defined by:

$$\mathbf{Z}_i^{(r)} = \delta^{(r)}\left(\mathbf{W}^{(r)}\mathbf{Z}_i^{(r-1)} + \sum_{j\in\mathcal{N}_i}\mathbf{Z}_j^{(r-1)} \odot h^{(r)}(e_{i,j})\right), \tag{7.12}$$

where $h^{(r)}$ is a neural network to embed edge features $e_{i,j}$ between the ith and j^{th} atoms, and $\odot$ is the element-wise multiplication. By exploiting the edge features of the chemical bondings, CKCN was able to accurately predict mechanical and physical properties of the molecular structures on a large benchmark dataset [17].

7.3.5 Crystal Graph Convolutional Neural Network

Crystal graph convolutional neural network (CGCNN) [18] is a graph neural network that is specially designed to process the graph-structured data of the crystal structures. CGCNN also exploits the edge features of the chemical bondings as well as the node features of the atoms to extract meaningful information from the crystal structures. The node aggregation scheme specified to the crystal structures is called crystal graph convolution in CGCNN. The rth crystal graph convolution layer is defined by:

$$\mathbf{Z}_i^{(r)} = \phi^{(r)}\left(\mathbf{Z}_i^{(r-1)} + \sum_{j \in \mathcal{N}_i} h_f(\mathbf{s}_{i,j}) \odot h_g(\mathbf{s}_{i,j})\right), \tag{7.13}$$

where $\mathbf{s}_{i,j}$ is a concatenated vector of $\mathbf{Z}_i^{(r-1)}$, $\mathbf{Z}_j^{(r-1)}$ and $\mathbf{e}_{i,j}$. In the crystal graph convolution, two embedding networks h_f and hg are introduced to generate latent features of the nodes when considering the edge features of the chemical bondings.

In the original paper on CGCNN, seven materials properties were predicted by CGCNN from the crystal structures in large benchmark datasets. About 30,000 materials from the Materials Project database [48] were used to train and evaluate CGCNN. In the experimental evaluation, CGCNN predicted formation energy, band gap, bulk modulus, shear modulus, and Poisson's ratio with comparable errors to the classical density functional theory (DFT) [49]. In addition to the original CGCNN, a structural variant of CGCNN was proposed to improve the prediction capabilities [50]. Although CGCNNs were designed for the graph data of the crystal structures, they can be applied to predict molecular properties without the modifications in their implementations.

7.4 Graph-based Machine Learning for Molecular Interactions

So far, we have explained graph-based machine learning for a single molecular graph. However, the chemical properties from the molecule-to-molecule interactions are common and important in analyzing the chemical processes and the chemical products in many scientific applications [35, 36]. As we discussed in Section 7.2.5, the data of the molecular interactions can be abstracted as a list of the molecular graphs, i.e., the dataset of the molecular interactions is given by:

$$\mathcal{D} = \left\{\left(\left\{G_{1,1},\ldots,G_{1,K}\right\}, y_1\right),\ldots,\left(\left\{G_{N,1},\ldots,G_{N,K}\right\}, y_N\right)\right\}, \tag{7.14}$$

where K is the number of molecules involved in the molecular interactions, and y is the target chemical properties from the molecular interactions. In this section, we will explore two primitive architectures to handle the multiple graphs for understanding design principles of graph neural networks for multiple input graphs.

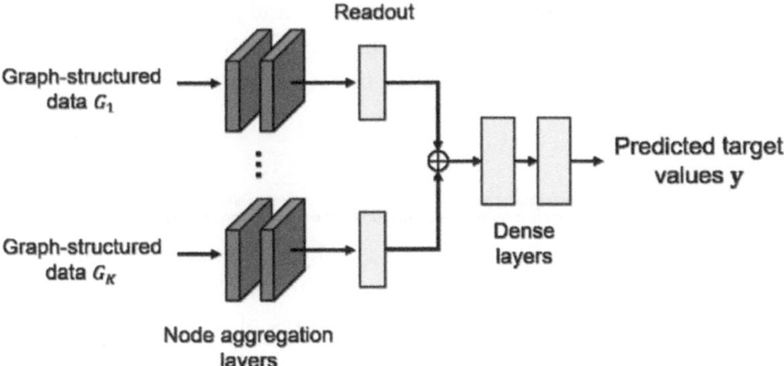

Figure 7.6 The architecture of the graph neural network with the vector concatenation to predict the target values from the graph interactions. The weights of the node aggregation layers can be shared or individually trained.

7.4.1 Vector Concatenation Approach to Prediction Molecule-to-Molecule Interactions

One of the most intuitive approaches to process K input graphs is to concatenate the vector-shaped graph embedding from individual node aggregation layers and the readout functions. Figure 7.6 illustrates the architecture of the graph neural network to predict target values from the molecular interactions described by K input graphs. In this architecture, the kth graph embedding of the input graph G_k is generated by the kth node aggregation layers and the readout function, and is formally given by:

$$\mathbf{s}_k = \phi_k\left(\mathbf{Z}_1^{(R)}, \ldots, \mathbf{Z}_{|\mathcal{V}_k|}^{(R)}\right) \tag{7.15}$$

where $\mathbf{s}_k$ is the readout function for G_k, $\mathcal{V}_k$ is a set of nodes of G_k, and R is the number of node aggregation layers in the graph neural network. After generating the K individual graph embeddings, a latent embedding that describes the entire molecular interactions is calculated by vector concatenation as:

$$\mathbf{s}^{(0)} = \mathbf{s}_1 \oplus \ldots \oplus \mathbf{s}_K. \tag{7.16}$$

Hence, the interaction embedding $\mathbf{s}^{(0)}$ is in $\mathbb{R}^{mK}$ for K graph embeddings $\mathbf{s}_k \in \mathbb{R}^{mK}$. Finally, the target values y of G_k are predicted by feeding the concatenated vector of the molecular interactions into the dense layers.

The concatenation-based architecture in Figure 7.6 is intuitive and easy to implement. Despite the simplicity in the architecture, a graph neural network based on the vector concatenation of the graph embeddings accurately predicted optical properties of the chromophores on an experimental database containing about 30,000 chromophore and solvent molecules with their optical properties [34]. The concatenation-based architecture can be extended to predict the target properties from the molecular interactions under specific environmental conditions by attaching the environmental conditions into the interaction vector $\mathbf{s}^{(0)}$.

7.4.2 Attention Map Approach for Interpretable Prediction of Molecule-to-Molecule Interactions

Although machine learning has shown state-of-the-art performances in various scientific applications, interpreting the reason for the prediction results is still an unresolved problem in machine learning because the prediction processes of most machine learning algorithms are black-box. However, in many scientific applications, we want to understand the reasons for the prediction results to gain insights into discovering new scientific knowledge. In particular, it is important to interpret the atomic interactions between the molecules in terms of the target molecular properties to understand the underlying principles of the chemical systems for discovering novel molecules.

For the interpretable predictions of the molecular interactions, a new graph neural network was proposed based on a concept of the attention map [37]. For two input molecules, G_1 and G_2, the attention map is defined as a matrix **M** where Mij indicates the attention score between the ith atom in G_1 and j^{th} atom in G_2. For example, the attention map of two graphs $G_1 = (\mathcal{V}_1, \mathcal{U}_1, \mathbf{X}_1 \mathbf{E}_1)$ and $G_2 = (\mathcal{V}_2, \mathcal{U}_2, \mathbf{X}_2 \mathbf{E}_2)$ is represented as a matrix $\mathrm{M} \in \mathbb{R}|\mathcal{V}_1| \times |\mathcal{V}_2|$. An (i,j) element $\mathrm{M}_{i,j}$ indicates the importance of the atomic interaction between the ith atom in G_1 and the jth atom in G_2. The attention score $\mathrm{M}_{i,j} = \phi(\mathbf{V}_i^{(R)}, \mathbf{U}_j^{(R)})$ is usually calculated by feeding the node features of the atoms, where ϕ is a neural network (called the attention network) to calculate the attention scores, $\mathbf{V}_i^{(R)}$ is the latent node embedding of ith node in G_1, $\mathbf{U}_j^{(R))}$ is the latent node embedding of jth node in G_2, and R is the number of node aggregation layers. Based on the attention map, the feature vector representing the molecular interactions is calculated by:

$$\mathrm{z} = \sum_{i=1}^{|V_1|} \sum_{j=1}^{|V_2|} \mathbf{M}_{i,j} \mathbf{V}_i^{(R)} \oplus \mathbf{U}_j^{(R)}, \tag{7.17}$$

where $\oplus$ is the vector concatenation. As shown in Equation 7.17, the pair-wise attention $\mathbf{M}_{i,j}$ can be regarded as a weight in the weighted sum of the atomic interaction between the ith atom in G_1 and the jth atom in G_2. Thus, the attention $\mathbf{M}_{i,j}$ can be interpreted as importanct to the atomic nteractions between the ith atom in G_1 and the jth atom in G_2.

Figure 7.7 illustrates the architecture of the graph neural network with the attention map to predict the target values from the molecular interactions. For two input graphs G_1 and G_2, the node

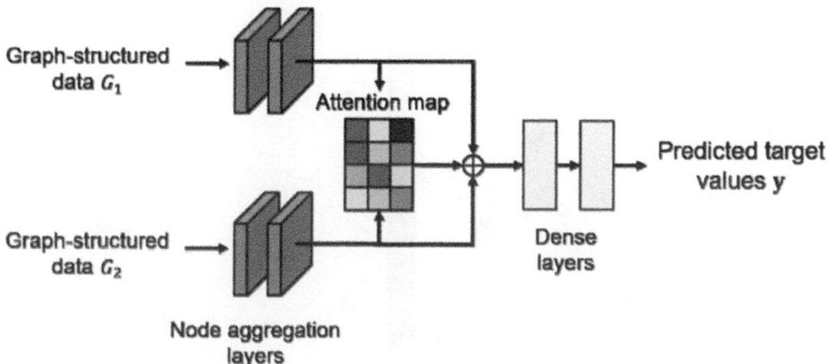

Figure 7.7 The architecture of the graph neural network with the attention map to predict the target values from the molecular interactions. Note that the molecular interactions are abstracted as a list of two input graphs $[G_1, G_2]$.

embeddings are calculated through the node aggregation layers. For each pair of the nodes in G_1 and G_2, the attentions are calculated by the dot product of the node embeddings or a trainable neural network to calculate attentions. Then, the feature vector of the molecular interaction is calculated by the weighted sum of the node embeddings with weights of the attentions. Finally, the target values $\mathbf{y}$ are predicted by entering the feature vector of the molecular interaction into the dense layers.

7.5 Representation Learning from Molecular Graphs

Although the graph-based representations of the molecular structures have achieved state-of-the-art performance in various chemical applications based on machine learning, the graph-based representations may not be optimal to describe the molecular structures. As an example, Figure 7.8 shows the data distribution of the molecular graphs on a public molecular dataset called the Estimated SOLubility (ESOL) dataset [25]. The ESOL dataset contains 1128 SMILES of the organic molecules with their aqueous solubilities. In the figure, each point is a molecular graph representing a molecular structure, and the color of the point indicates the value of the aqueous solubility of the molecule. In this example, the molecular graph is not an optimal method to represent molecular structures in terms of aqueous solubility because there is no clear relationship between the molecular structures and the aqueous solubilities as shown in the mixed data distribution.

Representation learning is a machine learning program to learn data representations that optimally describe the data in the given datasets. The most common method for representation learning is Autoencoder that generates latent and compact representations of the data [51]. Autoencoder is an unsupervised representation learning method that generates a general data representation independent of target values. By contrast, metric learning is a supervised

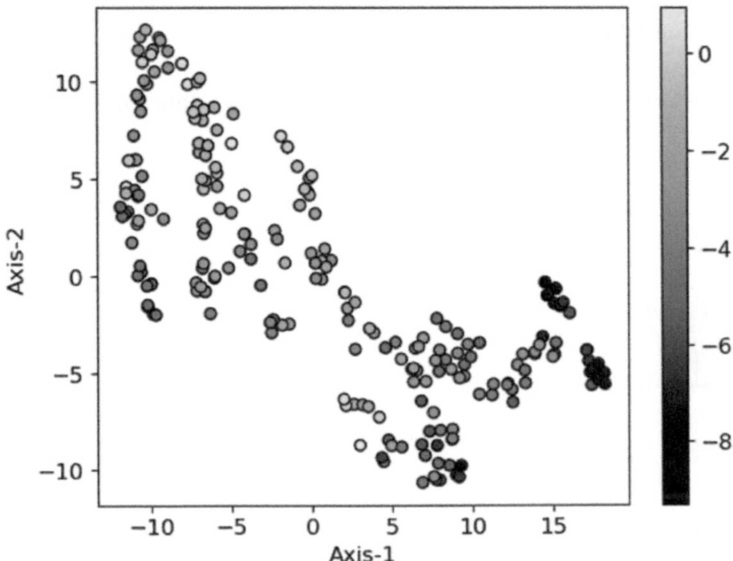

Figure 7.8 t-SNE visualization for data distribution of the molecular graphs on a public ESOL dataset [25]. Each point is a molecular graph representing a molecular structure, and the color of the point indicates the value of the aqueous solubility of the molecule.

representation learning method that generates a data representation explicitly specified to given target values [52, 53]. The effectiveness of the supervised representation learning in predicting molecular properties was demonstrated in various benchmark molecular datasets [53].

In the machine learning predictions with the representation learning, the prediction processes are performed through two training steps as:

- **Representation Learning**: A representation network f that transforms the original features into the latent features is trained to minimize or maximize the criteria for the representation learning. The training process of the representation networks can be unsupervised or supervised.
- **Regression (or Classification)**: A prediction model g is trained to minimize the prediction errors for the latent data representations $\mathrm{z} = f(\mathbf{x})$ rather than the original data representation x. Thus, the prediction model is optimized to predict the target values from the trained latent features of the data.

The main difference between the traditional regression (or classification) and the representation learning based regression (or classification) is that the data representation is trained with an explicitly defined embedding criteria in the representation learning based regression. The effectiveness of the representation learning based classification and regression have been widely demonstrated in the applications of computer vision [52, 54]. In the following sections, we will discuss the unsupervised and supervised representation learning methods and explore the representation learning methods for the molecular structures.

7.5.1 Unsupervised Representation Learning

Autoencoder and its variants are popular in representation learning of image data under an unsupervised manner. Autoencoders are also widely applied to various chemical applications to improve the prediction accuracies of the machine learning algorithms [55, 56]. For a given dataset $\mathcal{D} = \{\mathrm{x}_1, \ldots, \mathrm{x}_N\}$, the training problem of the autoencoders is usually defined by minimizing reconstruction loss as:

$$\theta^*, \mu^* = \operatorname{argmin}_{\theta,\mu} \sum_{i-1}^{N} \left\| \mathbf{x}_i - g(f(\mathbf{x}_i; \theta); \mu) \right\|_2^2, \tag{7.18}$$

where x_i is the input data in original feature space, $f(\mathbf{x}_i; \theta)$ is an encoder to generate the latent representation $\mathrm{z}_i = f(\mathbf{x}_i; \theta)$ of the original data x_i, and $g(\mathrm{z}_i; \mu)$ is a decoder to restore x_i from z_i. In the encoding process, the new data representation z_i is generated so that informative and latent features of $\mathbf{x}_i$ are extracted in low-dimensional spaces. By minimizing the reconstruction loss, as shown in Equation 7.18, the encoder is trained to generate the compact and informative data representation of the original data.

7.5.2 Supervised Representation Learning

In contrast to the unsupervised learning, supervised learning aims to generate a latent representation of the data in terms of given target variables. With supervised representation learning, we can obtain the latent representations that well describe the relationships between the input and target data. Metric learning [52] is the most representative method of supervised representation learning. The goal of metric learning is to learn a parameterized distance function that precisely describes the similarities and dissimilarities of the data. In computer science, metric learning has been widely studied to process the image data [52, 57, 58]. For the classification tasks on a dataset $\mathcal{D} = \{(\mathbf{x}_1, y_1), \ldots, (\mathbf{x}_N, y_N)\}$, the supervised learning is commonly defined by the following triplet embedding problem [52] as:

$$\theta^* = \underset{\theta}{\arg\min} \sum_{\mathbf{x} \in \mathcal{X}} \max(d(\mathbf{x}, \mathbf{x}_+; \theta) - d(\mathbf{x}, \mathbf{x}_-; \theta) + \alpha, 0) \tag{7.19}$$

where $d(\mathbf{x}, \mathbf{x}_+; \theta) = \|f(\mathbf{x}) - f(\mathbf{x}_+)\|_2^2$, $d(\mathbf{x}, \mathbf{x}_-; \theta) = \|f(\mathbf{x}) - f(\mathbf{x}_-)\|_2^2$ for a trainable neural network $f(\mathbf{x}; \theta)$ called embedding network, and $a > 0$ is a pre-defined margin. Two sampled data $\mathbf{x}_+$ and $\mathbf{x}_-$ are positive and negative samples of anchor data $\mathbf{x}$ respectively. Note that the anchor data is sequentially selected in D or a mini-batch in the training phase. Usually, the positive data is randomly selected from the data with the same label (target value) as the anchor data, whereas the negative data is randomly selected from the data with the different label to the anchor data. However, in addition to the random sampling, various sampling methods for positive and negative mining have been proposed to improve the embedding capabilities of the embedding networks. By training the embedding network to minimize the triplet embedding loss, we can get a feature space where the data with the same label is closely distributed, and the data with the different label is distributed in the different regions. In computer vision, the effectiveness of the supervised learning in image classification tasks is experimentally demonstrated in extensive benchmark datasets and various training environments [52, 57, 58].

Although metric learning has been successfully applied to various applications of computer vision, most existing metric learning methods are not applicable to the chemical applications on the regression problems because the existing metric learning methods are only defined on the classification problems. In other words, most existing metric learning methods cannot handle the continuous target values in their representation learning. In computer vision, log-ratio loss [54] was proposed to extend the applicability of metric learning into the continuous target values. The log-ratio loss L_{lr} is defined for the embedding network f as:

$$L_{lr} = \left\{ \log \frac{\|f(\mathrm{x}) - f(\mathrm{x}_+)\|_2^2}{\|f(\mathrm{x}) - f(\mathrm{x}_-)\|_2^2} - \log \frac{\|y - y_+\|_2^2}{\|y - y_-\|_2^2} \right\}^2, \tag{7.20}$$

where (x_+, y_+) and (x_-, y_-) are the selected samples from the entire training dataset or a mini-batch, and where $\|f(\mathbf{x}) - f(\mathbf{x}_+)\|_2^2 < \|f(\mathbf{x}) - f(\mathbf{x}_-)\|_2^2$ for all data. In metric learning, the model parameters of the embedding network were optimized to minimize the log-ratio loss. As shown in the log-ratio loss, the embedding network is trained to generate the data embeddings so that the distances between the data embeddings have the same values as the distances between the target values. However, the general applicability of the log-ratio loss is limited due to the numerical instability of the gradients in the training of the embedding network [53].

To overcome the numerical instability of the log-ratio loss, smoothed log-ratio loss [53] was proposed, and its effectiveness were empirically demonstrated on various molecular datasets. The smoothed log-ratio loss is defined by:

$$L_{lr} = \left\{ \psi \frac{\|f(\mathbf{x}) - f(\mathbf{x}_+)\|_2^2}{\|f(\mathbf{x}) - f(\mathbf{x}_-)\|_2^2} - \psi \frac{\|y - y_+\|_2^2}{\|y - y_-\|_2^2} \right\}^2, \tag{7.21}$$

where $\psi = -exp(\mathbf{x} + 1)$ is an approximated logarithm that has smooth gradients around zero. Furthermore, the smooth log-ratio loss generalizes the embedding networks to the graph neural networks beyond the convolutional neural networks [2]. Note that the encoder *f* should be defined as a network project that codes the data in a domain of the mathematical graphs into m-dimensional real valued space R^m for the unsupervised representation learning of the graphs [59]. With the embedding networks based on graph neural networks, the smooth log-ratio loss was applied to learn latent representations of the molecular from their molecular structures. Then, the latent molecular representations were used to predict the target molecular properties. In the experimental evaluations of the regression tasks, the conventional machine learning algorithms that employed the molecular representations of the embedding network *f* optimized to minimize the smooth log-ratio loss significantly outperformed the state-of-the-art graph convolutional network [53]. Specifically, the prediction errors of the machine learning algorithms reduced by 16.03%–48.99% in predicting aqueous solubilities [25], hydration free energies [26], octanol/water distribution coefficient [27], and atomization energies [23, 60].

7.6 Python Implementations

Python[4] is a high-level general-purpose programming language and most popular in machine learning implementation. Also, various state-of-the-art methods have been published in open source repositories with the python implementations. In addition to its popularity, Python is easy and efficient to implement software because it supports user-friendly programming environments and various third-party packages through PyPI.[5] For these reasons, Python will be a good choice to implement the machine learning applications for your first research work. In this section, we will present several examples of Python implementations for the molecular graphs and the graph-based machine learning for the chemical applications.

Figure 7.9 presents the workflow of the graph-based machine learning with the Python implementations. The workflow consists of two steps for data conversion and machine learning. In the data conversion, the chemical data representing the molecular structures is converted into the molecular graphs by Python packages for data-driven chemistry, such as RDKit[6] and Pymatgen.[7] After the data conversion, the graph neural networks are defined and trained through deep learning frameworks such as PyTorch.[8] In the following sections, we will explore several Python implementations of the machine learning workflow for the graph-based prediction based on the molecular structures.

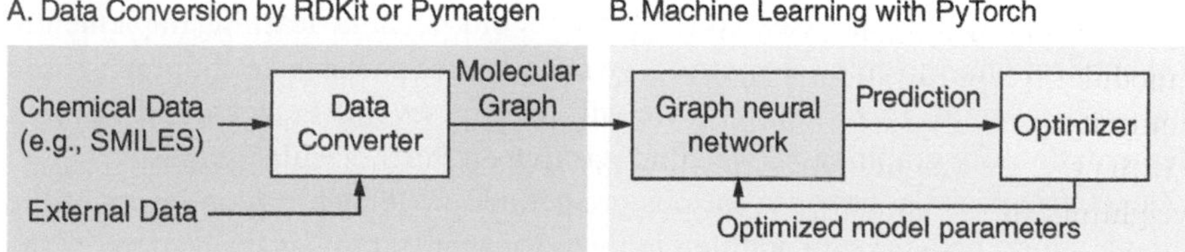

Figure 7.9 Overview of the Python implementation of the graph-based machine learning for the molecular and crystal structures. The implementation of the graph-based machine learning consists of two steps: 1) data conversion to represent the chemical data as the machine-readable data; and 2) machine learning based on graph neural networks.

7.6.1 Data Conversion: Molecular Structures to Molecular Graphs

The first step of the graph-based machine learning for chemical applications is to convert the chemical data into the machine-readable numerical data. One of the challenging parts in this data conversion is to generate the molecular graphs from the XYZ and SMILES representations of the molecules. However, various open-source cheminformatics libraries to support various data-driven research are publicly available. A well-known library called RDKit provides various data conversion methods to generate molecular object from SMILES, mol file, molecular image, and so on. For example, the SMILES representation of the melatonin "CC(=O)NCCC1=CNc2c1cc(OC)cc2" can be converted into the Python Mol object by the following Python code based on RDKit.

```
from rdkit import Chem
mol = Chem.MolFromSMILES('CC(=O)NCCC1=CNc2c1cc(OC)cc2')
```

For the XYZ file, two-step data conversion is required: 1) XYZ file to SMILES; and 2) SMILES to the Mol object. All Python modules for the two-step data conversion are provided in RDKit.

After generating the Mol object, you can construct the molecular graph $G = (\mathcal{V}, \mathcal{U}, \mathbf{X}, \mathbf{E})$ by extracting the atomic information and the adjacency matrix from the Mol object. To feed the molecular graphs into the machine learning models, PyTorch Geometric library[9] can be used in your Python implementations. You can generate PyTorch Geometric Data object by passing the node-feature matrix, the bond information, and the edge-feature matrix as:

```
from torch_geometric.data import Data
g = Data(x=atom_feats, edge_index=bonds, edge_attr=bond_feats)
```

where atom_feats is the node-feature matrix, bonds is a set of the paired indices of the atoms constructing a chemical bond, and bond_feats is the edge-feature matrix. If you generate the data object from your molecular structures, everything needed for data preparation in the graph-based machine learning is done.

7.6.2 Machine Learning: Deep Learning Frameworks for Graph Neural Networks

The second step of the graph-based machine learning for chemical applications is to train the graph-based prediction models for the generated molecular graphs. To this end, we will use PyTorch which is one of the most popular deep learning frameworks in machine research. Also, we can employ various third-party packages based on PyTorch for our machine learning implementations for chemical applications. PyTorch Geometric is an open-source package for the graph-based machine learning with PyTorch. In this section, we will present several PyTorch and PyTorch Geometric implementations for graph-based machine learning on the molecular graphs.

The PyTorch Geometric package provides various pre-defined node aggregation layers of the state-of-the-art graph neural networks. In addition to the node aggregation layers, it supports the readout functions and the pooling methods for extracting the graph-level representations from the node embeddings. For these reasons, we can efficiently implement the graph neural networks without low-level programming, by employing the PyTorch and PyTorch Geometric packages. For example, we can construct GCN for molecule-to-property prediction by declaring the pre-implemented network layers and stacking them as the following Python code.

In the above Python code, a graph neural network with the two graph convolutional layers and the following two dense layers are implemented to predict the target values from the molecular

```
import torch.nn as nn
import torch.nn.functional as F
from torch_geometric.nn import CGConv
from torch_geometric.nn import global_mean_pool

class GCN(nn.Module):
    def __init__(self, dim_in, dim_out):
        super(GCN, self).__init__()
        self.gc1 = GCNConv(dim_in, 128)
        self.gc2 = GCNConv(128, 128)
        self.fc1 = nn.Linear(128, 128)
        self.fc2 = nn.Linear(128, dim_out)

    def forward(self, g):
        h = F.relu(self.gc1(g.x, g.edge_index))
        h = F.relu(self.gc2(h, g.edge_index))
        hg = global_mean_pool(g, g.batch)
        hg = F.relu(self.fc1(hg))
        out = self.fc2(hg)
        return out
```

graphs. To generate the molecule-level representation from the atom embeddings, the readout function denoted by global_mean_pool is implemented in the forward method of the GCN class. The readout function is placed between the GCNConv and nn.Linear layers. The input argument g in the forward method is the PyTorch Geometric Data object generated from the XYZ file or SMILES in the data conversion step.

The next implementation of the graph-based machine learning is to optimize the model parameters of the graph neural networks for the training dataset. In this step, the molecular structures should be converted into the PyTorch Geometric Data objects, and the dataset of the molecular structures should be defined as a list of the PyTorch Geometric Data objects. For these datasets, the training step of the graph neural network can be written as the following Python code.

```
import torch
from torch_geometric.data import DataLoader
from sklearn.metrics import mean_absolute_error
from sklearn.metrics import r2_score

loader_train = DataLoader(dataset_train, batch_size=64)
loader_test = DataLoader(dataset_test, batch_size=64)

model = GCN(dim_in=32, dim_out=1).cuda()
optimizer = torch.optim.Adam(model.parameters())
loss = torch.nn.L1Loss()

for i in range(0, 1000):
    train(model, loader_train, optimizer, loss)

y_pred = predict(model, loader_test)
mae_test = mean_absolute_error(y_test, y_pred)
r2_test = r2_score(y_test, y_pred)
```

The first work of the training is to construct a data loader object for the datasets. In this implementation, two data loader objects loader_train and loader_test are constructed for the training and test datasets, respectively. After that, the graph neural network is defined as GCN that predicts the 1D target values from the 32-dimensional input features. The dimensionalities of the input and the target variables can be adjusted to your dataset by passing the arguments of the implemented GCN class. The optimizer and loss are the pre-defined classes in PyTorch for the training of the neural networks. The model parameters are optimized in the for loop during 1000 iterations. The train method should be manually implemented according to the graph neural network, but the PyTorch and PyTorch Geometric packages provide the code block for the training methods of the graph neural networks. After the training, the trained graph neural network is evaluated on the test dataset. Note that you should prepare the test dataset to evaluate the generalization capabilities of your machine learning model. The predict method also should be manually implemented with the code blocks of the PyTorch and PyTorch Geometric packages. Finally, the evaluation metrics provided in Scikit-learn[10] package are used to measure the prediction performances of the graph neural network for the unseen molecular structures.

7.6.3 Pymatgen for Crystal Structures

In addition to the molecular structures, the periodic crystal structures are popular in chemical data and the chemical applications. The chemical and physical properties of the crystal structures can also be predicted by machine learning implementations described in Section 7.6.2 without additional implementations, if the crystal structures are converted into the mathematical graphs. In this section, we present a Python implementation to convert the crystal structures into the mathematical graphs for machine learning in materials science.

Pymatgen is an open-source Python package for materials analysis. It provides highly flexible classes for the representation of molecule and crystalline systems. Also, it supports various input/output formats of the chemical data, such as VASP, ABINIT, CIF, and XYZ. The crystal structures stored by CIF, XYZ, and JSON files can be converted into the machine-readable mathematical graphs by the Structure model of Pymatgen. The following Python code converts the crystal structure in the "crystal.cif" file into the Python crystal object.

```
from pymatgen.core.structure import Structure
crystal = Structure.from_file('crystal.cif')
```

The converted crystal object contains chemical and physical properties, geometric information, and periodic structure of the crystal. By extracting the atomic and geometric features, we can generate the attributed graph $G = (\mathcal{V}, \mathcal{U}, \mathbf{X}, \mathbf{E})$ from the chemical data.

After the conversion, the crystal structures can be used for machine learning based on the Python implementations in Section 7.6.2, without additional implementations for the crystal structures. Furthermore, the composite data of the molecular and crystal structures can also be employed for machine learning based on the RDKit and Pymatgen packages, because the molecular and crystal structures that are heterogeneous in chemical data are converted into the mathematical graphs. For this reason, the graph-based machine learning is popular in materials science as well as molecular science.

7.7 Graph-based Machine Learning for Chemical Applications

Graph-based machine learning with the molecular graphs has shown state-of-the-art performances in various chemical applications [29, 34, 37]. The most intuitive application of the graph-based machine learning is to predict molecular properties from molecular structures. In addition to the property prediction, graph-based machine learning has been widely applied to predict and analyze the chemical reactions [37], drug discovery [30], and so on. In this section, we will explore the graph neural networks and the graph-based molecular representations for each chemical application.

7.7.1 Message Passing Neural Network to Predict Physical Properties of Molecules

Message passing is a method of node embedding for graph neural networks. Conceptually, message passing is similar to the atomic interactions in the molecules because it determines the state of the atom based on the neighborhood atoms and the environment of the atom. Message passing neural network (MPNN) [17] is a graph neural network to predict molecular properties from the molecular structures with the message passing between the atoms. MPNN predicted molecular properties on the QM9 dataset [24, 33] containing 130 k small organic molecules with 13 properties for each molecule. The molecular structures represented by SMILES were converted into the molecular graph $G = (\mathcal{V}, \mathcal{U}, \mathbf{X})$ without the bond features. The atomic features consist of atom type, atomic number, accepter type, donor type, aromaticity, hybridization, and number of bound hydrogens. In experimental evaluations, MPNN outperformed the conventional machine learning algorithms and showed an accurate prediction of HOMO-LUMO gaps of the molecules.

In addition to the single model and dataset, the prediction capabilities of the graph neural networks were comprehensively investigated on various benchmark molecular datasets [28]. The graph neural networks achieved better prediction accuracies than the conventional machine learning algorithms on 17 benchmark molecular datasets of quantum mechanics, physical chemistry, biophysics, and physiology. The evaluation results of the graph neural networks on the various benchmark datasets with different molecular properties showed the effectiveness of the graph neural networks and the graph representations of the molecular structures in predicting molecular properties.

7.7.2 Scale-Aware Prediction of Molecular Properties

To predict the graph-level properties, the graph-level embeddings should be calculated from the node and edge embeddings. To this end, the readout functions are placed between the node aggregation and dense layers in the graph neural networks. However, the readout functions can distort the scale information of the graph. This information loss is crucial in chemical applications because the scale information of the molecular structures such as molecular weight is important to predict the molecular properties. For example, the molecular weight is directly related to the aqueous solubility of the molecules in many cases. The scale distortion problem by the readout function is caused by the mean-based aggregation scheme of the readout functions. In the most popular mean-based readout functions, the graph-level embeddings are generated from the node embeddings as:

$$\mathbf{q} = \frac{1}{|\mathcal{V}|} \sum_{i=1}^{|\mathcal{V}|} \mathbf{Z}_i^{(K)}, \tag{7.22}$$

where $\mathbf{Z}^{(K)}$ is the node embeddings of the last node aggregation layers. As shown in the definition, the atom embeddings are divided by the number of atoms, and the information about the number of atoms and scale of the graph will disappear. For this reason, the graph neural networks with the mean-based readouts sometimes fail to predict the molecular properties related to the molecular scale [39].

To overcome the limitation of the mean-based readout functions, the readout function was investigated analytically and empirically [39]. The main results of the investigation are that the unnormalized readout function is preferred to preserve the scale information of the molecular graphs. For example, the mean-based readout is normalized because the weight of each atomic embedding is $\frac{1}{|\mathcal{V}|}$ and the sum of the weights is one. In contrast to the mean-based readout, the suggested unnormalized readout [39] has the weights in (0, 1) and their sum can exceed one. By employing the unnormalized readout, the prediction errors were reduced from 0.668 to 0.356 in predicting aqueous solubility in log(mol/L). In addition to the aqueous solubility, the performance improvements in machine learning were demonstrated on various benchmark molecular datasets. Thus, the nunormalized readout can be considered in the implementation of graph neural networks if the target molecular properties were related to the scale of the molecules, such as the molecular weight and the number of rings.

7.7.3 Prediction of Optimal Properties From Chromophore-Solvent Interactions

Recently, graph-based machine learning has been applied to predict the chemical and physical properties from the molecular interactions beyond the single molecular structure [34, 37]. As a successful application, a graph neural network to predict seven optical properties of 30,095 pairs of the chromophores and the solvents was proposed in chemical science [34]. In the paper, a pair of the chromophore and the solvent is abstracted as a pair of two molecular graphs of them. Then, the molecular graphs are separately embedded through two GCNs. Finally, the optical properties of the chromophores in the solvent are predicted by entering the concatenated vector of the graph embeddings of the molecular graphs into the dense prediction layers. The prediction process to predict target optical property y can be formally defined as:

$$y = f(g_c(G_c) \oplus g_s(G_s)), \tag{7.23}$$

where f is the dense prediction layer, g_c is GCN for the molecular graphs of the chromophores G_c, g_s is GCN for the molecular graphs of the solvents G_S, and $\oplus$ is the vector concatenation operator.

To describe the molecular structures as the molecular graphs, they defined the feature vectors of the atoms, as shown in Figure 7.10. In the actual implementation of the graph-based machine learning, the molecular structures of the chromophores and the solvents are converted into the molecular graphs with the node-feature matrix containing the atomic features in the figure. Based on the molecular graphs and the proposed interaction network in Equation 7.23, the seven optical properties were predicted with the prediction accuracies in R^2 scores of 0.714–0.926 [34]. In particular, the proposed neural network predicted the first absorption peak position of the chromophores with the R^2 score of 0.926. Therefore, it can be considered that the graph-based machine learning for predicting the first absorption peak position of the chromophore and solvent pairs has already reached the state of real-world practical applications.

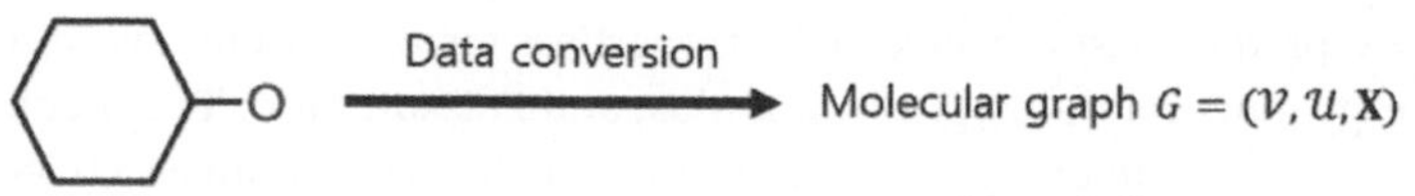

Data conversion → Molecular graph $G = (\mathcal{V}, \mathcal{U}, \mathbf{X})$

$\mathbf{X} =$

Element					# of H			# of atoms			Aromaticity	Hybridization			Ring	Formal charge		
C	N	O	...	Ge	0	...	4	0	...	5	ARM	sp	...	sp3d2	R	-4	...	4
1	0	0	...	0	1	...	0	0	...	0	1	0	...	0	1	0	...	0
1	0	0	...	0	0	...	0	0	...	0	1	0	...	0	1	0	...	0
1	0	0	...	0	0	...	0	0	...	0	1	0	...	0	1	0	...	0
1	0	0	...	0	0	...	0	0	...	0	1	0	...	0	1	0	...	0
1	0	0	...	0	0	...	0	0	...	0	1	0	...	0	1	0	...	0
1	0	0	...	0	0	...	0	0	...	0	1	0	...	0	1	0	...	0
0	0	1	...	0	0	...	0	0	...	0	0	0	...	0	0	0	...	0

Figure 7.10 Selected elemental attributes to define the atom-feature vectors in generating the molecular graphs. The atom-feature vectors are generated as a sparse vectors with binary values from seven categorical features of the elements.

7.7.4 Drug Discovery with Reinforcement Learning

The fundamental goal of machine learning in chemical science is to discover novel molecules with desired properties. That is, we want to devise a machine learning algorithm that automatically assembles the molecular graph of the desired target values. The most popular research field of machine learning to search the molecular structures is drug discovery. To search optimal graph data from the machine learning predictions, reinforcement learning [61] can be applied to the graph-based machine learning. Reinforcement learning is a search method based on the Markov decision process to discover optimal input that maximizes reward in the game [62].

In machine learning, graph convolutional policy network (GCPN) [30] was proposed to search optimal molecular graph for drug discovery. GCPN is an integrated algorithm of GCN and reinforcement learning. GCPN performs reinforcement learning to optimize the predicted value of GCN and provides the optimized molecular structures of the desired properties. In the experimental evaluations to discover the molecular structures of desired aqueous solubilities, GCPN found the molecular structures of the desired target values with a 100% probability. In addition to GCPN, various machine learning algorithms to discover novel molecular structures were proposed with machine learning methodologies of autoencoders [63], generative adversarial networks [64], and tree algorithms [65].

7.7.5 Graph Neural Networks for Crystal Structures

In addition to the molecules, the inorganic crystals form a large field of chemical applications, such as sensor device, battery, and solar cell materials. To predict physical properties from the crystal structures, several graph neural networks for periodic systems have been proposed in physics and materials science. The graph neural networks showed comparable accuracies with a classical density functional theory (DFT) [49] on benchmark materials datasets [18].

CGCNN [18] is the most popular graph neural network for crystal structures. The node aggregation layer of CGCNN is implemented based on crystal graph convolution that was designed to extract latent information from the crystal structures. The performance evaluation of CGCNN in predicting materials properties was conducted on a large open Materials Project Database.[11]

In the evaluation, CGCNN predicted seven materials properties: formation energy, absolute energy, band gap, Fermi energy, bulk modulus, shear modulus, and Poisson ratio. The prediction errors of CGCNN in predicting formation energy and band gap were 0.039 eV/atom and 0.388 eV, respectively. The overall prediction errors of CGCNN were comparable with DFT [18].

MEGNet [29] is also a graph neural network for the crystal structures. To improve the prediction accuracy, MEGNet exploits state attributes of the molecular and crystal structures as well as the local information of the atom and bond attributes. The state attribute was introduced to feed global information of the entire crystal structure into the node aggregation layers. By exploiting the global information from the entire crystal structures, MEGNet outperformed CGCNN in predicting formation energy, band gap, bulk modulus, and shear modulus on the Materials Project database.

7.8 Conclusion

In this chapter, we explored the basic machine-readable representations of the chemical data to the advanced graph-based representation of the molecular structures. The graph-based representation of the molecules is a useful method to describe the chemical characteristics and the structural information of the molecules. With the molecular graphs, graph neural networks have achieved state-of-the-art prediction accuracies over the conventional calculation- and simulation-based methods in various chemical applications. In addition to the conceptual description of graph neural networks, we presented the Python implementations of graph neural networks based on PyTorch and PyTorch Geometric libraries, and it will be helpful in developing your machine learning model for target chemical applications. However, existing molecular graphs are not complete and sometimes overlook the geometric information between the atoms in the molecular structures. As future research, novel graph-based molecular representations that can fully preserve the geometric information of the molecules should be investigated.

Notes

1 https://mendeleev.readthedocs.io/en/stable.
2 https://www.rdkit.org.
3 https://www.rdkit.org.
4 https://www.python.org.
5 https://pypi.org.
6 https://www.rdkit.org.
7 https://pymatgen.org.
8 https://pytorch.org.
9 https://pytorch-geometric.readthedocs.io/en/latest.
10 https://scikit-learn.org.
11 https://materialsproject.org.

References

1 Rosenblatt, F. (1958). The perceptron: a probabilistic model for information storage and organization in the brain. *Psychol. Rev.* 65: 386–408. doi:10.1037/h0042519.
2 LeCun, Y., Boser, B., Denker, J.S. et al. (1989). Backpropagation applied to handwritten zip code recognition. *Neural Comput.* 1 (4): 541–551.

3 Quinlan, J.R. (1986). Induction of decision trees. *Mach. Learn.* 1: 81–106. ISSN 0885-6125. doi:10.1023/A: 1022643204877.

4 Lee, J., Seko, A., Shitara, K. et al. (2016). Prediction model of band gap for inorganic compounds by combination of density functional theory calculations and machine learning techniques. *Phys. Rev. B Condens. Matter* 93: 115104. doi:10.1103/PhysRevB.93.115104.

5 Chen, X., Huang, L., Xie, D. et al. (2017). Egbmmda: extreme gradient boosting machine for mirna-disease association prediction. *Cell Death and Dis.* 31 (9): 3564–3572. https://doi.org/10.1038/s41419-017-0003-x. https://www.nature.com/articles/s41419-017-0003-x.

6 Zhuo, Y., Tehrani, A.M., and Brgoch, J. (2018). Predicting the band gaps of inorganic solids by machine learning. *J. Phys. Chem. Lett.* 9 (7): 1668–1673. doi:10.1021/acs.jpclett.8b00124.

7 Kim, H.W., Lee, S.W., Na, G.S. et al. (2021). Reaction condition optimization for non-oxidative conversion of methane using artificial intelligence. *React. Chem. Eng.* 6: 235–243. doi: 10.1039//D0RE00378F.

8 Na, G.S., Jang, S., and Chang, H. (2021). Predicting thermoelectric properties from chemical formula with explicitly identifying dopant effects. *NPJ Comput. Mater.* 7 (1): 106. ISSN 2057-3960. doi:10.1038/s41524-021-00564-y.

9 Quito, M., Monterola, C., and Saloma, C. (2001). Solving *N*-body problems with neural networks. *Phys. Rev. Lett.* 86: 4741–4744. doi:10.1103/PhysRevLett.86.4741. https://link.aps.org/doi/ 10.1103/PhysRevLett.86.4741.

10 Do, K., Tran, T., and Venkatesh, S. (2017). Matrix-centric neural networks. *CoRR*, abs/1703.01454. http://arxiv.org/abs/1703.01454.

11 Hirsch, R. (2004). *Exploring Colour Photography: A Complete Guide*. London: Laurence King. ISBN 9781856694209.

12 Duvenaud, D.K., Maclaurin, D., Iparraguirre, J. et al. (2015). Convolutional networks on graphs for learning molecular fingerprints. *Advances in Neural Information Processing Systems (NIPS)* 28.

13 Yang, M., Tao, B., Chen, C. et al. (2019). Machine learning models based on molecular fingerprints and an extreme gradient boosting method lead to the discovery of jak2 inhibitors. *J. Chem. Inf. Model.* 59 (12): 5002–5012. doi:10.1021/acs.jcim.9b00798. https://doi.org/10.1021/acs.jcim.9b00798.

14 Choi, H. (2018). Deep learning in nuclear medicine and molecular imaging: current perspectives and future directions. *Nucl. Med. Mol.* 52 (2): 109–118. doi: 10.1007/s13139-017-0504-7. https://pubmed.ncbi.nlm.nih.gov/29662559.

15 Prates, M., Avelar, P.H.C., Lemos, H. et al. (2019). Learning to solve np-complete problems: a graph neural network for decision TSP. In: *Proceedings of the AAAI Conference on Artificial Intelligence (AAAI)*, 33(1): 4731–4738. doi:10.1609/aaai.v33i01.33014731. https://ojs.aaai.org/index.php/AAAI/article/view/4399.

16 Ying, R., He, R., Chen, K. et al. (2018). Graph convolutional neural networks for web-scale recommender systems. In: *Proceedings of the 24th ACM SIGKDD International Conference on Knowledge Discovery and Data Mining (KDD)*, 974–983. Association for Computing Machinery. ISBN 9781450355520. doi:10.1145/3219819.3219890. https://doi.org/10.1145/3219819.3219890.

17 Justin Gilmer, S.S., Schoenholz, P.F., Riley, O.V. et al. (2017). Neural message passing for quantum chemistry. In: *Proceedings of the 34th International Conference on Machine Learning (ICML)*, 1263–1272. ICML'17. JMLR.org.

18 Xie, T. and Grossman, J.C. (2018). Crystal graph convolutional neural networks for an accurate and interpretable prediction of material properties. *Phys. Rev. Lett.* 120: 145301. doi:10.1103/PhysRevLett.120.145301. https://link.aps.org/doi/10.1103/PhysRevLett.120.145301.

19 Na, G.S., Kim, H.W., and Chang, H. (2020). Costless performance improvement in machine learning for graph-based molecular analysis. *J. Chem. Inf. Model* 60 (3): 1137–1145. doi:10.1021/acs.jcim.9b00816. https://doi.org/10.1021/acs.jcim.9b00816.

20 Zang, C. and Wang, F. (2020). *MoFlow: An Invertible Flow Model for Generating Molecular Graphs*, 617–626. Association for Computing Machinery. ISBN 9781450379984. https://doi.org/10.1145/3394486.3403104.

21 Na, G.S., Jang, S., Lee, Y.-L. et al. (2020). Tuplewise material representation based machine learning for accurate band gap prediction. *J. Phys. Chem. A* 124 (50): 10616–10623. doi:10.1021/acs.jpca.0c07802. https://doi.org/10.1021/acs.jpca.0c07802.

22 Weininger, D. (1988). Smiles, a chemical language and information system. *J. Chem. Inf. Comput. Sci.* 28: 31–36.

23 Blum, L.C. and Reymond, J.-L. (2009). 970 million druglike small molecules for virtual screening in the chemical universe database GDB-13. *J. Am. Chem. Soc.* 131: 8732–8733.

24 Ramakrishnan, R., Dral, P.O., Rupp, M. et al. (2014). Quantum chemistry structured and properties of 134 kilo molecules. *Sci. Data* 1 (1): 1–7

25 Delaney, J.S. (2004). ESOL: estimating aqueous solubility directly from molecular structure. *J. Chem. Inf. Comput. Sci.* 44: 1000–1005.

26 David, L., Mobley, and Guthrie, P. (2014). Freesolv: a database of experimental and calculated hydration free energies, with input files. *J. Comput. – Aided Mo. Des.* 28: 711–720.

27 Wenlock, M. and Tomkinson, N. (2015). Experimental *in vitro* DMPK and physicochemical data on a set of publicly disclosed compounds. https://www.ebi.ac.uk/chembl/document-report-card/CHEMBL3301361.

28 Wu, Z., Ramsundar, B., Feinberg, E.N. et al. (2018). Moleculenet: a benchmark for molecular machine learning. *Chem. Sci.* 9: 513–530.

29 Chen, C., Weike, Y., Zuo, Y. et al. (2019). Graph networks as a universal machine learning framework for molecules and crystals. *Chem. Mater.* 31 (9): 3564–3572. doi: 10.1021/acs.chemmater.9b01294. URLhttps://doi.org/10.1021/acs.chemmater.9b01294.

30 You, J., Liu, B., Ying, R. et al. (2018). Graph convolutional policy network for goal-directed molecular graph generation. In: *Proceedings of the 32nd International Conference on Neural Information Processing Systems (NIPS)*, NIPS'18. 6412–6422.

31 Kipf, T.N. and Welling, M. (2017). Semi-supervised classification with graph convolutional networks. In: *International Conference on Learning Representations (ICLR).*

32 Velickovic, P., Cucurull, G., Casanova, A. et al. (2018). Graph attention networks. In: *International Conference on Learning Representations (ICLR).*

33 Ruddigkeit, L., Ruud, V.D., Blum, L.C. et al. (2012). Enumeration of 166 billion organic small molecules in the chemical universe database gdb-17. *J. Chem. Inf. Model* 52: 2864–2875.

34 Joung, J.F., Han, M., Hwang, J. et al. (2021). Deep learning optical spectroscopy based on experimental database: potential applications to molecular design. *JACS Au* 1 (4): 427–438. doi:10.1021/jacsau.1c00035. https://doi.org/10.1021/jacsau.1c00035.

35 Grambow, C.A., Pattanaik, W.H., and Green, L. (2020). Reactants, products, and transition states of elementary chemical reactions based on quantum chemistry. *Sci. Data* 7 (1): 137. May. ISSN 20524463. doi:10.1038/s41597-020-0460-4. https://doi.org/10.1038/s41597-020-0460-4.

36 Joung, J.F., Han, M., Jeong, M. et al. (2020). Experimental database of optical properties of organic compounds. *Sci. Data* 7 (1): 295. ISSN 2052-4463. doi:10.1038/s41597-020-00634-8. https://doi.org/10.1038/s41597-020-00634-8.

37 Pathak, Y., Laghuvarapu, S., Mehta, S. et al. (2020). Chemically interpretable graph interaction network for prediction of pharmacokinetic properties of drug-like molecules. *Proceedings of the AAAI Conference on Artificial Intelligence (AAAI)* 34 (1): 873–880. doi:10.1609/aaai.v34i01.5433. https://ojs.aaai.org/index.php/AAAI/article/view/5433.

38 Keyulu, X., Weihua, H., Leskovec, J. et al. (2019). How powerful are graph neural networks? In: *International Conference on Learning Representations (ICLR).*

39 Na, G.S., Kim, H.W., and Chang, H. (2020). Scale-aware graph-based machine learning for accurate molecular property prediction. In: *2020 IEEE International Conference on Big Data (Big Data).* 196–203. doi:10.1109/BigData50022.2020.9377905.

40 Vinyals, O., Bengio, S., and Kudlur, M. (2016). Order matters: sequence to sequence for sets. In: *4th International Conference on Learning Representations, ICLR.*

41 Lee, J.B., Rossi, R., and Kong, X. (2018). Graph classification using structural attention. In: *ACM SIGKDD Conference on Knowledge Discovery and Data Mining (KDD).*

42 Ying, R., You, J., Morris, C. et al. (2018). Hierarchical graph representation learning with differentiable pooling. In: *Conference on Neural Information Processing Systems (NIPS).*

43 Keyulu, X., Chengtao, L., Tian, Y. et al. (2018). Representation learning on graphs with jumping knowledge networks. In: *International Conference on Machine Learning (ICML).*

44 Vaswani, A., Shazeer, N., Parmar, N. et al. (2017). Attention is all you need. *Advances in Neural Information Processing Systems (NIPS)*, 30

45 Devlin, J., Chang, M.-W., Lee, K. et al. (2019). BERT: pre-training of deep bidirectional transformers for language understanding. *NAACL-HLT* 1: 4171–4186. doi:10.18653/v1/N19-1423.

46 Wang, F., Jiang, M., Qian, C. et al. (2017). Residual attention network for image classification. In: *2017 IEEE Conference on Computer Vision and Pattern Recognition (CVPR)*, 6450–6458. doi:10.1109/ CVPR.2017.683.

47 Simonovsky, M. and Komodakis, N. (2017). Dynamic edge-conditioned filters in convolutional neural networks on graphs. In: *IEEE Conference on Computer Vision and Pattern Recognition (CVPR)*, 29–38. doi: 10.1109/CVPR.2017.11.

48 Jain, A., Ong, S.P., Hautier, G. et al. (2013). Commentary. The materials project: a materials genome approach to accelerating materials innovation. *APL Mater.* 1 (1): 011002. doi:10.1063/1.4812323.

49 Kohn, W. and Sham, L.J. (1965). Self-consistent equations including exchange and correlation effects. *Phys. Rev.* 140: A1133–A1138. doi:10.1103/PhysRev.140.A1133. https://link.aps.org/doi/10.1103/PhysRev.140.A1133.

50 Park, C.W. and Wolverton, C. (2020). Developing an improved crystal graph convolutional neural network framework for accelerated materials discovery. *Phys. Rev. Mater.* 4 (6): 063801.

51 Baldi, P. (2012). Autoencoders, unsupervised learning, and deep architectures. *Proceedings of Machine Learning Research (JMLR)* 27: 37–49.

52 Weinberger, K.Q., Blitzer, J., and Saul, L.K. (2009). Distance metric learning for large margin nearest neighbor classification. In: *Conference on Neural Information Processing Systems (NIPS).*

53 Na, G.S., Chang, H., and Kim, H.W. (2020). Machine-guided representation for accurate graph-based molecular machine learning. *Phys. Chem. Chem Phys.* 22 (33): 18526–18535.

54 Kim, S., Seo, M., Laptev, I. et al. (2019). Deep metric learning beyond binary supervision. In: *IEEE Conference on Computer Vision and Pattern Recognition (CVPR).*

55 Gomez-Bombarelli, R., Wei, J.N. et al. (2018). Automatic chemical design using a data-driven continuous representation of molecules. *ACS Cent. Sci.* 4 (2): 268–276. doi:10.1021/acscentsci.7b00572.

56 Liu, Q., Allamanis, M., Brockschmidt, M. et al. (2018). Constrained graph variational autoencoders for molecule design. In: *Proceedings of the 32nd International Conference on Neural Information Processing Systems (NIPS)*, NIPS'18, 7806–7815. Red Hook, NY: Curran Associates Inc.

57 Hadseel, R., Chopra, S., and Yann, L. (2006). Dimensionality reduction by learning an invariant mapping. In: *IEEE Conference on Compute Vision and Pattern Recognition (CVPR).*

58 Sohn, K. (2016). Improved deep metric learning with multi-class n-pair loss objectives. In: *Conference on Neural Information Processing Systems (NIPS).*

59 Kipf, T.N. and Welling, M. (2016). Variational graph auto-encoders. *NIPS Workshop on Bayesian Deep Learning.*

60 Rupp, M., Tkatchenko, A., Muller, K.-R. et al. (2012). Fast and accurate modeling of molecular atomization energies with machine learning. *Phys. Rev. Lett.* 108: 058301.

61 Sutton, R.S. and Barto, A.G. (2018). *Reinforcement Learning: An Introduction.* Cambridge, MA: A Bradford Book. ISBN 0262039249.

62 Silver, D., Huang, A., Maddison, C.J. et al. (2016). Mastering the game of Go with deep neural networks and tree search. *Nature* 529 (7587): 484–489. ISSN 0028-0836. doi:10.1038/nature16961.

63 Lim, J., Ryu, S., Kim, J.W. et al. (2018). Molecular generative model based on conditional variational autoencoder for *de novo* molecular design. *J. Cheminformatics* 10 (1): 31. ISSN 1758-2946. doi:10.1186/s13321-018-0286-7. https://doi.org/10.1186/s13321-018-0286-7.

64 Chenthamarakshan, V., Das, P., Hoffman, S.C. et al. (2020). Cogmol: target-specific and selective drug design for Covid-19 using deep generative models. *Advances in Neural Information Processing Systems* 33: 4320–4332.

65 Jin, W., Barzilay, R., and Jaakkola, T. (2018). Junction tree variational autoencoder for molecular graph generation. In: *International Conference on Machine Learning*, 2323–2332.

8

Extracting Metal-Organic Frameworks Data from the Cambridge Structural Database

Aurelia Li, Rocio Bueno-Perez, and David Fairen-Jimenez

Adsorption & Advanced Materials Laboratory (AAML), Department of Chemical Engineering & Biotechnology, University of Cambridge, Cambridge, United Kingdom

8.1 Introduction

Metal-organic frameworks (MOFs) are generally described as compounds coordinatively assembled from metal clusters or nodes (also called secondary building units or SBUs) and organic ligands in a building block approach (Figure 8.1) [1]. Their modularity has brought them increasing attention in the last two decades, making them one of the fastest-growing subsets of extended structures [2–7]. In addition, their relatively straightforward synthesis has enabled the experimental design of evermore customized structures. Their large range of pore sizes, geometries, internal surface areas (up to 8000 m^2/g) [8], and pore volumes are the reason MOFs are considered for a variety of applications, ranging from gas storage [9–13], separation [14–18], catalysis [19–21], to drug delivery [22–26] and bio-imaging [23,24,27]. At this point, some questions that have arisen in our community include: i) How many MOFs have been synthesized? ii) What are their structure–property landscapes? and iii) Is there already a best-fit structure for a given application? To answer these questions, the computational MOF community has gradually turned to data-mining techniques, usually involving database probing and high-throughput screenings (HTS). Although crystallinity is not mentioned in their definition [28], in this chapter, we will consider crystalline MOFs only for two reasons: i) crystalline MOFs are the most extensively studied MOFs, and ii) their crystalline data are readily available, accessible, and convenient for many types of simulations requiring the knowledge of atom positions.

Indeed, when publishing a new crystal structure, authors are many times required to deposit the corresponding data with the Cambridge Structural Database (CSD), curated by the Cambridge Crystallographic Data Centre (CCDC) [29]; also, the CCDC looks for new structures in the literature to keep the CSD updated. The data consist in Crystallographic Information Files (CIFs) resulting from X-ray, neutron, and electron diffraction analyses, and containing, among other crystalline data, information such as the atom positions, and the unit cell lengths and angles [30]. The CSD is therefore the largest database of experimentally-obtained crystal structures, including organic and metal-organic. Each structure is identified with a unique CSD refcode composed of six letters and, if necessary, two digits. The number of deposited structures has greatly increased over the last 44 years, reaching the milestone of 1 million in 2019 [31]. Among these data are those of crystalline MOFs, and several separate attempts at extracting them have been made in the past [32–34]. One of the first and largest subsets extracted is that of Watanabe et al., who mined 30,000 MOFs from the CSD to identify promising structures for the separation of CO_2 and N_2. Unfortunately, the methods for mining the MOFs were not fully disclosed [32]. Later, using a set of labeled MOFs, Goldsmith et al. identified data features that indicate if a structure is a MOF [33]. Based on these features, 38,800

AI-Guided Design and Property Prediction for Zeolites and Nanoporous Materials, First Edition. Edited by German Sastre and Frits Daeyaert.

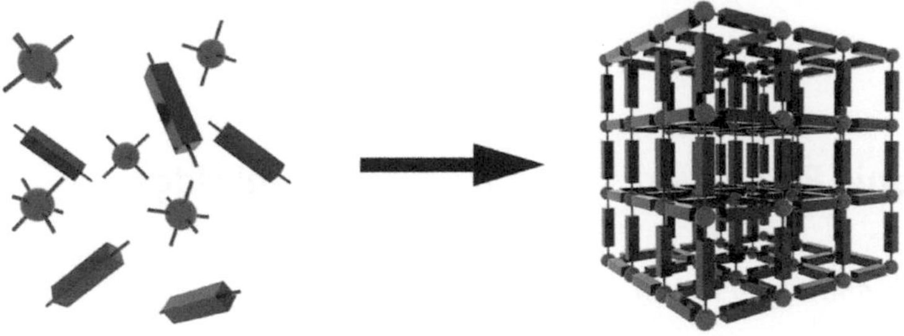

Figure 8.1 Metal-organic frameworks (MOFs) are assembled from metal clusters and organic linkers. *Source:* Reproduced from *Chem. Mater.*, 2017, 29, 7, 2618–2625. © 2017, American Chemical Society.

structures were further extracted from the CSD, and 20,700 were then successfully cleaned for simulations. Using the structure-property landscape of this dataset of structures, Goldsmith et al. were able to identify the theoretical limits of hydrogen storage. However, the first publicly available dataset of MOFs was published by Chung et al. Named Computation-ready, Experimental (CoRE) MOF database, this dataset grew from an initial 5000 curated structures in 2014 to over 14,000 in 2019 [34,35]. The MOF data in the CoRE MOF database were prepared to be simulation-ready. After having extracted an initial set of MOFs using ConQuest [36], search software implemented by the CCDC for the CSD, the structures were further cleaned: bound and/or unbound solvents were removed to mimic the experimental "activation" process where the solvents introduced during the solvothermal synthesis are removed under heat and vacuum; missing hydrogens from the crystalline data were added; and disordered structures were either discarded or manually repaired.

The CoRE MOF database proved very useful and was quickly adopted by the computational MOF community for myriads of HTSs [37–41]. However, being focused on gas adsorption applications, it consists of porous three-dimensional (3D) MOFs only and is not suitable for the study of electrical conductivity [42] or luminescent [27] and magnetic [43] properties, for which porosity and 3D structures are not essential. It also requires frequent manual updates, as well as manual cleaning and processing without any version control. To offer a more global and free-to-process dataset, we teamed up with the CCDC [44] to build the first CSD-integrated, automatically, and quarterly updated MOF dataset: the CSD MOF subset. This new subset contains a wide variety of porous and non-porous, 1D, 2D, and 3D MOF-like structures, which added up to a total of 70,000 structures in 2016 (Figure 8.2) and more than 100,000 in 2021 [44]. To extract the data, the authors used ConQuest to develop seven criteria that filter the CSD database and return MOF-like structures. This subset is now available in the software package developed by the CCDC [36]. To aid the computational community with the preparation of structures for simulations, a non-disordered CSD MOF subset of ca. 55,000 MOFs was also released. By being fully integrated into the CSD, the CSD Python application programming interface (API) [29] naturally applies to the CSD MOF subset. As a result, several Python scripts also complement the database for users to: i) remove unbound and/or bound solvents on a single structure; ii) remove these solvents on a subset of structures; and iii) identify the framework dimensionality.

Avci et al. recently illustrated the importance of such regularly updated databases by investigating the newly added structures in the CSD MOF subset [45]. The authors analyzed the structures for carbon capture and hydrogen storage in the CSD MOF subset in 2018 (3857 structures) and 2020 (10,221 structures). The results showed that a larger number of MOFs among the updated dataset exceed the target adsorption performance; many MOFs even out-perform benchmark zeolites on specific adsorption metrics.

This chapter first presents the building process of the CSD MOF subset, before providing a general overview of its characterization and methods for the targeted classification of structural data stored therein. We also discuss the place of the CSD MOF subset among all the existing MOF

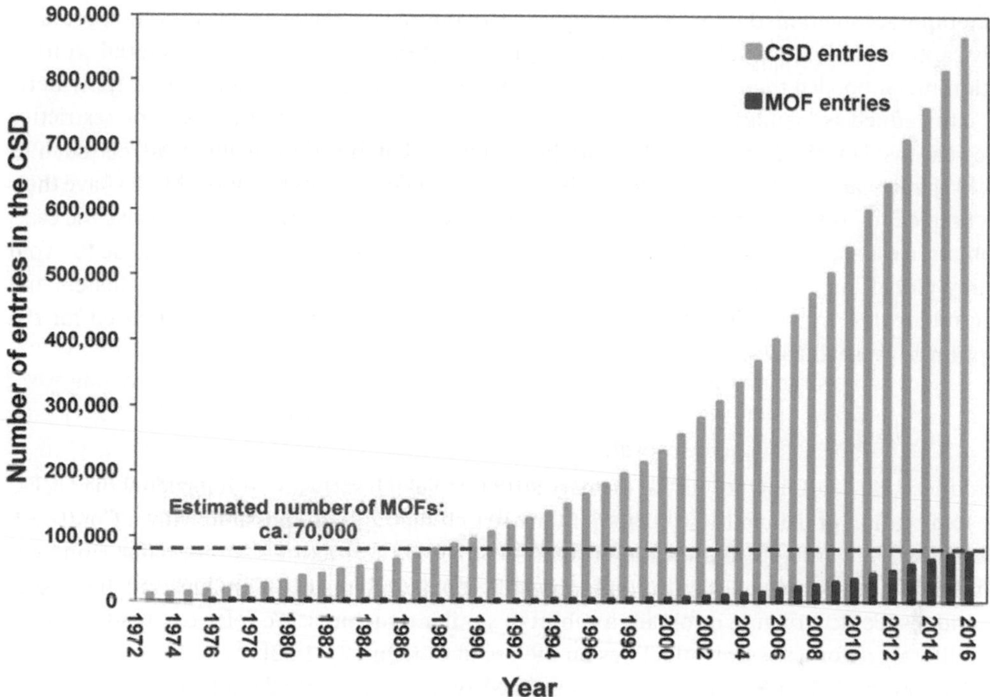

Figure 8.2 Evolution of the number of Cambridge Structural Database (CSD) entries and estimated number of deposited metal–organic framework (MOF) structures up to 2016. *Source:* Reproduced from *Chem. Mater.*, 2017, 29, 7, 2618–2625. © 2017, American Chemical Society.

databases. While the CSD MOF subset and CoRE MOF contain data of MOFs obtained experimentally, other databases provide theoretical structures that may not have been synthesized yet. The availability of a diverse set of MOF databases ensures that a large MOF landscape is covered and studied, provided the structural differences between the databases are well-understood.

8.2 Building the CSD MOF Subset

8.2.1 What Is a MOF?

To extract MOFs from the CSD, the first intuitive step is to define them. Unfortunately, this is not straightforward, as different researchers across different disciplines conceptualize MOFs differently, so much so that the International Union of Pure and Applied Chemistry (IUPAC) purposely kept a definition that is wide enough to account for all the possibilities: MOFs are "coordination polymer[s] (or, alternatively, coordination network[s]) with an open framework containing potential voids" [46]. A coordination polymer is defined here as "a coordination compound continuously extending in 1, 2, or 3 dimensions through coordination bonds," and a coordination network is a subset of the former: "a coordination compound extending, through coordination bonds, in 1 dimension, but with cross-links between two or more individual chains, loops or spiro-links, or a coordination compound extending through coordination bonds in 2 or 3 dimensions." MOFs are therefore coordination polymers, with potential porosity. The IUPAC even considers that, for a structure to be considered as a MOF, "no physical measurements of porosity or other properties are demanded *per se*" [28], which makes MOF distinction even more difficult.

As a temporary solution, the CCDC had suggested an informal MOF description, which could be directly captured in ConQuest [36]. In this guideline, a MOF was described as [a material with] "a transition metal bonded to an oxygen or nitrogen atom via a polymeric bond," where "polymeric bonds" are defined as "bonds between repeating units." However, this definition was too restrictive for two reasons: i) it excludes all MOFs containing metals that are not transition metals, such as alkalis [47], rare earths [48], and metalloids [49], among others; and ii) while most MOFs have their metals bonded to organic linkers via at least one nitrogen or oxygen atom, it is not always the case.

To obtain a dataset amidst the debates surrounding the definition of MOFs, we chose to avoid using any global interpretation and to describe a variety of linkages between metal clusters and organic molecules. In the following section, we present the software and methods used for the description of these linkages.

8.2.2 ConQuest

As introduced earlier, ConQuest is the primary structure search software developed by the CCDC for the exploration of data in the CSD [36]. Many search methods are possible within ConQuest, from drawing a substructure in three dimensions, to providing a structure's author name, and from combining different search queries to combining different search results. ConQuest can be used to either find a specific structure or obtain a subset of similar structures. We refer our readers to the relevant literature for extensive guidelines on the use of ConQuest [31,50].

The derivation of the CSD MOF subset combined two search methods: a textual information search that captured the periodic nature of MOFs; and the *Draw* function, which targeted specific linkage chemistries. The textual search looked for the keyword *catena* (declined as "catena-," "catena(" and "catena[" in the CIFs), a tag the CCDC uses for "polymeric" structures containing metal atoms. In the CSD, polymeric structures are defined as structures containing "polymeric" bonds. Essentially, the presence of polymeric bonds translates the fact that MOFs are extended structures. *Draw* enables users to draw entire molecules or fragments that should be contained in the targeted structure. By default, *Draw* queries are made in two-dimensional (2D) and the input is the targeted structure's chemical diagram. In this chapter, we will use indifferently the terms "2D diagram" and "chemical diagram" to refer to the chemical representation of the structures stored in the CSD. Figures 8.3a and 8.3c give two examples of such diagrams. ConQuest will then look through all the chemical diagrams in the CSD and find an exact match. However, designing a *Draw* query for MOFs is tricky. As extended structures, MOFs are only partially represented, with the help of the previously introduced polymeric bonds [51]. This means that there are different ways of placing the polymeric bonds and, in the absence of any convention, different ways of representing the same structure in a 2D diagram. Figure 8.3 shows two structures from one of the simplest configurations, where the carboxylate ligands are connected to zinc atoms. The blue circles highlight the linkages stored in the CSD, and in this case, there are at least two possible representations. The query used in Figure 8.3a to describe the linkage is intuitive but also too specific and will miss structures such as the one in Figure 8.3d.

The seven criteria, or *Draw* queries, built for the CSD MOF subset, are presented in Figure 8.4. These searches were made with version 5.37 of the CSD, including updates up to May 2016 and resulted in 69,666 MOFs. Each criterion represents a category of SBU-ligand linkage. To make sure we covered as many linkages as possible, the dataset was constantly compared to known existing MOFs and to the CoRE MOF database. Each time we found a known MOF that was not included in the CSD MOF subset, a new criterion for ConQuest was developed or an existing criterion was refined, until all known MOFs were included in the subset. The criteria are not all mutually

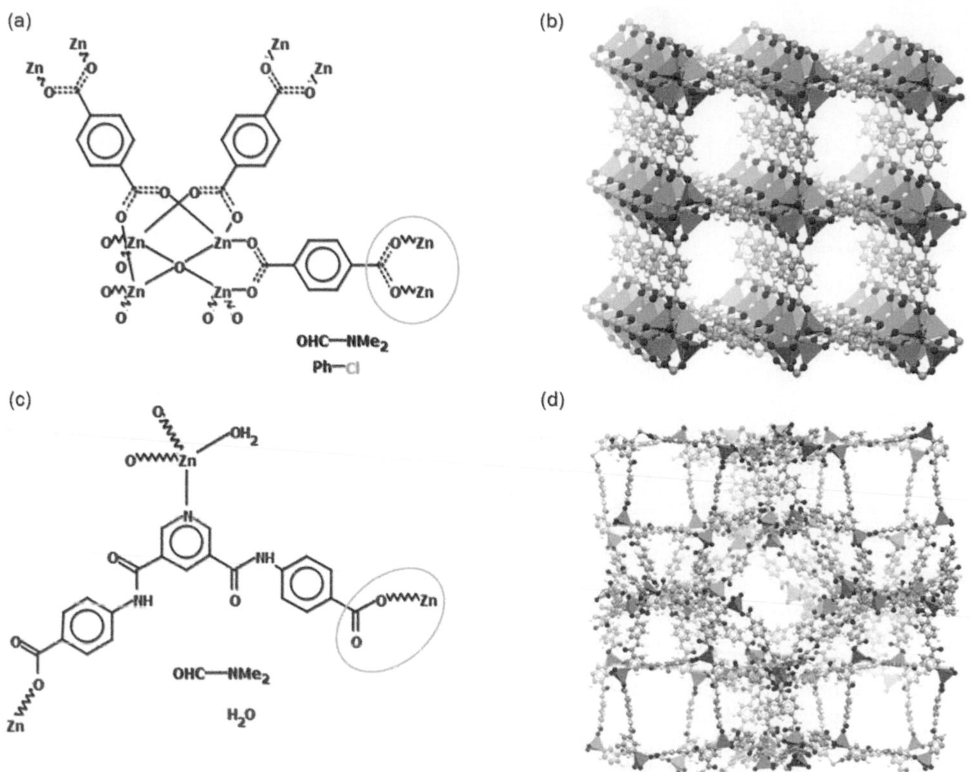

Figure 8.3 Example of two different ways of representing carboxylate ligands: (a) chemical diagram of SAHYIK; (b) 3D representation of SAHYIK: (c) chemical diagram of ADUROI; and (d) 3D representation of ADUROI. *Source: CrystEngComm*, 2020, 22, 7152–7161 from the Royal Society of Chemistry.

exclusive, as some structures can be found with several criteria. We describe each criterion in more detail in the following paragraphs.

Criterion 1 is based on carboxylate linkages that comprise the majority of MOFs synthesized today. Some of the most well-known examples are IRMOF-1, HKUST-1, UiO-66, and their derivatives. In their simplest forms, they contain 1,4-benzenedicarboxylic (BDC) linkers connected to zinc, copper, and zirconium oxide metal corners, respectively. There are three types of carboxylate-based linkers: i) those containing aromatic rings only; ii) those containing hydrocarbon chains only; and iii) those containing a mix of both. Figure 8.5 shows the tuning of criterion **1** from a simple BDC linker to a query that would return all carboxylate-based MOFs with cyclic hydrocarbon linkers. In its most intuitive form, metal atoms are connected to the BDC linkers (Figure 8.5a). To include carboxylate-derived linkers, the atoms of oxygen and carbon are replaced by QA, which could be any of C, N, O, P, B, or S (Figure 8.5b). To design the broadest criterion possible, we then paid special attention to the position of the polymeric bond, as explained previously. Similarly to the situation described in Figure 8.3, the criterion in Figure 8.5b was too restrictive. We, therefore, tuned it to the criterion shown in Figure 8.5c; this new criterion is broader without losing any of its specificity. The criterion is finally generalized to all types of rings and not limited to aromatic rings (Figure 8.5d). A similar procedure was carried out for MOFs with organic chains. Criterion **1** constitutes the majority of the database, returning 50,046 structures, i.e., 72% of the final database.

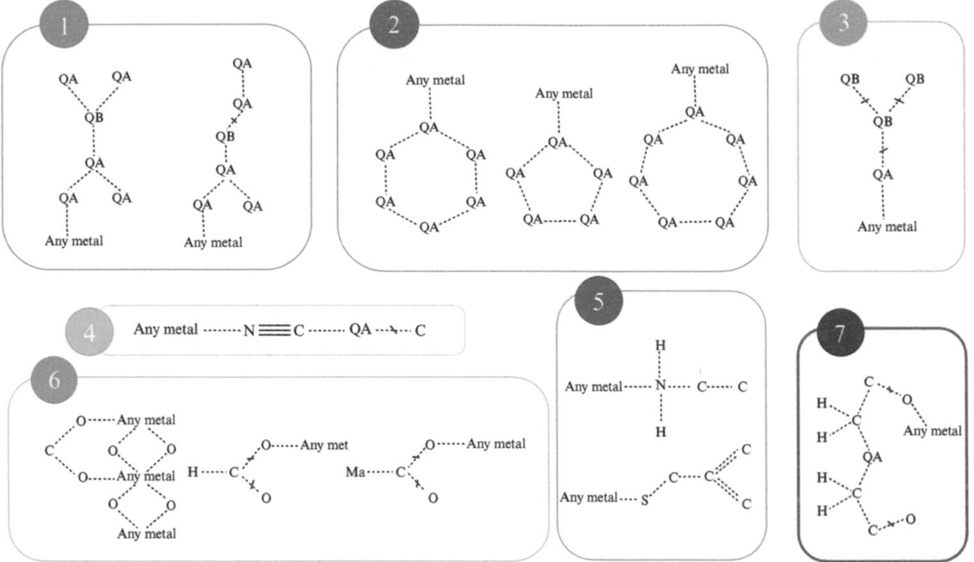

Figure 8.4 Summary of the seven criteria designed to build the Cambridge Structural Database (CSD) metal–organic framework (MOF) subset, where QA = O, N, P, C, B, and S. QB = N, P, B, S, and C and superscripts; "c" and "a" impose the corresponding atoms to be "cyclic" or "acyclic," respectively. Me denotes methyl groups. The dotted line refers to any of the bond types stored in the CSD (single, double, triple, quadruple, aromatic, polymeric, delocalized, and pi). The dotted line with the two lines through indicates a variable bond type (i.e., two or more of the options above). In these cases, the variable type is single, double, or delocalized. The union of these seven criteria combined with the presence of catena in the compound name found 69,666 MOF structures from version 5.37 of the CSD, including updates up to May 2016. *Source:* Reproduced with permission from *Chem. Mater.*, 2017, 29, 7, 2618–2625. © 2017, American Chemical Society.

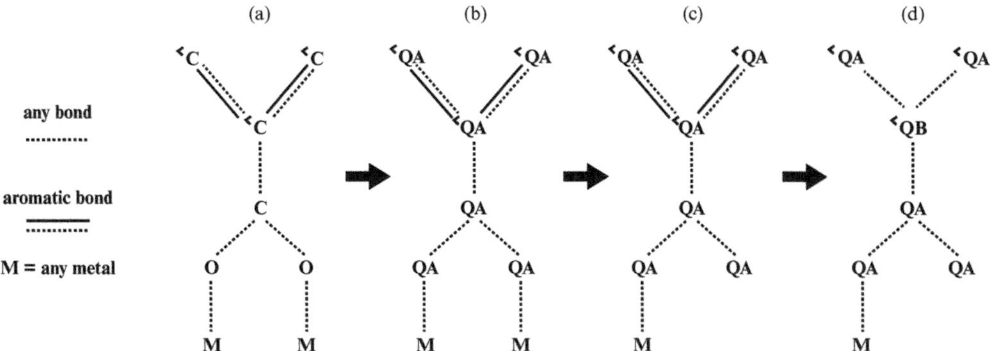

Figure 8.5 Evolution of Criterion 1 to describe metal–organic frameworks (MOFs) containing aromatic carboxylate-derived linkers: (a) describes carboxylate linkers; (b and c) both describe carboxylate-derived linkers; (c) yields results (b) does not return, due to the way ConQuest carries out its searches; and (d) final Criterion 1 extended to all types of bonds. QA = O, N, P, C, B, and S. QB = N, P, B, S, and C. Superscript "c" imposes the corresponding atom to be cyclic. *Source:* Reproduced with permission from *Chem. Mater.*, 2017, 29, 7, 2618–2625. © 2017, American Chemical Society.

Criterion 2 corresponds to another large group of MOFs containing cyclic organic compounds directly linked to at least one metal atom. Famous examples of this type are some ZIFs and FMOF-1 type of structures, where several metal atoms are directly linked to a ring. Additionally, each ring must have at least one carbon atom to prevent the inclusion of inorganic cyclic structures. Criterion **2** returns

Table 8.1 Summary of the number and percentage of structures found with each criterion.

Criterion	Number of structures	Percentage of structures
1	50,046	72
2	44,616	64
3	15,380	22
4	2167	3
5	832	1
6	3732	5
7	1808	3

The criteria are not mutually exclusive. See Figure 8.6 for a schematic representation of the overlaps.

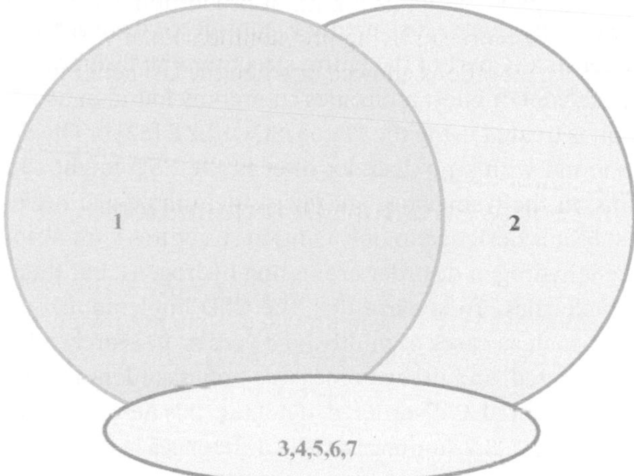

Figure 8.6 Schematic representation of the proportion and overlaps of structures found with each criterion. *Source:* Adapted with permission from *Chem. Mater.*, 2017, 29, 7, 2618–2625. © 2017, American Chemical Society

44,616 structures (or 64% of the final database), of which 30,481 structures are mutual with those found with criterion **1**. Together, criteria **1** and **2** yield 63,662 structures (i.e., 91% of the final database).

The remaining structures of the CSD MOF subset are obtained with **Criteria 3–7**, and correspond to less common MOF materials, such as structures with indirect metal–ring connections, with linkers containing cyanides, structures with metal atoms directly linked to nitrogen or sulfur atoms, with formate-derived linkers, and diglycolic acid-derived linkers. Structures containing linkers composed of cyanides were not explicitly included in the database, as structures where carbon is only present as a cyanide do not currently meet the criteria for being considered as organic by the CSD.

Table 8.1 summarizes the number and percentage of structures found with each criterion. Figure 8.6 presents a schematic view of the proportions and overlaps between the different criteria.

These seven criteria are now integrated into the CSD as filters, thereby automatically updating the CSD MOF subset as the CSD is updated. Importantly, these criteria are flexible and can be revised if required in the future, to reflect a new definition of MOFs, to add a class of MOFs that was not previously included, or to discard structures that should not be considered as MOFs.

It is worth noting here that the same structure can be represented several times in the CSD. Indeed, the CSD receives and records all experimental data corresponding to publications. Different research groups can therefore deposit the same structure, obtained and/or measured under different conditions. For instance, at least 50 entries correspond to HKUST-1 [34]. Therefore, the numbers presented in this section correspond to unique entries, but not necessarily to unique structures. We discuss in more detail the impact of these repeated occurrences later in this chapter.

8.3 The CSD MOF Subset

The CSD MOF subset is accessible via ConQuest. However, the entries in the CSD MOF subset may contain disorder. This may not be an issue when looking at one structure at a time, but becomes troublesome when screening a large number of structures. To provide a collection of structures that can be used in HTS calculations, a second subset, the CSD non-disordered MOF subset, was created to exclude any structures where the framework was found to be disordered. This subset is also available in ConQuest. This non-disordered subset should not be confused with a "no disorder" search in ConQuest. Indeed, among the many search possibilities provided in ConQuest, a no disorder filter can be applied to any query. A structure is considered disordered if it contains any non-hydrogen-related disorder in any part of the entire structure, the framework, and any other unmodeled molecule, such as solvent or guest molecules commonly found in MOF-like compounds where the disordered solvent is treated using the Platon/SQUEEZE [52] or Olex2/Mask [53] software. Therefore, a structure found with a no disorder filter in the CSD might still have missing or disordered hydrogen atoms, in the framework and/or in the unmodeled molecules. However, the non-disordered MOF subset is designed to only contain structures with absolutely no disorder within the framework, be it hydrogen disorder or missing hydrogens; but there might still be disorder in the unmodeled molecules. To achieve this, the CSD implemented an algorithm that: i) looks for disordered atoms, such as cases of multi-site disorder; ii) searches for the nearest neighboring atom that is non-disordered; and iii) examines this non-disordered atom: if it is part of the framework, the structure is considered as disordered, if not (i.e. it is near a solvent molecule), it is considered as non-disordered. Figure 8.7 summarizes the differences between the CSD MOF subset and the non-disordered MOF subset.

Cambridge Structural Database (CSD)
Disordered structures
Contains non-hydrogen related disorder
'Non-Disordered' structures
Might contain hydrogen-related disorder in any part of the entry
CSD MOF subset
'Non-disordered' CSD MOF subset
Entry might contain hydrogen-related disorder in unmodelled molecules only
Entry has no hydrogen-related disorder
Zero-disorder structures
Does not contain hydrogen-related disorder in any part of the entry

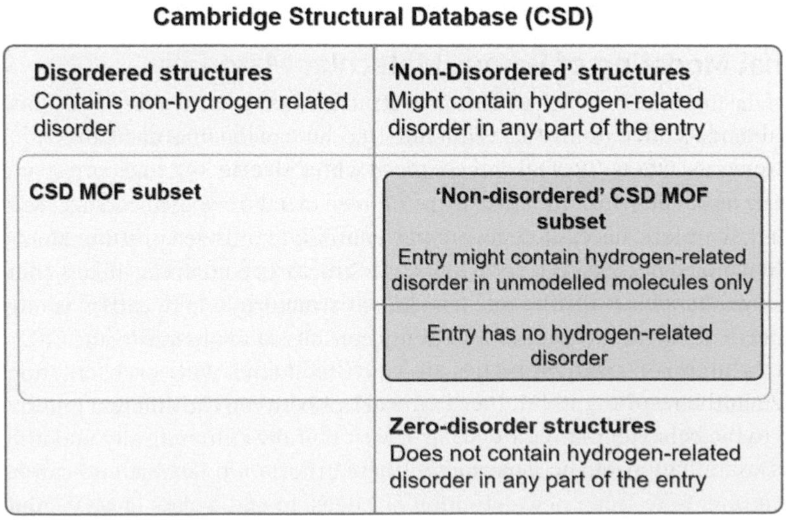

Figure 8.7 Organization of the metal–organic framework (MOF) data in the Cambridge Structural Database (CSD). *Source: CrystEngComm*, 2020, 22, 7152–7161 from the Royal Society of Chemistry.

8.3.1 Removing Solvents With the CSD Python API

Once the desired MOF subset is obtained from ConQuest, the data can be further explored with the CSD Python API. Along with the release of the CSD MOF subset, we published two Python scripts to remove bound and/or unbound solvent, the manuals of which can be found with the relevant publications [31, 44]. One script is to be used within Mercury [54], the visualization software developed by the CCDC, a single structure at a time. The other is to be used for the high-throughput cleaning of a subset. This latter algorithm takes a list of structures, a list of solvents if required, and outputs the desired CIFs [44]. It first looks for metal atoms present in the framework, then removes all bonds around them, before comparing the removed fragments to the provided list of solvents. If no list is provided by the user, the algorithm uses the default list of the 74 most common solvents curated by the CCDC. It is important to note that the provided script will remove both bound and unbound solvents. However, we recommend removing bound solvents only on structures where the removal of bound solvent will not cause structural collapse, such as structures containing Cu-Cu paddlewheels or structures similar to CPO-27/MOF-74. Indeed, structures such as HKUST-1 or CPO-27/MOF-74 are known to maintain their integrity upon activation, but this is not always the case. Some structures' stability critically depends on the presence of these bound solvents [55,56]. Moreover, it is not always correct to assume the bound solvents can be removed without any geometric optimization. In Section 8.5, we explain methods to identify HKUST-1-like and CPO-27/MOF-74-like structures. To only remove the unbound solvent, users can use an empty solvent file. Another quick, but often less reliable, way of removing unbound solvents with the API is to return only the heaviest weight component (`heaviest_component`) of an entry. The `heaviest_component` corresponds to the group of connected atoms in the entry with the highest molecular weight. Most of the time, the `heaviest_component` is the framework. However, there are many exceptions, and it is advised to also add a "polymeric" check with the `is.polymeric` attribute to make sure the heaviest weight component is indeed polymeric. If it is not, one of the substructures is polymeric by definition of the subset, and this substructure should be kept as the framework.

8.3.2 Adding Missing Hydrogens

We explained earlier that the non-disordered MOF subset does not include frameworks with missing hydrogens. What does "missing hydrogens" mean and why does this happen? In a CSD entry, only the atoms modeled from the original experimental data have coordinates and can be visualized in Mercury [54]. As hydrogen is the lightest atom, it is very difficult to detect with X-ray diffraction, which is why hydrogen atoms are sometimes not found in the original data, and therefore not modeled. However, to make sure the structure makes chemical sense, these hydrogen atoms are still accounted for in the CSD and are referred to as "siteless hydrogens." They do not appear in the original CIF but are taken into account in the chemical diagrams and therefore in search queries. Figure 8.8 shows the example of RUBTAK01, one of many entry versions of UiO-66 in the CSD. RUBTAK01 is part of the non-disordered MOF subset and has siteless hydrogens. The `add_hydrogen` function is available in the API to obtain the coordinates of these siteless hydrogens. It calculates the most likely bond lengths and angles based on the structure's local geometry. The added hydrogen atoms will then appear in the CIF.

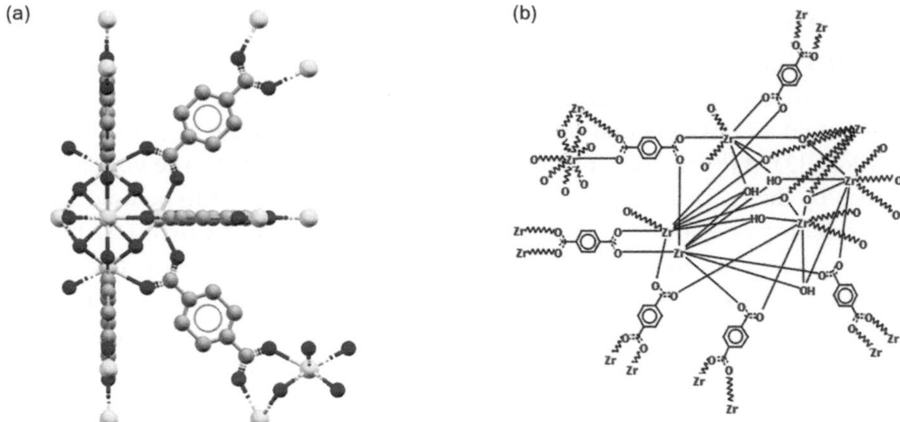

Figure 8.8 Example of a structure from the non-disordered metal–organic framework (MOF) subset with siteless hydrogen atoms: UiO-66; Cambridge Structural Database (CSD) refcode: RUBTAK01: (a) 3D visualization of a repeating unit in Mercury; and (b) 2D (hydrogen-depleted) diagram available in Mercury. *Source: CrystEngComm*, 2020, 22, 7152–7161 from the Royal Society of Chemistry.

8.4 Textural Properties of MOFs and Their Evolution

The CSD MOF subset contains most of the MOFs synthesized worldwide, if not all. What do these MOFs look like? To answer this question, we used the open-source software, Zeo++ [57], to geometrically characterize MOFs of interest in a high-throughput manner. The calculated properties include the largest cavity diameter (LCD), the pore limiting diameter (PLD), the void fraction (ratio of pore volume over the volume of the framework), and the gravimetric and volumetric surface area. Figure 8.9 presents the concepts of LCD and PLD. The bulk material is represented in gray and the porous area in white. The LCD is the size of the largest sphere that can fit into a cavity, and the PLD is the smallest opening of the channel that a molecule can diffuse through (Figure 8.9).

Before computationally characterizing the selected MOFs, the residual solvent inside the frameworks were first removed. We used the library of the 74 common solvent molecules previously introduced to identify the solvents present in the subset; 88% of the MOFs in the CSD MOF subset have solvents, of which 52% are unbound and 48% are bound solvents. Water is by far the most common bound or unbound solvent, present in 47,478 MOF structures. Among the nonaqueous bound solvents, 33 of the 73 solvents are present, DMF being the most common by a large margin,

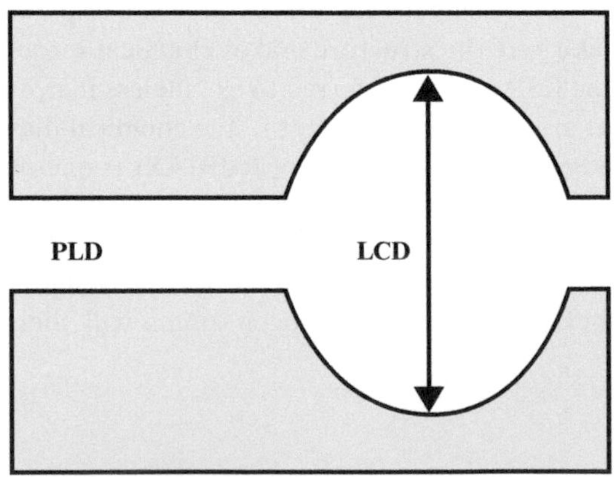

Figure 8.9 Schematic 2D illustration of largest cavity diameter (LCD) and pore limiting diameter (PLD).

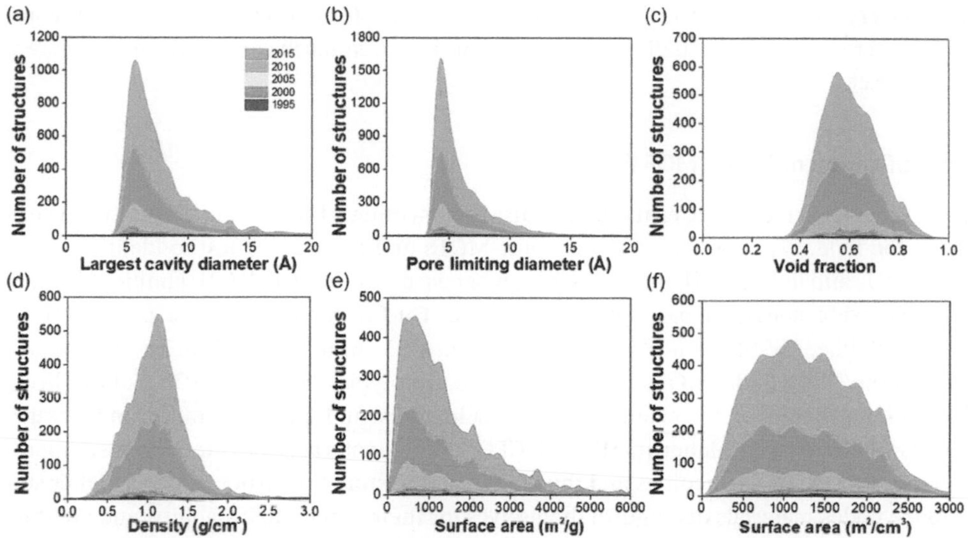

Figure 8.10 Histograms comparing geometric properties for all the porous metal–organic frameworks (MOFs) in the Cambridge Structural Database (CSD) MOF subset from 1995 to 2015: (a) largest cavity diameter (LCD); (b) pore limiting diameter (PLD); (c) void fraction; (d) density; (e) gravimetric surface area; and (f) volumetric surface area. All family-property relationships of the 8253 porous MOFs presented in this work can be found online at http://aam.ceb.cam.ac.uk/mof-explorer/CSD_MOF_subset. *Source: Chem. Sci.*, 2020, 11, 8373–8387 from the Royal Society of Chemistry.

present in 1509 MOF structures. We removed the unbound solvents from all the structures and the bound solvents from a total of 739 previously identified materials containing Cu-Cu paddle-wheels and CPO-27/MOF-74-like structures [44].

Using Zeo++, we then identified the subset of porous structures. MOFs here are considered porous if they are accessible to a nitrogen probe of 1.86 Å [44]. Thus, of the 55,000 non-disordered structures in the non-disordered MOF subset, ca. 8200 materials are porous. We applied Zeo++ to this subset of structures, the result of which is presented in Figure 8.10. This plot shows the distribution of the LCD, PLD, void fraction, density, and volumetric and gravimetric surface area of this subset of porous MOFs from 1995 to 2015. Only a few structures were known before 2000. However, the number of MOFs dramatically increased during the 21st century, translating the growing drive toward the exploration of MOFs. Most structures have pore sizes smaller than 10 Å and surface areas lower than 2000 m^2g. A possible explanation is the use of relatively cheap and commercially available short linkers, such as terephthalic acid. This range of pore size is also optimal for many gas storage and separation applications. As new synthesis methods are continuously being designed, the introduction of longer linkers and more complex SBUs have continued increasing during the past decade [58], leading to more structures with larger pore sizes and higher surface areas.

8.5 Classification of MOFs

One of the advantages of the CSD MOF subset is the possibility of directly using CSD tools to mine more information. In this section, we present some useful classifications obtained solely with these tools: most common MOF families (i.e., MOFs that are structurally similar), surface functionalizations, chiral MOFs, MOF channel dimensionality, and MOF network dimensionality. We also touch upon the crystal quality of the CSD MOF subset. All the information presented here is compiled in an interactive data visualization website at http://aam.ceb.cam.ac.uk/mof-explorer/CSD_MOF_subset.

8.5.1 Identification of Target MOF Families

Some MOFs are considered classic examples because their synthesis brought about significant contributions to the field. We chose six of these iconic MOFs and used ConQuest to identify MOFs with similar SBUs and linkages. These MOFs form what we define here as "MOF families." The six groups are: Zn-oxide nodes, in particular IRMOF-like, Cu-Cu paddlewheels (e.g., HKUST-1), MOF-74/CPO-27-like, ZIF-like, and Zr-oxide nodes (e.g., UiO-66) materials. The Zn-oxide nodes materials, and the IRMOF-like MOFs in particular, are prototypical MOFs. Cu-Cu paddlewheels are well-known for their open metal sites (OMS), which are useful for gas separation and storage. The number of OMS is even higher in MOF-74/CPO-27-like materials. ZIFs tend to form topologies similar to those found in zeolites and thus inherit interesting properties, such as higher stability. Zr-oxide nodes materials distinguish themselves for their remarkable stability, and are thus promising in gas adsorption and catalysis. It is important to note that the choice of these families is arbitrary, and more classes of materials can be extracted. The methods used here are the same as those used for building the CSD MOF subset, with two exceptions: i) the searches are carried out within the CSD MOF subset itself; and ii) the queries are more specific and use a combination of search results, where some queries are specifically designed to discard unwanted structures. These latter queries are labeled later as "must not have" and indicated in red in Figure 8.11, whereas the queries that target the desired structures are referred to as "must have" and indicated in green. Indeed, as much as MOFs have a vague definition and some leeway is advisable to reflect this gray area, well-known MOFs have very distinct features. The combination of "must have" and "must not have" queries guarantee a more accurate subset of structures.

Figure 8.11 provides a summary of the criteria developed for each MOF family. Some groups, such as IRMOF-like structures, only need a single "must have" criterion. Others require a combination of "must-have" and "must not have" queries. In particular, there are two types of "must have" combinations: those that correspond to an AND statement (i.e., the final structures result from the intersection of the hits returned by each query), and those that correspond to an OR statement (i.e., the final structures result from the union of the hits returned by each query). The AND statement is indicated with a dotted box surrounding the queries of interest in Figure 8.11. This is the case of the MOF-74/CPO-27-type structures. The absence of this dotted box signifies an OR statement. This is the case of the Zr-oxide-based family. Both statements can also be applied to the "must not have" queries; however, only OR was used in this case.

We explain in more detail here the derivation of the criteria designed for the family of Cu-Cu paddlewheel MOFs. This family is particularly interesting because of the different possible ways of representing the same paddlewheel. Figure 8.12a represents the diagram of a complete paddlewheel and its connection to the linker via the two oxygen atoms. An example of a structure captured by this search is given in Figure 8.12e. However, there are cases (such as the one represented in Figure 8.12f) where only half of the paddlewheel is represented. These structures are found using the diagram in Figure 8.12b, which corresponds to only a section of the paddlewheel. Note that we avoided adding extra oxygen atoms from the linker, as we found that keeping these atoms made the query too restrictive and returned fewer target structures. The two copper atoms are now bonded along the rotational axis of the paddlewheel. Other structures, such as those in Figures 8.12g and 8.12h have a "broken" representation of their paddlewheels. This situation is fixed with

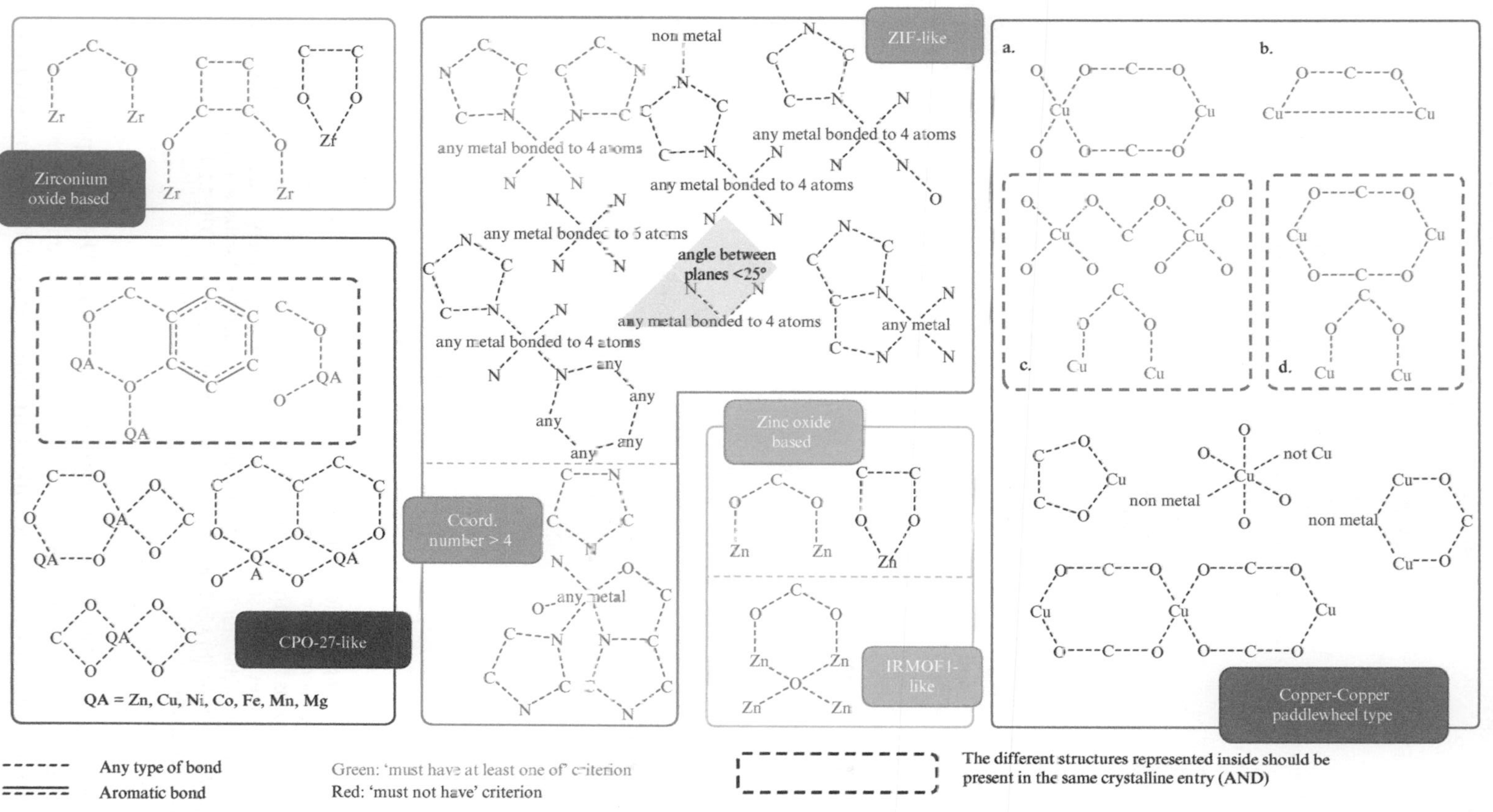

Figure 8.11 Criteria developed for the identification of metal–organic framework (MOF) families in the Cambridge Structural Database (CSD) MOF subset based on specific secondary building units and their connection to the organic linkers. The target MOF families are zirconium oxide, MOF-74/CPO-27-like, ZIF-like, zinc oxide, and IRMOF-like, as well as Cu-Cu paddle-wheeled materials: (a to d) diagrams used to look for structures containing Cu-Cu paddlewheels. The dotted box for (c and d) means the structures inside should be considered as one single query. The red diagrams are queries used to eliminate undesired structures. *Source: Chem. Sci.*, 2020, 11, 8373–8337 from the Royal Society of Chemistry.

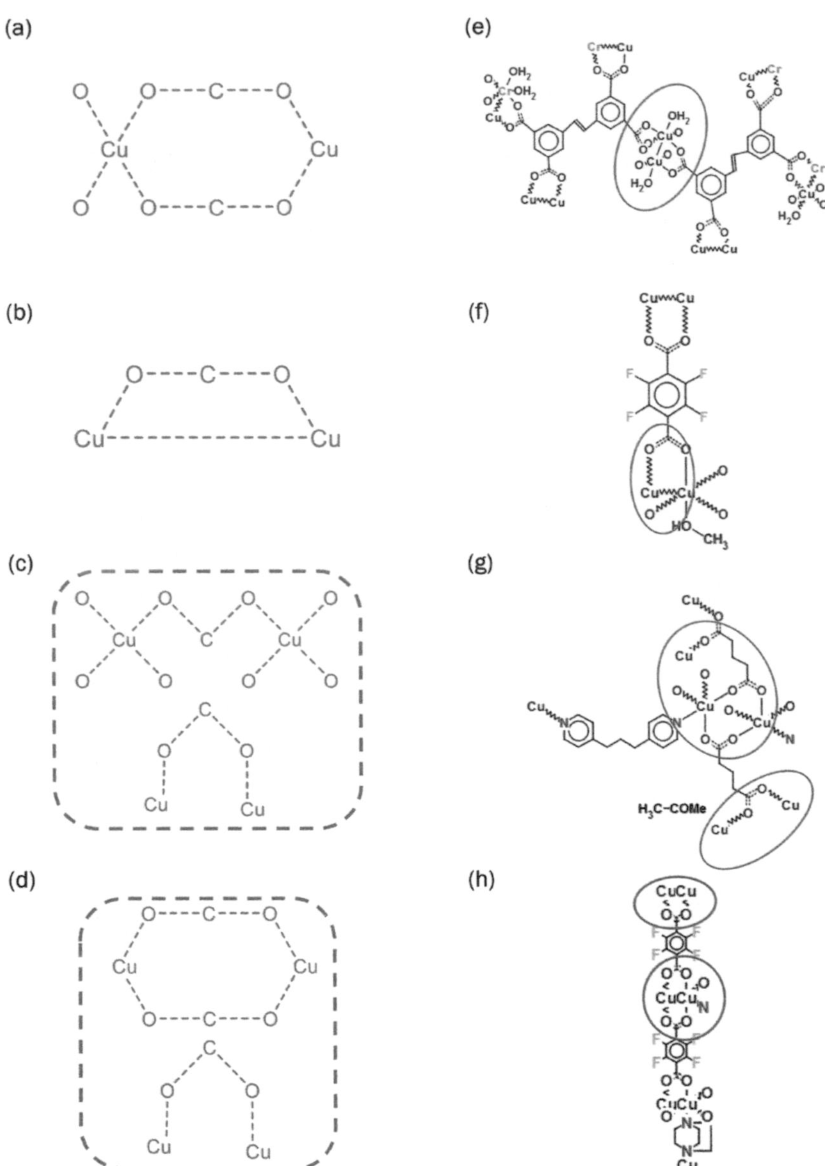

Figure 8.12 (a to d) Criteria developed to look for structures containing Cu-Cu paddlewheels; (e to h) example structures found using the criterion on the left. (a) returns 988 hits; (b) returns 611 hits, adds 178 to the list; (c) returns 716 hits, adds 248 to the list; and (d) returns 647 hits, adds 12 to the list. For (c and d), the dotted box means the structures inside should be considered as one single query. The blue circled areas show the parts that have been searched for in ConQuest. Cambridge Structural Database (CSD) refcodes; (e) ACUJOZ; (f) ACASUT; (g) ACAJOF; and (h) ACATAA. *Source:* Adapted from *Chem. Sci.*, 2020, 11, 8373–8387 from the Royal Society of Chemistry.

the queries in Figures 8.12c and 8.12d. Both are comprised of two parts. The upper part in Figure 8.12c returns structures in which the represented paddlewheel is "broken." However, this also brings in other Cu-based structures where the linkers are linear; this is avoided by adding the lower part, which corresponds to the connection between the metal atoms and the linkers. The upper part of the diagram in Figure 8.12d is similar to the diagram in Figure 8.12a, but this time it is cyclic, and devoid of any oxygen atoms from the linkers bonded to the Cu atoms. With the lower part of the search criterion, the overall diagram from Figure 8.12d returns MOFs where the paddle-wheel and the linkages are represented separately in ConQuest. All in all, the four "must-have" queries result in 1426 structures, some of which are not of the target type. To filter out these unwanted structures, we included a set of "must not have" criteria according to a list of undesired structures (Figure 8.13). The combination of all the queries leads to a total of 1015 MOFs containing Cu-Cu paddlewheels.

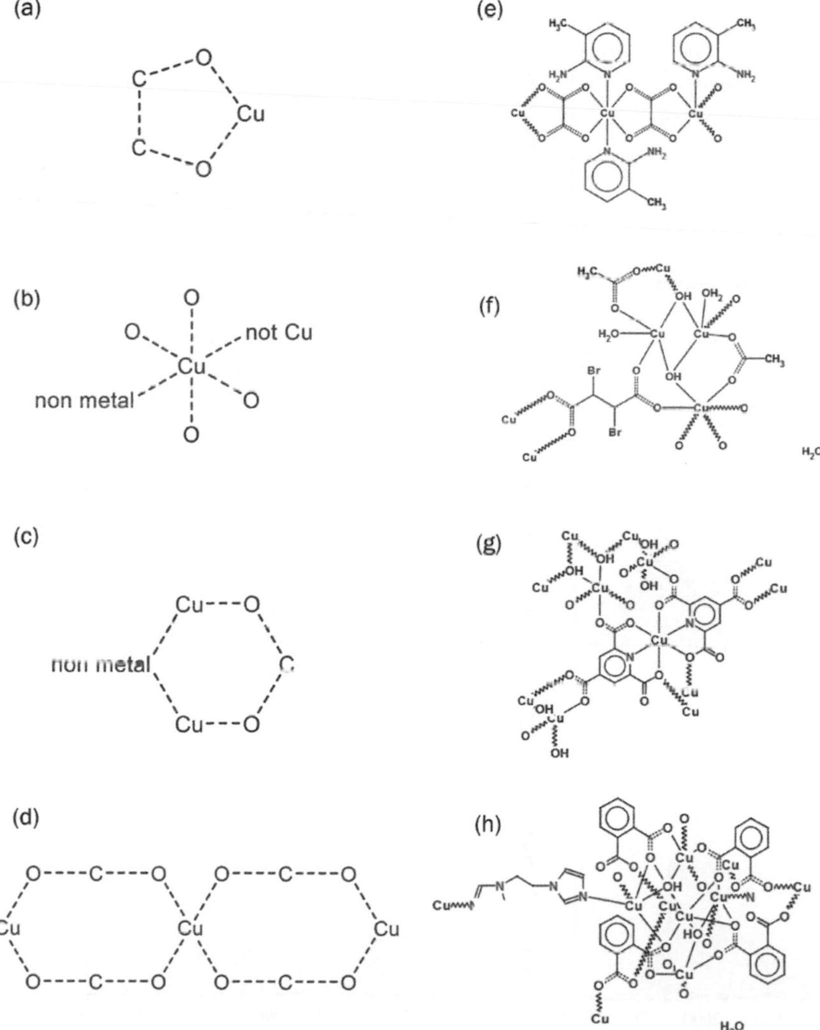

Figure 8.13 (a to d) Criteria used to eliminate undesired structures and the number of structures eliminated at each step; (e to h) examples of eliminated structures corresponding to the criteria on the left. (a) eliminates 128 hits; (b) eliminates 190 hits; (c) eliminates 88 hits; and (d) eliminates 5 hits. Cambridge Structural Database (CSD) refcodes: (e) ABOCUP; (f) AHEGIF; (g) AGUMAR; and (h) ASEWEB. *Source:* Adapted from *Chem. Sci.*, 2020, 11, 8373–8387 from the Royal Society of Chemistry.

Altogether, Zn-oxide and IRMOF-like materials account for 3187 structures, followed by 1015 for Cu-Cu paddlewheels, 274 for ZIFs, 108 for CPO-27-like structures, and 77 for Zr-oxide structures in the CSD 5.37 version from May 2016. It is interesting to note that MOFs from these six families only represent ca. 7% of the whole dataset. Figure 8.14 presents histograms that map the geometric properties of each MOF category. Zn-oxide MOFs, being the largest family, the corresponding structures cover the widest range of LCD, PLD, void fraction, density, and surface area. Despite this wide coverage, a few modes can be observed: LCDs of 5, 9, 10, and 16 Å, PLDs of 5, 7, and 10 Å, void fractions around 0.6 and 0.85, densities around 0.5 and 1.0 g/cm^3, volumetric

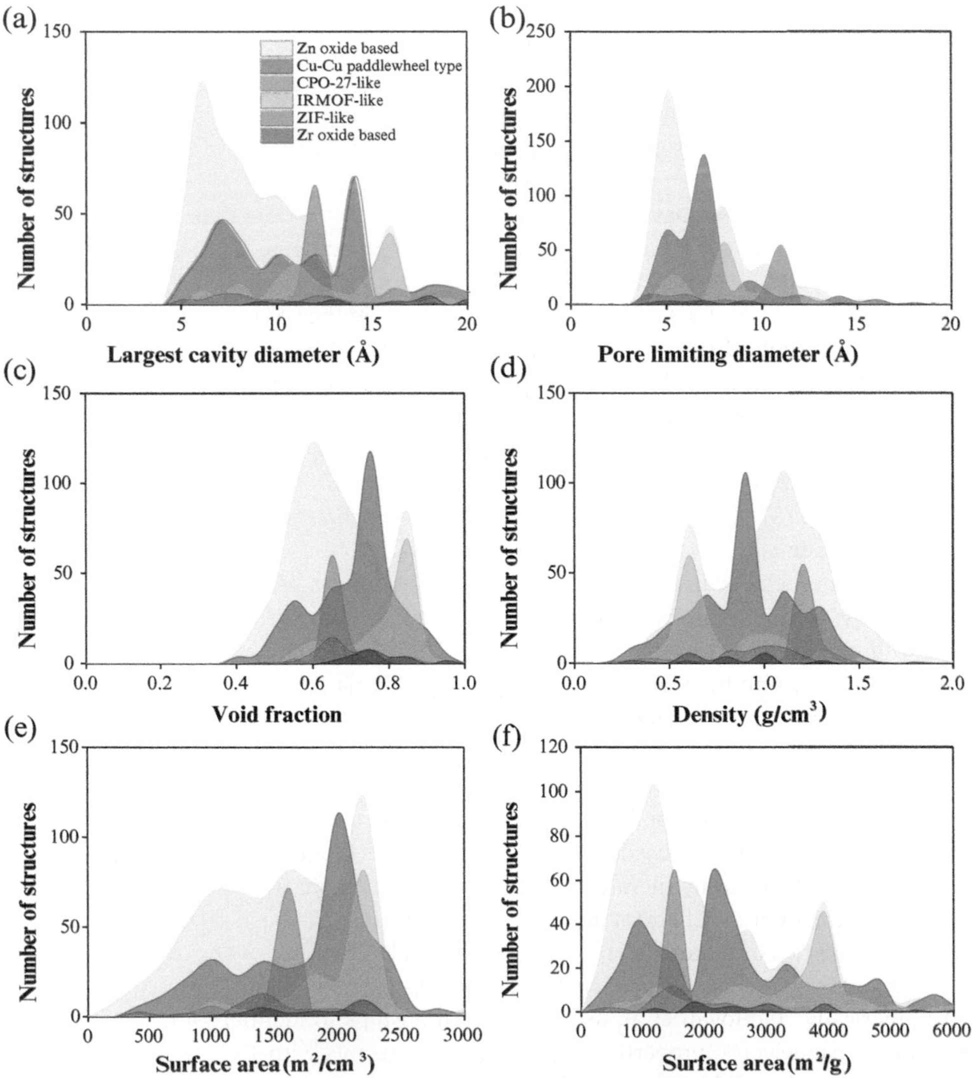

Figure 8.14 Histograms showing the geometric properties for each MOF family identified in the Cambridge Structural Database (CSD) metal–organic framework (MOF) subset: (a) largest cavity diameter (LCD); (b) pore limiting diameter (PLD); (c) void fraction; (d) density; (e) volumetric surface area; and (f) gravimetric surface area. *Source:* Adapted from *Chem. Sci.*, 2020, 11, 8373–8387 from the Royal Society of Chemistry.

surface areas of 1000, 1500, and 2200 m^2/cm^3, and gravimetric surface areas of 1000, 1500, 1800, and 4000 m^2/g. In particular, the IRMOF-like materials fit perfectly one mode in each of these structural properties: LCD of 16 Å, PLDs of 7 Å, void fractions of 0.85, densities around 0.5 g/cm^3, volumetric surface area of 2200 m^2/cm^3, and gravimetric surface area of 4000 m^2/g. MOFs containing Cu-Cu paddlewheels also show distinct modes despite covering a range of values: LCDs around 7, 9, 12, and 14 Å, PLDs around 5 and 7 Å, void fractions around 0.5 and 0.7, densities around 0.9 g/cm^3, volumetric surface areas around 1000, 1400, and 2000 m^2/cm^3, and gravimetric surface areas around 1000 and 2000 m^2/g mostly. CPO-27/MOF-74-like structures present one distinct mode for each property: LCDs of 12 Å, PLDs of 11 Å, void fractions of 0.65, densities of 1.2 g/cm^3, volumetric surface areas of 1500 $m^{2/}cm^3$, and gravimetric surface areas of 1100 m^2/g. ZIF-like structures and Zr-oxide MOFs are in significantly smaller numbers with values covering a wide spectrum for most properties. However, ZIF-like structures do show a distinct peak of the void fraction at 0.65, volumetric surface area of 1200 m^2/cm^3, and gravimetric surface area of 1100 m^2/g. As for Zr-oxide MOFs, most structures have void fractions around 0.7, three modes of densities (0.6, 0.7, and 1.0 g/cm^3), and volumetric surface areas around 2200 m^2/cm^3. These histograms thus show that the chosen classification was able to capture MOFs with a range of differing behaviors.

8.5.2 Identification of Surface Functionalities in MOFs

Surface functionalization is critically important when fine-tuning the chemical and physical properties of a MOF. Scientists have rationally incorporated chemical functionalities with pre- or post-synthetic engineering techniques for applications ranging from carbon capture [59, 60], gas separation and sensing [61–63], catalysis [64, 65], to light harvesting [66] and optical luminescence [67]. In this section, we look at methods to identify structures with specific functional groups: alkyls (methyl, ethyl, propyl and alkyls containing more than 4 carbon atoms), alkoxys (methoxy, ethoxy, propyloxy), halogens (-F, -Cl, -Br), and polar functionalities ($-NH_2$, $-NO_2$, -CN, -COOH, -OH). Figure 8.15 summarizes the combination of ConQuest queries used to target these functionalized MOFs. However, this time, we also used the CSD Python API to ensure that the searched fragments are only present in the main framework and not in a solvent molecule. Figure 8.16 presents the distribution of the geometrical properties obtained for the different polar groups. Conversely to the families classification, these histograms show less distinctive behaviors for structures with different functional groups. This indicates that the SBUs are a primary driver of a structure's characteristics, while the effects of the functional groups are relatively secondary. However, we encourage our readers to explore the combined effects of families and functional groups on our data explorer: http://aam.ceb.cam.ac.uk/mof-explorer/CSD_MOF_subset.

8.5.3 Identification of Chiral MOFs

MOF chirality is extremely useful for niche catalytic and enantio-separation applications [14,68–70]. Here, a MOF is considered chiral either when it contains chiral atoms in the structure or when it presents a chiral crystal packing; 4504 structures were found to contain S/R-chiral atoms and 6859 structures are in Sohncke-chiral space groups; and 2010 structures had both chiral atoms and a chiral space group. Note that we focused on R/S chirality only. Structures with metal lambda/delta or axially-chiral structures were therefore not accounted for. Figure 8.17 shows the geometric properties of 1911 chiral and porous structures. These MOFs represent almost a quarter of the 8253 porous MOFs. The distribution of geometrical properties is similar, where the majority of chiral structures have pores smaller than 10 Å and surface area values of less than 2000 m^2/g. However,

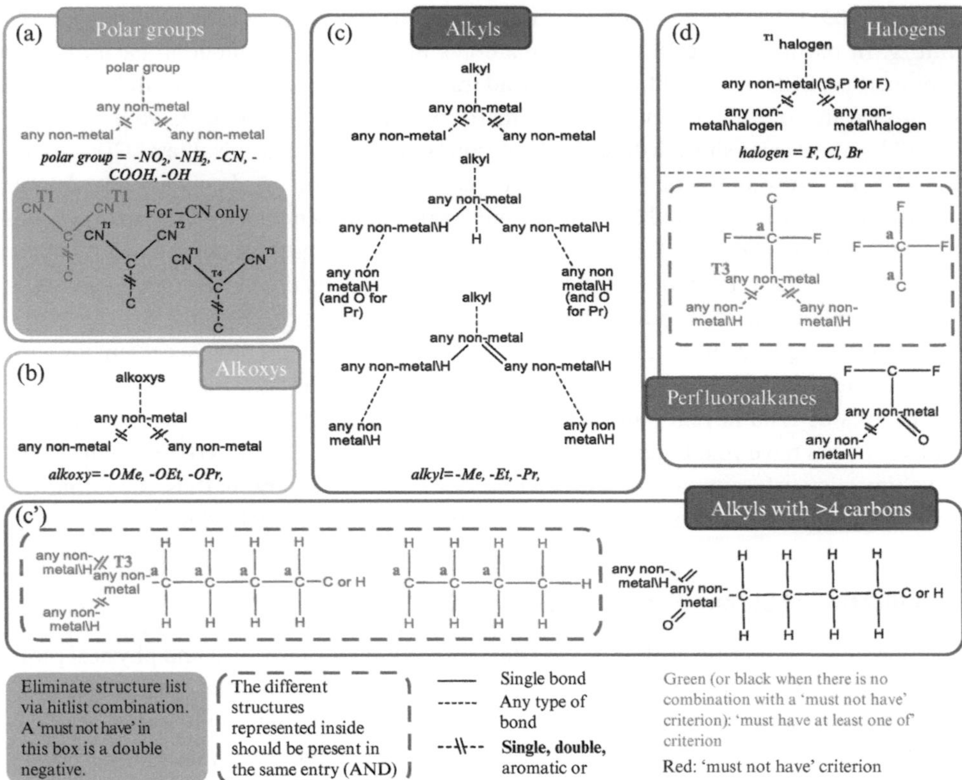

Figure 8.15 Criteria developed to identify metal–organic frameworks (MOFs) with common functionalities in the Cambridge Structural Database (CSD) MOF subset: (a) polar groups (-NH_2, -NO_2, -CN, -COOH, and -OH). For the -CN case, the red box represents queries that target dicyanides that are chosen to be eliminated. This dicyanide search is obtained via a combination of one "must-have" query and two "must not have" queries. The green diagram is thus an overall negative and the red diagrams are double negatives; (b) alkoxys (methoxy, ethoxy, propyloxy); (c) alkyls (methyl, ethyl, propyl); (c') alkyls with more than 4 carbon atoms on the left); and (d) halogens (-F, -Cl, -Br), and structures with perfluoroalkane groups. The variable bonds are all the same type for queries within the gray dotted box: single, double, aromatic, or delocalized. For the three queries outside of the gray dotted box, the variable bonds are either aromatic or delocalized. *Source: Chem. Sci.*, 2020, 11, 8373–8387 from the Royal Society of Chemistry.

only about 5% of the chiral MOFs are non-porous. This observation possibly suggests that researchers have been actively looking for porous structures that are also chiral. This might be connected to the fact that more than 90% of chiral MOFs were synthesized in the 21st century when interest in using MOFs for their potential in catalytic applications and enantiomeric resolution started growing.

8.5.4 Porous Network Connectivity and Framework Dimensionality

Understanding a MOF's porous network connectivity or dimensionality (also referred to as percolation or channel dimensionality) is crucial in determining its applicability in certain adsorption applications. For instance, depending on the channel size and shape, 1D-channeled MOFs can be

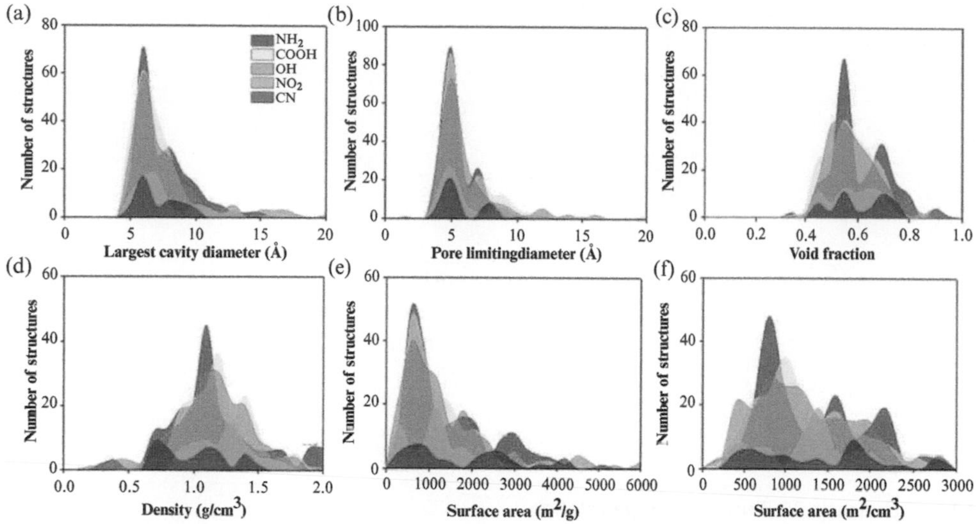

Figure 8.16 Histograms showing geometric properties distribution in metal–organic frameworks (MOFs) with different polar groups: (a) largest cavity diameter (LCD); (b) pore limiting diameter (PLD); (c) void fraction; (d) density; (e) gravimetric surface area; and (f) volumetric surface area. *Source:* Adapted from *Chem. Sci.*, 2020, 11, 8373–8387 from the Royal Society of Chemistry.

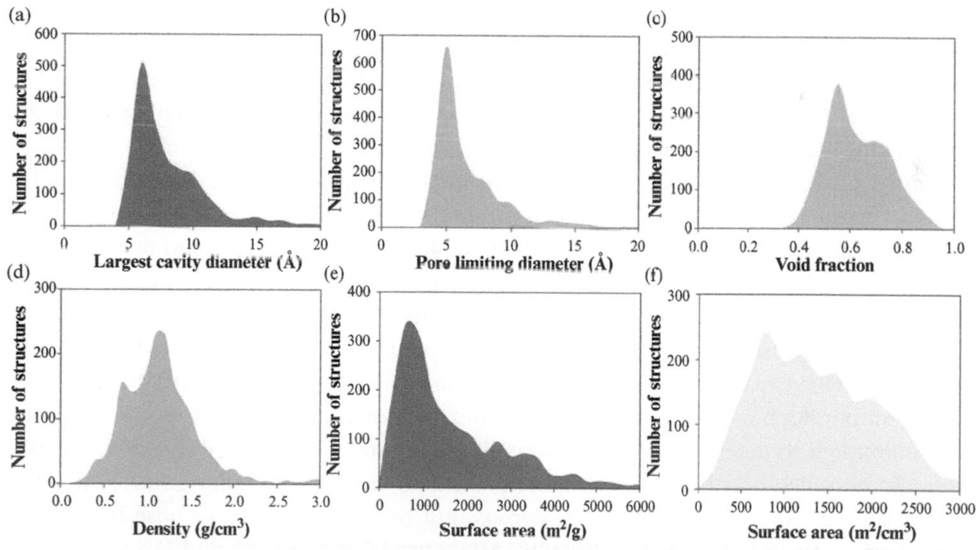

Figure 8.17 Histograms of the geometric properties of 1911 chiral structures with non-zero gravimetric surface area in the Cambridge Structural Database (CSD) metal–organic framework (MOF) subset: (a) largest cavity diameter (LCD); (b) pore limiting diameter (PLD); (c) void fraction; (d) density; (e) gravimetric surface area; and (f) volumetric surface area. *Source: Chem. Sci.*, 2020, 11, 8373–8387 from the Royal Society of Chemistry.

Figure 8.18 Schematic 2D illustration of percolation.

highly selective for the separation of hydrocarbons [71–73]. Figure 8.18 presents different cases of percolation (1D, 2D, 3D, and 0D). An isolated pore is shown in the top right corner. This pore is non-accessible and would correspond to 0D percolation. The rest of the schematic presents two porous channels crossing each other, thus forming a system of 2D channels, or 2D percolation. If another channel perpendicular to this page was connected to these two channels, the system would be 3D-percolated. The channel dimensionality is not to be confused with the overall structure dimensionality, which we will discuss in the next paragraph. To determine the dimensionality of porous networks, the void space and the connected components are mapped with mesh/grid-based propagation techniques [74–77]. We used here Poreblazer [77], freely available software for the structural characterization of materials, to determine the percolation of all 8253 porous MOFs. The results are shown in Figure 8.19a, where 86% of the structures are 1D-percolated, 9% are 2D-percolated, and 4% are 3D-percolated.

As mentioned previously, percolation is not to be confused with framework dimensionality, which is also crucial for choosing a MOF for a given application. While 3D porous MOFs are essential for gas adsorption and electrically conductive [42], luminescent [27] and magnetic [43] MOFs are often 1D or 2D. To identify a structure's framework dimensionality, we published another Python script that heavily relies on the CSD Python API. This algorithm first generates the smallest box containing the smallest repeating unit of the given structure. The latter is then expanded in all possible directions (the expanding points are given by the positions of the polymeric bonds) and a new smallest-containing box is created. The ratio of dimensions of the final box versus the initial box is then computed to determine the directions the structure has expanded. From the 55,000 MOFs from the non-disordered subset, a number presented various occupancy issues [78]. After discarding these, we were left with ca. 5300 porous and non-porous MOFs whose framework dimensionalities are presented in Figure 8.19a: 40% MOFs are 1D, 29% are 2D, and 31% are 3D [78].

8.5.5 An Insight into Crystal Quality of Different MOF Families

This last section takes a closer look at the crystal quality in the CSD MOF subset. Using the CSD Python API, users can easily access the structures' crystallographic R-factors, which measure the discrepancy between the observed structure factor F_{obs} and the calculated structure factor, F_{calc}, upon crystal determination:

$$R = \frac{\sum \left\| F_{obs} \right| - \left| F_{calc} \right\|}{\sum \left| F_{obs} \right|} \tag{8.1}$$

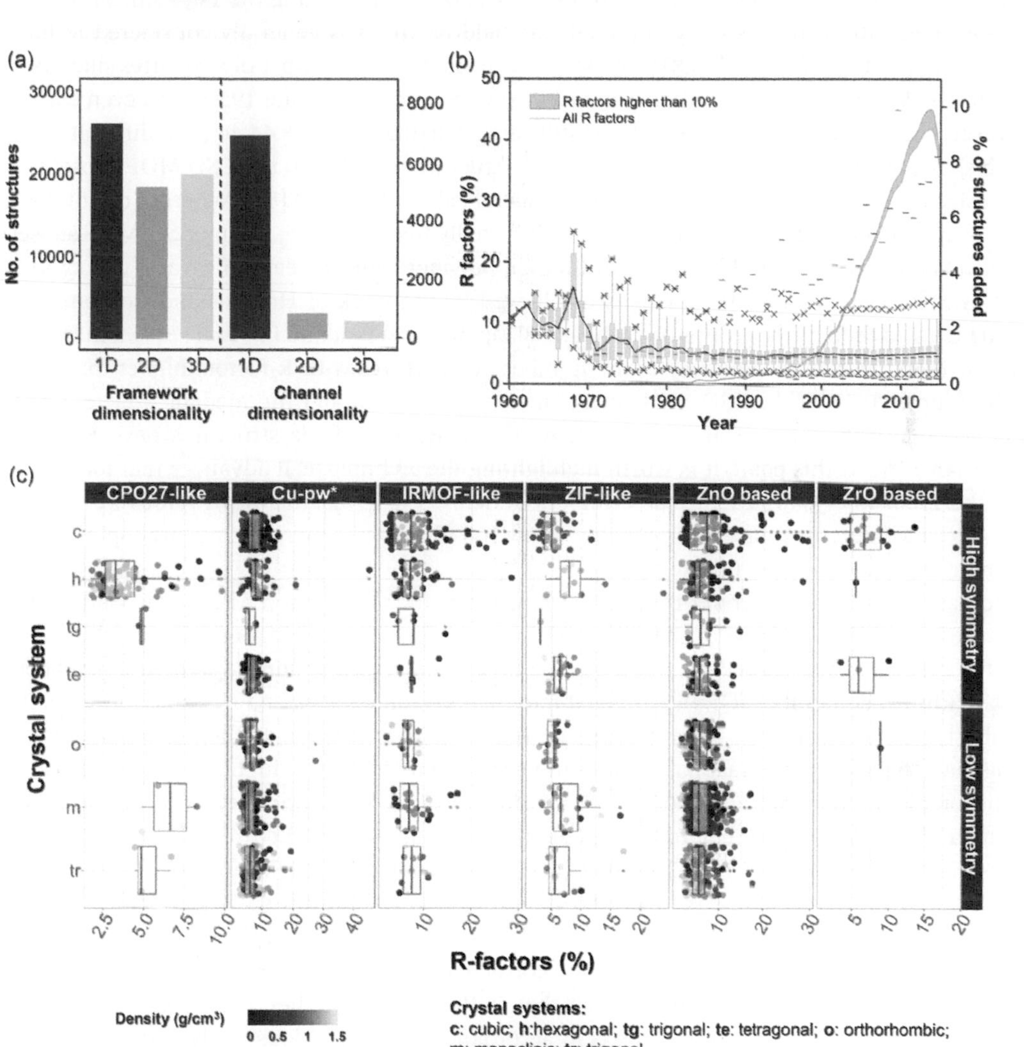

Figure 8.19 Analysis of metal–organic frameworks (MOFs) included in the Cambridge Structural Database (CSD): (a) Histograms of framework and channel/pore dimensionalities characterized for the 52,787 structures; (b) Non-cumulative evolution of R factors of the MOF subset from 1960 to 2015. Blue: boxplots of R-factors per year. Percentiles used: 1% (lower dash symbol), 25% (lower cross symbol), 50% (dash in the box), 75% (upper cross symbol), and 99% (upper dash symbol). A black line connects the means across all the boxes; the orange curve shows the percentage of structures added to the database per year. The orange area under the orange curve highlights the number of structures with an R-factor higher than 10%. (c) Distribution of R-factors and density across different MOF families and crystal systems of low or high symmetry. *Source: Chem. Sci.*, 2020, 11, 8373–8387 from the Royal Society of Chemistry.

where the sum extends over all the X-ray reflections measured and calculated. An R-factor of 0 corresponds to a perfect agreement between the experimental measurements and the predicted structure factors. R-factors higher than 10% are considered to reflect refinement models that may contain systematic errors [79]. Figure 8.19b shows the evolution of the R-factors of the MOFs from the CSD MOF subset from 1960 to 2015. The share of structures with R-factors higher than 10% is highlighted in orange. Each year has a box and whiskers plot, indicating the 1st, 25th, 50th, 75th, and 99th percentiles. Interestingly, although the field of MOFs is generally considered to have started in the last two decades [80,81], as shown by the increasing number of structures since 1995 in Figure 8.19b, the first coordination polymers were recorded in the late 1950s, and even earlier. However, because the definition of MOFs is still being debated today [28,44,46], it is difficult to tell which structure truly was the first MOF. The first structure recorded in the CSD MOF subset was synthesized in 1940 and is a sodium formate (CSD refcode: NAFORM) [82]. Whether or not it is a MOF is up to debate. In the meantime, it still marginally fits the criteria of the CSD MOF subset. ADINCU is the most "MOF-like" 3D coordination polymer from the early days and was synthesized by Saito et al. in 1959 [83]. This was followed by the work of Hoskins and Robson (with JARMEU), and then by the groups of Yaghi and Kitagawa. The timeline in Figure 8.19b, therefore, starts in 1960. Despite the increase of the number of structures with R-factors higher than 10%, reaching up to 0.7% of the CSD MOF subset in 2013, the mean and the median R-factors have remained relatively constant at around 5%. Additionally, 99% of the structures have R-factors lower than 12%. At this point, it is worth highlighting the technological advances that took place since the 1960s in crystal structure determination. Before the 1970s, the mean values were above 10%. A decade later, the R-factors dropped to below 10%, despite the increase in more sophisticated and large structures being synthesized [84].

The classification and other information extracted from the CSD Python API can be combined to further explore the landscape of the CSD MOF subset. For instance, Figure 8.19c combines the structures' previously obtained R-factors with their families, crystal systems, symmetries, and densities. Within each family, MOFs are divided into their crystal systems and a boxplot presents the distribution of their R-factors. The crystal systems are ordered according to their symmetry: cubic, hexagonal, trigonal, and tetragonal systems are considered as having "high symmetry," whereas orthorhombic, monoclinic, and triclinic are considered to be "low symmetry." Each point corresponds to a structure and is colored according to its density. This structure–property landscape shows that families such as CPO-27/MOF-74 and Zr-oxide MOFs crystallize in specific crystal systems, whereas others crystallize in all crystal systems, though not evenly so. For example, IRMOF-like structures tend to crystallize mainly in cubic or hexagonal systems and show higher R-factors in these systems. How these observations are correlated would, however, require more in-depth research. Nevertheless, Figure 8.19c suggests that, within all the families, low-density MOFs tend to form high symmetry structures. This is in accordance with a previous analysis carried out by Øien-Ødegaard et al. [79].

To fully understand the meaning of the R-factor, it is also important to take into account the structure refinements a structure went through. These refinements are often carried out with software such as SQUEEZE [52], and consist in identifying and including (or not including) the contribution of disordered solvent in the calculated structure factors when determining the crystal structure. This procedure allows users to mask the solvents and artificially correct the experimental values obtained from the X-ray diffraction patterns. Such methods interfere with the interpretation of the R-factors. Which structure provides the better information quality: the one with a lower R-factor after SQUEEZE was applied, or the one with a higher R-factor and an attempt to model all the disorder positions of the framework and/or guests? It should also be highlighted that the R-factor is only a convenient metric to quickly assess the quality of a crystal structure. It measures the agreement between the refined model and the experimental data, but does not take into account how meaningful the resulting structure is, chemically and physically, nor does it tell whether any use of solvent masking is appropriate or whether there are large residual electron density peaks.

8.6 The CSD MOF Subset Among All the MOF Databases

MOF databases such as the CSD MOF subset have become treasure troves in which computational researchers can carry out HTS to find the optimum structures for a given application. However, as highlighted earlier, several databases exist. The CSD MOF subset is an experimental database, that is, all the structural data included were obtained experimentally. To the best of our knowledge, all other experimental databases are derived from the CSD. The CoRE MOF was the first publicly available one [34]. In the density-derived electrostatic and chemical (DDEC) [85,86] dataset, charges were added to 2900 structures from the CoRE MOF database, allowing the study of adsorption cases where electrostatic interactions play a role. Theoretical structures are gathered in hypothetical databases. The hMOF dataset contains 138,956 hypothetical structures built from a "bottom-up" approach: each structure is generated from the recombination of 102 SBUs and organic linkers of available crystallographic data of existing MOFs [87]. The ToBaCCo (Topology-Based Crystal Constructor) database uses a "top-down" (or reverse topological) approach to focus on the diversity of possible MOF topologies [9]. The number of obtained structures, therefore, varies depending on the chosen topologies, 13,512 when considering 41 topologies. Boyd et al. used a similar approach to generate 300,000 structures from 46 topologies [88]. Finally, the Quantum MOF database (QMOF) gathers a mix of ca. 18,000 unique experimental and hypothetical structures for which the partial charges have been machine-learned [89]. Other studies use their own hybrid mix of data, the purpose being to widen the pool of potential candidates [90]. Evidently, the final calculated outcome depends heavily on the data source, and in particular on the variety and spread of the data and their quality.

Moosavi et al. compared the diversity of MOFs in several databases, CoRE MOFs being the only experimental one [91]. The authors computed revised autocorrelation functions (RACs) to obtain the correlations between heuristic atomic properties on a molecular graph, thus extracting information such as linker chemistry, metal chemistry, and functional groups [92]. Combined with the analysis of simple geometric descriptors, they found that the databases considered covered a significantly different chemical space. In particular, one experimental database was composed mostly of structures with small pores, while hypothetical databases each covered a different pore size region. This is a natural consequence of how the databases were built, as they were intended to cover different topological spaces. Another result of these artificially built structures is a more thorough coverage of pore geometry, linker chemistry, and functional groups spaces. However, the metal chemistry is significantly less varied and present. A simple explanation is that metal clusters in MOFs can only be known once they have been synthesized, as opposed to readily available lists of possible organic linkers and functional groups. Taking their analysis even further, Moosavi et al. applied machine learning to these databases and found that the resulting most important variables differed from one database to another. Metal chemistry is likely to be considered not important for certain applications if studied on hypothetical databases for instance. Finally, the authors show that the biases in the different databases mean that machine learning might not always be transferable, the best-case scenario being an algorithm trained on a more diverse set of structures and tested on a biased dataset. Through their diversity study, Moosavi et al. highlighted the potential differences between CoRE MOFs and a range of hypothetical MOFs, and the dangerous general conclusions we can draw when relying on biased datasets. In addition, while hypothetical databases aim to map out the unexplored regions of the MOF space, the existing ones are very restricted and heavily depend on the knowledge drawn from real structures. Unfortunately, the CSD MOF subset was not considered in this study. How different is it from the CoRE MOF?

To answer this question, Altintas et al. compared the two experimental databases (5109 structures from CoRE MOF version 2014 and a subset of 19,123 non-disordered MOFs from the CSD MOF subset version 5.37 May 2016, with solvents removed with the provided Python script) in the case of 3D MOFs, for methane and hydrogen adsorption [93]. They found 3490 structures in common in the two datasets, of which 387 (or 11%) had significantly different final gas uptakes. These differences are due to the different data cleaning procedures in the two databases. The authors also compared these uptakes with the original uptake measured in the corresponding papers, and found that neither database is in perfect agreement with the experimental results. However, it is tricky to compare these two databases, as the structures underwent different cleaning procedures. More precisely, it is unclear whether the solvents were removed from the appropriate set of structures in the CSD MOF subset, and the lack of version control in the CoRE MOF database is an obstacle in understanding the origin of the structural differences with the CSD data it was derived from. Daglar et al. followed a similar procedure to compare updated versions of the two databases (10,143 structures from CoRE MOF version 2019 and 79,943 non-disordered MOFs from the CSD MOF subset with updates up to November 2019, where the solvents were removed with the same provided Python script) for the adsorption of methane, hydrogen, carbon dioxide, and the separation of a 50:50 methane/hydrogen mixture [94]. The authors compared the chemical formulas of the 3543 structures in common and found that 69% of the MOFs had identical chemical formulas, as well as identical structural properties. Among the remnant 31%, the differences lie in: i) the calculated uptakes at low pressures, where gas–MOF interactions are the most influential; and ii) the nature of atoms that have been removed during data processing. Further atomic-level examination of the structures revealed that these differences are related to the removal of specific guest molecules, such as H_2, D_2, metal-O complexes, and bulky halogen groups.

Such reproducibility issues could be overcome with the integration of trackable workflows such as the Automated Interactive Infrastructure and Database for Computational Science (AiiDA) [95]. Ongari et al. demonstrated the use of AiiDA on a database of CURATED (Clean, Uniform, and Refined with Automatic Tracking from Experimental Database) covalent-organic frameworks (COFs) [96], for which the modifications and resulting structures obtained at each stage are accessible on the Materials Cloud platform [95]. The end goal of this infrastructure is to create a platform that is capable of matching experimentally-obtained materials with potential applications [97]. We expect such a tool to be groundbreaking, as it would bridge the gap between computational and experimental researchers.

Beyond the differences between the two experimental databases, it is also worth remembering that the data quality of the CSD depends highly on the deposited experimental data itself, even after being edited by the CCDC. Upon deposition, each CIF undergoes a thorough validation process, where it first has to pass a series of 539 detailed automated checks, the list of which is available online [98]. A chemical structure is then automatically assigned and its reliability is assessed based on the knowledge accumulated from all the other validated CSD structures [99]. Finally, a scientific editor manually checks the structure before its publication [100]. Errors are inevitable in such a complex and long procedure. Therefore, discrepancies can also occur between uptakes computed from the original CIF and the measured ones. We thus refer MOF researchers to the general guidelines given by the CCDC on good-quality CIFs [101].

8.7 Conclusions

In this chapter, we presented the CSD MOF subset, its building process, its general characteristics, and CSD methods to explore it further. While a license is needed to access the subset, ca. 10,000 3D structures are freely accessible via the MOF CIF Collection [102]. The Python scripts are provided as Supporting Information with the relevant publications [44,78]. The structure–property landscapes obtained from the targeted classification can be visualized online: http://aam.ceb.cam.ac.uk/mof-explorer/CSD_MOF_subset. With the CSD MOF subset and the tools provided here, users can run simulations where electrostatic interactions can be neglected. For example, Bucior et al. and Ahmed et al. carried out separate HTS for the storage of hydrogen (77 K, 100 bar–160 K, 5 bar, and 77 K, 5–100 bar, respectively) on a hybrid mix of data (ca. 50,000 and 500,000, respectively) [103,104]. Out of the four selected candidates, three were from the CSD MOF subset: MFU-4*l* (UPOZAB) [105], PCN-610/NU-100 (HABQUY [106]/GAGZEV [107]), and ZELROZ [108]. In particular, PCN-610/NU-100 surpassed the usable gravimetric capacity of the previous record holder under the same conditions (IRMOF-20): 10.0 wt.% vs. 5.7 wt.%. It even surpassed the 2020 target and the ultimate target set out by the Department of Energy by 124% and 55%, respectively [104]. However, analyses of the volumetric capacities revealed a ceiling of 40 g/L, which is more tricky to attain. In addition, the availability of the synthesis protocols in the CSD MOF subset was very convenient to then experimentally validate the simulated results. For the assignment of partial charges, multiple methods exist and we refer readers to the relevant literature [86, 109].

As explained in this chapter, designing an efficient search query is not straightforward. Without a usable definition of MOFs, it is also difficult to determine how accurate a set of queries is. While the criteria presented here helped in digging MOFs and MOF-like structures that have been deposited in the CSD in the past, more efforts can be made toward labeling future structure submissions. In fact, the CCDC has started assigning to some structures their common names (e.g., HKUST-1, MOF-5, etc.). This information can be searched for as a string in Conquest or with the CSD Python API.

We also mentioned earlier that identical and similar structures can be found several times in the CSD. While some researchers advocate for a database containing a single representation of a given MOF, the presence of similar structures can also be beneficial. Indeed, repeated occurrences can skew a data analysis, but the comparison of a variety of similar structures can also reveal interesting structural behaviors, such as flexibility. The CCDC currently groups similar structures under the same refcode family, composed of the same six-letter code but varying ending digits. For example, RUBTAK and RUBTAK01 are two similar entries of UiO-66. The similarity here is determined by the CSD editors, who use a set of different techniques, such as molecules overlay and powder patterns analyses, combined with the chemistry intended by the original authors. However, this is not always straightforward for MOFs, as two entries can have identical frameworks but with different guest molecules. These entries will therefore not be considered as part of the same family. To overcome this, Barthel et al. suggested comparing the structures' bond networks [110]. After analysing 502 CoRE MOFs, 15.5% were found redundant. Bucior et al. developed automated cheminformatics algorithms that systematically assign to each unique MOF a MOFid and a MOFkey [111]. The 2019 update of the CoRE MOF database uses a Python script to compare the

CIFs directly [35]. The open-source Python library for materials analysis Pymatgen had a StructureMatcher algorithm which uses a similar method [35, 112].

Taking a step back from CSD-specific improvements, the MOF field has seen an explosion of MOF data, MOF databases, and MOF simulation data, all without standardization. In addition, only a minority of the computational studies are confirmed with experimental validation. This lack of feedback loop is partly due to the poor communication and collaboration between computational and experimental researchers. While each research group is able to draw conclusions from analyzing simulated and/or experimental data, the analyses are scattered across publications. Therefore, it is time for the MOF field to have a holistic research framework in which researchers are encouraged to move from separate databases to a common knowledge base that incorporates all computational and experimental data. However, quality knowledge still requires quality and well-understood data. For experimental databases such as the CSD, this means that it is up to MOF researchers to decide on the data they use and on the data processing, depending on the studied applications. MOF scientists should, therefore, be equipped with the right tools to find and analyze data on their own. We hope this chapter is useful to the reader, who should now be able to not only directly use the CSD MOF subset, but also find their structures of interest in the CSD. Finally, the field of MOFs is constantly growing and changing, and so should the CSD MOF subset. We, therefore, encourage all users to send us and the CCDC any feedback and suggestions, so as to keep it as useful as possible to the wider MOF community and to capture and reflect the evolution of the field.

Acknowledgments

We thank the Royal Society of Chemistry for giving us permission to reproduce material from *CrystEngComm*, 2020, 22, 7152–7161 and *Chem. Sci.*, 2020, 11, 8373–8387. We thank the American Chemical Society for giving us permission to reproduce and adapt material from *Chem. Mater.*, 2017, 29, 7, 2618–2625.

References

1 Eddaoudi, M., Moler, D.B., Li, H. et al. (2001). Modular chemistry: secondary building units as a basis for the design of highly porous and robust metal–organic carboxylate frameworks. *Accounts of Chemical Research* 34: 319–330.

2 Li, H., Eddaoudi, M., O'Keeffe, M. et al. (1999). Design and synthesis of an exceptionally stable and highly porous metal-organic framework. *Nature* 402: 276.

3 Hoskins, B.F. and Robson, R. (1989).Infinite polymeric frameworks consisting of three dimensionally linked rod-like segments. *Journal of the American Chemical Society* 111: 5962–5964.

4 Maurin, G., Serre, C., Cooper, A. et al. (2017). The new age of MOFs and of their porous-related solids. *Chemical Society Reviews* 46: 3104–3107.

5 Kitagawa, S., Kitaura, R., and Noro, S. (2004). Functional porous coordination polymers. *Angewandte Chemie International Edition* 43: 2334–2375.

6 Cheetham, A.K., Férey, G., and Loiseau, T. (1999). Open-framework inorganic materials. *Angewandte Chemie International Edition* 38: 3268–3292.

7 Slater, A.G. and Cooper, A.I. (2015). Function-led design of new porous materials. *Science* 348: aaa8075.

8 Farha, O.K., Eryazici, I., Jeong, N.C. et al. (2012). Metal–organic framework materials with ultrahigh surface areas: is the sky the limit? *Journal of the American Chemical Society* 134: 15016–15021.

9 Gomez-Gualdron, D.A., Colon, Y.J., Zhang, X. et al. (2016). Evaluating topologically diverse metal-organic frameworks for cryo-adsorbed hydrogen storage. *Energy & Environmental Science* 9: 3279–3289.

10 Murray, L.J., Dinca, M., and Long, J.R. (2009). Hydrogen storage in metal–organic frameworks. *Chemical Society Reviews* 38: 1294–1314.

11 Getman, R.B., Bae, Y.-S., Wilmer, C.E. et al. (2012). Review and analysis of molecular simulations of methane, hydrogen, and acetylene storage in metal–organic frameworks. *Chemical Reviews* 112: 703–723.

12 He, Y., Zhou, W., Qian, G. et al. (2014). Methane storage in metal–organic frameworks. *Chemical Society Reviews* 43: 5657–5678.

13 Mason, J.A., Oktawiec, J., Taylor, M.K. et al. (2015). Methane storage in flexible metal–organic frameworks with intrinsic thermal management. *Nature* 527: 357–361.

14 Van de Voorde, B., Bueken, B., Denayer, J. et al. (2014). Adsorptive separation on metal–organic frameworks in the liquid phase. *Chemical Society Reviews* 43: 5766–5788.

15 Li, J.-R., Sculley, J., and Zhou, H.-C. (2012). Metal–organic frameworks for separations. *Chemical Reviews* 112: 869–932.

16 Furukawa, H., Cordova, K.E., O'Keeffe, M. et al. (2013). The chemistry and applications of metal–organic frameworks. *Science* 2013: 341.

17 Moghadam, P.Z., Ivy, J.F., Arvapally, R.K. et al. (2017). Adsorption and molecular siting of CO_2, water, and other gases in the superhydrophobic, flexible pores of FMOF-1 from experiment and simulation. *Chemical Science* 8: 3989–4000.

18 Bobbitt, N.S., Mendonca, M.L., Howarth, A.J. et al. (2017). Metal–organic frameworks for the removal of toxic industrial chemicals and chemical warfare agents. *Chemical Society Reviews* 46: 3357–3385.

19 Lee, J., Farha, O.K., Roberts, J. et al. (2009). Metal–organic framework materials as catalysts. *Chemical Society Reviews* 38: 1450–1459.

20 Zhang, T. and Lin, W. (2014). Metal–organic frameworks for artificial photosynthesis and photocatalysis. *Chemical Society Reviews* 43: 5982–5993.

21 Rogge, S.M.J., Bavykina, A., Hajek, J. et al. (2017). Metal–organic frameworks for artificial photosynthesis and photocatalysis. *Chemical Society Reviews* 46: 3134–3184.

22 Teplensky, M.H., Fantham, M., Li, P. et al. (2017). Temperature treatment of highly porous zirconium-containing metal–organic frameworks extends drug delivery release. *Journal of the American Chemical Society* 139: 7522–7532.

23 Horcajada, P., Chalati, T., Serre, C. et al. (2010). Porous metal–organic-framework nanoscale carriers as a potential platform for drug delivery and imaging. *Nature Materials* 9: 172–178.

24 Della Rocca, J., Liu, D., and Lin, W. (2011). Nanoscale metal–organic frameworks for biomedical imaging and drug delivery. *Accounts of Chemical Research* 44: 957–968.

25 Li, P., Modica, J.A., Howarth, A.J. et al. (2016). Toward design rules for enzyme immobilization in hierarchical mesoporous metal–organic frameworks. *Chem.* 1: 154–169.

26 Abánades Lázaro, I., Haddad, S., Sacca, S. et al. (2017). Selective surface PEGylation of UiO-66 nanoparticles for enhanced stability, cell uptake, and pH-responsive drug delivery. *Chem.* 2: 561–578.

27 Miller, S.E., Teplensky, M.H., Moghadam, P.Z. et al. (2016). Metal–organic frameworks as biosensors for luminescence-based detection and imaging. *Interface Focus* 6.

28 Batten, S.R., Champness, N.R., Chen, X.-M. et al. (2013). *Pure and Applied Chemistry* 85: 1715.

29 Groom, C.R., Bruno, I.J., Lightfoot, M.P. et al. (2016). The Cambridge Structural Database. *Acta Crystallographica Section B* 72: 171–179.

30 Brown, I.D. and McMahon, B. (2002). CIF: the computer language of crystallography. *Acta Crystallographica Section B* 58: 317–324.

31 Li, A., Bueno-Perez, R., Wiggin, S. et al. (2020). Enabling efficient exploration of metal–organic frameworks in the Cambridge Structural Database. *CrystEngComm.* 22: 7152–7161.

32 Watanabe, T. and Sholl, D.S. (2012). Accelerating applications of metal–organic frameworks for gas adsorption and separation by computational screening of materials. *Langmuir* 28: 14114–14128.

33 Goldsmith, J., Wong-Foy, A.G., Cafarella, M.J. et al. (2013). Theoretical limits of hydrogen storage in metal–organic frameworks: opportunities and trade-offs. *Chemistry of Materials* 25: 3373–3382.

34 Chung, Y.G., Camp, J., Haranczyk, M. et al. (2014). Computation-ready, experimental metal–organic frameworks: a tool to enable high-throughput screening of nanoporous crystals. *Chemistry of Materials*26: 6185–6192.

35 Chung, Y.G., Haldoupis, E., Bucior, B.J. et al. (2019). Advances, updates, and analytics for the computation-ready, experimental metal–organic framework database: CoRE MOF 2019. *Journal of Chemical & Engineering Data* 64: 5985–5998.

36 Bruno, I.J., Cole, J.C., Edgington, P.R. et al. (2002). New software for searching the Cambridge Structural Database and visualizing crystal structures. *Acta Crystallographica Section B* 58: 389–397.

37 Qiao, Z., Zhang, K., and Jiang, J. (2016). *In silico* screening of 4764 computation-ready, experimental metal–organic frameworks for CO_2 separation. *Journal of Materials Chemistry A* 4: 2105–2114.

38 Qiao, Z., Xu, Q., and Jiang, J. (2018). High-throughput computational screening of metal-organic framework membranes for upgrading of natural gas. *Journal of Membrane Science* 551: 47–54.

39 Li, S., Chung, Y.G., and Snurr, R.Q. (2016). High-throughput screening of metal–organic frameworks for CO_2 capture in the presence of water. *Langmuir* 32: 10368–10376.

40 Ahmed, A., Liu, Y., Purewal, J. et al. (2017). Balancing gravimetric and volumetric hydrogen density in MOFs. *Energy & Environmental Science* 10: 2459–2471.

41 Simon, C.M., Kim, J., Gomez-Gualdron, D.A. et al. (2015). The materials genome in action: identifying the performance limits for methane storage. *Energy & Environmental Science* 8: 1190–1199.

42 Campbell, M.G., Sheberla, D., Liu, S.F. et al. (2015). Cu_3(hexaiminotriphenylene)2: an electrically conductive 2D metal–organic framework for chemiresistive sensing. *Angewandte Chemie International Edition* 54: 4349–4352.

43 Canepa, P., Chabal, Y.J., and Thonhauser, T. (2013). When metal organic frameworks turn into linear magnets. *Physical Review B* 87: 094407.

44 Moghadam, P.Z., Li, A., Wiggin, S.B. et al. (2017). Development of a Cambridge Structural Database subset: a collection of metal–organic frameworks for past, present, and future. *Chemistry of Materials* 29: 2618–2625.

45 Avci, G., Erucar, I., and Keskin, S. (2020). Do new MOFs perform better for CO_2 capture and H_2 Purification? Computational screening of the updated MOF database. *ACS Applied Materials & Interfaces* 12: 41567–41579.

46 Batten, S.R., Champness, N.R., Chen, X.-M. et al. (2012). Coordination polymers, metal-organic frameworks and the need for terminology guidelines. *CrystEngComm.* 14: 3001–3004.

47 Forgan, R.S., Smaldone, R.A., Gassensmith, J.J. et al. (2012). Nanoporous carbohydrate metal–organic frameworks. *Journal of the American Chemical Society* 134: 406–417.

48 White, K.A., Chengelis, D.A., Zeller, M. et al. (2009). Nanoporous carbohydrate metal–organic frameworks. *Journal of the American Chemical Society* 30: 4506–4508. https://pubs.rsc.org/en/content/articlelanding/2009/cc/b909658b

49 Xie, M.-H., Yang, X.-L., Zou, C. et al. (2011). An SnIV–porphyrin-based metal–organic framework for the selective photo-oxygenation of phenol and sulfides. *Inorganic Chemistry* 50: 5318–5320.

50 CCDC (2019). *ConQuest User Guide and Tutorials*. www.ccdc.cam.ac.uk/support-and-resources/ccdcresources/ConQuest-UserGuide.pdf (accessed August 2019).

51 CCDC (2019). "I find searching for polymeric structures difficult and I get results I don't fully understand. Can you explain how polymeric structures are defined so I can tailor my searches to be more effective?" www.ccdc.cam.ac.uk/support-and-resources/support/case/?caseid=f75281ce-d3fe-472e-8438-7b6cdea5accb (accessed August 2019).

52 Spek, A. (2015). PLATON SQUEEZE: a tool for the calculation of the disordered solvent contribution to the calculated structure factors. *Acta Crystallographica Section C* 71: 9–18.

53 Dolomanov, O.V., Bourhis, L.J., Gildea, R.J. et al. (2009). OLEX2: a complete structure solution, refinement and analysis program. *Journal of Applied Crystallography* 42: 339–341.

54 Macrae, C.F., Sovago, I., Cottrell, S.J. et al. (2020). Mercury 4.0: from visualization to analysis, design and prediction. *Journal of Applied Crystallography* 53: 226–235.

55 Howarth, A.J., Peters, A.W., Vermeulen, N.A. et al. (2017). Best practices for the synthesis, activation, and characterization of metal–organic frameworks. *Chemistry of Materials* 29: 26–39.

56 Calahorro, A.J., Salinas-Castillo, A., Fairen-Jimenez, D. et al. (2015). Long lifetime photoluminescence emission of 3D cadmium metal–organic frameworks based on the 5-(4-pyridyl) tetrazole ligand. *Inorganica Chimica Acta* 427: 131–137.

57 Willems, T.F., Rycroft, C.H., Kazi, M. et al. (2012). Algorithms and tools for high-throughput geometry-based analysis of crystalline porous materials. *Microporous and Mesoporous Materials* 149: 134–141.

58 Li, P., Vermeulen, N.A., Malliakas, C.D., Gómez-Gualdrón, D.A. et al. (2017). Bottom-up construction of a superstructure in a porous uranium-organic crystal. *Science* 356: 624–627.

59 Sumida, K., Rogow, D.L., Mason, J.A. et al. (2012). Carbon dioxide capture in metal–organic frameworks. *Chemical Reviews* 112: 724–781.

60 Flaig, R.W., Osborn Popp, T.M., Fracaroli, A.M. et al. (2017). The chemistry of CO_2 capture in an amine-functionalized metal–organic framework under dry and humid conditions. *Journal of the American Chemical Society* 139: 12125–12128.

61 Cohen, S.M. (2012). Postsynthetic methods for the functionalization of metal–organic frameworks. *Chemical Reviews* 112: 970–1000.

62 Deria, P., Mondloch, J.E., Karagiaridi, O. et al. (2014). Beyond post-synthesis modification: evolution of metal–organic frameworks via building block replacement. *Chemical Society Reviews* 43: 5896–5912.

63 Moghadam, P.Z., Fairen-Jimenez, D., and Snurr, R.Q. (2016). Efficient identification of hydrophobic MOFs: application in the capture of toxic industrial chemicals. *Journal of Materials Chemistry A* 4: 529–536.

64 Bernales, V., Ortuño, M.A., Truhlar, D.G. et al. (2018). Computational design of functionalized metal–organic framework nodes for catalysis. *ACS Central Science* 4: 5–19.

65 Zhu, L., Liu, X.-Q., Jiang, H.-L. et al. (2017). Metal–organic frameworks for heterogeneous basic catalysis. - *Chemical Reviews* 117: 8129–8176.

66 So, M.C., Wiederrecht, G.P., Mondloch, J.E. et al. (2015). Metal–organic framework materials for light-harvesting and energy transfer. *Chemical Communications* 51: 3501–3510.

67 Cui, Y., Yue, Y., Qian, G. et al. (2012). Luminescent functional metal–organic frameworks. *Chemical Reviews* 112: 1126–1162.

68 Peng, Y., Gong, T., Zhang, K. et al. (2014). Engineering chiral porous metal–organic frameworks for enantioselective adsorption and separation. *Nature Communications* 5: 4406.

69 Navarro-Sánchez, J., Argente-García, A.I., Moliner-Martínez, Y. et al. (2017). Peptide metal–organic frameworks for enantioselective separation of chiral drugs. *Journal of the American Chemical Society.* 139: 4294–4297.

70 Ma, L., Falkowski, J.M., Abney, C. et al. (2010). A series of isoreticular chiral metal–organic frameworks as a tunable platform for asymmetric catalysis. *Nature Chemistry* 2: 838.

71 Herm, Z.R., Wiers, B.M., Mason, J.A. et al. (2013). Separation of hexane isomers in a metal–organic framework with triangular channels. *Science* 340: 960–964.

72 Torres-Knoop, A., Krishna, R., and Dubbeldam, D. (2014). Separating xylene isomers by commensurate stacking of p-Xylene within channels of MAF-X8. *Angewandte Chemie International Edition* 53: 7774–7778.

73 Holcroft, J.M., Hartlieb, K.J., Moghadam, P.Z. et al. (2015). Carbohydrate-mediated purification of petrochemical. *Journal of the American Chemical Society* 137: 5706–5719.

74 Haranczyk, M. and Sethian, J.A. (2010). Automatic structure analysis in high-throughput characterization of porous materials. *Journal of Chemical Theory and Computation* 6: 3472–3480.

75 Haldoupis, E., Nair, S., and Sholl, D.S. (2011). Pore size analysis of >250,000 hypothetical zeolites. *Phys. Chem. Chem. Phys.* 13: 5053–5060.

76 Coudert, F.-X. and Fuchs, A.H. (2016). Computational characterization and prediction of metal–organic framework properties. *Coord. Chem. Rev.* 307: 211–236.

77 Sarkisov, L. and Harrison, A. (2011). Computational structure characterisation tools in application to ordered and disordered porous materials. *Molecular Simulation* 37: 1248–1257.

78 Moghadam, P.Z., Li, A., Liu, X.-W. et al. (2020). Targeted classification of metal–organic frameworks in the Cambridge Structural Database (CSD). *Chemical Science* 11: 8373–8387.

79 Oien-Odegaard, S., Shearer, G.C., Wragg, D.S. et al. (2017). Pitfalls in metal–organic framework crystallography: towards more accurate crystal structures. *Chemical Society Reviews* 46: 4867–4876.

80 Li, H., Eddaoudi, M., Groy, T.L. et al. (1998). Establishing microporosity in open metal—organic frameworks: gas sorption isotherms for Zn(BDC) (BDC = 1,4-benzenedicarboxylate). *Journal of the American Chemical Society* 120: 8571–8572.

81 Kondo, M., Yoshitomi, T., Matsuzaka, H. et al. (1997). Three-dimensional framework with channeling cavities for small molecules: $\{[M2(4,4'\text{-bpy})3(NO_3)_4]\cdot xH_2O\}n$ (MCo, Ni, Zn). *Angewandte Chemie International Edition in English* 36: 1725–1727.

82 Zachariasen, W.H. (1940). The crystal structure of sodium formate, $NaHCO_2$. *Journal of the American Chemical Society* 62: 1011–1013.

83 Kinoshita, Y., Matsubara, I., Higuchi, T. et al. (1959). The crystal structure of bis(adiponitrilo) copper(I) nitrate. *Bulletin of the Chemical Society of Japan* 32: 1221–1226.

84 Strutt, N.L., Fairen-Jimenez, D., Iehl, J. et al. (2012). Incorporation of an A1/A2-difunctionalized pillar[5]arene into a metal–organic framework. *Journal of the American Chemical Society* 134: 17436–17439.

85 Manz, T.A. and Sholl, D.S. (2010). Chemically meaningful atomic charges that reproduce the electrostatic potential in periodic and nonperiodic materials. *Journal of Chemical Theory and Computation* 6: 2455–2468.

86 Nazarian, D., Camp, J.S., and Sholl, D.S. (2016). A comprehensive set of high-quality point charges for simulations of metal–organic frameworks. *Chemistry of Materials* 28: 785–793.

87 Wilmer, C.E., Leaf, M., Lee, C.Y. et al. (2011). Large-scale screening of hypothetical metal–organic frameworks. *Nature Chemistry* 4: 83.

88 Boyd, P.G. and Woo, T.K. (2016). A generalized method for constructing hypothetical nanoporous materials of any net topology from graph theory. *CrystEngComm* 18: 3777–3792.

89 Rosen, A.S., Iyer, S.M., Ray, D. et al. (2021). Machine learning the quantum-chemical properties of metal–: organic frameworks for accelerated materials discovery. *Matter* 4: 1578–1597.

90 Banerjee, D., Simon, C.M., Plonka, A.M. et al. (2016). Metal–organic framework with optimally selective xenon adsorption and separation. *Nature Communications* 7: 11831.

91 Moosavi, S.M., Nandy, A., Jablonka, K.M. et al. (2020). Metal–organic framework with optimally selective xenon adsorption and separation. *Nature Communications* 11: 4068.

92 Janet, J.P. and Kulik, H.J. (2017). Resolving transition metal chemical space: feature selection for machine learning and structure–property relationships. *The Journal of Physical Chemistry A* 121: 8939–8954.

93 Altintas, C., Avci, G., Daglar, H. et al. (2019). An extensive comparative analysis of two MOF databases: high-throughput screening of computation-ready MOFs for CH_4 and H_2 adsorption. *Journal of Materials Chemistry A* 7: 9593–9608.

94 Daglar, H., Gulbalkan, H.C., Avci, G. et al. (2021). Effect of metal–organic framework (MOF) database selection on the assessment of gas storage and separation potentials of MOFs. *Angewandte Chemie International Edition* 60: 7828–7837.

95 Pizzi, G., Cepellotti, A., Sabatini, R. et al. (2016). AiiDA: automated interactive infrastructure and database for computational science. *Computational Materials Science* 111: 218–230.

96 Ongari, D., Yakutovich, A.V., Talirz, L. et al. (2019). Building a consistent and reproducible database for adsorption evaluation in covalent–organic frameworks. *ACS Central Science* 5: 1663–1675.

97 Ongari, D., Talirz, L., and Smit, B. (2020). Too many materials and too many applications: an experimental problem waiting for a computational solution. *ACS Central Science* 6: 1890–1900.

98 IUCr (2020). Details of checkCIF/PLATON tests. http://journals.iucr.org/services/cif/datavalidation.html (accessed July 2020).

99 Bruno, I.J., Shields, G.P., and Taylor, R. (2011). Deducing chemical structure from crystallographically determined atomic coordinates. *Acta Crystallographica Section B* 67: 333–349.

100 Holgate, S. (2020). CSD data curation: the human touch. https://www.ccdc.cam.ac.uk/Community/blog/CSD-data-curation-the-human-touch (accessed July 2020).

101 CCDC (2020). The CCDC CIF deposition guidelines. https://www.ccdc.cam.ac.uk/Community/depositastructure/cif-deposition-guidelines (accessed August 2020).

102 Li, A., Perez, R.B., Wiggin, S. et al. (2021). The launch of a freely accessible MOF CIF collection from the CSD. *Matter* 4: 1105–1106.

103 Bucior, B.J., Bobbitt, N.S., Islamoglu, T. et al. (2019). Energy-based descriptors to rapidly predict hydrogen storage in metal–organic frameworks. *Molecular Systems Design & Engineering* 4: 162–174.

104 Ahmed, A., Seth, S., Purewal, J. et al. (2019). Exceptional hydrogen storage achieved by screening nearly half a million metal-organic frameworks. *Nature Communications* 10: 1568.

105 Denysenko, D., Grzywa, M., Tonigold, M. et al. (2011). Elucidating gating effects for hydrogen sorption in MFU-4-type triazolate-based metal–organic frameworks featuring different pore sizes. *Chemistry – A European Journal* 17: 1837–1848.

106 Yuan, D., Zhao, D., Sun, D. et al. (2010). An isoreticular series of metal–organic frameworks with dendritic hexacarboxylate ligands and exceptionally high gas-uptake capacity. *Angewandte Chemie International Edition* 49: 5357–5361.

107 Farha, O.K., Özgür Yazaydın, A., Eryazici, I. et al. (2010). *De novo* synthesis of a metal–organic framework material featuring ultrahigh surface area and gas storage capacities. *Nature Chemistry* 2: 944–948.

108 Rankine, D., Avellaneda, A., Hill, M.R. et al. (2012). Control of framework interpenetration for *in situ* modified hydroxyl functionalised IRMOFs. *Chemical Communications* 48: 10328–10330.

109 Hamad, S., Balestra, S.R.G., Bueno-Perez, R. et al. (2015). Atomic charges for modeling metal–organic frameworks: why and how. *Journal of Solid State Chemistry* 223: 144–151.

110 Barthel, S., Alexandrov, E.V., Proserpio, D.M. et al. (2018). Distinguishing metal–organic frameworks. *Crystal Growth & Design* 18: 1738–1747.

111 Bucior, B.J., Rosen, A.S., Haranczyk, M. et al. (2019). Identification schemes for metal–organic frameworks to enable rapid search and cheminformatics analysis. *Crystal Growth & Design* 19: 6682–6697.

112 Ong, S.P., Richards, W.D., Jain, A. et al. (2013). Python materials genomics (pymatgen): a robust, open-source python library for materials analysis. *Computational Materials Science* 68: 314–319.

9

Data-Driven Approach for Rational Synthesis of Zeolites and Other Nanoporous Materials

Watcharop Chaikittisilp

Research and Services Division of Materials Data and Integrated System (MaDIS), National Institute for Materials Science (NIMS), Tsukuba, Japan

9.1 Introduction

Zeolites, metal–organic frameworks, and mesoporous solids are representative nanoporous materials that have found their central place in several industrial applications [1–5]. Their applications range from traditional uses, such as adsorption, separation, and catalysis, to emerging areas, including sensing, electronics, and biomedicine. In the past few decades, progress in nanoporous materials has been accelerated by advances in experimental, analytical, and computational methods, thereby bringing the materials to practical applications. However, to bring the materials from laboratory to commercialization, in principle, it starts from materials searching, through optimization of materials properties and performance, for example, by tuning structures and compositions, to large-scale production. This process generally takes more than 10–15 years. To accelerate the development of materials, computational calculations and advanced characterization techniques have been employed to understand the synthesis–structure–property relationships. However, these relationships are sometime very complex and complicated.

As our society is changing at a rapid pace, with increasing demand for materials that can sustain us on our planet, the integration of materials science and data science, also known as materials informatics, has become an important research area. This also holds for nanoporous materials [6–8]. Conventionally, the development of nanoporous materials has heavily relied on an Edisonian trial-and-error approach based on chemical/materials intuition of human experts, which have been generally performed by alteration of experimental parameters, etc., a highly time-consuming approach. Such an exploratory search for the optimal materials is prohibitively experimentally expensive. To overcome this experimental challenge, a data-driven approach by applying data science techniques such as machine learning algorithms to experimental and computational data, enables us to extract the most significant material descriptors over the complex chemical/materials space with high dimension and massive entries, which is sometimes very difficult for humans to handle. In particular, the pattern recognition capability of machine learning can be exceptionally effective for nanoporous materials. This is because nanoporous materials consist of framework bodies (pore walls) and internal cavities (pore spaces), resulting in substantial interfaces in their confined spaces that can affect their properties and functions. Simultaneous optimization of pore walls and pore spaces becomes multidimensional and causes complex problems, which are difficult to resolve by straightforward computational and experimental methodologies.

In this chapter, the applications of data science techniques for assisting (and accelerating) the synthesis of nanoporous materials are discussed. First, machine learning algorithms are applied to analyze the synthesis records of zeolites previously reported in the literature, to extract the

AI-Guided Design and Property Prediction for Zeolites and Nanoporous Materials, First Edition. Edited by German Sastre and Frits Daeyaert.

synthesis descriptors of zeolites. These synthesis descriptors are then linked to the structure descriptors of zeolites, to rationalize the synthesis–structure relationship and subsequently to suggest the synthesis parameters for selected zeolites. Second, by a similar approach, machine learning algorithms are used to extract the (most) influential materials descriptors of nanoporous electrocatalysts from the experimental records reported in the literature to rationalize the structure–property relationship, which subsequently suggests the better materials for targeted reactions. Lastly, an active learning scheme, based on Bayesian optimization, is described for optimization of synthesis parameters of nanoporous metals for electrochemical reactions. This approach can efficiently assist the experimental design in discovering the optimal materials, as it is revealed that the number of necessary experiments can be decreased substantially.

9.2 Rationalization of the Synthesis–Structure Relationship in Zeolite Synthesis: Application Machine Learning and Graph Theory to Zeolite Synthesis

Typically, zeolites are synthesized under hydrothermal conditions in the presence of silica, alumina (or sources of other substituting atoms such as boron and germanium), mineralizing agents (hydroxide or fluoride), and water [9–11]. It is well-accepted that all existing zeolites are crystallized as metastable phases, which are less energetically favorable than their stable counterparts. The formation of metastable materials is often controlled via kinetics-dominant pathways, where entropic effects can be more significant than enthalpic effects. In zeolite synthesis, organic structure-directing agents and hydrated inorganic cations (sometimes referred to as inorganic structure-directing agents) are employed to compensate for such an energy penalty, thereby stabilizing the metastable structures. As a result, synthesis of zeolites generally involves complicatedly interrelated synthesis parameters. Only slight alteration of the synthesis conditions such as types of structure-directing agents, chemical compositions of raw materials, and heating time and temperature, can lead to the formation of different zeolites. Their formation mechanisms have not been fully clarified yet because zeolites are formed through an intricate sequence of chemical reactions under hydrothermal conditions. Consequently, the discovery of new zeolites and the subsequent property optimization for targeted applications have relied heavily on a trial-and-error approach.

In the past two decades, there have been many attempts to understand the formation mechanism of zeolites [12–14], and to predict and optimize zeolite synthesis based on advances in theory and computation chemistry [15–17]. Still, it is nearly impossible to capture the whole process of zeolite synthesis. To this end, data science has recently been applied to extract chemical knowledge hidden behind the synthesis. To capture synthesis knowledge of experimentalists, Jensen et al. created a literature data extraction pipeline to extract synthesis experience of zeolites from the literature using natural language processing [18]. They then focused on the subset of extracted data to correlate synthesis conditions to framework properties of the synthesized Ge-containing zeolites, resulting in some knowledge on the chemical trends of successful synthesis. The same automatic literature data extraction was later applied to generalize relationships between zeolite products and organic structure-directing agents used in the synthesis [19]. They applied a principal component analysis to reduce the dimensionality of organic structure-directing agents into a two-dimensional (2D) space, providing a geometric parameter to efficiently compare the similarity of different organics used in the successful synthesis of zeolites. Their approach can be an alternative to computationally expensive workflows typically used for designing the organic structure-directing agents for targeted zeolites.

In organic-free synthesis of zeolites, zeolite formation can become more complicated, because the structure-direction ability of hydrated inorganic cations (i.e., inorganic structure-directing agents) is somehow not so selective, in comparison with organic structure-directing agents. To capture the chemical intuition of human experts, Muraoka et al. developed a computational workflow to correlate synthesis descriptors to structure descriptors of zeolites (Figure 9.1) [20]. The dataset collected from the literature contains 686 synthesis records with 23 different products (22 crystalline zeolitic phases and amorphous). It is noteworthy that the synthesis records resulting in the amorphous solids can somewhat represent the "unsuccessful" conditions. When working on data science techniques based on literature data, the lack or insufficient amount of data on unsuccessful (or failed) experiments can sometimes be problematic and can lead to inaccurate results. It is clear that the presence of data on (partially) unsuccessful experiments is of significance to extract chemical knowledge and intuition in materials synthesis, akin to the way that humans naturally learn and gain their experience from trial-and-error experiments [21–23].

Classification models are constructed to predict the synthesis outcome (i.e., zeolite phases or amorphous) from the synthesis conditions (i.e., synthesis descriptors) using four different classifiers: support vector machine, decision tree, random forest, and extreme gradient boosting (XGBoost). A support vector machine is based on the idea that training data that are nonlinearly separable in the original low-dimensional space can become separable in a higher-dimensional space. The support vector machine builds a hyperplane (or set of hyperplanes) in a higher-dimensional space to separate data with the largest distance to the nearest training data point.

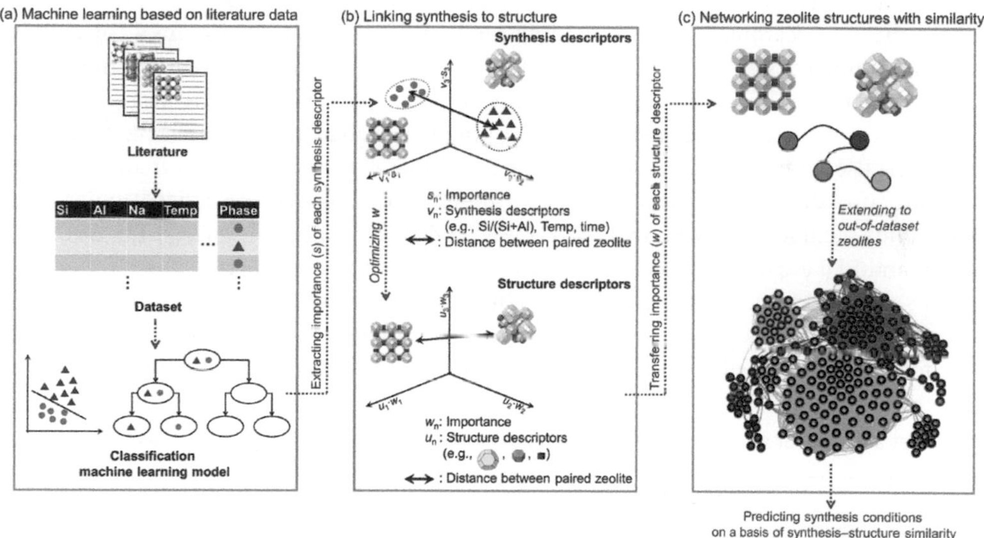

Figure 9.1 A computational workflow to correlate synthesis and structure descriptors in zeolites. (a) Classification machine learning models were constructed based on literature data for organic-free synthesis of zeolites. (b) Extraction of synthesis descriptors with their importance from the trained machine learning models and mapping the synthesizable domains of zeolites onto a multidimensional phase diagram. The distance between the synthesis conditions for each zeolite structure is measured and used to represent the synthesis similarity. Then, building units appearing in each zeolite structure are used as structure descriptors that can define the structural similarity based on the presence or absence of the building units. The distance between the zeolite structures representing the structural similarity is made to be close to the synthesis similarity (an arrow in the multidimensional phase diagram in the upper part in (b)) by tuning the weight (*w*) for each structure descriptor. (c) A network of zeolites is constructed based on the structural similarity.

The decision tree is a tree-like model of decisions, represented as chains or series of decisions by separating data based on decision rules (or criteria). Random forest uses an ensemble approach to transform decision trees to a forest by bagging ensemble, in which a subset of features is selected randomly to build a collection of trees (i.e., forest). XGBoost is a decision tree-based ensemble model based on boosting ensemble, where gradient boosting is optimized in parallel.

The first step is to select appropriate input features for the models. Chemical compositions of raw materials are considered as the most significant synthesis descriptor for zeolite synthesis, together with synthesis temperature and time. Aging conditions and source of raw materials, especially sources of silicon and aluminum, also highly affect the synthesis outcomes. The dataset is divided into a training set (80%) and a test set (20%). Four machine learning models are trained using different combinations of input features (i.e., synthesis descriptors) with appropriate hyperparameter tuning. Table 9.1 summarizes the test accuracy of different machine learning models.

Random forest and XGBoost models outperform support vector machine and decision tree, giving an approximate 80% test accuracy, regardless of the combinations of input features. The highest accuracy is gained when the random forest model is constructed based on all synthesis descriptors. However, in some literature, detailed synthesis conditions, including aging conditions and sources of raw materials, are not described properly. To this end, the XGBoost model using chemical compositions, synthesis temperature, and synthesis time as inputs is selected because of its reasonably high accuracy. The resulting XGBoost model can then be used to predict the synthesis outcomes and to quantify the likeliness of the formation of specific zeolite phases under a given condition.

To extract chemical knowledge from the constructed models, we need to interpret the results. In general, the machine learning models, including random forest and XGBoost, are a "black box" predictor with limited degree of interpretation. One approach to interpret the XGBoost model is to construct decision trees from the trained model. A decision tree built from the best performing XGBoost model is depicted in Figure 9.2. The resulting tree is first split based on the amount of sodium cations used in the synthesis. It should be noted that sodium is the most frequently used inorganic structure-directing agent in zeolite synthesis. Zeolite phases crystallized under thigh sodium contents include FAU, GIS, ANA, LTA, CAN, and SOD, while MFI, MOR, LTL, ERI, OFF, HEU, MER, and EDI are formed from the mixtures with lower sodium content.

The next splitting criterion is the amounts of aluminum, expressed as Si/(Si + Al) ratios. This criterion remarkably represents the chemical intuition of humans because the chemistry of aluminosilicate formation under silicon-rich and aluminum-rich conditions is greatly different. In the

Table 9.1 Test accuracy of machine learning models using different combinations of synthesis descriptors.

	Synthesis descriptor				
Model	**Base**[a]	**Base + Aging**	**Base + Al source**	**Base + Si source**	**Base + Aging + Al source + Si source**
Support vector machine	0.66	0.69	0.70	0.65	0.71
Decision tree	0.76	0.77	0.76	0.72	0.72
Random forest	0.80	0.79	0.79	0.77	0.82
XGBoost	0.80	0.80	0.79	0.80	0.79

[a]Base descriptors include chemical compositions, synthesis temperature, and synthesis time.

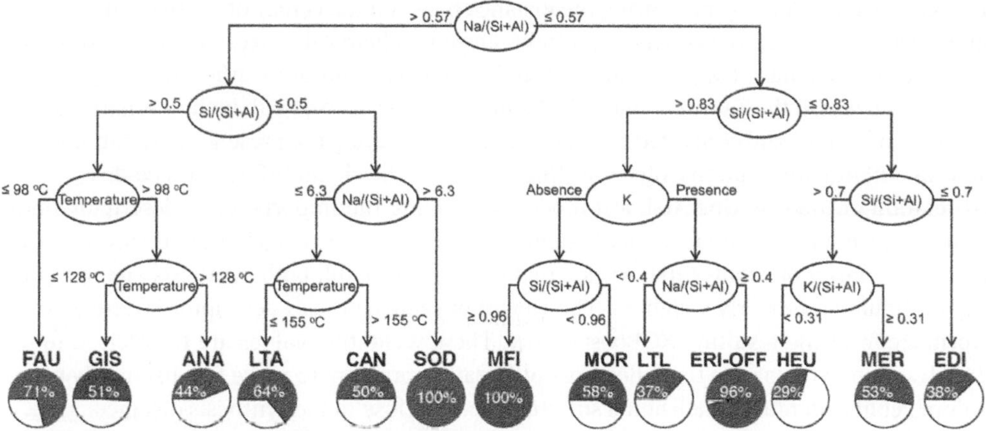

Figure 9.2 A decision tree (with a depth of 4) built from the trained XGBoost model with the highest accuracy. The depth of the complete tree is 12. The percentages shown under the zeolite framework codes are the fractions of the dominant phases presenting in the deeper branches. The remaining phases are other zeolite phases and amorphous product.

high sodium side, decision is made at the Si/(Si + Al) ratio of 0.5, corresponding to the Si/Al ratio of 1. Interestingly, the Si/(Si + Al) ratio of 0.5 (or the Si/Al ratio of 1) is the upper limit for aluminum to be incorporated into the zeolite framework, on a basis of the Löwenstein's rule [24], which describes the absence of Al–O–Al bonds in aluminosilicate zeolites because of their energetically unfavorability. More precisely, Löwenstein suggests that whenever two aluminate tetrahedra are jointed together, at least one of them must become five- or six-coordinated. It is noteworthy that the Al–O–Al bonds in zeolite frameworks can be energetically stable if their negative charges are counter-balanced by high-charge density cations such as proton and lithium, as suggested by theoretical calculations [25].

When Si/(Si + Al) > 0.5, FAU, GIS, and ANA are dominant phases. Their formations depend on the synthesis temperature, with FAU and ANA being the dominant phases at the lowest and highest temperatures, respectively. This can relate to the zeolite phase variation with their synthesis temperature, according to Ostwald's step rule [26,27]. This rule explains a consequence of growth kinetics (or the formation of multiple metastable phases in sequence), starting from the least stable, until reaching the most stable phase. When linking the thermodynamic stability to the synthesis temperature, the less stable phase often forms at lower temperatures. In fact, the Gibbs free energies of aluminosilicate FAU, GIS, and ANA zeolites at identical chemical compositions are in the following order: FAU > GIS > ANA. These results again suggest the effectiveness of machine learning to extract the chemical knowledge hidden behind the data.

When Si/(Si + Al) ≤ 0.5, LTA, CAN, and SOD dominate. LTA and CAN are then separated from SOD by the amount of sodium. Synthesis temperature is a splitting criterion for LTA and CAN, as the formation of LTA is favorable at lower temperatures. Computational calculations also reveal that the Gibbs free energy of LTA is higher than that of CAN, again indicating the LTA-to-CAN transformation according to Ostwald's step rule. Similar interpretation based on thermodynamic stability and chemical compositions can also be made for the zeolite phases formed under the lower sodium composition (the right side in Figure 9.2) [20].

In zeolite synthesis, it has been observed that there are some empirical relations between synthesis conditions and structures of zeolite products [28,29]. Correlating the structure–synthesis similarity may be useful as a guideline for designed synthesis of zeolite [30]. The synthesis similarity can be quantified by measuring the distance between the center of synthesis descriptors (or conditions) of each pair of zeolites in multidimensional chemical spaces; while the structural similarity can be calculated based on a similarity index such as Tanimoto, Carbo, and Hodgkin similarity indices, in which the Tanimoto index is the most common. To calculate the similarity index, the structures of zeolites are converted to binary vectors, indicating the presence (1) or absence (0) of structural building units, often called the "fingerprint" in the chemoinformatics field.

Interpretation of the resulting XGBoost model by deriving the importance of descriptors indicates that all the descriptors are not equally significant. Therefore, to correlate synthesis similarity to structure similarity of zeolites that are synthetically meaningful, both synthesis and structure descriptors should be properly weighted. The weight of each synthesis descriptor is derived from its importance from the resulting XGBoost model. These weighting values are transferred to the structure descriptors by solving the following optimization problem to make the distance between each pair of zeolites in multidimensional structural spaces close to the synthesis distance:

$$\underset{w_i}{\text{minimize}} \sum_{i} \sum_{i \neq j} \left[(sv_i - sv_j) - (w_i u_i - w_j u_j) \right]$$

where i represents the zeolite structure of interest, j is the paired structure, s is the importance of the synthesis descriptors derived from the XGBoost model, v_i is the central value of the synthesis descriptors in the zeolite i, u_i is the binary vector indicating the presence (1) or absence (0) of structural building units in the structure i, and w_i is the weighting value of the structural building units.

With the properly weighted structure descriptors, the structure similarity network of zeolites can be built to be synthetically meaningful. Groups of zeolites can be clustered by partitioning the network using modularity optimization [31], expressing the densely connected nodes into the community. Similarity networks of zeolites constructed with equally important structure descriptors and with properly weighted structure descriptors are substantially different, indicating the effects of weighting. In the similarity network shown in Figures 9.3, zeolite structures are

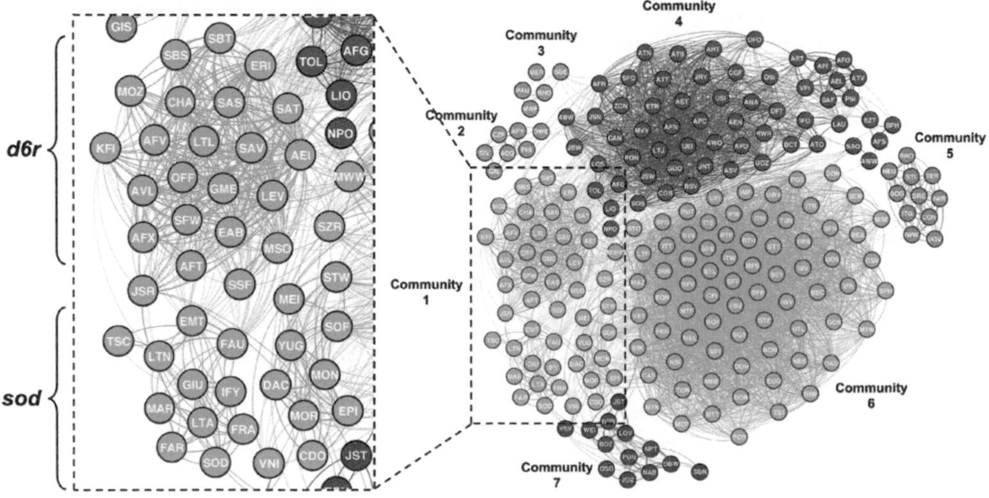

Figure 9.3 A similarity network (right) for the zeolite structures where the communities 1–7 are identified using a clustering algorithm based on the modularity optimization. A magnified network focusing on the community 1 is shown on the left.

expressed as nodes, in which the layout of the nodes reflects the structural similarity. There are seven communities present in the network with proper weighting. The formed communities show several similarity features, both chemically and structurally. For example, community 1 represents zeolites that can be formed in an aluminum-rich composition, while community 6 is dominated by silicon-rich zeolites. Looking more closely at community 1, the lower part is populated with zeolites having the *sod* building unit, whereas zeolites having the *d6r* unit are concentrated in the upper part. Overall, starting solely from the synthesis results reported in the literature for 22 zeolites, techniques in data science, including machine learning and networking, can extract some chemical knowledge hidden behind the data, and can transfer such knowledge to all existing zeolites, including those outside the initial dataset, providing an opportunity to identify the synthesis conditions, which may have never been investigated or reported in literature, of targeted zeolites including the hypothetical structures.

Although similarity of molecules and materials can be estimated from a binary vector using a similarity index, not all of the similarity information can be extracted. Zeolite structures can be considered as networks (or graphs), where silicon atoms (and other tetrahedral atoms) are treated as nodes and oxygen atoms are edges. As a result, instead of a similarity index, graph isomorphism can be considered to provide more comprehensive information on similarity of zeolite structures. One approach to identify and quantify isomorphisms among graphs (i.e., zeolite structures) is to calculate the graph's dissimilarity measure (*D* score) [32], which has been applied to evaluate the structure similarity of zeolites and subsequently to explain and predict topotactic conversion of 2D layered precursors to three-dimensional (3D) zeolites [33] and interzeolite transformations [34].

9.3 Extraction of the Structure–Property Relationship in Nanoporous Nitrogen-Doped Carbons: Dealing with the Missing Values in Literature Data

In Section 9.2, application of machine learning algorithms to zeolite synthesis was described. All zeolites have been synthesized under a similar hydrothermal method and the zeolite community has reported the synthesis conditions of zeolite in the literature in a similar manner. However, when similar approaches are applied to recognize the structure–property relationship based on literature data, missing values will always be problematic. It is hard to construct complete datasets of materials data collecting from the literature. Therefore, appropriate ways to handle the missing values in literature data is essential.

In general, there are three mechanisms of missing data considered to identify why they are missing [35]. It is useful to know the mechanisms of missing data because selection of the most appropriate method to handle the missing data can depend on the mechanisms. It should be noted that identification of the actual mechanism of the missing data is impossible because the actual values of missing data are never known. We can only compare the distribution of the variables/features with and without missing values.

i) *Missing Completely at Random (MCAR)*
   The missing data occur completely at random if the probability of their absence is the same for all features. The occurrence of these missing data is therefore not related to any observed or unobserved characteristics. If data are missing completely at random, removal of the data that contain missing values does not bias the datasets. However, missing completely at random is rare.

ii) *Missing at Random (MAR)*
If the occurrence of missing data is dependent on other features/variables but independent from themselves, this mechanism is known as missing at random. Because of the dependency, other observed data can be used to predict the missing values. There is no systematic difference between the available data and missing data.

iii) *Missing Not at Random (MNAR)*
Missing not at random is a mechanism when the missing data is dependent on the unobserved or unreported values themselves. The available data and missing data are systematically different.

The missing values are often present in the datasets of materials data, especially those that are extracted from the literature. Before these datasets can be analyzed by any machine learning techniques, the missing values must be handled properly. The simplest way to handle the missing values is deletion to eliminate the missing data. This can only be done if there are a small number of missing values, in comparison with the whole data. Deletion can be made by deleting rows (listwise deletion), deleting columns, and pairwise deletion. The other way to handle the missing values is filling out the missing values, for example, by mean, mode, or median. If the missing values exist over the whole datasheet, data imputation techniques based on machine learning algorithms should be considered. The frequently used imputation techniques include *k*-nearest neighbor [36], random forest [37], and XGBoost [38]. It should be noted that it is impossible to determine the accuracy or correctness of the data imputation and the way that the missing values are handled because the actual values are unknown. One method can outperform other strategies for certain datasets, but may be less effective on other datasets. Here, an example on handling the missing values in the iterature data reported for nanoporous materials is described [39].

Electrochemical catalytic oxygen reduction reaction (ORR) is an essential reaction in fuel cells and metal–air batteries [40,41]. Over decades, many attempts have been made to explore efficient ORR catalysts as an alternative to expensive metals, typically used in commercial fuel cells. One of the promising candidates is nanoporous carbons. Doping the carbon frameworks with other heteroatoms such as boron, nitrogen, phosphorus, and sulfur has been reported as a strategy to enhance the ORR catalytic performance, partly due to the modified electronic structure of carbon frameworks. In order to find better doped carbon materials for ORR applications, experimental data of metal-free, nitrogen-doped nanoporous carbons reported for ORR are collected from 50 publications. In the dataset, there are 123 entries representing different carbon materials as ORR catalysts. Table 9.2 lists 13 materials features identified to describe the carbon materials. Not surprisingly, there are many missing values, depending on the literature, in the dataset.

To impute the missing values, the MissForest algorithm [37], a non-parametric method based on random forest, is selected. This algorithm has been used to handle the missing values in bioinformatics and clinical data, showing reasonable effectiveness. Data imputation is performed on the whole dataset (i.e., applying to all 13 features and 123 catalyst entries). After imputing the data, machine learning algorithms can be applied to extract the relationship between materials structures (i.e., input variables) and catalytic properties (i.e., output variables). Among four output features, the onset potential is selected because it contains the lowest number of missing values. Nitrogen content, surface area, degree of graphitization, and fractions of pyrrolic, graphitic, pyridinic, and pyridinic–oxide nitrogen species are selected as input variables.

Regression models between the onset potential and seven input variables are constructed by using *k*-nearest neighbor regression (KNR), support vector regression (SVR), random forest regression (RFR), and XGBoost regression (XBR). The imput dataset is divided into a training set (70%)

Table 9.2 Input and output features (variables) in the dataset for nanoporous nitrogen-doped carbons reported in the literature for oxygen reduction reaction.

Feature (variable)	Number of missing values	Maximum value	Minimum value	Average value
Input variable (x)				
Carbon content (atomic %)	51	97.03	68.58	87.68
Oxygen content (atomic %)	21	13.1	0	5.86
Nitrogen content (atomic %)	52	23.9	0.96	5.99
Surface area ($m^2\,g^{-1}$)[a]	16	2986	35.7	856
Degree of graphitization (–)[b]	75	2.25	0.77	1.05
Fractions of pyrrolic nitrogen (%)[c]	43	74.7	0	17.3
Fractions of graphitic nitrogen (%)[c]	43	81.8	0	39.1
Fractions of pyridinic nitrogen (%)[c]	43	61.0	5.3	34.9
Fractions of pyridinic–oxide nitrogen (%)[c]	43	54.12	0	8.64
Output variable (y)				
Onset potential (V vs. RHE)	11	1.1	0.67	0.87
Half-wave potential (V vs. RHE)	48	0.9	0.26	0.74
Electron transfer number (–)	34	4.2	1.1	3.4
Diffusion limiting current density ($mA\,cm^{-2}$)	54	6.1	2.0	4.4

[a]Specific Brunauer–Emmett–Teller (BET) surface area calculated from nitrogen adsorption isotherms.
[b]Reported as a ratio of disorder and graphitic carbons as determined by Raman spectroscopy (I_D/I_G ratio).
[c]Percentages of nitrogen atoms with different local environments as estimated from XPS spectroscopy.

and a test set (30%). For KNR and SVR, values of the variables in both training and test sets must be standardized. Hyperparameters of the models are tuned by the three-fold cross-validation against the training set with a grid search over a range of candidate values to minimize root mean squared error (RMSE). Prediction accuracy of the constructed regression models is compared using RMSE and mean absolute error (MAE):

$$RMSE = \sqrt{\frac{1}{n}\sum_{i=1}^{n}(y_i - \hat{y}_i)^2}$$

$$MAE = \frac{1}{n}\sum_{i=1}^{n}\left|y_i - \hat{y}_i\right|$$

where y_i is the actual value, $\hat{y}_i$ is the predicted value, and n is the number of entries/observations.

Figure 9.4 shows correlations between the onset potentials that are reported in literature and that are predicted by the regression models. XBR and RFR outperforms KNR and SVR, judging by their lower RMSE and MAE values against the test data. However, the XBR model seems to be overfitted. As a result, the RFR model is selected for further interpretation.

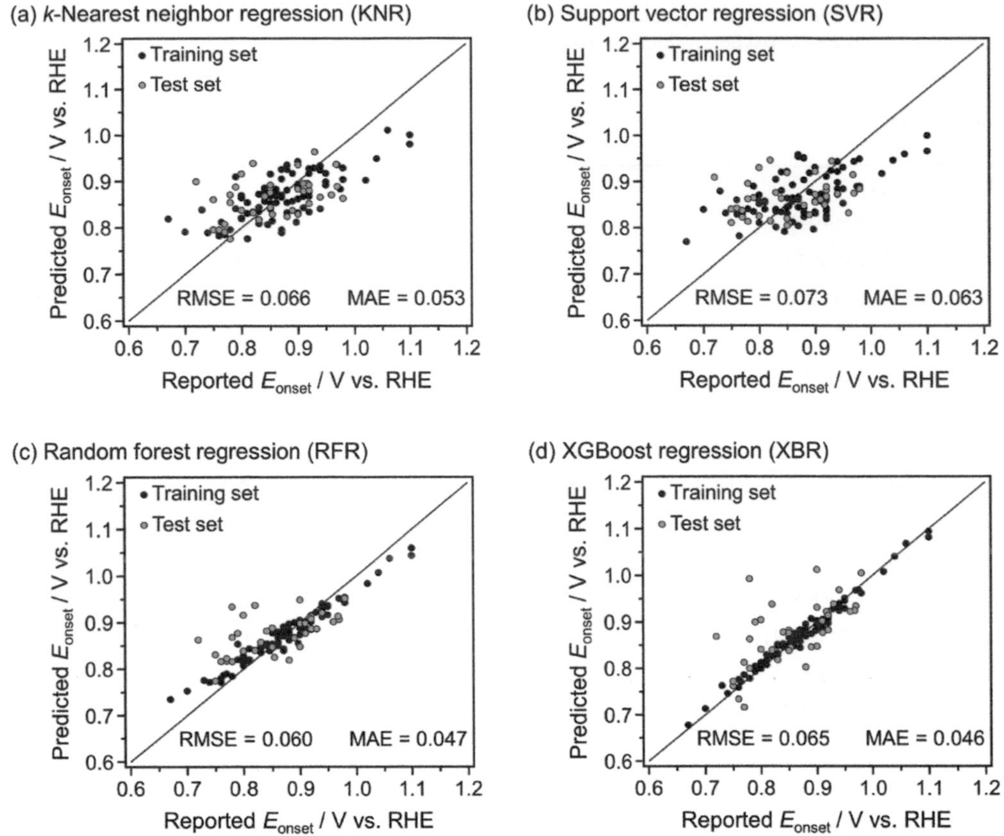

Figure 9.4 Correlations between the reported onset potentials and the values predicted from seven input variables (materials descriptors) using the trained (a) *k*-nearest neighbor regression (KNR); (b) support vector regression (SVR); (c) random forest regression (RFR); and (d) XGBoost regression (XBR).

As explained in Section 9.2, the machine learning models, including random forest, are black-box predictors. One approach to interpreting the results of these models is to extract the importance of input variables, although the interpretation degree of this approach is limited. Based on the built RFR model, the importance of seven input variables is derived. As shown in Figure 9.5a, bulk properties of the carbon materials, namely, nitrogen content and surface area, are more influential against the onset potential, in comparison with local structure of carbon frameworks (i.e., degree of graphitization) and atomic natures of doping nitrogen species.

Correlations between the onset potential and the top three influential input variables are plotted in Figures 9.5b–d. Better ORR catalysts should possess higher onset potentials. It is clearly seen that there are optimal ranges for nitrogen content and surface area to maximize the onset potential, while such a range is hardly noticed in the degree of graphitization (I_D/I_G ratio). For nitrogen-doped carbons to be used as an ORR catalyst, nitrogen content and surface area should be in the ranges of 3–5 at% and 900–1300 m^2/g, respectively. Some experimentalists may hypothesize that increases in nitrogen content and surface area of carbon materials would enhance the ORR catalytic performance. However, the results from the RFR model suggest that nitrogen-doped carbon ORR catalysts have optimal ranges of nitrogen content and surface area, at least in terms of the onset potential. To confirm these results, further literature surveys on nitrogen-doped carbon

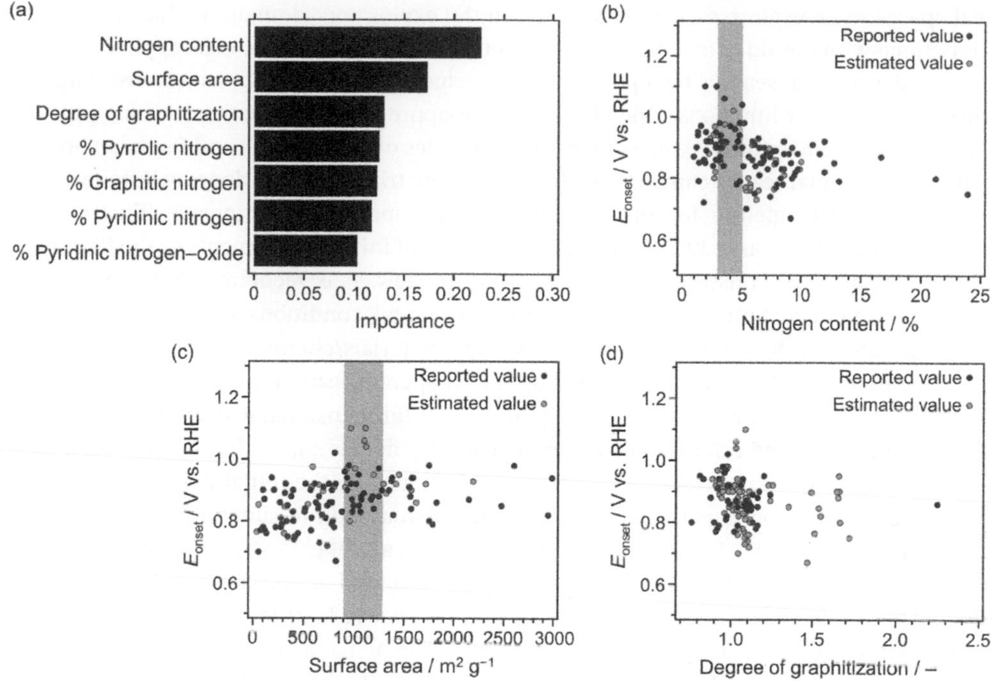

Figure 9.5 (a) Importance of the materials descriptors derived from the best performing RFR model. Relationships between (b) nitrogen content; (c) surface area; and (d) degree of graphitization versus the onset potential.

ORR catalysts is made, particularly regarding the carbon materials that are not included in the original dataset. It is found that several carbon materials exhibiting excellent ORR performance possess nitrogen content and surface area within the optimal ranges, including defect-rich nitrogen-doped graphene nanomesh [42].

The missing values of experimental data, especially those that are collected from the literature, are always the problem when the collected data are processed through any machine learning models. The accuracy or validity of the imputed data is always questionable because the true values are unknown. However, with proper data imputation, the machine learning models can give some quantitative trends that may be useful as a guideline for materials design. Generally, most of all materials properties can affect the performance of materials. However, their influences vary in extent. To prepare the best performing materials, simultaneous optimization of all materials properties is required. However, this optimization is almost impossible, or requires thousands of experiments. Proper analysis of the machine learning results, even based on the imputed data, can provide some guidelines, and may help experimentalists to decrease a number of necessary experiments.

9.4 Acceleration of Experimental Exploration of Nanoporous Metal Alloys: An Active Learning Approach

There is no doubt that the machine learning techniques are useful and powerful tools to accelerate materials discovery. By proper use and analysis, the machine learning models can provide guidelines for design of synthesis conditions of materials. This can substantially narrow the materials/

chemical spaces to be explore experimentally. Once the exploratory domain of objective experiments is defined with the aid of machine learning, optimization algorithms can be applied to accelerate the experimental search by optimizing the experimental conditions toward targeted structures, properties, or functions. One of the efficient approaches is an active learning pipeline coupling with Bayesian optimization. By repeating the cycles of experiments and Bayesian optimization, the experimental search can be much faster than the trial and error done by humans.

An active learning pipeline for optimization of experimental conditions is illustrated in Figure 9.6. Initially, the dataset, which can come from a set of initial experiments or the literature, is needed as an input to construct machine learning models. It is recommended to use some Design-of-Experiment techniques to decide the experimental conditions in the initial dataset because the initial data should be random and cover the materials/chemical space of interest. For example, the initial conditions may be selected by Latin hypercube sampling, which is a statistical method for generating a near-random sampling over a multidimensional distribution [43]. If the starting dataset is small and not random, machine learning models should be constructed to provide a reasonable number of virtual data, to minimize any bias in the initial dataset. In the next step, Bayesian optimization is performed based on the virtual data to suggest the experimental conditions for the next cycle. These experimental conditions are likely to produce materials with better targeted properties. Then, experiments are carried out based on the suggested experimental conditions. The obtained data are added to the dataset. This active learning cycle is repeated until the property is optimized or reaches the targeted value. The active learning has been applied to several classes of materials, including polymeric materials as adhesives [44], mixed oxides for batteries application [45], permanent magnets [46], and nanoporous metallic alloys [47].

Nanoporous metals and metal alloys have been studied intensively in the past few decades for electrochemical applications such as electrochemical catalysis and sensing [48–50]. Experimental search of chemical compositions of multi-metallic alloys (e.g., ternary and quaternary alloys) for targeted properties is an exceptionally time-consuming exploratory process due to a nearly infinite number of possible combinations. With the active learning pipeline, such experimental exploration can be accelerated. As a model system, optimization of compositions of trimetallic PtPdAu alloys toward electrocatalytic activity of methanol oxidation is described. In practical direct methanol fuel cells, Pt is used as a catalyst to oxidize methanol. When alloying Pt with Pd or Au, catalytic

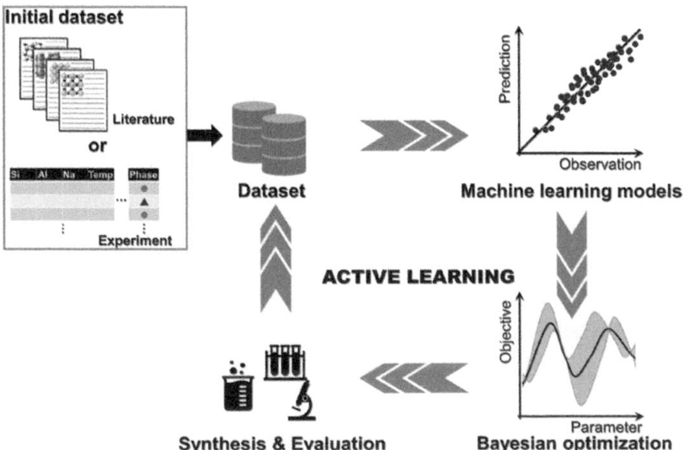

Figure 9.6 An active learning scheme for optimization of experimental conditions.

activity or durability can be improved [51–53]. To find the optimal composition of PtPdAu alloys, thousands of experiments may be required. For example, if the fraction of each metal is altered from 0 to 100% with an increment of 1%, the exploration space contains 5151 different compositions.

In this work, the compositional ratios of each metal in the starting precursors are an input variable, while the volume current densities of methanol oxidation reaction using the resulting alloys as catalysts are an objective variable. In the initial experimental set, the experimental conditions (compositions of each metal) are decided by experimentalists based on their knowhow and intuition. The compositions cover the wide composition space of interest (Figure 9.7a). The initial dataset containing 17 experiments is fed to the random forest model for modelization (input: compositional ratios, output: electrocatalytic activity). The random forest model is then used for randomly spanning of the compositional space. This step allows creating a virtual dataset devoid of biases that would otherwise impair the objective selection of conditions with potentially increased activity.

The virtual dataset is given to the Bayesian optimization to be used as a basis for the training of a Gaussian process and the selection of proposals for the later experiments through an acquisition function scoring. Bayesian optimization is a probabilistic methodology to optimize an unknown function globally, *f*, which is often non-convex, non-linear, highly dimensional, and noisy [54]. Bayesian optimization finds the maxima of an objective function *f* (i.e., maximizing the volume current densities):

$$f \equiv f(x_{Pt} + x_{Pd} + x_{Au})$$

where x_{Pt}, x_{Pd}, and x_{Au} are fractions of Pt, Pd, and Au in the starting precursors, respectively. x_i ranges from 0 to 1. It is worth mentioning that metal compositions in the starting precursors are used instead of compositions in the obtained alloys, because the precursor compositions are experimentally adjustable.

For adjustable variables, some constraints are applied to the constrained domain:

$$\mathbb{C} \equiv \{x_{Pt}, x_{Pd}, x_{Au}\}$$

where the alloy composition is stochiometric:

$$x_{Pt} + x_{Pd} + x_{Au} = 1$$

with

$$\Delta x_{Pt} = \Delta x_{Pd} = \Delta x_{Au} = 0.01$$

n Random forest models are trained using a leave-one-out cross-validation procedure to predict the volume current density from compositional ratios (x_{Pt}, x_{Pd}, x_{Au}) with an increment $\Delta x_{Pt}, \Delta x_{Pd}, \Delta x_{Au}$, where *n* is equal to a number of entries in the dataset (initially $n = 17$). The random forest regression (as a first surrogate model) is used here as a proxy to the initial dataset in order to guide the Gaussian process regressor (as a second surrogate model) to produce more accurate predictions and widen its view of the compositional space reachable by interpolation, as the Gaussian process

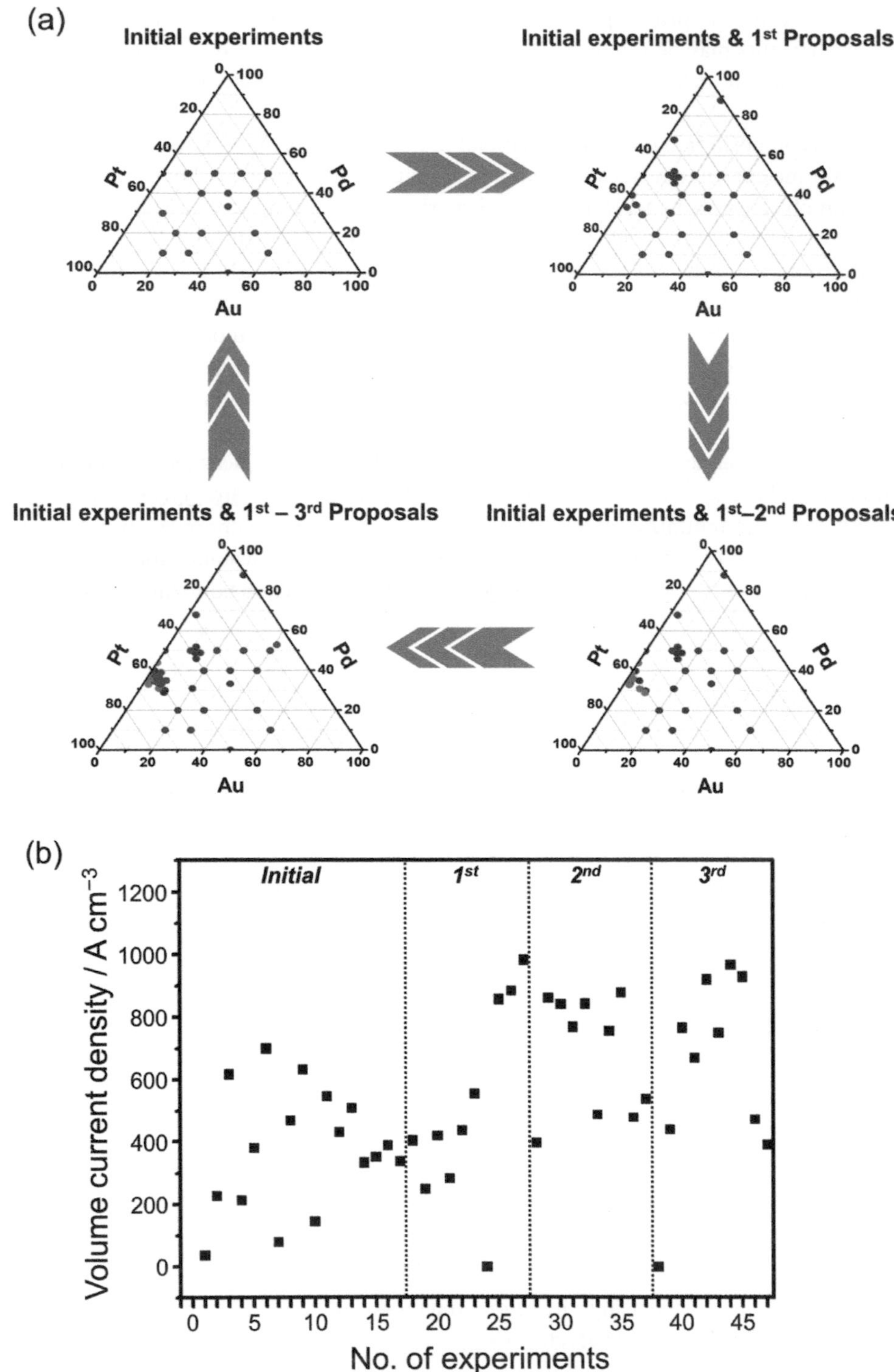

Figure 9.7 (a) Ternary diagrams of chemical compositions of metal precursors Pt, Pd, and Au in the electrolyte solutions: initial dataset (in red) and after first (in blue), second (in green), and third (in pink) Bayesian optimization cycles. (b) Peak current volume density of all of the resulting mesoporous trimetallic PtPdAu films.

is more robust to close extrapolation. Note that if a number of data in the initial dataset are large enough, use of two surrogate models may not be necessary because the Gaussian process alone as a surrogate model for Bayesian optimization should be sufficient.

A set of top ten proposed compositional ratios (a number limited to what is experimentally feasible in a limited amount of time) with the largest acquisition function score is created. These top-ten ratios are then experimentally checked. The experimental feedback is added to the initial dataset, which is fed again to the random forest model for re-training and to gain in novel valuable information. The rest is cycled again and again until the objective is reached.

As can be seen in Figure 9.7a, the compositional ratios in the first proposal are dominated in a few areas, which may imply that the optimization process tries to find some possible compositional space with maximal activity. The current densities of the first proposed compositions are much higher than those of the initial compositions (Figure 9.7b). After repeating the active learning cycles, the compositional ratios in the third proposal are limited in a narrow area. The current densities of the obtained alloys are almost saturated; therefore, the active learning cycle is stopped (Figure 9.7). The current density of the best performing product is enhanced by at least 30%, in comparison with some of the best PtPdAu alloys reported in literature so far for electrocatalytic methanol oxidation [47,55,56]. The compositional ratios of metal precursors can be found by performing only 47 experiments; 1% of total possible compositions, clearly indicating the effectiveness of the active learning pipeline for accelerating the experimental exploratory. It is worth mentioning that the composition yielding the highest current density is Pt:Pd:Au of 59:40:1. Introduction of metallic components at a very low concentration is not usual in the preparation of nanoporous metal alloys, although it is a common practice in the bulk alloy community. This result reveals that the active learning may broaden the exploration space beyond the common thoughts in scientific fields.

9.5 Summary

Over a few decades, progress in synthesis of zeolites and other nanoporous materials has been made in more rationalized ways based on chemical/materials intuition of humans, combined with advances in analytical and computational chemistry. Given the fact that synthesis of zeolites and other nanoporous materials are highly complex, as slight changes in synthesis conditions and protocols can lead to products with different structures and properties, techniques in data science can be particularly useful to solve such complex problems in materials synthesis. A workflow that combines experiments with computations and artificial intelligence would pave a new way for rational synthesis of nanoporous materials.

To further accelerate the discovery of practically useful nanoporous materials, data science techniques should be combined with high-throughput experimentations for realizing autonomous synthesis. It is noteworthy that autonomous synthesis is not automatic, because intelligent design of the workflow and proper use of data science techniques are essential. By combing high-throughput experimentations and robotics with appropriate models and algorithms in data science, such as the active learning pipeline with Bayesian optimization, autonomous synthesis of nanoporous materials can be developed. To gain experimental feedback within a reasonable timeframe, high-throughput experiments for materials characterizations and property evaluations must also be developed. Although there is still a lot of research to be done, a data-driven scheme with high-throughput experimentations would potentially become a common practice in materials research in the near future, including in the field of nanoporous materials.

Acknowledgments

The author thanks Japan Society for the Promotion of Science (JSPS) and Japan Science and Technology Agency (JST) for financial supports through a Grant-in-Aid for Scientific Research (B) (KAKENHI: JP20H02528) and a JST-ERATO Project (JPMJER2003), respectively.

References

1 Davis, M.E. (2002). Ordered porous materials for emerging applications. *Nature* 417: 813–821.
2 Chaikittisilp, W. and Okubo, T. (2017). Zeolite and zeolite-like materials. In: *Handbook of Solid State Chemistry*, 4e (ed. R. Dronskowski, S. Kikkawa, and A. Stein), 97–119. Weinheim: Wiley-VCH.
3 Horike, S., Shimomura, S., and Kitagawa, S. (2009). Soft porous crystals. *Nat. Chem.* 1: 695–704.
4 Ariga, K., Vinu, A., Yamauchi, Y. et al. (2012). Nanoarchitectonics for mesoporous materials. *Bull. Chem. Soc. Jpn.* 85: 1–32.
5 Slater, A.G. and Cooper, A.I. (2015). Function-led design of new porous materials. *Science* 348: aaa8075.
6 Moliner, A., Román-Leshkov, Y., and Corma, A. (2019). Machine learning applied to zeolite synthesis: the missing link for realizing high-throughput discovery. *Acc. Chem. Res.* 52: 2971–2980.
7 Jablonka, K.M., Ongari, D., Moosavi, S.M. et al. (2020). Big-data science in porous materials: materials genomics and machine learning. *Chem. Rev.* 120: 8066–8129.
8 Clayson, I.G., Hewitt, D., Hutereau, M. et al. (2020). High throughput methods in the synthesis, characterization, and optimization of porous materials. *Adv. Mater.* 32: 2002780.
9 Davis, M.E. and Lobo, R.F. (1992). Zeolite and molecular sieve synthesis. *Chem. Mater.* 4: 756–768.
10 Cundy, C.S. and Cox, P.A. (2003). The hydrothermal synthesis of zeolites: history and development from the earliest days to the present time. *Chem. Rev.* 103: 663–702.
11 Bellussi, G., Carati, A., Rizzo, C. et al. (2013). New trends in the synthesis of crystalline microporous materials. *Catal. Sci. Technol.* 3: 833–857.
12 Cundy, C.S. and Cox, P.A. (2005). The hydrothermal synthesis of zeolites: precursors, intermediates and reaction mechanism. *Micropor. Mesopor. Mater.* 82: 1–78.
13 Grand, J., Awala, H., and Mintova, S. (2016). Mechanism of zeolites crystal growth: new findings and open questions. *CrystEngComm.* 18: 650–664.
14 De Yoreo, J.J., Gilbert, P.U.P.A., Sommerdijk, N.A.J.M. et al. (2015). Crystallization by particle attachment in synthetic, biogenic, and geologic environments. *Science* 349: aaa6760.
15 Catlow, R., Bell, R., Cora, F. et al. (2005). Computer modelling of inorganic materials. *Annu. Rep. Prog. Chem., Sect. A* 101: 513–547.
16 Li, Y. and Yu, J. (2014). New stories of zeolite structures: their descriptions, determinations, predictions, and evaluations. *Chem. Rev.* 114: 7268–7316.
17 Van Speybroeck, V., Hemelsoet, K., Joos, L. et al. (2015). Advances in theory and their application within the field of zeolite chemistry. *Chem. Soc. Rev.* 44: 7044–7111.
18 Jensen, Z., Kim, E., Kwon, S. et al. (2019). Approach to zeolite synthesis enabled by automatic literature data extraction. *ACS Cent. Sci.* 5: 892–899.
19 Jensen, Z., Kwon, S., Schwalbe-Koda, D. et al. (2021). Discovering relationships between OSDAs and zeolites through data mining and generative neural networks. *ACS Cent. Sci.* 7: 858–867.
20 Muraoka, K., Sada, Y., Miyazaki, D. et al. (2019). Linking synthesis and structure descriptors from a large collection of synthetic records of zeolite materials. *Nat. Commun.* 10: 4459.

21 Raccuglia, P., Elbert, K.C., Adler, P.D.F. et al. (2016). Machine-learning-assisted materials discovery using failed experiments. *Nature* 533: 73–76.

22 Moosavi, S.M., Chidambaram, A., Talirz, L. et al. (2019). Capturing chemical intuition in synthesis of metal–organic frameworks. *Nat. Commun.* 10: 539.

23 Xie, Y., Zhang, C., Hu, X. et al. (2020). Machine learning assisted synthesis of metal–organic nanocapsules. *J. Am. Chem. Soc.* 142: 1475–1481.

24 Loewenstein, W. (1954). The distribution of aluminum in the tetrahedra of silicates and aluminates. *Am. Miner.* 39: 92–96.

25 Fletcher, R.E., Ling, S., and Slater, B. (2017). Violations of Löwenstein's rule in zeolites. *Chem. Sci.* 8: 7483–7491.

26 Van Santen, R.A. (1984). The Ostwald step rule. *J. Phys. Chem.* 88: 5768–5769.

27 Navrotsky, A., Trofymluk, O., and Levchenko, A.A. (2009). Thermochemistry of microporous and mesoporous materials. *Chem. Rev.* 109: 3885–3902.

28 Itabashi, K., Kamimura, Y., Iyoki, K. et al. (2012). Hypothesis for broadening framework types of zeolites in seed-assisted synthesis without organic structure-directing agent. *J. Am. Chem. Soc.* 134: 11542–11549.

29 Guo, P., Shin, J., Greenaway, A.G. et al. (2015). A zeolite family with expanding structural complexity and embedded isoreticular structures. *Nature* 524: 74–78.

30 Chaikittisilp, W. and Okubo, T. (2021). No more trial and error for zeolites. *Science* 374: 257–258.

31 Blondel, V.D., Guillaume, J.-L., Lambiotte, R. et al. (2008). Fast unfolding of communities in large networks. *J. Stat. Mech. Theory Exp.* 2008: P10008.

32 Schieber, T.A., Carpi, L., Díaz-Guilera, A. et al. (2016). Quantification of network structural dissimilarities. *Nat. Commun.* 8: 13928.

33 Xu, L., Choudhary, M.K., Muraoka, K. et al. (2019). Bridging the gap between structurally distinct 2D lamellar zeolitic precursors through a 3D germanosilicate intermediate. *Angew. Chem. Int. Ed.* 58: 14529–14533.

34 Schwalbe-Koda, D., Jensen, Z. et al. (2019). Graph similarity drives zeolite diffusionless transformations and intergrowth. *Nat. Mater.* 18: 1177–1181.

35 Jakobsen, J.C., Gluud, C., Wetterslev, J. et al. (2017). When and how should multiple imputation be used for handling missing data in randomised clinical trials: a practical guide with flowcharts. *BMC Med. Res. Methodol.* 17: 162.

36 Schmitt, P., Mandel, J., and Guedj, M. (2015). A comparison of six methods for missing data imputation. *J. Biomet. Biostat.* 6: 1000224.

37 Stekhoven, D.J. and Bühlmann, P. (2012). MissForest: non-parametric missing value imputation for mixed-type data. *Bioinformatics* 28: 112–118.

38 Zhang, X., Yan, C., Gao, C. et al. (2020). Predicting missing values in medical data via XGBoost regression. *J. Healthc. Inform. Res.* 4: 383–394.

39 Xia, W., Hou, Z., Tang, J. et al. (2022). Materials informatics-guided superior electrocatalyst: a case of pyrolysis-free single-atom coordinated with N-graphene nanomesh. *Nano Energy* 94: 106868.

40 Dai, L., Xue, Y., Qu, L. et al. (2015). Metal-free catalysts for oxygen reduction reaction. *Chem. Rev.* 115: 4823–4892.

41 Gewirth, A.A., Varnell, J.A., and Diascro, A.M. (2018). Nonprecious metal catalysts for oxygen reduction in heterogeneous aqueous systems. *Chem. Rev.* 118: 2313–2339.

42 Xia, W., Tang, J., Li, J. et al. (2019). Defect-rich graphene nanomesh produced by thermal exfoliation of metal–organic frameworks for the oxygen reduction reaction. *Angew. Chem. Int. Ed.* 58: 13354–13359.

43 Heltona, J.C. and Davis, F.J. (2003). Latin hypercube sampling and the propagation of uncertainty in analyses of complex systems. *Reliab. Eng. Syst. Saf.* 81: 23–69.

44 Pruksawan, S., Lambard, G., Samitsu, S. et al. (2019). Prediction and optimization of epoxy adhesive strength from a small dataset through active learning. *Sci. Technol. Adv. Mater.* 20: 1010–1021.

45 Harada, M., Takeda, H., Suzuki, S. et al. (2020). Bayesian-optimization-guided experimental search of NASICON-type solid electrolytes for all-solid-state Li-ion batteries. *J. Mater. Chem. A* 8: 15103–15109.

46 Lambard, G., Sasaki, T.T., Sodeyama, K. et al. (2022). Optimization of direct extrusion process for Nd-Fe-B magnets using active learning assisted by machine learning and Bayesian optimization. *Scr. Mater.* 209: 114341.

47 Nugraha, A.S., Lambard, G., Na, J. et al. (2020). Mesoporous trimetallic PtPdAu alloy films toward enhanced electrocatalytic activity in methanol oxidation: unexpected chemical compositions discovered by Bayesian optimization. *J. Mater. Chem. A* 8: 13532–13540.

48 Zhang, J. and Li, C.M. (2012). Nanoporous metals: fabrication strategies and advanced electrochemical applications in catalysis, sensing and energy systems. *Chem. Soc. Rev.* 41: 7016–7031.

49 Lim, H., Kani, K., Henzie, J. et al. (2020). A universal approach for the synthesis of mesoporous gold, palladium and platinum films for applications in electrocatalysis. *Nat. Protoc.* 15: 2980–3008.

50 Wang, T., Chutia, A., Brett, D.J.L. et al. (2021). Palladium alloys used as electrocatalysts for the oxygen reduction reaction. *Energy Environ. Sci.* 14: 2639–2669.

51 Liu, Y., Chi, M., Mazumder, V. et al. (2011). Composition-controlled synthesis of bimetallic PdPt nanoparticles and their electro-oxidation of Methanol. *Chem. Mater.* 23: 4199–4203.

52 Zhang, Z., Wang, Y., and Wang, X. (2011). Nanoporous bimetallic Pt–Au alloy nanocomposites with superior catalytic activity towards electro-oxidation of methanol and formic acid. *Nanoscale* 3: 1663–1674.

53 Xu, S., Hou, W., Jiang, R. et al. (2019). Regulating locations of active sites: a novel strategy to greatly improve the stability of PtAu electrocatalysts. *Chem. Commun.* 55: 13602–13605.

54 Packwood, D. (2017). *Bayesian Optimization for Materials Science*. Singapore: Springer.

55 Zhang, Y., Gu, Y., Lin, S. et al. (2011). One-step synthesis of PtPdAu ternary alloy nanoparticles on graphene with superior methanol electrooxidation activity. *Electrochim. Acta* 56: 8746–8751.

56 Chen, D., Luo, L.-M., Zhang, R.-H. et al. (2018). Highly monodispersed ternary hollow PtPdAu alloy nanocatalysts with enhanced activity toward methanol oxidation. *J. Electroanal. Chem.* 812: 90–95.

10

Porous Molecular Materials

Exploring Structure and Property Space with Software and Artificial Intelligence

Steven Bennett and Kim E. Jelfs

Department of Chemistry, Molecular Sciences Research Hub, Imperial College London, White City Campus, London, United Kingdom

10.1 Introduction

Porous molecular materials (PMMs) are a subclass of porous material, which lack the extended network formed by directional covalent or coordination bonds found in other porous materials such as metal–organic frameworks (MOFs) or covalent organic frameworks (COFs) [1]. Instead, PMMs are comprised of discrete molecular units held together by weaker non-covalent interactions. Recent developments in the field of PMMs have resulted in the discovery of several new classes of porous material [2], which offer great promise in the fields of heterogenous catalysis [3], molecular separations [4], and gas storage [5], among others [6]. Compared to their network counterparts, PMMs are rare [7], and their discovery is increasingly being contributed to by computational techniques, and, in more recent times, the outcomes of data-driven artificial intelligence (AI) methods [8]. In this chapter, we define the focus to be the computational discovery of PMMs, notably excluding porous network materials such as MOFs and zeolites. We also exclude coordination cages containing metal ions, focusing instead on their organic analogs. However, we will compare computational methods used across many material types, highlighting approaches that could be transferred to PMMs and shared challenges across the porous materials modeling field. The focus of this chapter is on the increasing role computational techniques are playing in rationalizing both structure and properties of PMMs, with an emphasis on how data-driven techniques are becoming increasingly prevalent within the field. Throughout, we discuss the role that AI is progressively playing in the *in silico* discovery and experimental realization of PMMs, by providing chemists with not only a better understanding of what to synthesize, but how to synthesize it. We highlight the challenges the PMM field must overcome to fully utilize the potential of AI, from computational predictions to the formidable challenge of fully autonomous discovery.

The serendipitous discovery of the porous molecular crystal Dianin's molecule in 1976 spurred interest in PMMs. This relatively simple molecule was able to achieve porosity by inefficient packing, resulting in the formation of hourglass-shaped columns running along a single axis in a material also termed an "organic zeolite" [9]. Compared with their network counterparts, PMMs have typically exhibited low porosity (~200–300 m^2/g) owing to the challenge of overcoming attractive intermolecular interactions that remove void space between molecules [7]. However, research over the last 15 years has resulted in the discovery of new PMMs, including several molecular materials featuring surface areas up to 1000 m^2/g [10]. The vast number of possible precursors that could be combined, in addition to the diverse range of accessible topologies, makes them exciting candidate materials for both computational and experimental chemists alike.

AI-Guided Design and Property Prediction for Zeolites and Nanoporous Materials, First Edition. Edited by German Sastre and Frits Daeyaert.

Although far fewer PMMs are known than their network material counterparts [7], they offer several advantages. Conventional porous materials, such as zeolites, provide limited scope for post-synthetic modifications due to their highly crystalline nature, while this is easily achieved in PMMs owing to their solution processability. PMMs' fully organic nature allows for facile modification, including functionalization, akin to post-synthetic modification methods in the MOF literature [11]; indeed, even minor chemical modifications can facilitate large changes in porosity [12]. One example of this is the reduction of the imine bonds present in the porous organic cage (POC) **CC3** to amines, resulting in the collapse of the internal cavity, and subsequent deterioration of porosity [13]. Many PMMs are also solution processable, easing both purification using standard organic chemistry techniques and subsequent processing steps [10]. Three prominent classes of PMMs embody these advantages: i) porous liquids [14]; ii) porous amorphous solids [15]; and iii) porous molecular crystals. Each achieve porosity by the presence of void space between discrete molecular units. However, porous molecular crystals must retain this void space upon desolvation by maintaining permeability to liquids or gases in the solid-state.

Owing to the discrete molecular nature of PMMs, porosity occurs from either extrinsic or intrinsic void spaces. Extrinsic porosity arises from inefficient packing of molecular components and refers to the void space generated between molecules. There have been several notable examples within the porous molecular crystal family, shown in Figure 10.1, which

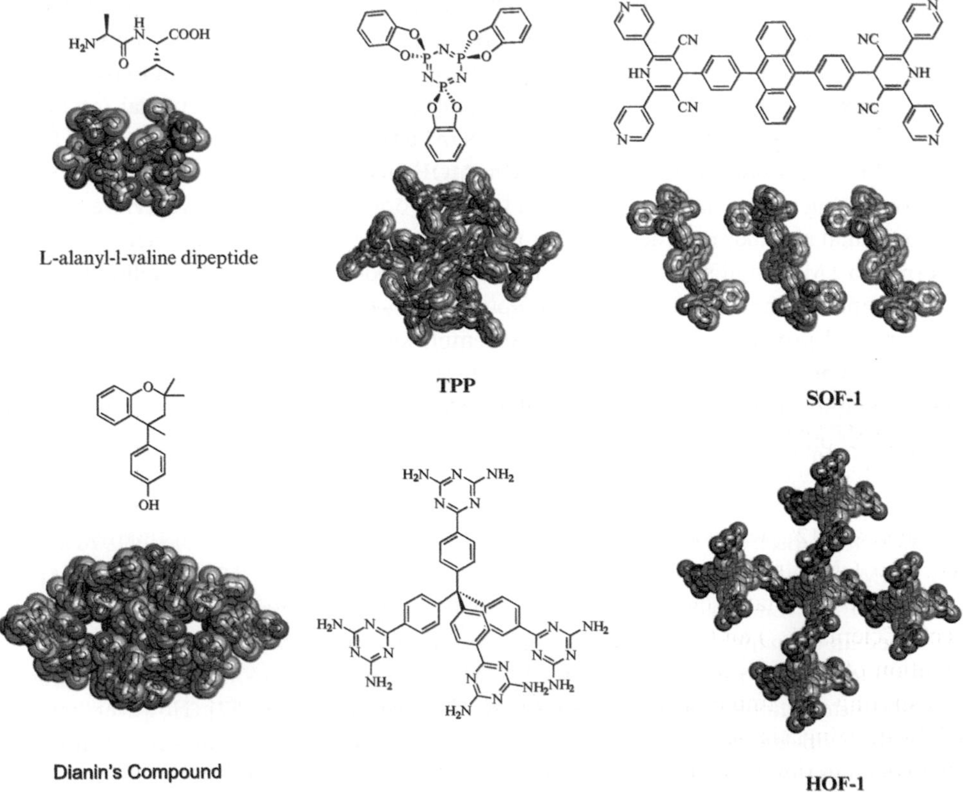

Figure 10.1 Structures of molecules that form porous molecular crystals accompanied by their unit cell structures. Oxygen is shown in red, nitrogen in blue, phosphorus in purple, and carbon in black. Hydrogen from all crystal structures and solvent molecules present in the **HOF-1** crystal structure have been omitted for clarity. Spheres represent the van der Waals surface of each atom.

are: tris(o-phenylenedioxy) phosphonitrile trimer (**TPP**), whose porosity has been extensively studied [16,17]; 4,4′,4″,4‴-tetra(4,6-diamino-s-triazin-2-yl)tetraphenylmethane (**HOF-1**) [18]; 9,10-bis(4-((3,5-dicyano-2,6-dipyridyl)dihydropyridyl)phenylï)anthracene (**SOF-1**) [19]; and several crystalline dipeptides [20]. Yet, this type of porosity is inherently challenging to predict and design because molecules preferentially pack efficiently within the solid-state. Instead, many PMMs offer an alternative, intrinsic porosity. Intrinsic porosity arises from a well-defined interior cavity within a molecule. Examples of macrocyclic molecules with intrinsic cavities include cyclodextrins [21], cucurbiturils [22], calixarenes [23], both urea and amine macrocycles [24,25], and, finally, POCs [26]. In each of these molecules, the internal cavity is accessible to guest molecules via cavity windows.

An intrinsically porous molecule with more than two windows is termed a cage. In a cage, window size is typically equal to or less than the diameter of the internal cavity, and, thus, can restrict the diffusion of large molecules into and out of the internal cavity. This makes them excellent candidates for potential applications such as selective guest adsorption or separations [26]. Recently, these materials have been able to achieve surface areas exceeding many PMMs that rely solely on extrinsic porosity, and, in recent times have achieved a surface area of 3758 m^2/g [27].

POCs are cage molecules that possess a permanent internal cavity. Typically, POCs are synthesized using well-established reversible dynamic covalent chemistry (DCC) reactions, of which imine condensation remains the most prevalent [28]. However, boronic ester condensation and alkyne metathesis are also frequently employed [27,29], widely extending the potential functional groups that will undergo the desired DCC reaction. POC synthesis primarily relies on the formation of a thermodynamic product, often forming well-defined topologies based on Platonic or Archimedean solids, shown in Figure 10.2 [30]. A naming convention for the most commonly formed POC topologies is described by Santolini et al., ascribing nomenclature to the topicity of precursors that form the cage and the number of individual precursor molecules incorporated into the discrete cage molecule [30]. However, prediction of topological outcome for a single set of precursors is non-trivial [31]. Mixtures of topologies, polymers, or a range of oligomeric products of varying sizes are often observed, complicating experimental analysis. Indeed, even if the desired topology is experimentally observed, there is no certainty that the cage will be shape-persistent, such that the cage is able to maintain the internal cavity that gives rise to intrinsic porosity. Even if the topological outcome could be accurately predicted through computation, there is no certainty that the precursors can be synthesized, arguably the most labor- and cost-intensive aspect of POC discovery [32,33].

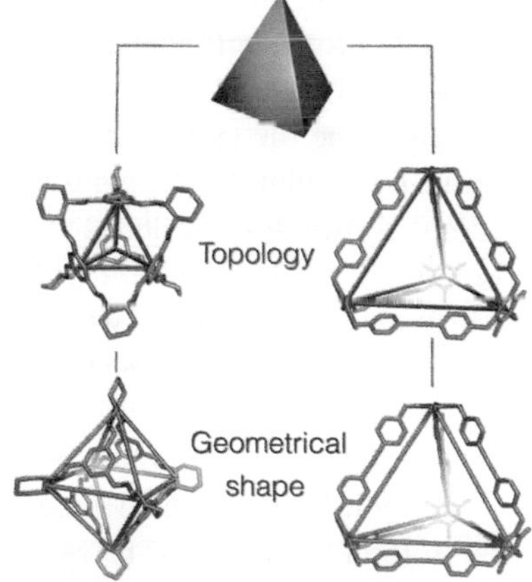

Figure 10.2 The topological and geometric shapes POCs can adopt. While two POCs may adopt the same topology, their underlying geometric shapes may differ depending on the flexibility of the building blocks that constitute each POC. *Source:* From [30], with permission from the Royal Society of Chemistry. Licensed under CC BY-SA 3.0.

Synthetic accessibility is frequently considered in the drug design field, and should ideally be a consideration in any computational material workflow [34]. However, whether or not a material, or material precursor, can be synthetically realized is challenging to predict computationally and, as a result, this factor is often neglected.

Furthermore, synthetic cost and yield within conventional materials development significantly deviate from that for pharmaceutical development, with materials requiring cheap commercially available precursors, or those accessible in several high-yielding reaction steps. Thus, computational methods devised to assess synthetic accessibility of novel pharmaceuticals are unsuitable for PMM precursor selection [33]. It is due to this unique combination of challenges that new POCs are commonly the product of serendipitous discovery and often the consequence of small iterative changes to precursors already well-known in the POC field. This form of discovery is slow and hinders the rate at which new POCs are discovered.

In recent years, there have been increasingly notable successes in using computational techniques to predict both the structure and properties of PMMs, most significantly in the discovery of POCs [35]. With the potential to accelerate materials discovery, computational tools are becoming essential to the discovery of new materials [36], and porous materials are no exception to this trend [8]. Ultimately, the purpose of any computational tool is to be able to predict, prior to any experimental synthesis or subsequent analysis, a property of interest. Promising candidate materials identified by computational predictions may be proposed to experimentalists; this significantly decreases materials discovery costs because computational prediction offers a more cost-effective and efficient route than material synthesis. However, simulations are often limited in accuracy due to the computational expense of performing *ab initio* or semi-empirical calculations, and instead rely on force field or coarse-grained approaches. These assumptions and simplifications are a prime feature of many high-throughput screening workflows [37], which use fast calculations to screen many candidates for those exhibiting a desired property.

Although screening is a powerful tool, the sheer number of potential PMMs afforded by the near-infinite combination of precursors renders it intractable, even considering the capabilities of modern-day computational calculations. Instead, the focus of chemists has been inverse design. For a desired property (or properties), can we design a molecule that will exhibit these properties? It is often the goal of chemists to develop relationships between structures of materials and their properties; however, identifying such relationships is often challenging. Specifically with respect to PMMs, small changes in structure can have large effects on the weak non-covalent interactions that determine the crystal structure, significantly altering bulk properties. AI techniques aim to capture these complex relationships using data, thereby facilitating faster property screening by eliminating the simplifications and limitations of conventional computational techniques.

Predictive machine learning (ML) techniques now play a vital role in chemistry, most notably the drug discovery process, where their application ranges from identifying retrosynthetic routes to predicting the outcome of chemical reactions [38,39]. These techniques are fuelled by extensive datasets and tasked with learning a relationship between numeric inputs and outputs. Given sufficient data, ML algorithms can develop non-linear, multivariable relationships which makes them most suitable in fields where large datasets are readily accessible and relationships are highly complex. In the PMM field, AI is well-suited to learn these complex structure–property relationships, avoiding approximations of conventional computational simulations. However, the few examples of PMMs limits available training data and restricts its utility; this is a recurring theme throughout this chapter. Despite this, there have been advances in the wider porous material field to overcome these challenges, including: i) the development of FAIR databases [40]; ii) novel ways of describing porous materials quantitatively; and iii) new ways in which ML algorithms have been translated from the computer science domain. It is with these incremental innovations that materials discovery is slowly beginning to transcend into a "fourth paradigm" [41], with an increased emphasis on data collection and the goal of uncovering the statistical patterns that exist within. However, the advantages of ML go beyond property prediction, as optimization algorithms are now able to predict the optimal next data point to obtain the largest performance increase.

When developing an ML model for materials screening, typically human intuition, which is subject to bias, guides the data collection process. This bias leads to the formation of densely clustered datasets that, when used to train a model, fail to generalize to unseen candidates. Naturally, humans are also biased toward successful outcomes, often neglecting or failing to report failures, especially when the cost of performing a reaction is high [42]. The growing accessibility of autonomous synthetic methods has the potential to eliminate this bias when combined with the potential of AI to propose the best next experiment [43], and has been used to discover new POCs [31]. With such developments, the ambitious concept of closed-loop discovery has been proposed in the materials community, and early implementations have been included [44]. Independent of human input, it is possible to envisage these techniques transforming many chemistry domains [44], including the PMM field. In this chapter, we will discuss: i) the computational methods used to model PMMs; ii) the use of AI techniques to predict PMM properties; iii) the treatment of synthetic accessibility within computational workflows; iv) several algorithms that have been used to systematically explore chemical and structural space; and, finally, iv) high-throughput synthesis methods that have been used to screen for new PMMs.

10.2 Computational Modeling of Porous Molecular Materials

At the core, computational models offer atomistic insight into the chemical phenomena underpinning experimental measurements. Often, the goal of these models is to identify the structure–property relationships governing bulk behavior. The porosity of PMMs is a fundamental characteristic of this class of materials and, therefore, necessitates computational characterisation. Since porosity can arise from the individual molecule (intrinsic porosity), or the packing of molecules in the solid-state (extrinsic porosity), PMM models featuring a range of length- and timescales are necessary; Figure 10.3 briefly canvases varying PMM models, ranging from an isolated molecular unit to periodic or amorphous structures. Although properties are often highly dependent on how those discrete units self-organize in the resulting PMM, the chemical behavior of the isolated molecule can often be extrapolated to the behavior of the extended structure [45,46]. This allows for a more computationally rigorous examination of the isolated unit using electronic structure methods,

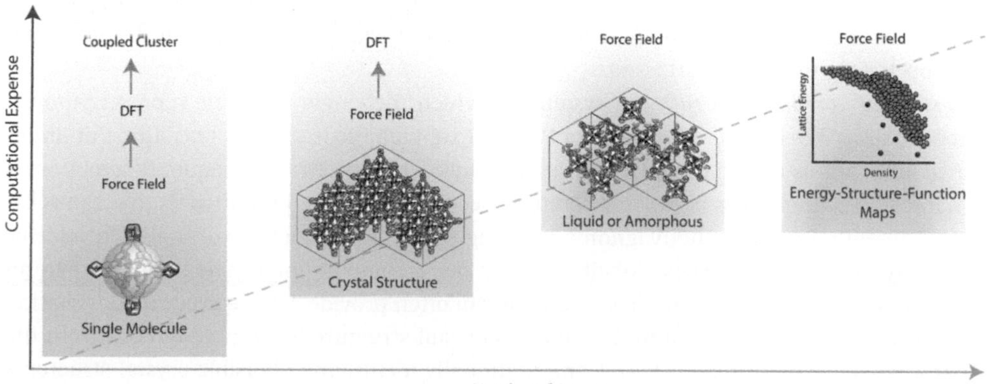

Figure 10.3 The qualitative computational expense against number of atoms of different methods frequently used to model porous molecular materials. As the number of atoms in the simulation grows, coupled cluster and DFT calculations quickly becoming intractable, leaving less accurate force field methods available to model the system. This is especially the case for generating ESF maps, where lattice energies of thousands of structures are typically required.

which can be used to probe individual interactions between molecules [47], and provide an insight into how those molecules may pack. Alternatively, less accurate computational methods, such as force fields, can provide useful heuristics for many properties, making them suitable for high-throughput screening studies of PMMs [48]. In this section, we will describe the various computational techniques used in the PMM field, including: i) porosity measures; ii) modeling discrete molecules; and iii) modeling crystalline and disordered structures.

10.2.1 Structure Prediction

Molecular construction and force fields: While identifying lowest energy conformations is a routine operation for small molecules [49], this is more challenging for PMM molecules. Their size and flexibility result in many independent degrees of freedom, which require conformational search algorithms to efficiently sample the extensive configurational space. Molecular mechanics can achieve this when combined with force field or electronic structure methods. However, force field choice is important; the universal force field [50] and COMPASS [51] inadequately reproduce the bond angles of imine bonds [52], a bond type frequently present in POCs. Parameterizing existing force fields is one solution [52]; however, the OPLS all-atom force field [53] has been shown to accurately reproduce POC structures and energetics [54]. With recent developments and additional parameters, OPLS has shown increasing accuracy on small-molecule solvation, protein-ligand binding benchmarks [55,56], and POCs [30]. While it is possible to propose plausible structures for large numbers of hypothetic PMMs using force field methods, the generation of the structures themselves is itself a demanding task.

Constructing cage molecule models provides an additional computational challenge, due to the vast number of precursors that can form a cage and the large number of potential topologies. The supramolecular toolkit (*stk*) [57,58] provides a solution to the automated construction and optimization of a variety of cage, framework, and polymer materials, allowing construction of three-dimensional (3D) molecules from two-dimensional (2D) precursor representations, such as SMILES or InChI. The utility of *stk* is multi-facetted: *stk* has been applied to the generation of cages [59], polymers [60,61], and metal-organic polyhedra [62], owing to its ability to systematically combine precursors in pre-defined defined topologies in a facile manner. Similarly, the software package *cgbind* provides autonomous capabilities for metallocages and host–guest systems [63]. The development of open-source computational tools for the construction, optimization, and analysis of PMMs has enabled high-throughput virtual screening of these systems [48], a task that would once have been a significant undertaking.

Crystal structure prediction (CSP): For systems in which the single molecule representation is insufficient to describe the bulk properties of PMMs, extended models of the crystalline or amorphous structure can instead be used. Chemical and physical properties are frequently dictated by the weak intermolecular forces between discrete molecules; therefore, bulk models that can reliably capture these interactions, wholly ignored by a single molecule model, are desirable to describe experimentally observed properties. Albeit typically more computationally demanding than an analysis of the single structural unit, these methods can often provide a more accurate description of porosity if the conformation of a molecule in its crystal structure is likely to deviate from the isolated ground state conformation [64]. Computationally identifying plausible crystal structures using CSP is a long-standing area of research and can now be routinely carried out for small drug-like molecules of interest in the pharmaceutical industry [65,66]. Significant complexity is introduced when identifying the crystal structure of PMMs due to the weak intermolecular forces that exist between individual molecular units, resulting often in many energetically similar crystal structures which could still structurally be quite diverse [67].

Despite the challenges of accurately predicting the bulk crystalline structure of PMMs, especially POCs, predicting plausible crystal structure geometries is possible using methods conventionally suited to small organic molecules [68]. The primary goal of CSP is to identify the lowest energy crystal structure of a molecule, while not discounting metastable states similar in energy to the thermodynamic minimum that may also be experimentally accessible. CSP is often performed in an iterative manner: i) thousands of initial possible structures are generated with large deviations in their confirmation; ii) the lattice energies are calculated, often using force field methods; and iii) high energy and crystallographically identical structures are removed. These steps are repeated with the goal of converging toward a low energy minima [69]. Increases in computational power has afforded the use of higher level calculations to allow energetic re-ranking on large numbers of possible structures [70]. Combined with more accurate DFT energy rankings on the lowest energy polymorphs, where the inclusion of dispersive interactions is key, successful CSP is no longer out-of-reach for many organic molecules.

Although currently restricted to tens of molecules [66], CSP has the potential to be accelerated by ML methods. The development of ML-derived force fields can accelerate energy ranking, replacing expensive final DFT calculations or inaccurate force field rankings [71]. To accurately describe long-range interactions, an ML correction approach has been applied to lower cost baseline methods, showcasing the ability to perform crystal relaxations at a significantly reduced computational cost [72]. While these methods are yet to be applied to the field of PMMs, it is clear that developments of these methods will help accelerate CSP.

Following the identification of a low-energy crystal structure, the primary task is to relate structures to properties, with the goal of targeting specific structures that will exhibit favorable properties. Visually mapping properties to structures results in an energy-structure-function (ESF) map, shown in Figure 10.4, and can help identify promising new candidate materials [67]. Typically, an ESF map shows the lattice energy landscape of a single molecule compared with the properties of interest for individual structures; this can be used to identify new porous materials by the existence of "spikes" on the landscape that correspond to low-energy structures with low density [67]. For six hydrogen-bonded organic frameworks known to exhibit porosity in the solid-state, energy-density maps demonstrated the large effect intermolecular hydrogen bonds and π–π interactions can have on structure and, thus, properties [73]. For POCs, where porosity can originate from the internal cavity, more structures on the landscape are predicted to be porous than molecules that rely on extrinsic porosity [67]. Typically requiring the generation and subsequent property evaluations of tens of thousands of potential structures, ESFs are computationally expensive to calculate [73]. Recently, the development of efficient optimization algorithms, which will be further discussed in Section 10.4, have reduced the required computational resources needed for these global structure searches by limiting expensive property calculations to only the most promising structures [74].

10.2.2 Modeling Porosity

For the individual molecular unit, porosity originating from the well-defined internal cavity of the PMM can be analyzed using computational software, provided an accurate, geometry-optimized structure is available. The importance of having a structure resembling the experimentally observed global (or local) minima is paramount; porosity models applied to high-energy unfeasible structures

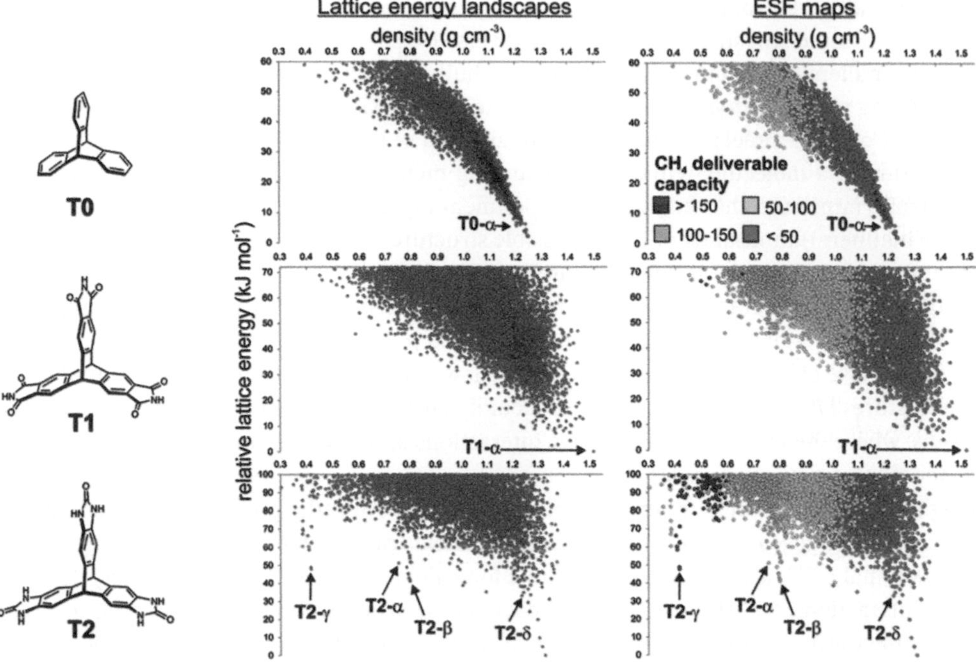

Figure 10.4 Energy-structure-function maps and lattice energy landscapes of three porous molecular materials adapted from the work of Day and Cooper [67]. These maps were able to predict the existence of four experimental isolable polymorphs of **T2**, one of which has favorable methane deliverable capacity (**T2**-γ). These maps show visually the relationship between stability and properties, which can be used to identify metastable polymorphs with favorable properties. *Source:* From [67] John Wiley & Sons. Licensed under CC BY-SA 4.0.

are unlikely to reproduce experimentally observed porosities [75]. Once an adequate structure of the discrete molecule is achieved using methods described in Section. 10.2.1, the void space encompassed the host molecule can be analyzed. Cavity sizes and window diameters can be calculated by sampling the internal cavity using a sphere of sampling points. Indeed, this approach is used by *pyWindow* [45], which uses vectors from the center of mass of the molecule joined to a sphere of sampling points surrounding the entire molecule to: i) identify individual windows; ii) calculate window diameters; and iii) calculate the diameter of the cavity. However, these approaches are limited to the cavity space accessible by a monatomic guest. A method of detecting whether a host molecule can encompass a guest was developed by Kravchenko et al., whose algorithm was able to calculate the accessible space for multi-atomic molecules [76]. Void space in the stationary host, defined as static porosity by Holden et al. [77], is the easiest to quantify computationally because it does not account for the effects of the dynamic behavior of the host molecule. Even if the internal cavity size is large enough to accommodate a guest molecule, host flexibility plays a key role in defining the timescale that windows are accessible to guest molecules; this is inherently neglected in the porosity measurement methods discussed thus far [77].

To account for porosity originating from host flexibility, demonstrated in Figure 10.5, termed dynamic porosity, MD simulations must be employed to probe the behavior of the host over time. The importance of host flexibility on porosity is exemplified in the experimental gas uptake measurements of the POC **CC2**. Despite possessing a permanent internal cavity, experimental gas uptake measurements for H_2 in **CC2** could only be reproduced by Holden et al. when performing grand canonical Monte Carlo simulations where the internal cavity was blocked, suggesting **CC2**

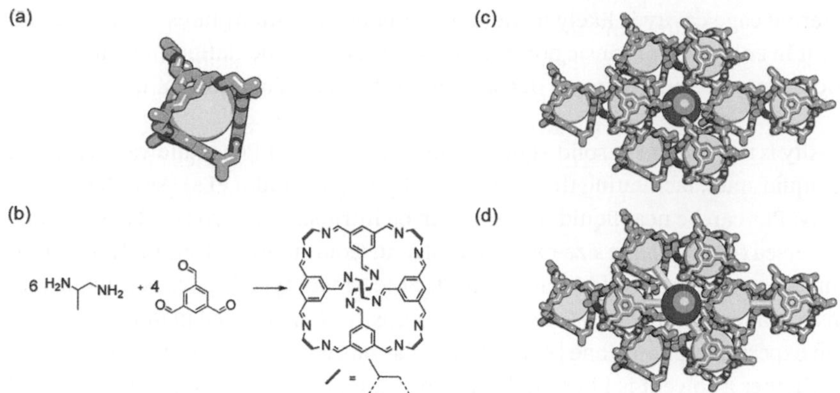

Figure 10.5 **CC2**: (a) formed by the imine condensation of 1,2-diaminopropane and 1,3,5-triformylbenzene; (b) possesses a permanent internal cavity in the crystal structure, shown in yellow. Window-to-arene packing results in the formation of 1D channels, shown in orange. However, these voids are inaccessible in the static structure, resulting in disconnected voids. (c) The connection of voids in **CC2**; (d) can be rationalized by considering the dynamic behavior of the host structure, or cooperative porosity resulting from interactions with guest molecules. *Source:* Adapted from [77].

is only dynamically porous to H_2 [77]. Meanwhile, experimental CO_2 uptakes indicated guests were able to access the internal cavity of **CC2**, a consequence of the cooperative porosity occurring due to interactions between CO_2 and the host structure. Indeed, a consideration of cooperative porosity is often required to describe the properties of many PMMs [78–81]. Cooperative porosity, originating from the influence of the guest on the host molecule, can explain unexpected diffusion behavior in materials that are predicted to be both dynamically and statically non-porous to a guest. Chen et al. used molecular dynamics (MD) to probe the interactions between SF_6 and **α-CC2** that resulted in cooperative porosity; SF_6 was able to access the internal cavity, overcoming the free energy barrier associated with passing through the window of **α-CC2** [82]. Static porosity measurements would be unable to account for most of this behavior, meaning a variety of computational techniques, in addition to some prior knowledge the underlying pore structure, is required to fully understand the origin of porosity.

10.2.3 Amorphous and Liquid Phase Simulations

PMMs have also been reported beyond the crystalline phase, as both amorphous solids and as porous liquids. The inherent disorder present in structure of amorphous and liquid structures requires additional modeling considerations. For example, it is challenging to identify the cause of porosity in amorphous phase PMMs (i.e., to what degree this is due to intrinsic porosity, low crystallinity, and inefficient packing), in addition to the challenges of modeling such disordered structures [83]. Interestingly, amorphous PMMs can exhibit increased porosity over their crystalline counterparts, suggesting that long-range order is not necessarily a requirement for high performing PMMs [83]. A computational workflow for amorphous structure prediction of several POCs has been developed and makes use of several MD steps to simulate the amorphization process [84]. Indeed, it is possible to use simulations to identify characteristic structural features that are likely to result in increased porosity. Abbott et al. were able to derive three design principles for increasing the porosity of amorphous organic molecules of intrinsic microporosity, based on simulations of 22 molecules: i) increased rigidity; ii) presence of bulky end groups; and iii) decreased planarity [85]. Similar structural relationships were developed via simulations of 11 amorphous cage structures; a large internal cage cavity is likely to yield a highly porous amorphous structure, and bulky end groups result in enhanced extrinsic porosity. However, these calculations also indicated that the amorphization process could increase density due to the interdigitation of discrete cage molecules [84].

Porous liquids: Porosity is not limited to solid-state; the internal cavity of POCs allows porosity to be maintained in the liquid state, facilitating the formation of porous liquids (PLs) [86]. PLs can be categorized accordingly: PLs can be neat liquids (type I), can be intrinsically porous substances dissolved (type II), or dispersed (type III) in a size-excluded solvent. Computational simulations, similar to those used to model amorphous solids, have been used to model type II porous liquids, and can provide an atomistic understanding into the origins of porosity of disordered materials, impossible to obtain through experimentation alone [87]. MD simulations with explicit solvent molecules present can indicate whether a solvent is likely to diffuse into, and, therefore, occlude the internal cavity of cages [88,89]. A more rigorous examination of solvent behavior can be achieved by analyzing the free energy profiles of solvent diffusion; a combination of umbrella sampling (US) and the weighted histogram analysis method (WHAM) [90] can be used to compute the free energy barrier of the solvent diffusing into a cage cavity [14]. One strategy to design PLs is with the addition of long hydrocarbon chains on the cage periphery, enhancing the solubility in organic solvents and reducing the melting point [91]. US and WHAM techniques are able to probe the behavior of periphery chains, determining whether the chains are likely to occupy the cage cavity, thus rendering the liquid non-porous [87]. Alternatively, mixtures of precursors can result in the formation of "scrambled" cages [92], whose structural disorder results in greater solubility, preventing crystallization [14]. Simulating disordered materials is challenging, especially providing an accurate representation of solvent molecules, and is often limited to a handful of candidates selected using chemist intuition.

Currently, these methods are intractable for high-throughput screening of promising candidates and are instead typically employed for post-synthetic rationalization. Although the number of known PLs is small, recently, high-throughput experimentation, which will be further discussed in Section 10.5, has been able to explore far more cage-solvent combinations than could be done manually [93]. It is possible to envisage ML-based interatomic potentials being employed to accelerate the structure prediction process of both liquids and amorphous material, providing the accuracy of quantum chemical calculations at a fraction of the computational cost [94].

Throughout the PMM field, simulations have proved useful not only for post-rationalization of properties, but also their prediction. The development of high-throughput techniques has forever changed the role of computation. Computational screening is now able to eliminate molecules with unfavorable properties and can subsequently strengthen the relationship between experimental efforts and larger computational simulations. The increasing tractability of large-scale computation afforded by advancements in computing hardware allows for larger, more detailed models, incrementally progressing toward the ultimate goal of bringing chemists closer to the ability to perform inverse design.

10.3 Data-Driven Discovery: Applying Artificial Intelligence Methods to Materials Discovery

In the last few decades, ML has seen rising popularity due to high-profile cases such as AlphaFold [95] and GPT-3 [96]. ML can be broadly classified as either supervized (data labels provided), or unsupervized (data labels are not provided) learning (Figure 10.6). Supervised learning uses labels

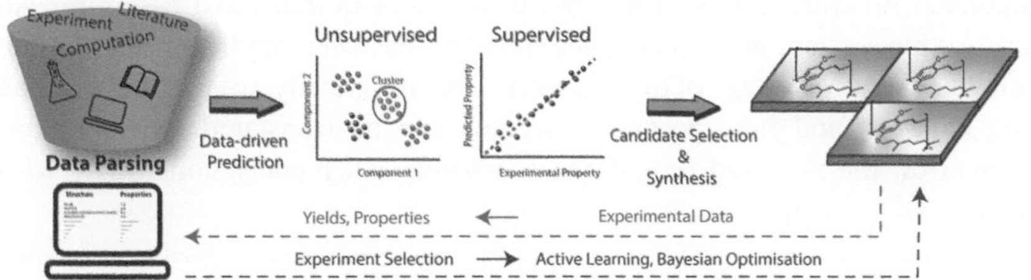

Figure 10.6 A closed-loop materials discovery workflow using data-driven techniques. Initially, data for predictive models are obtained from experiments, scientific literature, or computational calculations, which must be parsed into a machine-readable format. Given sufficient training data, supervized or unsupervized learning algorithms can be used to develop new structure–property relationships that can be used to predict the properties of candidate materials. Such models can be iteratively improved by expanding the initial dataset with data from experiments (red line) or can be used to efficiently acquire experimental data using active learning or Bayesian optimization (blue line).

to iteratively update functions that map a vector to an output. Conversely, unsupervised learning learns relationships without data labels, identifying hidden patterns from the structure of datasets. Models such as random forests, support-vector machines, neural networks (NNs), and even simplistic models such as linear regression fit within supervised learning. Meanwhile, clustering techniques such as k-means, t-SNE, and DBSCAN [97], in addition to some architectures of NN, fit within the unsupervised learning category. All these methods have found widespread use in materials discovery [36,98].

Neural networks: Of the available ML models, NNs have achieved significant popularity in the last decade. NNs are comprised of multiple layers of interconnected nodes, which conceptually mimic neurons in the brain. Composed of sequences of interlinked nodes, called hidden layers, NNs aim to approximate a function between input and output by applying nonlinear transformations at each node in the hidden layers. Inputs are propagated throughout the layers until the output layer is reached, where a prediction error is calculated. The weights in the network are then updated by calculating the gradient of the error function with respect to the weights of each node, a process known as backpropagation. Training is an iterative process; multiple forward and backward propagation steps are often required before the model's weights can converge toward optimal values.

Although the concept of NNs is not new [99], there has been a rapid increase in research interest in expanding and generating new NN architectures. Known as deep learning (DL), this field uses NNs with large numbers of hidden layers trained on extensive datasets for predictive tasks, outperforming humans at many tasks [100]. In materials chemistry, deep neural networks (DNNs) have been shown to excel at property prediction [36]; however, it is the customizability and modularity of DL techniques, when used in combination with other methods, which infers the largest advantages. When combined with reinforcement learning methods, DNNs excel at tasks from video games to drug design [101,102]. Many disciplines have made use of DL techniques, chemistry being no exception; in the drug discovery field [103], deep learning algorithms have now found widespread use in molecular generation [104], biological activity prediction [105], and reaction prediction [106]. Recently, the porous materials field has begun to utilize some of the recent advances in DL techniques, notably the work of Kim et al. on using generative adversarial networks to propose plausible new zeolite structures [107]. Their utility in the field of PMMs is hindered by a lack of data, often restricting DL to domains with large accessible datasets. However, the complexity in the generated function, and vast number of parameters that must be optimized, demand large datasets. In this section, we will describe how ML has been applied to the field of PMMs, highlighting challenges at each stage of the ML workflow. Specifically, we will focus on the generation of training datasets, and how porous molecules can be translated into a vector. Throughout, we comment on the uses and limitations of various ML models, concluding with recent applications in DL.

10.3.1 Training Data Generation

One of the most significant challenges associated with predicting PMM properties is the curation of a suitable training dataset. Ideally, a training dataset would contain few errors and contain enough datapoints to train a model with suitable accuracy for a predictive task. The amount of required training data depends on the complexity of predictive problem, and the complexity and number of parameters of the ML model. Generally, more training data will result a better performing model; however, simpler ML methods will often reach a point of diminishing returns, where additional data have little effect on model performance. For example, in the field of natural language processing, where textual data is ubiquitous across the Internet, extremely complex models can be trained, such as GPT-3 with 175 billion parameters [96].

For materials discovery, experimentally derived data is often prohibitively costly to obtain, and can be difficult to reproduce due to variations in conditions and sample purities between experiments. Synthesis of a novel PMM can often take months [26], a more time-consuming endeavor than image labeling, and there are only a few hundred literature reports of PMMs to date [108].

10.3.1.1 Hypothetical Structure Datasets

Hypothetical structure datasets are supplementing experimental datasets with computational data is a strategy that is proving increasingly popular to grow training datasets. In the field of MOFs [109], COFs [110], and zeolites [111], structural generation algorithms have resulted in the creation of hypothetical structure databases comprised of thousands of structures. Typically, these methods exploit the fixed geometric constraints of inorganic systems, involving the placement of building block units in pre-defined topologies [112,113]. In Chapter 12, an algorithm used to build structures based on geometric constraints is described in further detail. For MOFs, grand canonical Monte Carlo simulations are frequently used to evaluate hypothetical MOF structures for gas storage [109], and fit predictive ML models [114,115].

The absence of hypothetical structure databases for PMMs has limited ML application. Of the few examples, Turcani et al. created a database of 63,472 POCs, the largest known hypothetical POC database. The generation of such a large number of hypothetical POC structures was possible using *stk* [57], exploiting the well-defined topologies of POCs [30], and identifying plausible geometries using MD simulations [116]. Using this database, the authors were able to generate both classification and regression random forest models capable of predicting POC structure and cavity size, achieving a minimum accuracy of 0.86 when predicting whether a cage would remain shape-persistent. While models trained and tested on a single DCC reaction type performed well, the models struggled to generalize to cages formed from reactions outside of the training set. Predicting cavity size was also challenging, with an average mean absolute error (MAE) of all models of 1.5 Å. Despite these limitations, it was possible to identify several promising new cage candidates featuring cavity sizes exceeding 40 Å and high symmetry, as well as several design principles. Structural features, such as bridgehead atoms, double bonds, and fewer rotatable bonds present in precursors are more likely to constitute a shape-persistent cage. However, the assumption that

MD simulations can identify low energy conformers may not always be valid; instead, experimentally-observed structures can be used for training data generation.

10.3.1.2 Experimental Structure Datasets

One alternative source of structures are the many experimentally-obtained crystal structures contained within the CSD [117]. Evans et al. applied an autonomous screening workflow to identify 481 potential PMMs out of the 150,000 experimental structures in the CSD, a hit rate of 0.3% [118]. For each structure, they used the porosity analysis tool Zeo++ [119] to assess the pore volume accessible to a helium guest. Following subsequent structure optimization, a support vector machine was trained to classify porous and non-porous molecules, and was able to achieve 69.7% accuracy with a single molecular descriptor corresponding to the van der Waals surface area of the molecule. The small number of porous molecules contained within the CSD limits its utility for training ML models [118]; thus, hypothetical structures often remain the only viable option to generate ML training data.

It is evident that the lack of curated datasets is a significant drawback for constructing ML models of PMMs. While computation can extend datasets, approximations and simplifications in the calculations will persist in the training data, and large datasets can only be created at significant computational expense. Another viable, albeit challenging, alternative is to harness the data available in the scientific literature for dataset construction.

10.3.1.3 Extraction of Data From Scientific Literature

Natural language processing, the application of computational techniques to allow computers to process human language, has been used to extract data from the scientific literature [120,121]. The formulaic way in which scientific literature is often reported (figures and tables) is exploited by open-source software such as ChemDataExtractor [122], which has been used to extract both structures and properties of materials [122]. In the materials literature, word embeddings have been used to encode vector representations of text, allowing the relationship between words to be expressed numerically. Tshitoyan et al. used these relationships to identify inorganic materials with promising thermoelectric properties, supporting their results with DFT calculations [123]. Despite successes in other fields of materials chemistry [121], the literature remains primarily untapped for PMMs. Nonformulaic chemical naming conventions and complex chemical structures adds additional complications to text mining, making linking chemical structures and corresponding properties difficult [121]. Furthermore, the scientific literature rarely reports failed outcomes or materials without exceptional properties, creating bias in chemical datasets [42]. Training a decision tree on both a dataset comprised of both successful and unsuccessful reaction outcomes has been shown to increase model accuracy, outperforming human predictions [42].

10.3.1.4 Data Augmentation and Transfer Learning

Often, even with the multitude of approaches used to create a suitable training set in chemistry, only small amounts of data are accessible in the PMM domain. The cost and time associated with expanding these PMM datasets requires the use of models that can attain high predictive power given a small number of training points. One solution is data augmentation, which involves artificially enlarging the dataset with transformations of the existing data. Wang et al. combined predicted X-ray diffraction data with noise from experimental X-ray diffraction patterns from MOF samples to artificially enlarge their dataset, with the goal of training a convolutional neural network able to identify MOFs [124]. In the field of molecular property prediction, where encoded SMILES strings are common input representations, one popular strategy to extend dataset size is to generate multiple valid SMILES strings for a single molecule [125]. Successful augmentation strategies are specific to each problem domain; an augmentation strategy that worked for one task is unlikely to work for others [126].

Another approach is the use of transfer learning, in which a model is pre-trained on a larger dataset from a similar domain, followed by fine-tuning the weights of an NN on a smaller dataset directly related to the problem domain [127,128]. Alternatively, the output of an NN can be used as input for a model able to achieve better performance on smaller datasets, such as a random forest [129]. Averaging the prediction of multiple models [130], or incorporating cheap property estimates into the descriptor set [131], has reduced the bias that small datasets generate in the final prediction. Once a suitable dataset has been established, we must choose an appropriate method of encoding the dataset into a machine-readable format.

10.3.2 Descriptor Construction and Selection

As all ML models require numeric input, we must select a suitable method of quantifying a molecule into numeric vectors. For property prediction tasks, the goal of a descriptor is to encapsulate chemical information into a vector for a model to establish a relationship between numeric input and output. In former quantitative structure–activity relationship development, popular descriptors were typically easily calculable molecular, structural characteristics, such as molecular formula, or, more frequently, 2D descriptors based on the underlying atom and bond connectivity of a molecule [132]. Molecular fingerprints, a vector encoding of the unique substructures present in a molecule, are a popular 2D encoding developed for substructure searching and structure comparison [133]. These descriptors are typically inexpensive to calculate, and several software libraries have been created for this purpose [134,135]. In many pharmaceutical studies, 2D descriptors, such as molecular fingerprints, are sufficient for many predictive tasks, including toxicity and solubility [136]. In the field of porous materials, where descriptor choice is vast (Figure 10.7), even simple descriptors, such as the use of molecular fingerprints constructed solely on precursors of POCs, afford random forest models that can accurately predict whether a POC will remain shape-persistent [116].

10.3.2.1 Local Environment Descriptors

While there are a diverse range of descriptors that can be used for ML prediction, the properties of porous materials are dictated by complex *inter*molecular interactions, which descriptors limited to encoding *intra*molecular bonding fail to describe. Thus, descriptors based on the underlying local coordinate environments of each atom can be used, encoding the spatial information of the

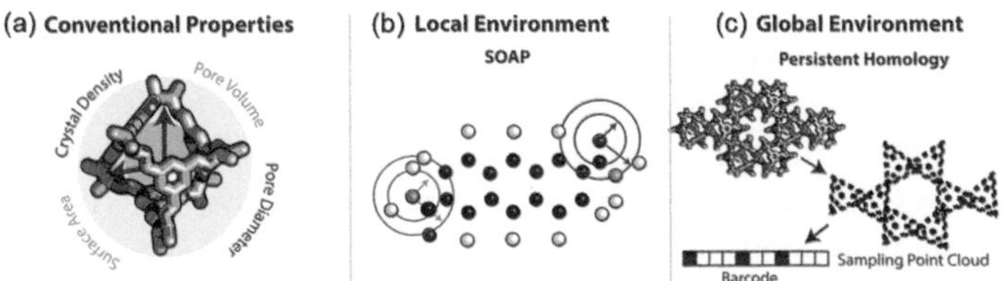

Figure 10.7 An overview of descriptors used to encode porous materials. Conventional descriptors are: (a) calculated porosity measures of the material, which sometimes fail to encode key characteristics of the void space. Instead, the smooth overlap of atomic positions (SOAP) kernel; (b) can be used to encode local atomic environments. Where porosity originates from long-range order, persistent homology (c) can encode defining topological features.

molecule. The smooth overlap of atomic positions (SOAP) kernel is one such embedding and is able to quantize the local environment information for every atom on a molecule [137]. Indeed, 3D descriptors, such as the SOAP kernel, can be used to distinguish between and predict properties of polymorphs, which can have vastly different properties [73]. In the wider materials community, Musil et al. used the SOAP-REMatch kernel to predict the charge mobility and energy, hence stability, of pentacene and azapentacenes polymorphs; the model predicts energies within the accuracy of DFT [138]. Relying on the ability to perform CSP to generate potential polymorphs, this method is primarily suited to rigid molecules, such as the typically conjugated structures of organic photovoltaics [139].

However, such local environments embeddings are not only useful inputs for ML models, but can be used for the generation and subsequent analysis of ESF maps. Zhao et al. used the SOAP kernel to distinguish between the structures of six rigid porous molecular crystals, generating ESF maps from the crystal energy landscape of each molecule. By applying unsupervised learning to these descriptors, the generated ESF maps were mapped to a 2D embedding, categorizing structures according their similarities [73].

10.3.2.2 Global Environment Descriptors

Due to the highly dependent nature of porosity on 3D structure, there has been a significant interest in developing descriptors based on encoding void space into lower-dimensional vectors, capturing information about the *global* atomic structure. Conventional porosity-based descriptors, as discussed in Section 10.1, such as pore diameter, pore volume, and surface area, are unable to capture the complex geometries of void spaces. Sturluson et al. used singular value decomposition to embed cavity information from images of 74 POCs into a lower-dimensional vector [140]. t-SNE plots showed how the salient characteristics of the POC cavities were maintained in this lower-dimensional embedding, able to cluster POCs by the geometry of their internal cavity. Using the method of persistent homology from mathematics, Lee et al. describe an approach to encode crystal structures of porous molecular crystals into a unique barcode, able to distinguish between different crystal structures in a lower dimensional space [141]. The persistent homology encoding, detailed in Chapter 12, proved better at distinguishing between materials with different pore shapes compared with the SOAP kernel, which was instead affected by local changes in atomic environments [142]. Using this representation, Moosavi et al. used kernel ridge regression (KRR) to estimate the methane storage capabilities of different crystal structures of three triptycene-based molecules, using the database of hypothetical structures generated using CSP methods by Pulido et al. [142,143]. Even with only 600 structures for their training data, the developed KRR model was able to achieve an MAE on 7 v STP/v when comparing the predicting methane deliverable capacity to simulations [142].

While the choice of descriptors is vast for porous materials, often certain descriptors remain better suited for certain predictive tasks; a 2D fingerprint encoding is insensitive to structural changes within the molecule, and is, therefore, unsuitable to differentiate between polymorphs. Descriptor selection can be challenging, as it is often unknown which descriptor will generate the best mapping of structure to properties. More recently, DL has helped overcome the challenge of descriptor choice, by using a NN to *learn* its own representation of a molecule that best describes a property. Notable architectures include graph neural networks (GNNs) and variational autoencoders (VAEs), which are both able to create learnt representations. GNNs use the graph representation of a molecule, aggregating information from atom (node) and bond (edge) environments [144], while VAEs encode discrete molecular representations into a numerical vector, able to translate back from the learnt representation using a decoder DNN [104]. Recently, in the MOF community, where data is more accessible than their porous molecular counterparts, GNNs have been used to learn the best encodings of MOFs that will predict the band gap [145]. However, learnt representations require extensive datasets to outperform human-selected descriptors, hindering their transition into the PMM field.

It is clear AI has the potential to accelerate PMM design and property prediction, as it has done so already in the wider porous material field. By accelerating both structure and property prediction, ML has positioned itself to overcome the limitations of many existing computational techniques, achieving high accuracy predictions with very little computational expense. However, further work in producing reliable, error-free datasets is required to achieve its full potential.

10.4 Efficient Traversal of the Chemical Space of Porous Materials

The nature of materials discovery is often serendipitous, with discoveries originating from many experimental or computational trials. Experiments are often selected by chemists to prioritize candidates that are most likely to form a useful material. This, however, tends to confine discovery algorithms to well-explored regions of chemical space, hindering discovery. This is disadvantageous in cases where the ideal molecule, or topology, resides in less-explored regions. The development of exploratory algorithms to efficiently explore a defined chemical space is becoming increasingly popular in materials chemistry due to the ever-increasing size of databases. Typically, the goal of these algorithms is to identify a global minimum (or maximum) of a well-defined objective function as efficiently as possible. Objective functions can range from a single property value of interest to function containing multiple property values. Combined with ML methods, exploratory algorithms, shown in Figure 10.8, provide a powerful method of exploring chemical space efficiently. In this section we categorize these generative algorithms, highlighting several approaches that have proved fruitful in the endeavor of PMM discovery.

10.4.1 Evolutionary Algorithms

Evolutionary algorithms (EAs) offer an alternative approach to molecular generation. Unrestricted by training set, EAs utilize the principles of natural selection and evolution to incrementally propose new candidates that improve upon the fitness of the previous generation. In the MOF field, EAs have been employed to search for promising candidates with favorable adsorption properties [146,147]. For candidates that can be designed in a modular fashion with well-defined topologies, such as MOFs and many POCs, EAs are particularly well-suited as they are able to systematically exchange the precursors that constitute the underlying topology. Berardo et al. exploited the modular nature of POCs to design an EA able to explore the landscape of POCs, identifying several candidates with cavity sizes rarely reported in the literature. In their implementation, the structure of a POC was abstracted to genetic equivalents, such that a POC represented a member of a population, and its precursors represented the chromosomes of that member. By utilizing mutation, where a precursor (chromosome) was randomly interchanged for another from a pre-defined database, and crossover, in which POCs interchanged precursors, the EA was able to converge on a population of POCs that were the optimal solution to a pre-defined fitness function for that specific precursor library [148].

A similar approach was used to identify POCs that could be used to encapsulate fullerene, whose selective extraction is of significant research interest due to its applications across materials science [149]. By tuning the fitness function to include the fullerene binding energy, Miklitz et al.

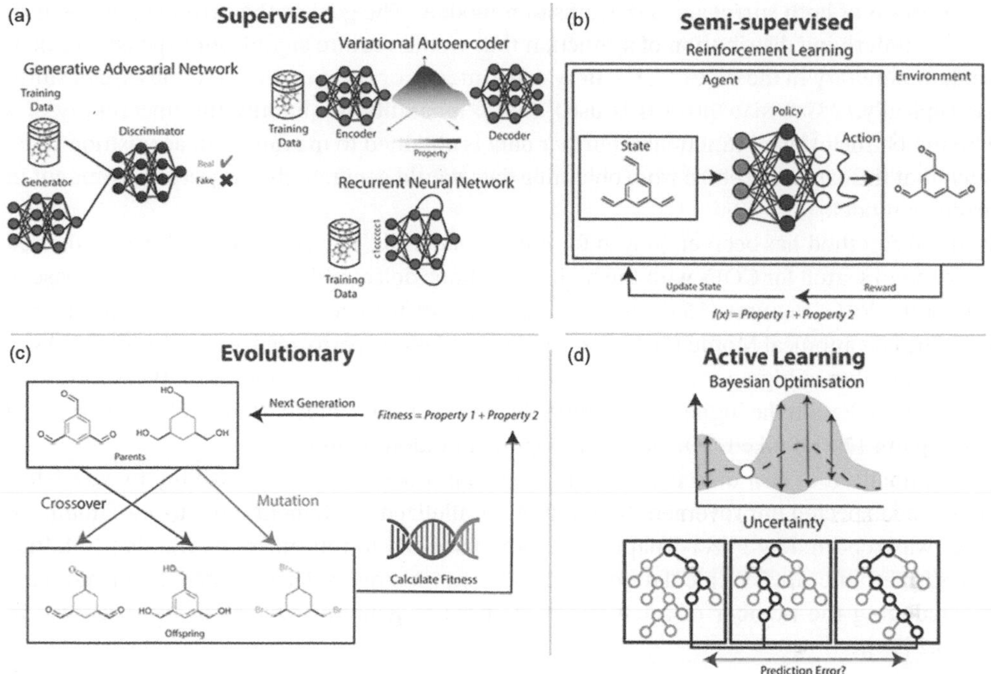

Figure 10.8 An overview of generative algorithms. Supervised generative algorithms (a) can be trained using extensive molecular datasets, often utilizing recurrent neural networks. Alternatively, molecular generation can be framed as a reinforcement learning problem (b), with the goal of training an agent able to take actions which will result in new molecules with desirable properties. If training data is costly to generate or scarce, evolutionary approaches (c) can be used instead, using concepts originating from natural selection to converge on a potential minimum for a pre-defined fitness function. However, if calculations are costly, active learning can be used to direct the training data generation procedure, with the goal of maximizing model performance in the fewest number of data points.

were able to identify several POCs that provided a solution to this specific application [150]. For these applications, generating and analyzing all possible combinations is unfeasible due to the vast number of combinations of precursors that can react to form a POC, and the vast number of topologies they can form. The search space of even a small database of 100 dialdehydes and 100 triamines, only considering three possible topologies, amounts to 30,000 possible combinations. Fortunately, EAs are particularly well-suited to finding solutions in enormous search spaces [151], making them particularly well-suited for molecular generation [152].

10.4.2 Reducing the Number of Experiments: Bayesian Optimization and Active Learning

In chemistry, the cost of experiments or even higher-level computational calculations can be prohibitively expensive, and, as a result, there has been significant interest in algorithms that minimize the number of experiments while still proposing high-quality solutions to often-multi-objective functions. Bayesian optimization (BO) is one such algorithm; from its original purpose of optimizing the hyperparameters of ML models [153], BO has been applied to reaction condition optimization [154], chemical design [155], and, more recently, porous material discovery [156]. Invariably,

BO is comprised of both surrogate and acquisition models. The goal of the surrogate model is to capture the underlying distribution of a function that would require significant expense to calculate, be that monetary in the case of laboratory experiments, or computational in the case of simulations. Typically, a Gaussian process is used due to its ability to quantify the uncertainty of a model to match the original functional. Further data is obtained to maximize an acquisition function, such that the expected return from obtaining data would generate the largest improvement in the surrogate model.

Indeed, this method has been employed for porous materials. For example, Deshwal et al. used this approach to search for COFs with the highest methane deliverable capacity from a database of 70,000 hypothetical structures [156]. Using BO, the authors were able to minimize the number of expensive grand canonical Monte Carlo adsorption simulations performed, outperforming EAs in terms of the number of high-performing candidates identified. In this work, the authors were able to identify the COF with the highest methane deliverable capacity after 110 simulations, compared with finding the 170th ranked COF after 500 steps of a random search.

In BO, acquisition of new data points is often a serial process, owing to inability to parallelize the often-used expected improvement function. A parallelized implementation, using Thompson sampling, was reported by Pyzer-Knapp et al., who used BO to minimize the number of lattice energy and methane uptake calculations to identify promising molecular crystals for methane capture, reducing the number of calculations required to generate ESF maps, as discussed in Section 10.2 [74].

Instead of generating a surrogate function to model an unknown objective function, uncertainties directly obtained from the supervised learning model can instead be used to select the next data point. Models containing inherent uncertainty in their prediction, such as random forest or Gaussian processes, can be used to point toward the next sample that contains the highest uncertainty in its prediction [157]. Alternatively, the difference between the prediction of multiple models can be quantified as an uncertainty [43,158]. These methods, collectively known as active learning, have been used to test new compounds in drug discovery for many years [159,160], but, more recently, have been used to develop ML potentials [43], and select the next synthesis to perform in high-throughput reaction screening [161].

10.4.3 Chemical Space Exploration with Deep Learning

More recently, DL techniques trained in a semi-supervised manner have been used to generate candidate molecules [162], most notably generative adversarial networks and VAEs. The former uses perturbations in a learnt latent space of molecules for generation, and the latter is comprised of competing generator and discriminator networks [104]. Ultimately, the goal of generative models is to learn an underlying distribution of the data, sampling from this distribution to produce new examples. Often, these models are biased toward the generation of molecules with desirable properties, achieved by incorporating a reward function during the training process [163]. The inclusion of recurrent neural networks, known to perform well at handwriting and speech recognition tasks [164,165], form an integral part of the architecture of the generative algorithm due to their ability to generate chemically valid SMILES. The ability of these methods to both generate molecules and develop functions mapping molecules to properties allow for the possibility of inverse design. Alternatively, reinforcement learning aims to train an agent to perform the best action in its environment to maximize a pre-defined reward function. In chemistry, reinforcement learning has been used for reaction condition selection [166], and drug discovery [101], using an NN, termed the policy, to predict the best action to take.

While generative models have found use in the inverse design of inorganic materials [167], generation of new zeolite architectures [107], and inorganic crystal structure prediction [168], it is challenging to apply these techniques to the discovery of PMMs due to the lack of training data. Typically, these approaches struggle to predict properties and, therefore, generate molecules outside of their training set distribution, restricting them to domains in chemistry where larger datasets are accessible. Instead, EAs and active learning methods are a promising alternative for exploring chemical space; not requiring an initial training set, active learning methods can provide an efficient, non-biased approach to generating training data, be its origins experimental or computational.

10.5 Considering Synthetic Accessibility

Using a combination of exploratory algorithms combined with ML, it is now possible to screen vast numbers of hypothetical molecules for promising properties. However, there is no certainty that those molecules are synthetically accessible, which can complicate the transition between computational prediction and experimental realization. For organic materials, often formed from robust reactions of precursors, the synthesis of the precursors themselves is often the most challenging and time-consuming part of the synthesis [32]. Following precursor synthesis, there are additional considerations: i) whether those precursors will combine to form a product with the desired topology; ii) whether that material will pack in a way in that the properties are maintained, and; finally, iii) how to fabricate the material into a device. Considering all these factors computationally is extremely challenging. As a result, they are largely neglected in many computational workflows, often relying on the ability and intuition of experimental scientists to create the final material [32].

The precursors of PMMs are typically small organic molecules with functional groups compatible with the DCC reaction that will form the final material [28]. Commercially-available and enumerated compound databases, such as the Enamine REAL [169], or ZINC database [170], can provide a library of precursors for computational screening. However, fluctuating chemical costs can prohibit the utility of these libraries. Recently, enumeration of compound libraries using synthetically feasible transformations has successfully resulted in the generation of molecules accessible via known reaction routes [171–173]. Alternatively, a computational prediction of synthetic accessibility could be used to selectively eliminate unsynthesizable molecules. For small molecules, a computational prediction of synthetic accessibility can be achieved using four main methods: i) using cheminformatic methods to quantify structural complexity [174]; ii) data-driven approaches using extensive reaction databases [175]; iii) modeling chemists' intuition [176,177]; and iv) retrosynthesis planning tools [178]. Typically, these methods are quick to calculate (with the exception of generating a full retrosynthesis route) and provide a single metric which can be incorporated into a multi-objective function [34].

Many of these methods are tailored to predict the synthetic accessibility of small, drug-like molecules of importance in the pharmaceutical industry, whose synthetic requirements differ from the materials industry. However, these heuristic measures can be applied to precursors of POCs. Bennett et al. trained an RF classification model on 12,553 molecules labeled as easy- and difficult-to-synthesize by expert chemists in the field of organic synthesis, with the goal of using the final model to eliminate cage precursors that would be challenging to synthesize, the workflow of which is shown in Figure 10.9 [33]. Using their model, in addition to two other computational methods of assessing synthetic accessibility, the authors were able to identify seven shape-persistent cages with a cavity size of larger than 16 Å whose precursors were predicted to be easy-to-synthesize. By removing difficult-to-synthesize molecules, computational efforts can be targeted toward

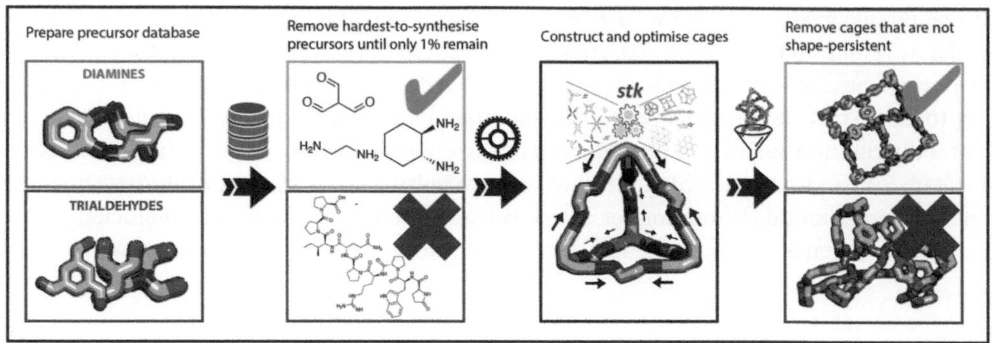

Figure 10.9 The workflow used by Bennett et al. for selectively eliminating difficult-to-synthesize POC precursors, targeting computational efforts toward POCs whose precursors are predicted to be easy-to-synthesize. *Source:* Reproduced from [33], with permission from American Chemical Society. © 2021.

precursors that *are* synthesizable in the lab, targeting subsequent computationally demanding simulations on only synthesizable candidates. Ideally, we would not only be proposed with a numerical measure of synthetic accessibility, but also a viable route to that compound. Computer-assisted retrosynthesis tools are now able to propose viable pathways to a molecule [179–181], even improving upon established synthetic routes [182]. Undoubtedly, automated retrosynthesis planners will play an increasing role in the synthesis of precursors for materials development.

Although it is possible to identify synthesizable precursors, there is no certainty that the precursors will undergo the required reaction to form a product with the desired topology. While formation energies of POCs can indicate the likely thermodynamic outcome [31,183], they cannot predict the formation of mixtures, or even oligomeric or polymeric products. Computational methods also cannot accurately describe the effect of reaction conditions, such as solvent choice or temperature, on cage formation; these methods often neglect the kinetic pathways that could result in a different reaction outcome [184]. Indeed, the presence of solvent can result in a crystal structure that deviates from the predicted global minimum. McMahon et al. were able to rationalize the experimentally observed high-energy crystal structure of **CC1** by using Monte Carlo methods to insert solvent molecules into possible crystal structures, providing an explanation for the solvamorphic behavior of **CC1** [185]. The computational demands of modeling reaction conditions makes ML an attractive alternative; now, models can predict optimal temperature, catalyst, and solvent for many organic reactions [186]. It is possible to envisage ML becoming increasingly prevalent in every aspect of the PMM discovery pipeline, from precursor selection to material fabrication. Given sufficient data, ML can make predictions of synthetic outcomes that would be too computationally expensive to model using conventional methods.

10.6 Closing the Loop: How Can High-Throughput Experimentation Feed Back into Computation?

Naturally, experimental chemists are more inclined to pursue reactions that will achieve a successful outcome or generate an interesting result. This process is typically slow and hinders novel discovery due to the small incremental modifications chemists perform on known molecules [47,187]. With the ability to systematically react potentially hundreds of precursors, automated synthesis methods allow chemists to test a far greater number of candidates, accelerating serendipitous discovery by providing greater opportunities to navigate unexplored areas of chemical space. Indeed, high-throughput automated synthesis techniques can help overcome many of the challenges associated with generating large datasets for ML [188]. While automated synthetic methods are mostly in their infancy, even today, the possibility exists to combinatorically react vast numbers of precursors and perform subsequent analyses. Combined with global optimization methods, as described in Section 10.4, exploratory algorithms are increasingly becoming a replacement for chemist intuition and can propose future experiments using a non-biased and systematic methodology.

10.6.1 High-Throughput and Autonomous Experimentation

Although the complexity of the reactions that can be performed using automated experimentation is currently limited, existing methods excel at systematically combining many different substances, which can be used to explore different reaction conditions by altering concentration, catalyst, or solvent [188,189]. Automated robotic platforms can be used to automate many syntheses, albeit limited by the challenges associated with being compatible with the diverse range of solvents used in chemistry and handling solids [93,190]. For POCs, experimentation and, consequently, experimental discovery is typically slow; this is exemplified by the average of three POCs discovered each year between 2009 and 2018 by Greenaway et al. [31]. However, using automated synthesis methods combined with computation, the authors were able to increase this number by 33, substantially increasing the number of discovered POCs [31]. A similar automated approach was applied to the formation of multi-component systems by combining tri- and di-topic amines and tri-topic aldehydes with the goal of generating an topology resembling an organic cage pot, exploiting the ability of precursors to self-sort into cages with three components [191]. This systematic exploration of precursor combination resulted in the discovery of **OCP3**, an organic cage pot with windows similar in geometry to **CC3**. Recently, Kearsey et al. were able to automate the discovery of type II PLs by combining precursors that resulted in the formation of scrambled cages [93]. Subsequent high-throughput solubility testing was performed on 29 scrambled cages to identify compatible solvents that would be size-excluded from the cage's internal cavity. While these studies do not incorporate ML in the workflow, they highlight how robotic synthesis can systematically navigate across a pre-defined reaction space, typically designed by the intuition of the chemist. However, for POCs, the time and cost associated with precursor synthesis remains the bottleneck to autonomous applications.

Combined with global optimization methods, it is possible to envisage a fully autonomous workflow with algorithms assisting chemists in not only *how* to synthesize a substance, but *what* to synthesize next. Currently, "closed-loop" experimental workflows that feed directly back into ML algorithms are scarce in the literature. This is likely due to the challenge of combining synthesis, analysis, model updating, and prediction in the same workflow [192]. Indeed, to date, there are no examples applying closed-loop discovery to PMMs, and there are few examples in the wider chemical literature. In one example, Christensen et al. used a fully autonomous discovery workflow to optimize the yield of a Suzuki–Miyaura coupling reaction by exploring different phosphine ligands and reaction conditions [193]. Their automated system was achieved by integrating a Chemspeed platform, Internet-accessible Agilent HPLC-UV system, a BO algorithm to propose the next set of conditions, and the use of the authors' experiment scheduling software, ChemOS [194]. In total, 23 categorical three continuous parameters were explored, a parameter space impossible to fully explore manually, achieving 65% yield of the desired product in 118 iterations. Similar autonomous workflows have been applied to the discovery of organic photovoltaics [195], thin films [196], and reaction discovery [197].

Autonomous discovery can help reduce the bias chemists inherently included in experiment selection [198]. However, it is currently limited by prohibitive equipment cost and limited integration capabilities for most equipment. Although research is ongoing for robotic platforms that can use conventional non-automated laboratory equipment and 3D-printed universal chemical reactors [199,200], more developments are required in the field before they can be applied to PMMs.

10.7 Conclusions

Computation and, more recently, AI is playing an increasing role in the wider porous materials community. These methods are now able to accelerate many workflows, including both structure and property prediction on a range of length- and timescales. With the ability to predict properties more efficiently than conventional computational techniques and with similar accuracy, ML techniques will continue to grow in popularity in the porous materials field as the amount of accessible data grows.

Computational methods are now reliably able to predict structures for the discrete molecular units that constitute PMMs, and even predict plausible bulk structures for crystalline and disordered PMMs. While ML use in the PMM field is currently hindered by a lack of training data, computation, combined with high-throughput screening, can help overcome this challenge. Indeed, ML can increase the accuracy of simulations using ML-potentials derived from electronic structure calculations. While these potentials have currently only been applied to the simulation of MOFs [201], they will no doubt find utility in PMM crystal or disordered structure prediction.

Recent developments in geometric descriptors have allowed researchers to capture more detailed structural information about porous materials, providing the ability to easily distinguish between different conformations of the same molecule, and to map these structures to properties. With a growth in training data, it is possible to foresee more deep learning methods applications applied to PMMs, including the ability to learn more detailed representations directly with neural networks.

To facilitate this transition to a more data-driven approach, a greater emphasis needs to be placed on reporting both successful and unsuccessful reactions in a reproducible and machine-readable format. High-throughput synthetic methods can help achieve this by systematically screening candidates more efficiently than a typical researcher. While many high-throughput techniques are currently in their infancy, the growing accessibility of open-source platforms will encourage other researchers to adopt these platforms and find solutions to the shortcomings of these methods, most notably the ability to combine synthesis and analysis in a single autonomous workflow.

While there are still a significant number of challenges to overcome, it is with the combination of high-throughput experimentation, ML prediction, and efficient global optimization techniques that it is now possible to envisage a future of truly autonomous materials discovery entirely algorithmically fuelled.

References

1 Tian, J., Thallapally, P.K., and McGrail, B.P. (2012). Porous organic molecular materials. *CrystEngComm* 14 (6): 1909.

2 Bennett, T.D., Coudert, F.-X., James, S.L. et al. (2021). The changing state of porous materials. *Nat. Mater.* 20 (9): 1179–1187.

3 Jiang, S., Cox, H.J., Papaioannou, E.I. et al. (2019). Shape-persistent porous organic cage supported palladium nanoparticles as heterogeneous catalytic materials. *Nanoscale* 11 (31): 14929–14936.

4 Song, Q., Jiang, S., Hasell, T. et al. (2016). Porous organic cage thin films and molecular-sieving membranes. *Adv. Mater.* 28 (13): 2629–2637.

5 Deegan, M.M., Dworzak, M.R., Gosselin, A.J. et al. (2021). Gas storage in porous molecular materials. *Chemistry* 27 (14): 4531–4547.

6 Slater, A.G. and Cooper, A.I. (2015). Porous materials: function-led design of new porous materials. *Science* 348 (6238): aaa8075.

7 Holst, J.R., Trewin, A., and Cooper, A.I. (2010). Porous organic molecules. *Nat. Chem.* 2 (11): 915–920.

8 Jablonka, K.M., Ongari, D., Moosavi, S.M. et al. (2020). Big-data science in porous materials: materials genomics and machine learning. *Chem. Rev.* 120 (16): 8066–8129.

9 Barrer, R.M. and Shanson, V.H. (1976). Dianin's compound as a zeolitic sorbent. *J. Chem. Soc. Chem. Commun.* 9: 333.

10 Budd, P.M., Ghanem, B.S., Makhseed, S. et al. (2004). Polymers of intrinsic microporosity (PIMs): robust, solution-processable, organic nanoporous materials. *Chem. Commun. (Camb).* 21 (2): 230–231.

11 Kalaj, M. and Cohen, S.M. (2020). Postsynthetic modification: an enabling technology for the advancement of metal–organic frameworks. *ACS Cent. Sci.* 6 (7): 1046–1057.

12 Wang, H., Jin, Y., Sun, N. et al. (2021). Post-synthetic modification of porous organic cages. *Chem. Soc. Rev.* 50 (16): 8874–8886.

13 Liu, M., Little, M.A., Jelfs, K.E. et al. (2014). Acid- and base-stable porous organic cages: shape persistence and pH stability via post-synthetic "tying" of a flexible amine cage. *J. Am. Chem. Soc.* 136 (21): 7583–7586.

14 Giri, N., Del Pópolo, M.G., Melaugh, G. et al. (2015). Liquids with permanent porosity. *Nature* 527 (7577): 216–220.

15 Cooper, A.I. (2017). Porous molecular solids and liquids. *ACS Cent. Sci.* 3 (6): 544–553.

16 Allcock, H.R. and Siegel, L.A. (1964). Phosphonitrilic compounds. III: 1, molecular inclusion compounds of Tris(o-phenylenedioxy)phosphonitrile trimer. *J. Am. Chem. Soc.* 86 (23): 5140–5144.

17 Sozzani, P., Comotti, A., Simonutti, R. et al. (2000). A porous crystalline molecular solid explored by hyperpolarized xenon. *Angew. Chem. Int. Ed. Engl.* 39 15): 2695–2699.

18 He, Y., Xiang, S., and Chen, B. (2011). A microporous hydrogen-bonded organic framework for highly selective C_2H_2/C_2H_4 separation at ambient temperature. *J. Am. Chem. Soc.* 133 (37): 14570–14573.

19 Yang, W., Greenaway, A., Lin, X. et al. (2010). Exceptional thermal stability in a supramolecular organic framework: porosity and gas storage. *J. Am. Chem. Soc.* 132 (41): 14457–14469.

20 Soldatov, D.V., Moudrakovski, I.L., and Ripmeester, J.A. (2004). Dipeptides as microporous materials. *Angew. Chem. Int. Ed. Engl.* 43 (46): 6308–6311.

21 Alsbaiee, A., Smith, B.J., Xiao, L. et al. (2016). Rapid removal of organic micropollutants from water by a porous β-cyclodextrin polymer. *Nature* 529 (7585): 190–194.

22 Zhang, G. and Mastalerz, M. (2014). Organic cage compounds: from shape-persistency to function. *Chem. Soc. Rev.* 43 (6): 1934–1947.

23 Atwood, J.L., Barbour, L.J., and Jerga, A. (2002). Storage of methane and freon by interstitial van der Waals confinement. *Science* 296 (5577): 2367–2369.

24 Dewal, M.B., Lufaso, M.W., Hughes, A.D. et al. (2006). Absorption properties of a porous organic crystalline apohost formed by a self-assembled bis-urea macrocycle. *Chem. Mater.* 18 (20): 4855–4864.

25 Chaix, A., Mouchaham, G., Shkurenko, A. et al. (2018). Trianglamine-based supramolecular organic framework with permanent intrinsic porosity and tunable selectivity. *J. Am. Chem. Soc.* 140 (44): 14571–14575.

26 Tozawa, T., Jones, J.T.A., Swamy, S.I. et al. (2009). Porous organic cages. *Nat. Mater.* 8 (12): 973–978.

27 Zhang, G., Presly, O., White, F. et al. (2014). A permanent mesoporous organic cage with an exceptionally high surface area. *Angew. Chem. Int. Ed. Engl.* 53 (6): 1516–1520.

28 Little, M.A. and Cooper, A.I. (2020). The chemistry of porous organic molecular materials. *Adv. Funct. Mater.* 30 (41): 1909842.

29 Avellaneda, A., Valente, P., Burgun, A. et al. (2013). Kinetically controlled porosity in a robust organic cage material. *Angew. Chem. Int. Ed. Engl.* 52 (13): 3746–3749.

30 Santolini, V., Miklitz, M., Berardo, E. et al. (2017). Topological landscapes of porous organic cages. *Nanoscale* 9 (16): 5280–5298.

31 Greenaway, R.L., Santolini, V., Bennison, M.J. et al. (2018). High-throughput discovery of organic cages and catenanes using computational screening fused with robotic synthesis. *Nat. Commun.* 9 (1): 2849.

32 Szczypiński, F.T., Bennett, S., and Jelfs, K.E. (2020). Can we predict materials that can be synthesised? *Chem. Sci.* 12 (3): 830–840.

33 Bennett, S., Szczypiński, F.T., Turcani, L. et al. (2021). Materials precursor score: modeling chemists' intuition for the synthetic accessibility of porous organic cage precursors. *J. Chem. Inf. Model.* 61 (9): 4342–4356.

34 Gao, W. and Coley, C.W. (2020). The synthesizability of molecules proposed by generative models. *J. Chem. Inf. Model.* 60 (12): 5714–5723.

35 Greenaway, R.L. and Jelfs, K.E. (2021). Integrating computational and experimental workflows for accelerated organic materials discovery. *Adv. Mater.* 9: e2004831.

36 Butler, K.T., Davies, D.W., Cartwright, H. et al. (2018). Machine learning for molecular and materials science. *Nature* 559 (7715): 547–555.

37 Clayson, I.G., Hewitt, D., Hutereau, M. et al. (2020). High throughput methods in the synthesis, characterization, and optimization of porous materials. *Adv. Mater.* 32 (44): e2002780.

38 Yang, X., Wang, Y., Byrne, R. et al. (2019). Concepts of artificial intelligence for computer-assisted drug discovery. *Chem. Rev.* 119 (18): 10520–10594.

39 Struble, T.J., Alvarez, J.C., Brown, S.P. et al. (2020). Current and future roles of artificial intelligence in medicinal chemistry synthesis. *J. Med. Chem.* 63 (16): 8667–8682.

40 Wilkinson, M.D., Dumontier, M., Aalbersberg, I.J.J. et al. (2016). The FAIR guiding principles for scientific data management and stewardship. *Sci. Data.* 3 (1): 160018.

41 Agrawal, A. and Choudhary, A. (2016). Perspective. Materials informatics and big data: realization of the "fourth paradigm" of science in materials science. *APL Mater.* 4: 053208.

42 Raccuglia, P., Elbert, K.C., Adler, P.D.F. et al. (2016). Machine-learning-assisted materials discovery using failed experiments. *Nature* 533 (7601): 73–76.

43 Smith, J.S., Nebgen, B., Lubbers, N. et al. (2018). Less is more: sampling chemical space with active learning. *J. Chem. Phys.* 148 (24): 241733.

44 Kusne, A.G., Yu, H., Wu, C. et al. (2020). On-the-fly closed-loop materials discovery via Bayesian active learning. *Nat. Commun.* 11 (1): 5966.

45 Miklitz, M. and Jelfs, K.E. (2018). pywindow: Automated structural analysis of molecular pores. *J. Chem. Inf. Model.* 58 (12): 2387–2391.

46 García, I.G. and Haranczyk, M. (2020). Toward crystalline porosity estimators for porous molecules. *CrystEngComm* 22 (43): 7242–7251.

47 Jones, J.T.A., Hasell, T., Wu, X. et al. (2011). Modular and predictable assembly of porous organic molecular crystals. *Nature* 474 (7351): 367–371.

48 Berardo, E., Greenaway, R.L., Miklitz, M. et al. (2020). Computational screening for nested organic cage complexes. *Mol. Syst. Des. Eng.* 5 (1): 186–196.

49 Hawkins, P.C.D. (2017). Conformation generation: the state of the art. *J. Chem. Inf. Model.* 57 (8): 1747–1756.

50 Rappe, A.K., Casewit, C.J., Colwell, K.S. et al. (1992). UFF, a full periodic table force field for molecular mechanics and molecular dynamics simulations. *J. Am. Chem. Soc.* 114 (25): 10024–10035.

51 Sun, H. (1998). COMPASS: an *ab initio* force-field optimized for condensed-phase applications. Overview with details on alkane and benzene compounds. *J. Phys. Chem. B.* 102 (38): 7338–7364.

52 Holden, D., Jelfs, K.E., Cooper, A.I. et al. (2012). Bespoke force field for simulating the molecular dynamics of porous organic cages. *J. Phys. Chem. C Nanomater. Interfaces.* 116 (31): 16639–16651.

53 Jorgensen, W.L., Maxwell, D.S., and Tirado-Rives, J. (1996). Development and testing of the OPLS all-atom force field on conformational energetics and properties of organic liquids. *J. Am. Chem. Soc.* 118 (45): 11225–11236.

54 Jelfs, K.E., Eden, E.G.B., Culshaw, J.L. et al. (2013). *In silico* design of supramolecules from their precursors: odd-even effects in cage-forming reactions. *J. Am. Chem. Soc.* 135 (25): 9307–9310.

55 Lu, C., Wu, C., Ghoreishi, D. et al. (2021). OPLS4: improving force field accuracy on challenging regimes of chemical space. *J. Chem. Theory. Comput.* 17 (7): 4291–4300.

56 Roos, K., Wu, C., Damm, W. et al. (2019). OPLS3e: extending force field coverage for drug-like small molecules. *J. Chem. Theory. Comput.* 15 (3): 1863–1874.

57 Turcani, L., Tarzia, A., Szczypiński, F.T. et al. (2021). *stk*: an extendable python framework for automated molecular and supramolecular structure assembly and discovery. *J. Chem. Phys.* 154 (21): 214102.

58 Turcani, L., Berardo, E., and Jelfs, K.E. (2018). *stk*: a python toolkit for supramolecular assembly. *J. Comput. Chem.* 39 (23): 1931–1942.

59 Greenaway, R.L. and Jelfs, K.E. (2020). High-throughput approaches for the discovery of supramolecular organic cages. *ChemPlusChem* 85 (8): 1813–1823.

60 Bai, Y., Wilbraham, L., Slater, B.J. et al. (2019). Accelerated discovery of organic polymer photocatalysts for hydrogen evolution from water through the integration of experiment and theory. *J. Am. Chem. Soc.* 141 (22): 9063–9071.

61 Heath-Apostolopoulos, I., Wilbraham, L., and Zwijnenburg, M.A. (2019). Computational high-throughput screening of polymeric photocatalysts: exploring the effect of composition, sequence isomerism and conformational degrees of freedom. *Faraday Discuss* 215 : 98–110.

62 Tarzia, A., Lewis, J.E.M., and Jelfs, K.E. (2021). High-throughput computational evaluation of low symmetry Pd2 L4 cages to aid in system design. *Angew. Chem. Int. Ed. Engl.* 60 (38): 20879–20887.

63 Young, T.A., Gheorghe, R., and Duarte, F. (2020). cgbind: a python module and web app for automated metallocage construction and host–guest characterization. *J. Chem. Inf. Model.* 60 (7): 3546–3557.

64 Thompson, H.P.G. and Day, G.M. (2014). Which conformations make stable crystal structures? Mapping crystalline molecular geometries to the conformational energy landscape. *Chem. Sci.* 5 (8): 3173–3182.

65 Reilly, A.M., Cooper, R.I., Adjiman, C.S. et al. (2016). Report on the sixth blind test of organic crystal structure prediction methods. *Acta. Crystallogr. B Struct. Sci. Cryst. Eng. Mater* 72 (4): 439–459.

66 Price, S.L. (2014). Predicting crystal structures of organic compounds. *Chem. Soc. Rev.* 43 (7): 2098–2111.

67 Day, G.M. and Cooper, A.I. (2018). Energy-structure-function maps: cartography for materials discovery. *Adv. Mater.* 30 (37): e1704944.

68 Pyzer-Knapp, E.O., Thompson, H.P.G., Schiffmann, F. et al. (2014). Predicted crystal energy landscapes of porous organic cages. *Chem. Sci.* 5 (6): 2235–2245.

69 Bowskill, D.H., Sugden, I.J., Konstantinopoulos, S. et al. (2021). Crystal structure prediction methods for organic molecules: state of the art. *Annu. Rev. Chem. Biomol. Eng.* 12 (1): 593–623.

70 Zhang, P., Wood, G.P.F., Ma, J. et al. (2018). Harnessing cloud architecture for crystal structure prediction calculations. *Cryst. Growth. Des.* 18 (11): 6891–6900.

71 Unke, O.T., Chmiela, S., Sauceda, H.E. et al. (2021). Machine learning force fields. *Chem. Rev.* 121 (16): 10142–10186.

72 Wengert, S., Csányi, G., Reuter, K. et al. (2021). Data-efficient machine learning for molecular crystal structure prediction. *Chem. Sci.* 12 (12): 4536–4546.

73 Zhao, C., Chen, L., Che, Y. et al. (2021). Digital navigation of energy-structure-function maps for hydrogen-bonded porous molecular crystals. *Nat. Commun.* 12 (1): 817.

74 Pyzer-Knapp, E.O., Chen, L., Day, G.M. et al. (2021). Accelerating computational discovery of porous solids through improved navigation of energy-structure-function maps. *Sci. Adv.* 7 (33): eabi4763.

75 Addicoat, M.A. (2021). Chapter 2: Structure prediction of porous materials. In: *Computer Simulation of Porous Materials* (ed. K.E. Jelfs), 27–78. Cambridge, UK: TheTheThe Royal Society of Chemistry.

76 Kravchenko, O., Varava, A., Pokorny, F.T. et al. (2020). A robotics-inspired screening algorithm for molecular caging prediction. *J. Chem. Inf. Model.* 60(3): 1302–1316.

77 Holden, D., Chong, S.Y., Chen, L. et al. (2016). Understanding static, dynamic and cooperative porosity in molecular materials. *Chem. Sci.* 7 (8): 4875–4879.

78 Bassanetti, I., Bracco, S., Comotti, A. et al. (2018). Flexible porous molecular materials responsive to CO_2, CH_4 and Xe stimuli. *J. Mater. Chem. A Mater. Energy Sustain.* 6 (29): 14231–14239.

79 Hasell, T., Miklitz, M., Stephenson, A. et al. (2016). Porous organic cages for sulfur hexafluoride separation. *J. Am. Chem. Soc.* 138 (5): 1653–1659.

80 Verploegh, R.J., Nair, S., and Sholl, D.S. (2015). Temperature and loading-dependent diffusion of light hydrocarbons in ZIF-8 as predicted through fully flexible molecular simulations. *J. Am. Chem. Soc.* 137 (50): 15760–15771.

81 Jackson, E., Miklitz, M., Song, Q. et al. (2019). Computational evaluation of the diffusion mechanisms for C8 aromatics in porous organic cages. *J. Phys. Chem. C Nanomater. Interfaces.* 123 (34): 21011–21021.

82 Chen, L., Che, Y., Cooper, A.I. et al. (2021). Exploring cooperative porosity in organic cage crystals using *in situ* diffraction and molecular simulations. *Faraday Discuss* 225 (0): 100–117.

83 Tian, J., Thallapally, P.K., Dalgarno, S.J. et al. (2009). Amorphous molecular organic solids for gas adsorption. *Angew. Chem. Int. Ed. Engl.* 48 (30): 5492–5495.

84 Evans, J.D., Huang, D.M., Hill, M.R. et al. (2015). Molecular design of amorphous porous organic cages for enhanced gas storage. *J. Phys. Chem. C Nanomater. Interfaces.* 119 (14): 7746–7754.

85 Abbott, L.J., McKeown, N.B., and Colina, C.M. (2013). Design principles for microporous organic solids from predictive computational screening. *J. Mater. Chem. A Mater. Energy Sustain.* 1 (38): 11950.

86 O'Reilly, N., Giri, N., and James, S.L. (2007). Porous liquids. *Chem. Eur. J.* 13 (11): 3020–3025.

87 Melaugh, G., Giri, N., Davidson, C.E. et al. (2014). Designing and understanding permanent microporosity in liquids. *Phys. Chem. Chem. Phys.* 16 (20): 9422–9431.

88 Greenaway, R.L., Holden, D., Eden, E.G.B. et al. (2017). Understanding gas capacity, guest selectivity, and diffusion in porous liquids. *Chem. Sci.* 8 (4): 2640–2651.

89 Deng, Z., Ying, W., Gong, K. et al. (2020). Facilitate gas transport through metal-organic polyhedra constructed porous liquid membrane. *Small* 16 (11): e1907016.

90 Kumar, S., Rosenberg, J.M., Bouzida, D. et al. (1995). Multidimensional free-energy calculations using the weighted histogram analysis method. *J. Comput. Chem.* 16 (11): 1339–1350.

91 Giri, N., Davidson, C.E., Melaugh, G. et al. (2012). Alkylated organic cages: from porous crystals to neat liquids. *Chem. Sci.* 3 (6): 2153.

92 Jiang, S., Jones, J.T.A., Hasell, T. et al. (2011). Porous organic molecular solids by dynamic covalent scrambling. *Nat. Commun.* 2 (1): 207.

93 Kearsey, R.J., Alston, B.M., Briggs, M.E. et al. (2019). Accelerated robotic discovery of type II porous liquids. *Chem. Sci.* 10 (41): 9454–9465.

94 Zubatiuk, T. and Isayev, O. (2021). Development of multimodal machine learning potentials: toward a physics-aware artificial intelligence. *Acc. Chem. Res.* 54 (7): 1575–1585.

95 Senior, A.W., Evans, R., Jumper, J. et al. (2020). Improved protein structure prediction using potentials from deep learning. *Nature* 577 (7792): 706–710.

96 Brown, T.B., Mann, B., Ryder, N. et al. (2020). Language models are few-shot learners [Preprint] arXiv:2005.14165v4 [cs.CL]. Available from http://arxiv.org/abs/2005.14165 (accessed 31 August 2021).

97 Ester, M., Kriegel, H.-P., Sander, J. et al. (1996). A density-based algorithm for discovering clusters in large spatial databases with noise. In: *Proceedings of the Second International Conference on Knowledge Discovery and Data Mining*, 226–231. AAAI Press. (KDD'96).

98 Bennett, S., Tarzia, A., and Zwijnenburg, M.A. (2020). Chapter 12: Artificial intelligence applied to the prediction of organic materials. In: *Machine Learning in Chemistry* (ed. H.M. Cartwright), 280–310. Cambridge, UK: The Royal Society of Chemistry.

99 McCulloch, W.S. and Pitts, W. (1943). A logical calculus of the ideas immanent in nervous activity. *Bulletin of Mathematical Biophysics* 5 (4): 115–133.

100 LeCun, Y., Bengio, Y., and Hinton, G. (2015). Deep learning. *Nature* 521 (7553): 436–444.

101 Popova, M., Isayev, O., and Tropsha, A. (2018). Deep reinforcement learning for *de novo* drug design. *Sci. Adv.* 4 (7): eaap7885.

102 Mnih, V., Kavukcuoglu, K., Silver, D. et al. (2015). Human-level control through deep reinforcement learning. *Nature* 518 (7540): 529–533.

103 Chen, H., Engkvist, O., Wang, Y. et al. (2018). The rise of deep learning in drug discovery. *Drug. Discov. Today.* 23 (6): 1241–1250.

104 Gómez-Bombarelli, R., Wei, J.N., Duvenaud, D. et al. (2018). Automatic chemical design using a data-driven continuous representation of molecules. *ACS Cent. Sci.* 4 (2): 268–276.

105 Rifaioglu, A.S., Atas, H., Martin, M.J. et al. (2019). Recent applications of deep learning and machine intelligence on *in silico* drug discovery: methods, tools and databases. *Brief Bioinform* 20 (5): 1878–1912.

106 Coley, C.W., Jin, W., Rogers, L. et al. (2019). A graph-convolutional neural network model for the prediction of chemical reactivity. *Chem. Sci.* 10 (2): 370–377.

107 Kim, B., Lee, S., and Kim, J. (2020). Inverse design of porous materials using artificial neural networks. *Sci. Adv.* 6 (1): eaax9324.

108 Evans, J.D., Jelfs, K.E., Day, G.M. et al. (2017). Application of computational methods to the design and characterisation of porous molecular materials. *Chem. Soc. Rev.* 46 (11): 3286–3301.

109 Colón, Y.J., Gómez-Gualdrón, D.A., and Snurr, R.Q. (2017). Topologically guided, automated construction of metal–organic frameworks and their evaluation for energy-related applications. *Cryst. Growth. Des.* 17 (11): 5801–5810.

110 Martin, R.L., Simon, C.M., Smit, B. et al. (2014). *In silico* design of porous polymer networks: high-throughput screening for methane storage materials. *J. Am. Chem. Soc.* 136 (13): 5006–5022.

111 Deem, M.W., Pophale, R., Cheeseman, P.A. et al. (2009). Computational discovery of new zeolite-like materials. *J. Phys, Chem. C*113 (51): 21353–21360.

112 Wilmer, C.E. and Snurr, R.Q. (2014). Large-scale generation and screening of hypothetical metal–organic frameworks for applications in gas storage and separations. *Top Curr. Chem.* 345: 257–289.

113 Martin, R.L. and Haranczyk, M. (2014). Construction and characterization of structure models of crystalline porous polymers. *Cryst. Growth. Des.* 14 (5): 2431–2440.

114 Shi, Z., Yang, W., Deng, X. et al. (2020). Machine-learning-assisted high-throughput computational screening of high performance metal–organic frameworks. *Mol. Syst. Des. Eng.* 5 (4): 725–742.

115 Wang, R., Zhong, Y., Bi, L. et al. (2020). Accelerating discovery of metal–organic frameworks for methane adsorption with hierarchical screening and deep learning. *ACS Appl. Mater. Interfaces* 12 (47): 52797–52807.

116 Turcani, L., Greenaway, R.L., and Jelfs, K.E. (2019). Machine learning for organic cage property prediction. *Chem. Mater.* 31 (3): 714–727.

117 Groom, C.R., Bruno, I.J., Lightfoot, M.P. et al. (2016). The Cambridge Structural Database. *Acta. Crystallogr. B Struct. Sci. Cryst. Eng. Mater.* 72 (2): 171–179.

118 Evans, J.D., Huang, D.M., Haranczyk, M. et al. (2016). Computational identification of organic porous molecular crystals. *CrystEngComm* 18 (22): 4133–4141.

119 Willems, T.F., Rycroft, C.H., Kazi, M. et al. (2012). Algorithms and tools for high-throughput geometry-based analysis of crystalline porous materials. *Microporous Mesoporous Mater.* 149 (1): 134–141.

120 Huang, S. and Cole, J.M. (2020). A database of battery materials auto-generated using ChemDataExtractor. *Sci. Data.* 7 (1): 260.

121 Kononova, O., He, T., Huo, H. et al. (2021). Opportunities and challenges of text mining in aterials research. *iScience* 24 (3): 102155.

122 Swain, M.C. and Cole, J.M. (2016). ChemDataExtractor: a toolkit for automated extraction of chemical information from the scientific literature. *J. Chem. Inf. Model.* 56 (10): 1894–1904.

123 Tshitoyan, V., Dagdelen, J., Weston, L. et al. (2019). Unsupervised word embeddings capture latent knowledge from materials science literature. *Nature* 571 (7763): 95–98.

124 Wang, H., Xie, Y., Li, D. et al. (2020). Rapid identification of X-ray diffraction patterns based on very limited data by interpretable convolutional neural networks. *J. Chem. Inf. Model.* 60 (4): 2004–2011.

125 Zhang, Y., Wang, L., Wang, X. et al. (2021). Data augmentation and transfer learning strategies for reaction prediction in low chemical data regimes. *Org. Chem. Front.* 8 (7): 1415–1423.

126 Fortunato, M.E., Coley, C.W., Barnes, B.C. et al. (2020). Data augmentation and pretraining for template-based retrosynthetic prediction in computer-aided synthesis planning. *J. Chem. Inf. Model.* 60 (7): 3398–3407.

127 Yuan, Q., Santana-Bonilla, A., Zwijnenburg, M.A. et al. (2020). Molecular generation targeting desired electronic properties via deep generative models. *Nanoscale* 12 (12): 6744–6758.

128 Smith, J.S., Nebgen, B.T., Zubatyuk, R. et al. (2019). Approaching coupled cluster accuracy with a general-purpose neural network potential through transfer learning. *Nat. Commun.* 10 (1): 2903.

129 Jiang, J., Wang, R., Wang, M. et al. (2020). Boosting tree-assisted multitask deep learning for small scientific datasets. *J. Chem. Inf. Model.* 60 (3): 1235–1244.

130 Vanpoucke, D.E.P., van Knippenberg, O.S.J., Hermans, K. et al. (2020). Small data materials design with machine learning: when the average model knows best. *J. Appl. Phys.* 128 (5): 054901.

131 Zhang, Y. and Ling, C. (2018). A strategy to apply machine learning to small datasets in materials science. *NPJ Comput. Mater.* 4 (1): 1–8.

132 Cherkasov, A., Muratov, E.N., Fourches, D. et al. (2014). QSAR modeling: where have you been? Where are you going to? *J. Med. Chem.* 57 (12): 4977–5010.

133 Willett, P. (2006). Similarity-based virtual screening using 2D fingerprints. *Drug. Discov. Today* 11 (23–24): 1046–1053.

134 Moriwaki, H., Tian, Y.-S., Kawashita, N. et al. (2018). Mordred: a molecular descriptor calculator. *J. Cheminform.* 10 (1): 4.

135 Mauri, A. (2020). alvaDesc: a tool to calculate and analyze molecular descriptors and fingerprints. In: *Methods in Pharmacology and Toxicology (ed. K. Roy)*, 801–820. New York: Springer US.

136 Gao, K., Nguyen, D.D., Sresht, V. et al. (2020). Are 2D fingerprints still valuable for drug discovery? *Phys. Chem. Chem. Phys.* 22 (16): 8373–8390.

137 Bartók, A.P., Kondor, R., and Csányi, G. (2013). On representing chemical environments. *Phys. Rev. B Condens. Matter.* 87 (18): 184115.

138 Musil, F., De, S., Yang, J. et al. (2018). Machine learning for the structure-energy-property landscapes of molecular crystals. *Chem. Sci.* 9 (5): 1289–1300.

139 Yang, J., De, S., Campbell, J.E. et al. (2018). Large-scale computational screening of molecular organic semiconductors using crystal structure prediction. *Chem. Mater.* 30 (13): 4361–4371.

140 Sturluson, A., Huynh, M.T., York, A.H.P. et al. (2018). Eigencages: learning a latent space of porous cage molecules. *ACS Cent. Sci.* 4 (12): 1663–1676.

141 Lee, Y., Barthel, S.D., Dłotko, P. et al. (2017). Quantifying similarity of pore-geometry in nanoporous materials. *Nat. Commun.* 8 (1): 15396.

142 Moosavi, S.M., Xu, H., Chen, L. et al. (2020). Geometric landscapes for material discovery within energy-structure-function maps. *Chem. Sci.* 11 (21): 5423–5433.

143 Pulido, A., Chen, L., Kaczorowski, T. et al. (2017). Functional materials discovery using energy-structure-function maps. *Nature* 54 (7647): 657–664.

144 Duvenaud, D., Maclaurin, D., Aguilera-Iparraguirre, J. et al. (2015). Convolutional networks on graphs for learning molecular fingerprints. In: *Proceedings of the 28th International Conference on Neural Information Processing System*, vol 2, 2224–2232. Cambridge, MA: MIT Press. (NIPS'15).

145 Fung, V., Zhang, J., Juarez, E. et al. (2021). Benchmarking graph neural networks for materials chemistry. *NPJ Comput. Mater.* 7 (1): 1–8.

146 Chung, Y.G., Gómez-Gualdrón, D.A., Li, P. et al. (2016). *In silico* discovery of metal-organic frameworks for precombustion CO_2 capture using a genetic algorithm. *Sci. Adv.* 2 (10): e1600909.

147 Bao, Y., Martin, R.L., Simon, C.M. et al. (2015). *In silico* discovery of high deliverable capacity metal–organic frameworks. *J. Phys. Chem. C Nanomater. Interface* 119 (1): 186–195.

148 Berardo, E., Turcani, L., Miklitz, M. et al. (2018). An evolutionary algorithm for the discovery of porous organic cages. *Chem. Sci.* 9 (45): 8513–8527.

149 Anthopoulos, T.D., Kooistra, F.B., Wondergem, H.J. et al. (2006). Air-stable *n*-channel organic transistors based on a soluble C84 fullerene derivative. *Adv. Mater.* 18 (13): 1679–1684.

150 Miklitz, M., Turcani, L., Greenaway, R.L. et al. (2020). Computational discovery of molecular C60 encapsulants with an evolutionary algorithm. *Commun. Chem.* 3 (1): 10.

151 Henault, E.S., Rasmussen, M.H., and Jensen, J.H. (2020). Chemical space exploration: how genetic algorithms find the needle in the haystack. *Peer J Physical Chemistry* 2 (e11): e11.

152 Jensen, J.H. (2019). A graph-based genetic algorithm and generative model/Monte Carlo tree search for the exploration of chemical space. *Chem. Sci.* 10 (12): 3567–3572.

153 Snoek, J., Larochelle, H., and Adams, R.P. (2012). Practical Bayesian optimization of machine learning algorithms. In: *Proceedings of the 25th International Conference on Neural Information Processing System*, vol 2, 2951–2959. Red Hook, NY: Curran Associates Inc. (NIPS'12).

154 Shields, B.J., Stevens, J., Li, J., Parasram, M. et al. (2021). Bayesian reaction optimization as a tool for chemical synthesis. *Nature* 590 (7844): 89–96.

155 Griffiths, R.R. and Hernández-Lobato, J.M. (2020). Constrained Bayesian optimization for automatic chemical design using variational autoencoders. *Chem. Sci.* 11 (2): 577–586.

156 Deshwal, A., Simon, C., and Doppa, J. (2021). Bayesian optimization of nanoporous materials. *Mol. Syst. Des. Eng.* 6 (12): 1066–1086.

157 Rouet-Leduc, B., Hulbert, C., Barros, K. et al. (2017). Automatized convergence of optoelectronic simulations using active machine learning. *Appl. Phys. Lett.* 111 (4): 043506.

158 Deringer, V.L., Bartók, A.P., Bernstein, N. et al. (2021). Gaussian process regression for materials and molecules. *Chem. Rev.* 121 (16): 10073–10141.

159 Fujiwara, Y., Yamashita, Y., Osoda, T. et al. (2008). Virtual screening system for finding structurally diverse hits by active learning. *J. Chem. Inf. Model.* 48 (4): 930–940.

160 Warmuth, M.K., Liao, J., Rätsch, G. et al. (2003). Active learning with support vector machines in the drug discovery process. *J. Chem. Inf. Comput. Sci.* 43 (2): 667–673.

161 Eyke, N.S., Green, W.H., and Jensen, K.F. (2020). Iterative experimental design based on active machine learning reduces the experimental burden associated with reaction screening. *React Chem Eng* 5 (10): 1963–1972.

162 Elton, D.C., Boukouvalas, Z., Fuge, M.D. et al. (2019). Deep learning for molecular design: a review of the state of the art. *Mol. Syst. Des. Eng.* 4 (4): 828–849.

163 De Cao, N. and Kipf, T. (2018). MolGAN: an implicit generative model for small molecular graphs [Preprint]. arXiv:1805.11973v1 [stat.ML] (accessed 31 August 2021). Available from http://arxiv.org/abs/1805.11973.

164 Yu, Y., Si, X., Hu, C. et al. (2019). A review of recurrent neural networks: LSTM cells and network architectures. *Neural Comput* 31 (7): 1235–1270.

165 Goehring, T., Keshavarzi, M., Carlyon, R.P. et al. (2019). Using recurrent neural networks to improve the perception of speech in non-stationary noise by people with cochlear implants. *J. Acoust. Soc. Am.* 146 (1): 705.

166 Zhou, Z., Li, X., and Zare, R.N. (2017). Optimizing chemical reactions with deep reinforcement learning. *ACS Cent. Sci.* 3 (12): 1337–1344.

167 Dan, Y., Zhao, Y., Li, X. et al. (2020). Generative adversarial networks (GAN) based efficient sampling of chemical composition space for inverse design of inorganic materials. *Npj Comput. Mater.* 6 (1): 1–7.

168 Kim, S., Noh, J., Gu, G.H. et al. (2020). Generative adversarial networks for crystal structure prediction. *ACS Cent. Sci.* 6 (8): 1412–1420.

169 Grygorenko, O.O., Radchenko, D.S., Dziuba, I. et al. (2020). Generating multibillion chemical space of readily accessible screening compounds. *iScience* 23 (11): 101681.

170 Irwin, J.J. and Shoichet, B.K. (2005). ZINC: a free database of commercially available compounds for virtual screening. *J. Chem. Inf. Model.* 45 (1): 177–182.

171 Gottipati, S.K., Sattarov, B., Niu, S. et al. (2020). Learning to navigate the synthetically accessible chemical space using reinforcement learning. In: *Proceedings of the 37th International Conference on Machine Learning* (ed. H.D. Iiiand A. Singh), 3668–3679. PMLR. (*Proceedings of Machine Learning Research*, vol. 119).

172 Korovina, K., Xu, S., Kandasamy, K. et al. (2020). ChemBO: Bayesian optimization of small organic molecules with synthesizable recommendations. In: *Proceedings of the Twenty Third International Conference on Artificial Intelligence and Statistics* (ed. S. Chiappa and R. Calandra), 3393–3403. PMLR. (*Proceedings of Machine Learning Research*, vol. 108).

173 Bradshaw, J., Paige, B., Kusner, M.J. et al. (2019). A model to search for synthesizable molecules. In: *Advances in Neural Information Processing Systems* (ed. H. Wallach, H. Larochelle, A. Beygelzimer, et al.). Red Hook, NY: Curran Associates, Inc.

174 Ertl, P. and Schuffenhauer, A. (2009). Estimation of synthetic accessibility score of drug-like molecules based on molecular complexity and fragment contributions. *J. Cheminform.* 1 (1): 8.

175 Coley, C.W., Rogers, L., Green, W.H. et al. (2018). SCScore: synthetic complexity learned from a reaction corpus. *J. Chem. Inf. Model.* 58 (2): 252–261.

176 Baba, Y., Isomura, T., and Kashima, H. (2018). Wisdom of crowds for synthetic accessibility evaluation. *J. Mol. Graph. Model.* 80: 217–223.

177 Bonnet, P. (2012). Is chemical synthetic accessibility computationally predictable for drug and lead-like molecules? A comparative assessment between medicinal and computational chemists. *Eur J. Med. Chem.* 54: 679–689.

178 Shen, Y., Borowski, J.E., Hardy, M.A. et al. (2021). Automation and computer-assisted planning for chemical synthesis. *Nature Reviews Methods Primers* 1 (1): 1–23.

179 Genheden, S., Thakkar, A., Chadimová, V. et al. (2020). AiZynthFinder: a fast, robust and flexible open-source software for retrosynthetic planning. *J. Cheminform.* 12 (1): 70.

180 Mo, Y., Guan, Y., Verma, P. et al. (2020). Evaluating and clustering retrosynthesis pathways with learned strategy. *Chem. Sci.* 12 (4): 1469–1478.

181 Schwaller, P., Petraglia, R., Zullo, V. et al. (2020). Predicting retrosynthetic pathways using transformer-based models and a hyper-graph exploration strategy. *Chem. Sci.* 1 (12): 3316–3325.

182 Klucznik, T., Mikulak-Klucznik, B., McCormack, M.P. et al. (2018). Efficient syntheses of diverse, medicinally relevant targets planned by computer and executed in the laboratory. *Chem* 4 (3): 522–532.

183 Zhu, G., Liu, Y., Flores, L. et al. (2018). Formation mechanisms and defect engineering of imine based porous organic cages. *Chem. Mater.* 30 (1): 262–272.

184 Briggs, M.E. and Cooper, A.I. (2017). A perspective on the synthesis, purification, and characterization of porous organic cages. *Chem. Mater.* 29 (1): 149–157.

185 McMahon, D.P., Stephenson, A., Chong, S.Y. et al. (2018). Computational modelling of solvent effects in a prolific solvatomorphic porous organic cage. *Faraday Discuss.* 211 (0): 383–399.

186 Gao, H., Struble, T.J., Coley, C.W. et al. (2018). Using machine learning to predict suitable conditions for organic reactions. *ACS Cent. Sci.* 4 (11): 1465–1476.

187 Hasell, T., Culshaw, J.L., Chong, S.Y. et al. (2014). Controlling the crystallization of porous organic cages: molecular analogs of isoreticular frameworks using shape-specific directing solvents. *J. Am. Chem. Soc.* 136 (4): 1438–1448.

188 Mennen, S.M., Alhambra, C., Allen, C.L. et al. (2019). The evolution of high-throughput experimentation in pharmaceutical development and perspectives on the future. *Org. Process. Res. Dev.* 23 (6): 1213–1242.

189 Hamilton, P., Sanganee, M.J., Graham, J.P. et al. (2015). Using PAT to understand, control, and rapidly scale up the production of a hydrogenation reaction and isolation of pharmaceutical intermediate. *Org. Process. Res. Dev.* 19 (1): 236–243.

190 Cui, P., McMahon, D.P., Spackman, P.R. et al. (2019). Mining predicted crystal structure landscapes with high throughput crystallisation: old molecules, new insights. *Chem. Sci.* 10 (43): 9988–9997.

191 Greenaway, R.L., Santolini, V., Pulido, A. et al. (2019). From concept to crystals via prediction: multi-component organic cage pots by social self-sorting. *Angew. Chem. Int. Ed. Engl.* 58 (45): 16275–16281.

192 Stein, H.S. and Gregoire, J.M. (2019). Progress and prospects for accelerating materials science with automated and autonomous workflows. *Chem. Sci.* 10 (42): 9640–9649.

193 Christensen, M., Yunker, L.P.E., Adedeji, F. et al. (2021). Data-science driven autonomous process optimization. *Commun. Chem.* 4 (1): 1–12.

194 Roch, L.M., Häse, F., Kreisbeck, C. et al. (2020). ChemOS: an orchestration software to democratize autonomous discovery. *PLoS One* 15 (4): e0229862.

195 Langner, S., Häse, F., Perea, J.D. et al. (2020). Beyond ternary OPV: high-throughput experimentation and self-driving laboratories optimize multicomponent systems. *Adv. Mater.* 3 (14): e1907801.

196 MacLeod, B.P., Parlane, F.G.L., Morrissey, T.D. et al. (2020). Self-driving laboratory for accelerated discovery of thin-film materials. *Sci. Adv.* 6 (20): eaaz8867.

197 Granda, J.M., Donina, L., Dragone, V. et al. (2018). Controlling an organic synthesis robot with machine learning to search for new reactivity. *Nature* 559 (7714): 377–381.

198 Jia, X., Lynch, A., Huang, Y. et al. (2019). Anthropogenic biases in chemical reaction data hinder exploratory inorganic synthesis. *Nature* 573 (7773): 251–255.

199 Kitson, P.J., Marie, G., Francoia, J.-P. et al. (2018). Digitization of multistep organic synthesis in reactionware for on-demand pharmaceuticals. *Science* 359 (6373): 314–319.

200 Burger, B., Maffettone, P.M., Gusev, V.V. et al. (2020). A mobile robotic chemist. *Nature* 583 (7815): 237–241.

201 Eckhoff, M. and Behler, J. (2019). From molecular fragments to the bulk: development of a neural network potential for MOF-5. *J. Chem. Theory. Comput.* 15 (6): 3793–3809.

11

Machine Learning-Aided Discovery of Nanoporous Materials for Energy- and Environmental-Related Applications

Archit Datar[1], *Qiang Lyu*[2], *and Li-Chiang Lin*[1,3]

[1] *William G. Lowrie Department of Chemical and Biomolecular Engineering, The Ohio State University, Columbus, Ohio, USA*
[2] *School of Materials Science and Engineering, China University of Petroleum (East China), Qingdao, Shandong, China*
[3] *Department of Chemical Engineering, National Taiwan University, Taipei, Taiwan*

11.1 Introduction

11.1.1 Nanoporous Materials

Nanoporous materials are materials with pore sizes similar to that of common gas molecules (i.e., <2 nm) [1]. They can be classified as crystalline or non-crystalline, where non-crystalline materials include activated carbons, single-walled carbon nanotubes (SWCNTs), etc., while crystalline materials include metal-organic frameworks (MOFs), covalent organic frameworks (COFs), zeolites, etc. Crystalline materials have particularly drawn attention over the past decade. According to the data from Web of Science, the number of publications on zeolites, MOFs, and COFs has been increasing every year over the past decade (Figure 11.1). An interesting feature of these materials is that they can be synthesized with a large porosity and a high internal surface area, with the highest reported to date being DUT-60 having a surface area of 7800 m^2/g [2]. The porosities and surface areas of these materials have substantially increased in the last few decades [3]. Another interesting aspect of these materials is their large tunability. For instance, in the case of MOFs, a variety of metal clusters can be combined with a multitude of organic linkers to result in a wide range of materials. This has led to the synthesis of over tens of thousands of MOFs with many more hypothesized. The Computation-Ready Experimental (CoRE) MOF database [4, 5] reports ~14,000 synthesized materials, while the BW-DB hypothetical MOF database reports ~300,000 materials [6]. Similarly, over 570 experimental COFs have been reported and their structures are made available for computation [7, 8]. While approximately 250 zeolites with distinct topologies have been reported and included in the International Zeolite Association (IZA) database [9], orders of magnitude more have been predicted [10]. Thus, as we can expect, studying these structures purely experimentally could likely be prohibitive, thus necessitating the need for large-scale and efficient screening approaches.

11.1.2 History and Development

The first use of porous materials dates back at least to World War I, when Edward Harrison developed the large box respirator (LBR) containing activated charcoal to protect soldiers from chemical warfare [11]. In the 1940s, extensive work was done on the crystalline inorganic zeolitic frameworks and aluminosilicates gels [12, 13]. An interesting development in this field was the combination of metal clusters with organic linkers to form crystalline networks of MOFs in the 1990s by Yaghi et al. with the synthesis of the earliest MOFs such as MOF-2 and MOF-5 [14, 15].

AI-Guided Design and Property Prediction for Zeolites and Nanoporous Materials, First Edition. Edited by German Sastre and Frits Daeyaert.

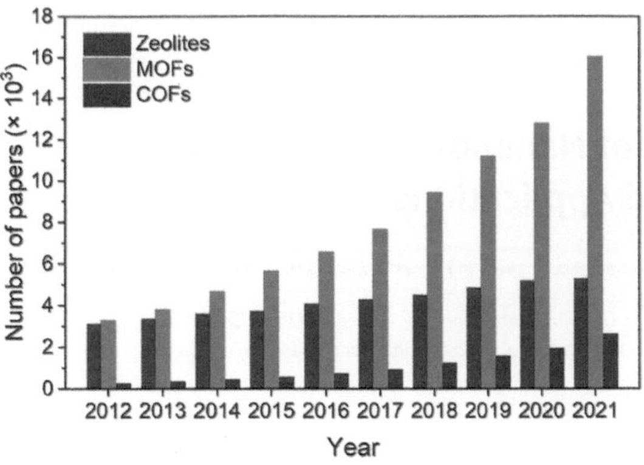

Figure 11.1 The number of zeolite-, MOF-, COF-related papers published each year from 2012 to 2021. Web of Science data on 21 March 2022.

More recently, the synthesis of COFs was achieved for the first time in 2005 with a two-dimensional (2D) form (COF-1 and COF-5) and in 2007 with a three-dimensional (3D) form (COF-105 and COF-108) [16, 17]. A compilation of the development of nanoporous materials is provided by Bennett et al. [13], while a detailed review of the development of MOFs and COFs has been provided by Freund et al. [18]. MOFs and COFs together, commonly known as reticular materials, have many interesting and desirable properties, as outlined by Freund et al. [18]. These are: i) precise control over topologies; ii) high porosity and tunable pore sizes; iii) ability to perform post-synthetic modifications; iv) (multivariate (MTV)) structures to create heterogeneity in the materials; v) easy characterization by X-ray diffraction; and vi) scalable synthesis. Their development is now paving the way for commercialization with over 27 companies working on MOF synthesis at the pilot scale [19].

11.1.3 Gas Separation and Storage Applications

Gas separation is the process to transform a gaseous mixture into its constituent components, while gas storage is the process of accumulating a gas and later releasing it for a particular purpose [1]. An example of a separation process is Xe/Kr gas separation, while an example of gas storage can be storing methane in the fuel tank of a vehicle [20]. While there are several methods to effect such separation or storage operations, porous materials have drawn increasing attention as promising candidates due to their aforementioned properties [21]. The large surface areas and porosities enable the materials to store large amounts of gases, while the ability to finely control their topology can provide high selectivity. Examples of separations using nanoporous materials include separating N_2/O_2 in air, N_2/CH_4, noble gases such as Kr/Xe, and CO_2 from flue gases [22–27]. Similarly, gas storage in nanoporous materials has been studied for energy-related applications such as H_2 storage [28], CH_4 storage [29], and storage of other hydrocarbons [30–32], for medical applications such as NO [33, 34], CO [35], and O_2 storage [36], as well as environmental applications such as CO_2 [37], SO_2 [38–40], and NH_3 storage [41]; the readers are referred to extensive reviews on its various applications [1, 42, 43].

11.1.4 Large-Scale Computational Screening for Gas Separation and Storage

While the details of each operation vary, separation and storage operations are often performed using temperature/pressure swing adsorption (T/PSA). In these applications, we try to exploit the differences between the adsorption behavior of the adsorbent material at different T/P conditions to achieve the separation or storage [44]. In general, for optimal performance, the material must adsorb large amounts of the desired component at the adsorption condition and be able to desorb it at the desired condition to regenerate itself. The difference between the quantities adsorbed at the adsorption and desorption conditions are known as the deliverable capacity or working capacity. In most applications, the adsorption conditions are set in advance as part of the process. In some applications, the desorption conditions can be variable. Thus, if the desorption condition is fixed, the material must have low uptake at the desorption condition and if it is variable, milder regeneration conditions are desirable so that low-quality energy sources may be used. In any case, the overall performance of a separation/storage process can be highly sensitive to the adsorption properties of the adsorbent materials used, since it can significantly affect the process energetics and ultimately the process costs. Thus, having optimal materials is necessary for the viability of many separation and storage processes. Given the large materials space, as mentioned above, computational methods can represent a promising means for large-scale screening of materials to find the optimal ones.

A common strategy for screening tens of thousands of materials for a potential application has been to use computational approaches to compute geometric and energetic properties such as largest cavity diameter (LCD), pore volume, void fraction, and accessible surface area calculated by geometry-based analysis methods, as well as energetic parameters such as heats of adsorption and Henry's constants computed through employing Monte Carlo simulations [45–49]. For instance, one of the earliest large-scale screening studies was conducted by Lin et al. [50] to identify zeolites and zeolitic imidazolate frameworks (ZIFs) for carbon capture and storage having the so-called lowest parasitic energy. In this work, they computed Henry's coefficients and heats of adsorptions for hundreds of thousands of structures via an efficient algorithm which makes use of graphics processing units (GPUs).

However, there are limitations to such a brute force approach. As noted above, the tunable nature of materials such as MOFs can result in a nearly infinitely large materials space, and therefore computing relevant properties may be excessively demanding. Furthermore, for simpler molecules such as CH_4 or CO_2 (at high temperature), some relatively easy-to-compute, geometric and/or energetic properties may be related to the performance parameter of interest. Using carbon capture (i.e., separating CO_2 from a CO_2/N_2 mixture) as an example, the materials' Henry coefficient of CO_2, a relatively easy-to-compute energetic property, has been shown to correlate strongly with their separation performance (i.e., lower energy requirement [25, 50]) or smaller global warming potential from a lifecycle point of view [51]). However, this may not always hold for more complicated molecules such as water. In fact, this was recently observed during the study of MOFs for water harvesting from air by Datar et al. [52, 53]. In that case, geometric and energetic properties were unable to account for the complex adsorption behavior of water in MOF structures due to which some interesting materials were not highlighted in screening. For such cases, more complicated simulations may be necessary which may take up to tens of days per structure, making large-scale computations more challenging. Thus, data-driven methods are being increasingly explored so that calculations done only for a fraction of possible structure candidates may be used to develop insights and predictions for the behaviors of large sets of structures.

In the following sections, the concepts and background for data-driven approaches will be first introduced, including dimensionality reduction and machine learning (ML) models. Second, various aspects of data-driven approaches such as nanoporous structure datasets, identification of the materials feature space for screening studies, methods to search for optimal structures, and modeling interatomic and intermolecular interactions will be discussed. Finally, case studies concerning post-combustion CO_2 capture, methane storage, and hydrogen storage will be specifically reviewed.

11.2 Concepts and Background for Data-Driven Approaches

Before discussing some of the data-driven approaches commonly used for materials screening, we deem it insightful to first develop some intuition about them. This section is not intended to provide a rigorous foundation for the associated statistics and ML techniques, as there are many excellent resources available such as *An Introduction to Statistical Learning* by James et al., which is a widely read textbook [54]. This section is meant to provide an intuition of these to the readers who may not be well-acquainted with them. Readers having a background in statistics and ML are encouraged to skip to the next section.

11.2.1 Dimensionality Reduction

A nanoporous material can be characterized using a large number of features when being analyzed. This is especially true of computational analysis since a lot of information about a material can be extracted, particularly if localized features (local energy, local density, etc.) are used. Dimensionality reduction then becomes key for many reasons, and a critical one is the compression of the data. If the data have features that do not particularly add any new information, i.e., redundant features, having them in the dataset increases the costs of data storage and further computation. This can often occur when the features are correlated with each other; for instance, when a feature can be easily expressed as a linear combination of other features, an example of which is when the same quantities are presented in different units. Furthermore, dimensionality reduction can help map materials to a lower dimensional space in which they can be analyzed. One such application is the development of a recommender system to connect materials with applications to which they may be suited, as is further discussed in Section 11.3.

Singular value decomposition (SVD) and principal component analysis (PCA) are some of the most popular methods often employed in dimensionality reduction. While their math is related, their application is somewhat different. Let $\mathbf{X}$ be a centered (the mean of column is 0) $n \times p$ matrix where n is the number of samples and p the number of features considered. The idea behind SVD is that this matrix containing the sample and feature data can be written as shown in Equation (11.1), where $\mathbf{S}$ is a diagonal matrix containing singular values of the matrix arranged in descending order. The larger singular values indicate the more important directions in the data, while the smaller ones indicate the less important ones. A singular value of zero indicates that the information held in the corresponding direction is redundant and can be ignored. While ignoring smaller singular values can result in a loss of information, it can lead to significant gains in computational efficiency because a compressed form of the matrix can be processed, making the processing of data much more computationally efficient.

$$\mathbf{X} = \mathbf{USV}^{\mathrm{T}} \tag{11.1}$$

A related approach of dimensionality reduction is PCA. In this approach, the variance matrix is factored out, as shown below in Equation (11.2). In linear algebra, the variance of a matrix $\mathbf{X}$ is written as $\mathbf{X}^{\mathrm{T}}\mathbf{X}/(\mathrm{n}-1)$, which is a square matrix. Substituting Equation (11.1) into this expression, we can obtain a relation between SVD and PCA.

$$\mathbf{X}^{\mathrm{T}}\mathbf{X} = (\mathbf{USV}^{\mathrm{T}})^{\mathrm{T}}(\mathbf{USV}^{\mathrm{T}}) = \mathbf{VSU}^{\mathrm{T}}\mathbf{USV}^{\mathrm{T}} = \mathbf{VS}^2\mathbf{V} = \mathbf{V\Lambda V}^{\mathrm{T}} \tag{11.2}$$

The $\mathbf{V}$ matrix denotes the eigenvectors of the variance matrix of $\mathbf{X}$, while $\mathbf{\Lambda}$ denotes its eigenvalues. The data can be transformed to a different coordinate system formed by the eigenvectors. The location of the data in this new system defined by the eigenvectors can be given by the score matrix computed as $\mathbf{T} = \mathbf{XV}$. This provides an intuitive way to think about the spread of the data in $\mathbf{X}$. If we were to visualize the data in terms of the eigenvectors, the directions along which the scores of the data are more spread out are considered to be more important as they explain more of the variance in the data. The less important directions (smaller eigenvalues) can be ignored, reducing the dimensionality of the data. The key difference between SVD and PCA is that PCA requires an extra computational step to calculate the score as compared to SVD. However, this allows the intuitive visualization of data which is not possible in SVD. Thus, SVD is typically used for data compression, while PCA is used for data visualization.

11.2.2 Machine Learning Models

ML, a data-driven approach, has recently draw considerable attention in the materials community in general, including the nanoporous materials community. ML, basically, can detect patterns in the data which can be very useful in the case of nanoporous materials design and analysis. There are many types of ML models with applications such as developing structure–property relationships and identifying groups of similar structures, among many others. While there may be many models suited to these applications, in this chapter, we briefly introduce the concepts behind some of the more widely used methods in the materials discovery literature.

11.2.2.1 Linear Models

One of the most familiar concepts in statistics is the ordinary least squares regression (OLS). In this method, predictor and response variables are assumed to be linearly related through a set of parameters. These parameters are estimated by minimizing the sum of squares of the errors between the predicted and true values, as indicated in procedure (11.3). The training data ($\mathbf{X}$) can be represented as an $n \times p$ matrix, as discussed in the previous section, and the corresponding outputs are shown by the vector y, while the linear parameters are represented using the vector β.

$$\begin{aligned} & y = \mathbf{X}\beta + \epsilon \\ & \underset{\beta}{\text{minimize}} \left[(y - \mathbf{X}\beta)^T (y - \mathbf{X}\beta) \right] \end{aligned} \tag{11.3}$$

While this familiar concept is fairly straightforward to interpret, there are important assumptions that the model makes about the underlying distribution of the data. Linear regression is a classic example of a parametric method due to these, which must be kept in mind while applying this approach: i) *Linearity*: there exists a linear relationship between the predictors and the responses; ii) *Homoscedasticity*: all responses are drawn from a single distribution, thus their variance is the same; iii) *Normality*: all responses are drawn from a normal distribution; and iv) *Independence*: the data points are independent of each other. The foregoing assumptions must be tested when this model is applied to the data. Another key limitation of this method is that it becomes unstable when the predictor variables are correlated. This means that the predicted coefficients can drastically vary if some values are slightly changed, making the interpretation of the predicted parameters unreliable. Thus, it is necessary to test the collinearity of the various features before the model is trained and interpreted. Furthermore, an important operational concern is the scales of the various features involved. Since the β values depend on the magnitudes of features, the ones with larger values may lead to larger associated coefficients. The feature must therefore be scaled to ensure that the more important variables are assigned larger coefficients. As a result, the linear models, while simple to understand, require some pre-processing of the data and understandings to interpret the results accurately. While this chapter does not delve into the procedural details of these methods, the readers are referred to many excellent resources such as the textbook by Ramsey and Schafer [55]. Furthermore, Hollander et al. present an account of linear methods when the foregoing assumptions are relaxed [56].

A special problem also arises when the number of features is comparable to the number of observations. This can cause the model to overfit the training data and underperform on the test set. A solution to such a problem is to use regularization or penalize the parameters. While these penalties can take many forms, the squares and moduli are the two most common with their associated methods being linear absolute shrinkage and selection operator (LASSO) regression and ridge regression, respectively. With these, the loss function can be written as shown in Equation (11.4).

$$\underset{\beta}{\text{minimize}}\left[(y-\mathbf{X}\beta)^T(y-\mathbf{X}\beta)\right]+\lambda P(\beta) \tag{11.4}$$

where $P(\beta)=|\beta|$ for LASSO and $P(\beta)=\beta^T\beta$ for ridge regression, while λ is the tuning parameter. The key difference between these is that the LASSO can set the importance parameter for certain features to zero, unlike ridge regression.

11.2.2.2 Decision Trees and Random Forests

A decision tree is a sequence of decisions to divide the data based on certain features in order to optimize the target function. Decision trees can be applied to both classification and regression problems and the target function is decided accordingly. In this section, we will develop the intuition for classification problems, but it can be easily extended for regression problems. Let us consider the data matrix **X** from the previous section. Assuming that the goal is to train a decision tree to classify each sample as class Y or N, a decision tree would employ a top-down, greedy classification algorithm. More specifically, at the first decision node, the feature space is searched to identify the point at which splitting the data would result in the best separation. Once this is identified and the data divided accordingly, this step is subsequently repeated until the resulting split of the data have fewer than a certain number of observations, or the tree reaches a certain depth. This creates an algorithm for classifying the data as either Y or N.

Decision trees have several advantages. Noticeably, they are intuitive and easy to explain as they can be visualized and interpreted as an algorithm. Additionally, they can easily provide a notion for the most important variable in the feature space. However, a decision tree can overfit the data on which it is trained; i.e., the decision rules chosen by the *greedy* approach at each step can fit to the noise in the data rather than the trend. While techniques such as bagging and pruning can be used to mitigate this problem, a widely used alternative is the random forest model. A random forest is a collection of decision trees. A key difference is that in a single decision tree, we have the choice to split the data based on any feature, while for a random forest, this choice is restricted. The subset of features from which we can choose the decision boundary are chosen randomly from the set of all features which ensures that the trees created are different from each other. Thus, an expectation is that the random forest will not overfit to the training data. The predicted class of a sample is then decided by taking a majority vote among the trees in the forest.

Like decision trees, random forests can also provide variable importance scores based on how much the classification improved when the data were split using a certain feature. This informs which features in the data are important. Decision trees and random forests are nonparametric methods; i.e., they do not assume that the sample data are drawn from a certain distribution, unlike some other methods such as linear regression. Moreover, decision trees as such can produce complicated decision boundaries which need not be linear or quadratic, etc. These attributes make tree-based methods widely applicable, especially if the exact distribution of the data is not known.

An important concept gaining popularity in the literature associated with tree-based methods is boosting. While boosting is a general method which can be applied to a variety of ML models, it is most widely used for tree-based methods and is therefore discussed here. Boosting is similar to a random forest in that there are many trees involved. However, in a random forest, the trees are grown independently with the variables along which the feature space is split randomly restricted. On the other hand, when boosting is used, the trees are grown sequentially and each tree is grown to fit the residuals from fitting the previous tree model to the data. The advantages of boosted trees include better performance as compared to non-boosted models such as random forest models. Furthermore, they allow the use of trees with a lower depth (lower complexity) which means better interpretability. For instance, a tree depth of 1 means that it is an additive model.

11.2.2.3 Support Vector Machine

The Support Vector Machine (SVM) is a widely used ML model and has a wide range of applications. Again, the concept in the context of a binary classification will be discussed in the following, but the extension to regression is straightforward. The basic concept is that the decision boundary (plane which separates the classes) is computed such that its distance from the points closest to it; i.e., the support vectors, is maximized. The points that are misclassified with this decision boundary are assigned penalties which are accounted for in the loss which is then minimized. This idea is inherently different from a linear discriminant analysis (LDA), which is related to ordinary least squares linear regression, because LDA considers all data points for computing the decision boundary, while SVM only considers a subset of these points. Thus, even if the data points not on the margin were to be slightly changed, the resulting decision boundary would not be affected.

An important property of SVM is that the predictor function depends only on the distance of a point x from the support vectors (x_i). Details of this calculation are not discussed here. In linear coordinates, this can be represented as a dot product between the two ($\langle x, x_i \rangle$). However, more generally, it can be represented using a kernel $K(x, x_i)$, which can take the form of a linear, polynomial, radial, or exponential kernel. These choices correspond to decision boundaries of various shapes, and the nature of the expected decision boundary should dictate the choice of the kernel used. The great flexibility offered by the SVM has made it a useful algorithm.

11.2.2.4 Neural Networks

Neural networks are a class of ML models that are used in a wide range of science and engineering disciplines. Originally inspired by the working principle of the human brain, the neural networks relate a certain input to an output via a sequence or neurons "firing" or becoming activated. The activation of the neurons is controlled by the weights assigned to them, which are trained using data (training set). A practical attribute associated with neural networks is their ability to scale with the amount of data. It has been found that as the amount of training data increases, larger neural networks with increasing performance can be built, unlike other models such as SVMs, linear models, etc., where the performance plateaus out after a certain number of data points are reached. Of course, even in such models, additional features can be engineered, but often the user has limited knowledge about the effectiveness of these features to improve performance. On the other hand, neural networks can learn complex features from the data. This virtue of neural networks can become particularly useful when the response depends on complex relationships between input variables, which are not established such as in the case of image or speech data.

However, an important consideration with neural networks is that a large number of parameters are involved. For a vanilla L-layered dense neural network, with the lth layer having n^l neurons, the number of parameters to be trained is $\sum_{l=1}^{L} n^l(n^{l-1}+1)$, where $n^0 = p$ or the number of features in the input data as discussed before. Thus, even for a moderately-sized neural network, this can result in a large number of parameters which can lead to overfitting. The number of layers and number of neurons in each layer, together referred to as the neural network architecture, along with the number of iterations for training and the learning rate (how much the parameters are updated at each iteration), comprise the hyperparameters for a neural network. It is often observed that the performance of the neural network is strongly dependent on its hyperparameters and that optimizing the hyperparameters can lead to significant improvements in performance [57]. Thus, neural networks are particularly useful when the response depends on complex relationships between the predictor variables and when the network is designed and hyperparameters selected to ensure that overfitting is mitigated.

11.2.2.5 Unsupervised Learning

The foregoing discussion (Sections 11.2.2.1–11.2.2.4) is based upon methods in which the data are labeled, i.e., there is a target variable and its values are known for the data at hand. The techniques are part of a class known as supervised ML algorithms. In these, the model is trained to reproduce these known target values, and its performance is assessed on how well it reproduces these. However, there are situations in which the target variable is not available, or its values are not known. In such situations, a class of methods known as unsupervised learning methods, can still provide valuable insights from the available data.

A widely used example of unsupervised learning is dimensionality reduction problems, including PCA as described in Section 11.2.1, which enables us to determine the important directions along which data varies. Another important example of unsupervised learning is data clustering. Often, it is necessary to understand which data points in the sample are "similar" to and which ones are "different" from each other. Using this, we can understand how the data are distributed and determine if the data form distinct groups. An interesting extension of this idea is anomaly detection, which aims to understand if a given data point is very different from the rest of the sample. Knowing this can be extremely useful in fault detection and the detection of unusual and fraudulent activities, among other applications.

In general, unsupervised learning methods are used more in exploratory data analysis to understand the data. Furthermore, because the ground truth about the data is unknown, unsupervised learning is more subjective than supervised learning methods, since the results cannot be checked against any known values. However, they find use in a wide variety of applications.

11.3 Data-Driven Approaches

Data-driven approaches have the potential for studying materials on a large scale and for providing insights into their development. While there is a great diversity of research in many different areas, we highlight a few trends observed in the literature where these approaches can have a great impact. We describe these in the following sections and discuss a few case studies involving the application of data-driven methods for the large-scale screening of materials for the adsorption of gases.

11.3.1 Nanoporous Structure Datasets

A key component in the large-scale screening of nanoporous structures is the availability of a large number of structures in a computationally usable form. The information commonly required for such analyses is the atom types and their location in the structure. From the interatomic distances, their bonding can usually be inferred. However, in some cases, it can also be separately specified. For structures containing many molecules in the unit cell, it may become cumbersome to specify the location of each atom separately and wherever possible, the symmetry in the structure may be exploited for the structure to be visualized and used in simulations. The standard format of specifying the above-mentioned information is the crystallographic information framework (CIF) format, which was adopted by the International Union of Crystallography (IUCr) in 1991 and has been made available on their website [58–60].

The standardization of the file format has enabled the development of large databases of nanoporous materials. One of the earliest such databases was the Database of Zeolite Structures by the IZA Structure Commission (IZA-SC) [9]. First published in 1996 at ETH Zürich, it was based on two books, *Atlas of Zeolite Structure Types* and *Collection of Simulated XRD Powder Patterns for Zeolites* [61, 62]. While the original version contained only 96 zeolite framework types, the latest version to date contains 255 framework types [9]. As also noted, hypothetical zeolite databases containing millions of zeolites, such as that by Deem et al., have been reported in the literature [10, 63]. A large number of them have also been predicted to be thermodynamically accessible. However, the synthesis of zeolites has been a grand challenge. For instance, it involves the use of proper structural direction agents (SDAs) to help guide the synthesis. As such, the number of currently reported zeolites is much fewer than those that have been predicted.

Several hypothetical MOF datasets have also been reported in the literature, with the first being the hypothetical MOF (hMOF) database generated using the "Tinkertoy" algorithm by snapping MOF building blocks for 130,000 structures reported by Wilmer et al. [64]. Subsequently, the ToBaCCo database, comprising 13,000 structures, was reported by Gómez-Gualdrón et al., employing topology-based algorithms to generate the structures [65]. Boyd and Woo also used a topology-based algorithm to develop the BW-DB database with over 300,000 structures [6, 66]. While the details of these approaches are discussed in a detailed review by Boyd et al. [67], the central idea behind the topology-based approaches is to use pre-defined topologies (placement of nodes and edges in the topological network) and fit the secondary building units (SBUs) to them. That is, hypothetical MOF databases are typically constructed by using building blocks derived from already synthesized materials to enumerate all possibilities. This, as expected, results in a very large number of materials. Aside from these hypothetical databases, the synthesized MOFs are reported in the Cambridge Structural Database (CSD) [68]. This database contains over 90,000 MOFs. Importantly, experimentally refined crystal structures reported in the CSD may include solvent molecules and partially occupied/disordered atoms. Such features must be removed prior to computational simulations of fully activated, solvent-free structures. Goldsmith et al., who developed a graph-labeling algorithm for the removal of solvents, identified some mislabeled structures, and repaired some structures manually to develop a database of ~22,700 structures (~4000 with nontrivial internal porosity) [69]. Further improvements were made in the CoRE MOF database to provide 4764 materials. These improvements included the removal of solvents from unsaturated metal atoms, retention of charge-balancing ions to ensure neutrality, and retention of interpenetrated structures [4]. The CoRE MOF database was further updated in 2019 to constitute over 14,000 experimentally synthesized structures [5].

It should be noted that, while these databases can provide a large number of structures (over 500,000), it has been pointed out that they may not necessarily yield a diversity of structures

[64, 67]. For instance, the hMOF database with over 130,000 structures contains only 6 topologies [45, 70]. These biases in the chosen databases can lead to incorrect conclusions in large-scale screening studies. Moosavi et al. [64] illustrated this point by studying the prediction of low pressure CO_2 adsorption (0.15 bar) as well as CH_4 deliverable capacity (between 65 bar and 5.8 bar pressure) using three datasets: CoRE MOF 2019 [5], BW-20K [6, 66] (a subset of 20,000 randomly selected structures from the BW-DB), and ARABG-DB [71]. They trained random forest regression models on geometric and chemical descriptors to predict these quantities. The SHapley Additive exPlanations (SHAP) values for each of these predictors shown in Figure 11.2 are significantly different for these three datasets. For instance, in the study of CO_2 adsorption at low pressure, the metal chemistry appears as the most important factor for the CoRE MOF database, while pore geometry appears to be more important for the BW-DB. The functional groups appear to be more important for the ARABG-DB. Thus, they conclude that the underlying distributions of these datasets are inherently different leading to these inconsistent results. In this work, they also propose a framework incorporating MOF chemistry and geometric properties to quantify the diversity of the database being used. Thus, it is important to have databases for high-throughput screening of materials, and conclusions drawn through such studies must account for the diversity (or lack thereof) of the datasets used. Important developments in this direction have been discussed in this section.

11.3.2 Identifying Feature Space of Materials to Screen

A critical aspect of large-scale materials screening is the choice of features used. Due to the availability of computational methods, it is possible to utilize many different features and the choice of features ultimately used will depend on the property that is being modeled and predicted. A typical workflow of screening endeavors is to first compute the target property using high-fidelity molecular and/or *ab initio* calculations for an adequate number of structures, subsequently training a surrogate model to reproduce this property, employing this model to predict the target property for the desired dataset, then identifying candidates with the optimal value of the target property.

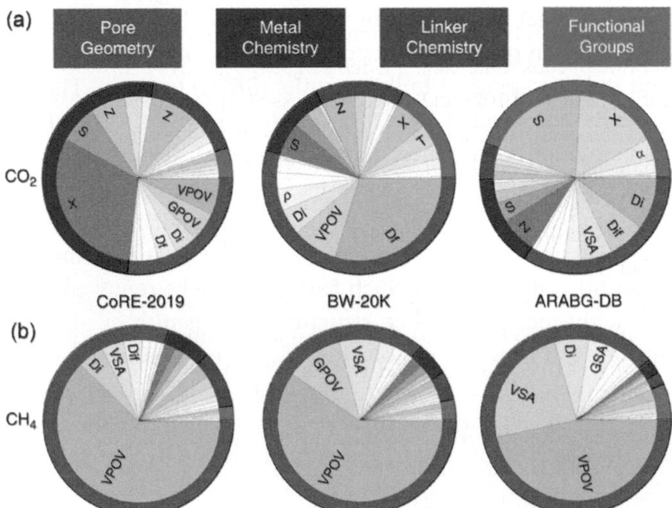

Figure 11.2 Database dependence of the importance of material characteristics or features. Pie charts showing the SHAP values (importance of variables) for (a) the low-pressure CO_2 adsorption and (b) CH_4 deliverable capacity. *Source:* From [64]. Springer Nature CC BY 4.0.

Data-driven approaches can therefore be useful if the target property is expensive to compute computationally, for instance, the adsorption step pressure of water adsorption in MOFs [52, 53, 72]. When the target property is expensive to compute, the number of structures for which it can be computed also tend to be limited. Therefore, it becomes critical to select features which can extract the relevant information from structures to build a useful model [73].

In the literature, simple geometric properties such as LCD, pore volume, void fraction, accessible surface area along with energetic parameters such as heats of adsorption and Henry's constants have been commonly employed in the screening of materials [45–49]. These features are attractive mainly for two reasons. First, they are intuitive and fairly easy to compute. The methods to compute those geometric features have been discussed in detail by Ongari et al. [74] and they have been implemented in, for example, the open-source Zeo++ software [75]. Energetic features such as heats of adsorption and Henry's constants are discussed in many sources, including the well-known book by Frenkel and Smit [76] and can be computed using, for example, the open-source RASPA package [77]. Second, some of these can be experimentally measured, thus providing a basis for comparison. However, despite their attraction, they may not always capture the information related to the target property and may be inadequate. For instance, in a recent study by Datar et al. [53], they attempted to study water adsorption in MOFs which is known to proceed through the formation of water clusters due to the extensive hydrogen bonding. The target property was the adsorption step relative pressure; i.e., the relative humidity at which the uptake is 50% of the maximum uptake. It was observed that, in general, energetic features such as heats of adsorption could predict the order of adsorption step pressure with larger heats of adsorption corresponding to a lower adsorption step relative pressure. Interestingly, however, some structures possessing high heats of adsorption also showed high adsorption step relative pressures. An instance of this occurred due to the heterogeneity in the structure; while the primary adsorption site was extremely favorable, the secondary and majority adsorption sites were relatively weak. Structural details such as this can often not be computed by an average property such as the heat of adsorption, and more advanced features are needed.

An alternative approach is to use measures of the interaction experienced by the guest molecule in the structure. Bucior et al. have demonstrated this for predicting H_2 deliverable capacity between 100 bar and 2 bar for over 50,000 experimentally synthesized MOFs from the Cambridge Crystallographic Data Centre (CCDC) database [78]. They sampled the energy landscape by overlaying a 3D grid on the simulation box and computing the energy experienced by the guest molecule (i.e., H_2 molecule in this work) at each grid point. These energies were then binned into a one-dimensional (1D) histogram and the frequency in each bin was used as the features. This was able to provide a simple yet descriptive parameter that could efficiently describe the adsorption chemistry of nanoporous materials. A LASSO model was then trained (on the training set), and the resulting model accurately predicted the deliverable capacity with a small root mean square error (RMSE) of 3 g/L on the test set for the given application. Recently, Cho et al. trained a convolutional neural network (CNN) to predict the adsorption isotherms of methane in zeolites [79]. Zeolite structures were represented by 3D voxels that contain geometric accessibility information. They found that the CNN trained can accurately reproduce the isotherm computed using grand canonical Monte Carlo (GCMC) simulations. While the results of both studies are promising, we note that these have been performed on single-site guest molecules such as methane. Further developments are required to apply these to more complicated molecules.

Another set of potentially informative features that has been explored in the literature involves graph-based feature descriptors such as the revised autocorrelations (RACs). RACs are correlations between atomic properties such as the Pauling electronegativity, nuclear charge, covalent radii, etc., on a graph. First introduced by Kulik et al. [80], RACs can be represented as shown in Equation (11.5). Here, the differences in the property P for each atom can be summed up for atoms i, j over the structure. In this equation, *start* refers to the atom from which we start computing this property while *scope* corresponds to the atoms over which this property is summed. Specifying these allows us to specifically sum these properties over specific regions of the framework such as the metal cluster. d_{ij} corresponds to the number of bonds between atoms i and j, while d corresponds to the desired depth over which the autocorrelation is to be computed. A higher d implies a more "global" average of the property, while a smaller d corresponds to a more local one. δ is the Kronecker's delta. The computation of the RACs is illustrated below in Figure 11.3.

$$a_{\text{scope}}^{\text{start}}\text{P}_{\text{d}}^{\text{diff}} = \sum_{\text{i}}^{\text{start}}\sum_{j}^{scope}(P_i - P_j)\delta(d_{ij}, d) \tag{11.5}$$

Moosavi et al. [64] used RACs for five atomic properties including atom identity (I), connectivity (T), Pauling electronegativity (χ), covalent radii (S), and nuclear charge (Z) to describe the chemistry of the MOF along with geometric features such as pore volumes and pore sizes to predict CO_2 uptake at 0.15 bar (low pressure) for MOF structures included in the CoRE MOF 2019 [5], BW-20K [6, 66] (a subset of 20,000 randomly selected structures from the BW-DB) and ARABG-DB [71] databases. For low pressure conditions, the chemistry of the structure is anticipated to become notably more important and indeed, for the CoRE MOF database, which is the more diverse, this was observed. Importantly, incorporating RACs as chemical features led to a Spearman rank correlation coefficient (SRCC) of above 0.83, indicating the ability of these features to capture useful chemical information and their usefulness in screening nanoporous materials.

In addition to the chemistry of the materials as discussed before, an interesting and intuitive feature to predict the adsorption performance of a nanoporous material is the size and shape of the

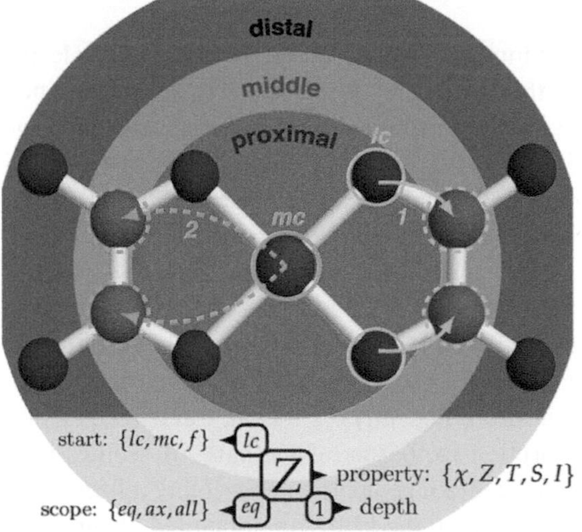

Figure 11.3 Schematic illustration of revised autocorrelations (RACs) in an iron octahedral complex represented in ball and stick (iron is brown, oxygen is red, and carbon is gray). Regions of the molecule used to classify descriptors are designated as proximal (metal and first coordination shell, in red), middle (second coordination shell, in green), and distal (third shell and beyond, in blue) throughout the text. Light green circles and arrows depict terms in a 2-depth RAC from the metal center (mc) and the light blue circles and arrows depict terms in a 1-depth RAC centered around the ligand (lc). *Source:* Reproduced from [80]. © 2017, American Chemical Society.

cages in the structures. While the foregoing methods can extract features from it for computation, the challenge remains to represent it in a lower dimensional space. This has been demonstrated by Sturlson et al. [81] for the case of 74 porous organic cages (POCs). While MOFs and COFs are extended networks of connected molecular building blocks by directional coordination and covalent bonds, respectively, POCs are the assembly/packing of porous organic cage-like molecules via non-covalent/non-coordination interactions [82]. They generate a 3D image of the cage using the presence/absence of framework molecules and embed it in a latent space using SVD. This space can be represented in terms of a few vectors which they refer to as "eigencages." They demonstrate that cages which are close in the latent space display similar Xe/Kr selectivity, indicating the ability of eigencages to depict the adsorption behavior of the cages.

We conclude this section by reiterating that the selection of features is important in the screening and development of nanoporous materials and that significant strides have been made in the literatures toward this end. This section does not aim to exhaustively introduce all possible features. To enhance the prediction accuracy, some ML works have further proposed some advanced features, and more details are discussed in those case studies presented in Section 11.4.

11.3.3 Methods to Search for Optimal Structures

In addition to the choice of features for screening and selection of optimal materials, another important aspect is the algorithms used for shortlisting. This becomes particularly important when high-fidelity simulations/experiments are expensive. The goal is to scan the materials space as efficiently as possible; i.e., perform the high-fidelity calculations for as few structures as possible, and as exhaustively as possible; and to predict performance for the most diverse set of materials to arrive at the optimal materials. A detailed review on this topic has been provided by Deshwal et al. [83].

The conventional approach to this problem has been one-shot supervised learning in which a set of structures is chosen, which is used to train a model to predict the desired property. This model is then applied to a larger set of structures and the best materials are predicted. The workflow of this approach is shown below in Figure 11.4, as summarized by Pollice et al. [84] and has been used by many researchers [73, 84–90]. While this approach has yielded promising results, it is limited by the training set chosen, as also mentioned; if the training set is not diverse, the conclusions can be inaccurate [64].

Another approach used has been that of genetic algorithms (GAs) [91]. In this approach, the features of candidates are assigned to them as chromosomes. An initial population of candidates is chosen, and each candidate is evaluated for its fitness to the task at hand. These candidates are

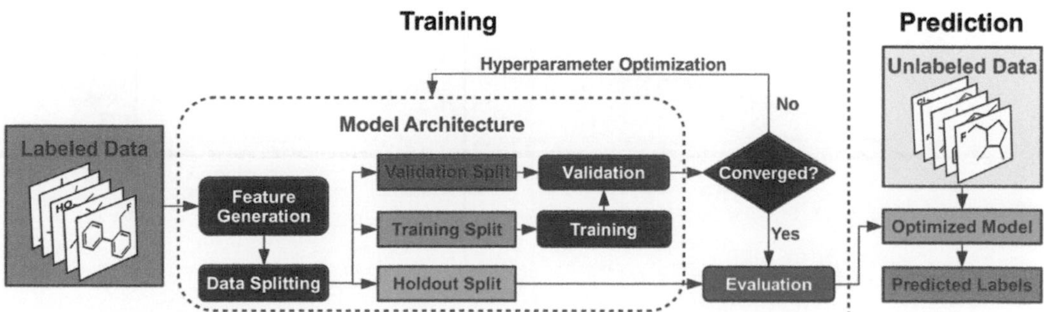

Figure 11.4 Workflow of one-shot supervised learning for optimal material selection. *Source:* Adapted from [84].

paired using the so-called tournament selection method [91] and a part of their chromosomes are exchanged through genetic operations and a new generation is created. This process is repeated until the process converges with the best candidate. The workflow of a typical GA can be seen from that reported by Chung et al. [92]. GAs have been used for a diverse range of applications such as screening MOFs for post-combustion carbon capture [93], methane storage [94], and finding optimal synthesis conditions [95]. While GAs are useful to find the optimal materials, they do not employ any models (structure-property relationship in this case). Thus, they can require many evaluations of the materials, thus making them inefficient.

This problem can be addressed using Bayesian optimization (BO) (See Figure 11.5 for the workflow). In this case, started with a prior idea of a model (called surrogate model), the structure and property of interest are related to provide the uncertainty with this prediction. An acquisition function quantifies the utility of evaluating a structure. Typically, structures which either have a high value of the desired property or a high amount of uncertainty are of interest when properly evaluated. For each structure that is evaluated, the surrogate model is updated with information about its structure–property relationship. This updated surrogate model is then used to select the next material that is to be evaluated using the acquisition function. This process is repeated until a structure with the optimal value of the desired property is found with a low amount of uncertainty. Deshwal et al. [83] applied BO to study methane adsorption in COFs. They found that the BO approach was able to identify the best material in 174 evaluations and the top 36% of the top 100 materials in 250 evaluations. They also found that the BO approach outperformed the one-shot supervised as well as the evolutionary search algorithms (specifically, the covariance matrix

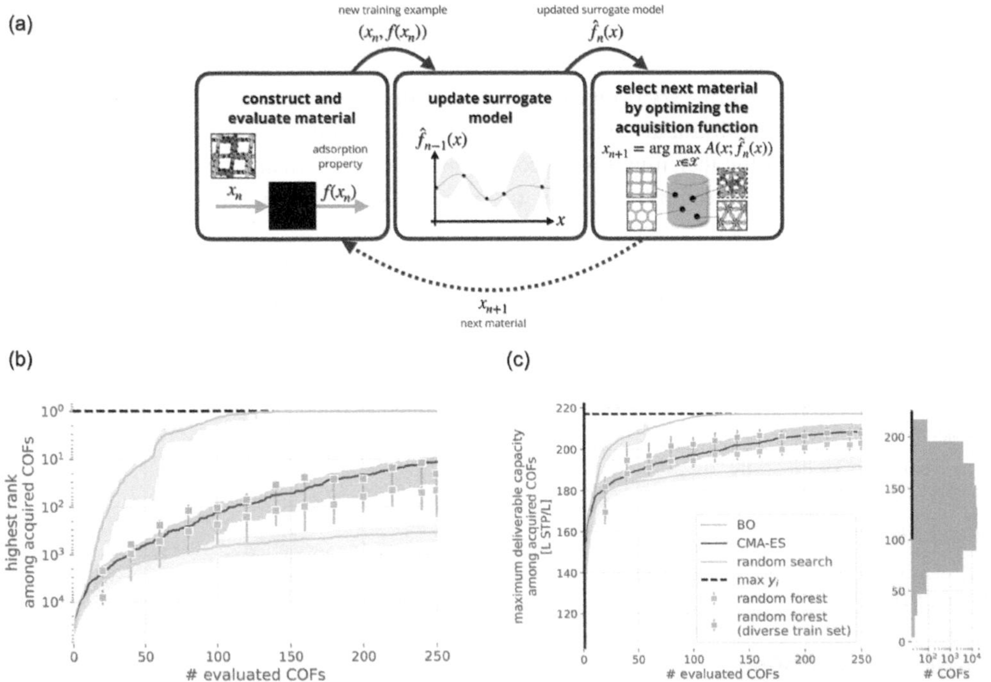

Figure 11.5 Bayesian optimization approach for materials search. (a) Workflow of the Bayesian optimization method. (b) and (c) Results for methane adsorption in the COF case study showing the ranks of the materials selected and their maximum deliverable capacity, respectively. *Source:* Figure reproduced from [83] Copyright (2021) Royal Society of Chemistry.

adaptation evolution strategy (CMA-ES)), as shown in Figure 11.5. The basic BO method suggests the evaluation of one structure at a time which can become cumbersome and disallow us from exploiting the power of parallel computation or experiments. However, evaluations can be performed in parallel using batch BO, a variant of this basic technique [96].

Overall, it is important to ensure that the method used to explore the materials is exhaustively and efficiently sampled. For cases where the researcher has an idea of the structure–property relationship, the BO method might be well-suited since it could balance the exhaustiveness of the search with its efficiency.

Furthermore, there are also applications in which it is necessary to estimate the applicability of a material for a certain application, as explored by Simon et al. [97]. In their work, they considered the dataset v9 on Materials Cloud [98] containing COFs and their reported applications. By "hiding" some of these data (i.e., treating them as blank), they developed models to predict those values by converting the material–property matrix to a low-rank matrix. Thus, they were able to recommend the best materials for a given application or, the best application for a given material. For a set of 16 properties they considered, their approach yielded a median Spearman rank correlation of >0.8 for the over 500 materials considered in their study. Thus, this approach could be easily translated to match materials with their applications.

11.3.4 Modeling Interatomic and Intermolecular Interactions

The application of accurate interatomic interactions is necessary to ensure that the simulations truly capture the behavior of gases in nanoporous materials. Some properties are more sensitive to these than others. For instance, we observed in a previous work that for water adsorption in a MOF, the pressure at which the isotherm shows a steep rise in uptake; i.e., the step pressure, can be strongly affected by the choice of the water model used in the simulation, while the maximum uptake remained relatively less affected by this choice [53]. Here, the various water models differ in their interaction parameters. For typical adsorption applications involving T/PSA, the step pressure is an important parameter, due to which the accurate modeling of interactions between atoms and molecules in the system is crucial.

Typically, the interactions have two components in the simulation: dispersion (van der Waals) interactions and Coulombic interactions. Many simulations in the literature employ generic force fields such as universal force field (UFF) [99] or Dreiding [100] to model the dispersion interactions [101]. The Coulombic interactions are modeled by assigning a set of partial charges to the atoms. The benchmark methods for assigning these charges to the framework atoms (host) involve electronic structure calculations to provide electronic density or electrostatic potential using methods such as density functional theory (DFT). We can assign point charges to the atoms to reproduce these using methods such as Mulliken [102], Hirschfield [103], iterative Hirschfield [104], and Bader [105] for the electronic densities and RESP [106], CHELPG [107], REPEAT [108], and DDEC for electrostatic potential [109–111]. The interaction parameters for guest molecules are often modeled separately and for a given molecule, a variety of models exist [112–114]. Due to the large number of atoms in typical nanoporous materials (often, a few hundred), assigning partial charges using the approaches discussed above often requires several hours on tens of cores, thus making it computationally expensive. For large-scale screening of materials, this may become impracticable. Due to this, many approximate methods to assign partial charges have been developed.

A well-known category of methods to this end are the charge equilibration (Qeq) methods such as those proposed by Rappé and Goddard [115]. We note that these are not ML methods in the strictest sense, but due to their extensive application in the literature, they are important to discuss. The general idea behind these methods is that the energy of an individual atom ($E_A(Q_A)$) is approximated by a Taylor series expansion centered around that of a reference partial charge ($E_A(Q_0)$) and involving the electronegativity (χ_A^0) and the electronic repulsion in the outer atomic orbital (J_{AA}^0) (also known as idempotential or atomic hardness), and as shown in Equation (11.6).

$$E_A(Q_A) = E_A(Q_0) + \chi_A^0 Q_A + \frac{1}{2} J_{AA}^0 Q_A^2 \tag{11.6}$$

The overall energy of the framework is given by the summation of these individual energies and their pairwise interactions, as shown in Equation (11.7).

$$E(Q_1, Q_2, \ldots, Q_N) = \sum_{A=1}^{N} (E_A + \chi_A^0 Q_A + \frac{1}{2} J_{AA}^0 Q_A^2 + \sum_{B>A}^{N} J_{AB} Q_A Q_B) \tag{11.7}$$

At equilibrium, the partial derivatives of the system energy with respect to the charges are equal (Equation (11.8) and the total charges on the molecule are constrained (Equation (11.9).

$$\chi_1 = \chi_2 = \ldots = \chi_N \tag{11.8}$$

$$Q_{tot} = \sum_{i=1}^{N} Q_i \tag{11.9}$$

The former condition is analogous to the equalization of chemical potentials during thermodynamic equilibria. Using these, we obtain a system of N equations with N unknowns, which can be solved exactly to provide the partial charges for the system. Several variants of this method have been proposed in the literature and a detailed review on this topic has been provided by Ongari et al. [101]. In this work, they used several variants to assign partial charges to a set of diverse MOFs and computed properties such as the H_2S heat of adsorption. They found that among these variants, the extended version of the Qeq (i.e., EQeq) method proposed by Wilmer et al. [116, 117] with the reference charges assigned to represent those of the common oxidation states and the parameters χ_A^0 and J_{AA}^0 obtained experimentally, provided the best agreement with the DDEC method with a Pearson coefficient of 0.756. They also noticed that these methods were often sensitive to the reference charges and parameters used, and further variants of the Qeq methods over the years did not provide a significant improvement in performance.

While the Qeq method and its variants seek to assign point charges to minimize the total energy of the MOF structure, an alternative approach has been the so-called connectivity-based atom contribution (CBAC) method first proposed by Xu and Zhong [118]. In this method, charges are assigned to framework atoms based on their bonding environment. They trained their model on 30 MOFs and validated them on 13 MOFs to reproduce the quantum mechanical (QM)-derived point charges which were available for those structures. They found that this method assigns reasonable point charges to the framework atoms and also results in similar isotherms for CO_2, CO, and N_2 as compared to the QM-derived charges. Further improvements were proposed by Zou et al. [119] through their mCBAC (multilayer CBAC) method in which they extended this method by assigning charges based on 2nd-layer connectivity in addition to the 0th- and 1st-layer connectivities (Figure 11.6a). A much larger database of more than 2000 MOFs with their DDEC charges assigned

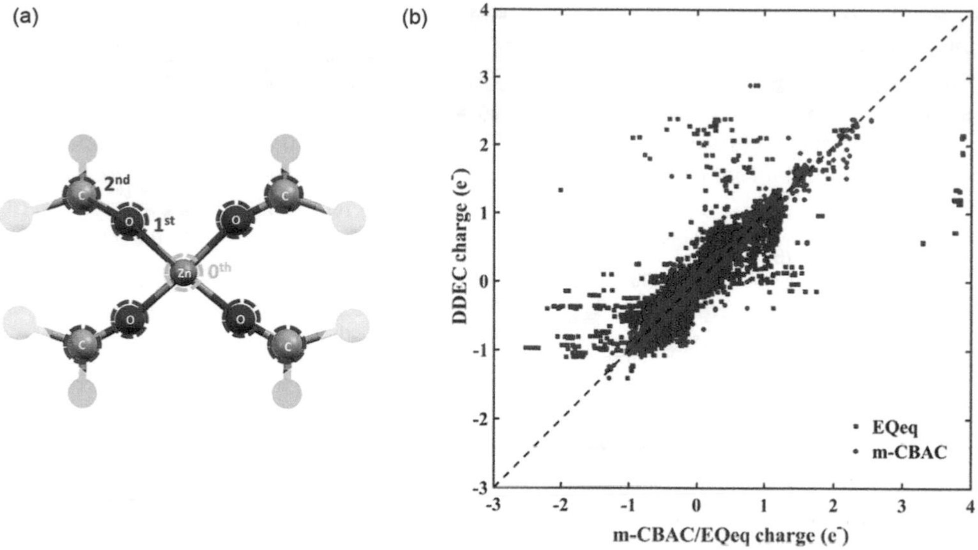

Figure 11.6 mCBAC method. (a) Illustration of the 0th-, 1st-, and 2nd layer connectivity. (b) Results for comparison between mCBAC and EQeq methods with DDEC charges as the benchmark. *Source:* Reproduced from [119]. © 2020, American Chemical Society.

by Nazarian et al. [120] was adopted as training input. They found that this method was able to reproduce the QM-derived charges more accurately than the EQeq method (Figure 11.6b). Furthermore, the importance of accurate charge assignments was demonstrated by computing the Henry's coefficients (K_H) for CO_2 adsorption in MOFs. The mCBAC approach provided a faithful representation of the K_H using DDEC charges with a Spearman rank correlation coefficient of 0.939, illustrating the importance of accurate charge assignments and also the data-driven approach by utilizing a large dataset.

Another recent development has been the use of graph-based neural networks for charge assignments in porous materials. The motivation for this arises from the idea that molecules can be intuitively represented as graphs (networks) and that the partial charges depend on the identity of the atom and its bonding environment [121]. While previous techniques such as CBAC [118] and mCBAC [119] use manually engineered features such as the 1st- or 2nd-layer connectivity, the neural networks can learn the appropriate features describing the bonding environment. Significant efforts have been made in this regard by Wang et al. [122], Unke et al. [123], and Raza et al. [121]. They trained message-passing neural networks (MPNNs) to predict high-fidelity partial atomic charges; for instance, Raza et al. [121] used the DDEC partial charges also computed by Nazarian et al. [120] for MOFs. At a high level, the graph is initialized with certain hidden states (related to the local bonding environment) for each atom. These states are then repeatedly updated based on the hidden states of their neighbors through which the bonding environment is "learned." After a sufficient number of updates, the node is said to have learned chemical bonding environment of the entire molecule. Subsequently, the partial charges are assigned to ensure that they represent the high-fidelity charges while ensuring charge neutrality. An important advantage of this technique for charge assignment is that the charge neutrality constraint can be enforced during model training, as opposed to other methods where the excess charges are adjusted to ensure neutrality. While the details of the algorithm can be found in Raza et al. (121), Figure 11.7 shows an overview of the MPNN used for charge assignment. Trained on 2250 MOFs, this method was

lengths of the edges of this network were then used to construct histograms which served as features to characterize the structure. They further employed the rationale that structurally similar structures are expected to exhibit similar adsorption properties. Thus, they searched for materials similar to those with high Henry coefficients or heats of adsorption. By doing this, they scanned the materials space efficiently, providing an enrichment of 65 times over random sampling; by scanning 500 zeolites similar to those with large heats of adsorption, they were able to identify 327 in the top 1%. As previously mentioned, the use of descriptive features can indeed help in efficient and accurate scanning of the materials space.

Yet another way to characterize the adsorbent structure is to use the familiar radial distribution functions (RDFs) of framework atoms. Fernandez et al. [130] used these features to develop quantitative structure–property relationships (QSPRs) by training support vector classifiers (SVC) on these features to predict the adsorption and desorption loadings and classifying them as "high-performing" or "low-performing". This QSPR model could recover 945 of the top 1000 MOFs in the test set while flagging only 10% of the whole library for compute intensive screening.

More recently, Krishnapriyan et al. have used features to encode the geometric, topological, and chemical information of a material to predict a variety of adsorption properties [131]. They have encoded the geometric information via commonly used features such as the LCD, pore volume, etc., while they have used a technique known as persistent homology to encode the topological features [132]. Treating the framework as a collection of spheres of uniform radius, as their radii are increased, certain cavities in the structure will start to disappear. For instance, if certain atoms form a loop, as their radii are increased, this loop ceases to exist. The longer a certain cavity persists, i.e., higher radius of the atoms, the more prominent it is (Figure 11.9). Pairs of the radii at which cavities form and collapse; i.e., birth-death, describe the topology of the structure. Additionally, they have represented the chemistry through word embeddings of chemical elements from a large corpus of abstracts in the literature. Using these features, they predicted various properties such as Henry's coefficients and uptakes of CO_2 and CH_4 at various pressures with higher R^2 values than the work of Fanourgakis et al., who used standard geometric features such as void fraction, LCD, surface area, etc. [133].

To date, a variety of features have been used to characterize nanoporous materials for carbon capture applications and various techniques have been employed to develop QSPR to efficiently screen the materials space for promising sorbents for post-combustion carbon capture. This

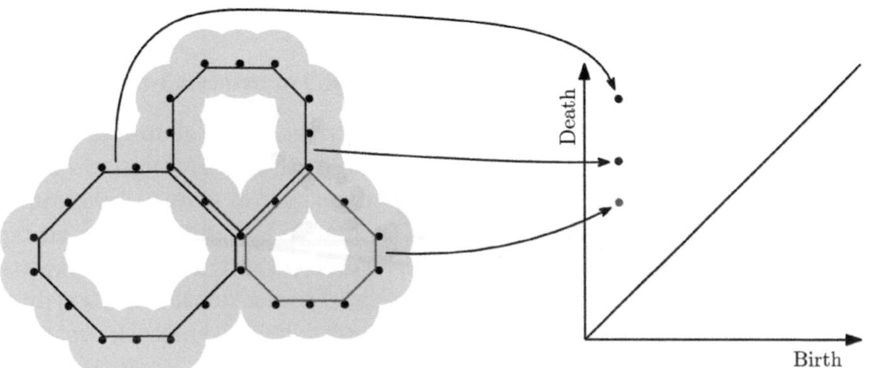

Figure 11.9 Persistent homology illustration. The atoms are represented as spheres and the "death" of the loops formed by them is represented in the panel showing the birth and death radii on the right. *Source:* From [131]. Springer Nature CC BY 4.0.

section does not aim to be exhaustive about these studies, and our intention is to give the readers an intuition of the challenges for materials used in post-combustion carbon capture and illustrate some techniques and strategies used to tackle these challenges. We note that, while most of the literature studies, including those discussed above, only considered a dry adsorbent material in their studies to evaluate the adsorbent performance, real-life CO_2 adsorption processes should inevitably be conducted in the presence of some amount of moisture; considering the moisture aspect is an important challenge that is being worked on in the literature using experiments, simulation studies, and data-driven synthesis [6, 134–139].

11.4.2 Methane Storage

To reduce the emission of carbon, relatively clean fuels, such as methane and hydrogen, have been considered as promising substitutes. A prominent challenge of these fuels in practical applications is their low-energy density per unit volume relative to traditional fuels. Therefore, developing efficient storage techniques has been a strategic direction in recent years. Specifically, adsorbent materials possessing adsorption uptakes that can exceed the target as suggested by the US Department of Energy (DOE) have drawn considerable attention in the past few years [140–145]. Notably, MOFs, materials of large porosity and high surface area as mentioned above, have been extensively studied for their potential in storing methane (this subsection) and hydrogen (the next subsection).

To date, numerous ML studies have focused on the methane storage using MOFs. In 2013, Fernandez et al. were among the first to use ML models to explore the design principle of MOFs for this purpose [146]. In their study, large-scale quantitative QSPR analysis was performed by using CH_4 adsorption data and structural features of ~130,000 hMOFs reported by Wilmer et al. [45]. Three ML regression models, including multilinear regression (MLR), DT, and SVM, were adopted for statistical analysis. Various widely adopted and rather simple geometrical descriptors (i.e., dominant pore diameter (the diameter of the pore that appears most frequently in the structure), the LCD, void fraction, framework density, volumetric surface area, and gravimetric surface area) were taken as input features in the ML models to predict the adsorption capacity. These ML models can all reasonably predict the CH_4 uptake, while nonlinear SVM models exhibited the best accuracy. SVM models were also found to be capable of better predicting the uptake at higher pressures (i.e., at 100 bar, $R^2 = 0.94$) than that at low pressures (i.e. at 35 bar, $R^2 = 0.85$; (Figure 11.10), suggesting those geometrical descriptors

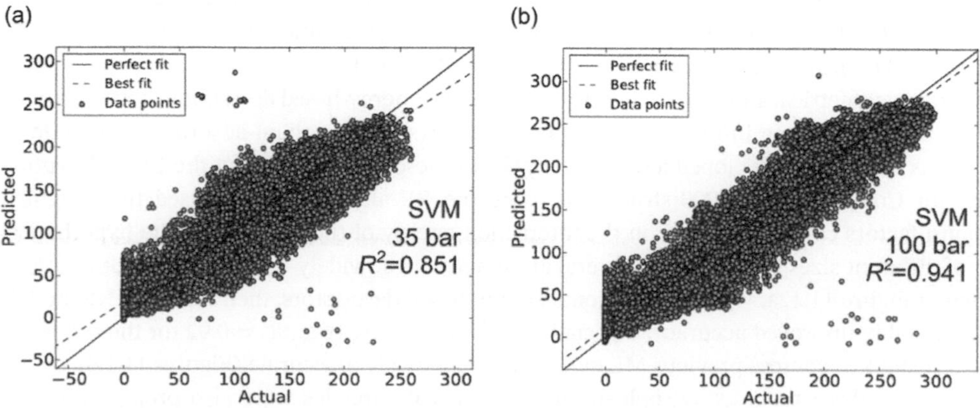

Figure 11.10 SVM-predicted vs. GCMC-computed CH_4 uptake of hMOFs at 35 bar (a) and 100 bar (b). *Source:* Reproduced from [146]. © 2013, American Chemical Society.

used were probably insufficient for depicting the CH_4 adsorption in the lower pressure regime. In addition, the outcome of the study indicated that MOFs with void fraction in the range of 0.40–0.50 and dominant pore diameter of >20 Å appear to result in the maximum CH_4 uptake.

To improve the prediction accuracy of CH_4 adsorption in the relatively lower pressure region (4.5, 2.5, and 0.5 bar), Fernandez et al. [147] further proposed a chemical descriptor: atomic property weighted radial distribution function (AP-RDF) based on electronegativity, polarizability, and van der Waals volume of atoms. The results showed that with the descriptor the prediction accuracy at 4.5 bar was noticeably improved to $R^2 = 0.83$ from $R^2 = 0.46$, emphasizing the importance of chemical descriptors for developing accurate ML models. Pardakhti et al. [148] further introduced various chemically intuitive descriptors (e.g., number of atomic type, metal/halogen to carbon ratio, nitrogen to oxygen ratio, and degree of electronegativity), along with some commonly employed structural features, to predict CH_4 uptake at 35 bar and 298 K for hMOFs using various regression algorithms (i.e., DT, Poisson regression, SVM, and RF). It was found that incorporating those chemical descriptors into ML models can greatly enhance the prediction accuracy. The RF model, as compared to the SVM model, exhibited an optimal prediction accuracy with an R^2 value of as high as 0.98 for predicting the volumetric-based CH_4 uptakes (Figures 11.11a and 11.11b). The RF model also identified that density, void fraction, surface area, pore diameter, metal to carbon ratio, and degree of unsaturation played a critical role in predicting the CH_4 uptake. Another work by Wu et al. [149] further incorporated Henry's coefficients, type of functional groups, atomic number density, and functional group number density as extended chemical descriptors for predicting CH_4 uptake of hMOF. Three different regression models (i.e., SVM, RF, and gradient boosting regression tree (GBRT)) were trained, and their prediction accuracy improved significantly with the newly introduced descriptors. Specifically, the GBRT model offered the best predictive power with $R^2 = 0.96$. The feature importance heat-map also elucidated that the additional descriptors of Henry coefficients as well as atom and functional group number density were important features in training the ML model. In a recent study by Fanourgakis et al. [133], atom counts were further categorized based on atom types in conventional force field (i.e., hybridization and connectivity of atoms). The atom type-based descriptors can effectively account for different chemistries of a specific atom, and hence can help describe the different chemical environments in nanoporous materials. The RF model coupled with the newly designed descriptors, with respect to other sets of descriptors as reported elsewhere [150], provided more accurate prediction for CH_4 uptake at different pressures, and much smaller training sets were needed (Figure 11.11c). With the generalizable atom type descriptors, interestingly, the transferability of resulting ML models was demonstrated for different classes of nanoporous materials. For instance, an RF model trained for MOFs can still make decently accurate predictions for COFs with $R^2 = 0.88$.

Aside from geometrical and simple chemical descriptors, energy-based descriptors have also been explored to provide more information regarding the physical adsorbent-adsorbate interactions. Fanourgakis et al. [151] developed a set of energy-based descriptors to predict the CH_4 adsorption capacity for CoRE MOFs (~4700 structures) using the RF algorithm. They used the weighted Boltzmann factors calculated based on the interaction energy of the MOFs with the hypothetical probes of different sizes, besides using several aforementioned, widely-used structural descriptors, as shown in Figure 11.12a. With the additional energy-based descriptors, their resulting RF models offered a notably improved accuracy, especially at a low pressure (e.g., $R^2 = 0.92$ for the test set at 5.8 bar and 298 K), whereas previous ML models were relatively inaccurate (Figures 11.12b, c).

Furthermore, descriptors using sophisticated topological barcodes were also proposed by Lee et al. [152]. They were implemented with various ML algorithms (i.e., RF, DT, kernel ridge

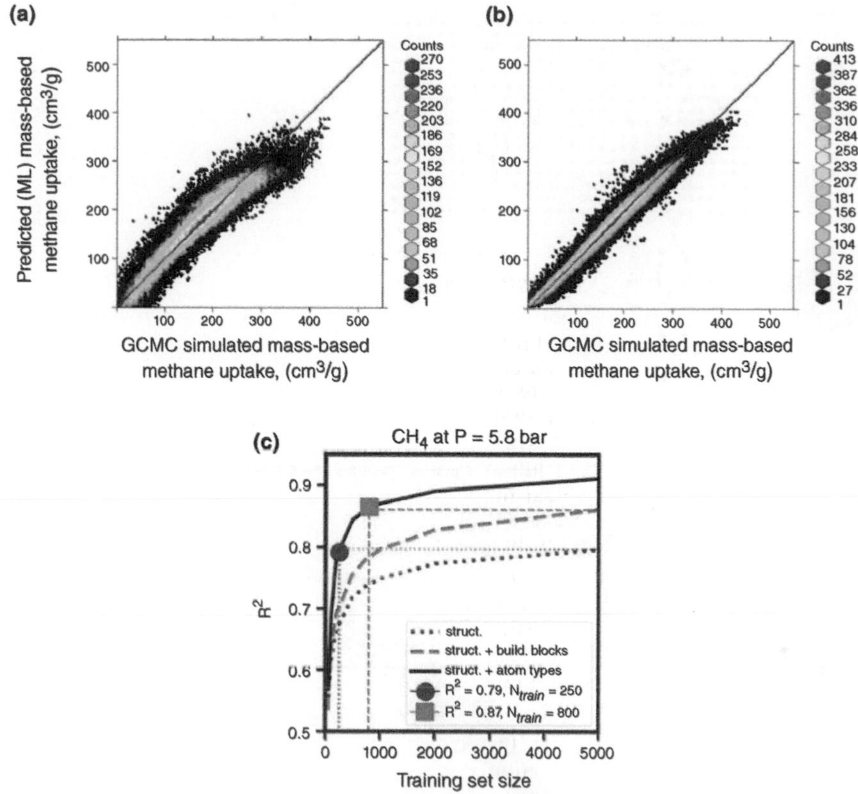

Figure 11.11 (a) SVM- and (b) RF-predicted CH_4 uptake values versus that computed by GCMC. (c) R^2 as a function of the training set size for the RF model. *Source:* (a) and (b) reproduced from [148]. American Chemical Society. (c) Reproduced from [133]. American Chemical Society.

regression (KRR), and SVM), to predict the deliverable capacity of CH_4 for both zeolites and MOFs, which were previously studied using GCMC simulations by Smit et al. [20, 153, 154]. Using the encoding approach, as shown in Figure 11.13, the prediction was not sensitive to the selection of the ML models. The KRR model exhibited the best accuracy (e.g., $R^2 = 0.72$ for zeolites and $R^2 = 0.88$ for MOFs). However, it should be noted that, even with the pore geometry barcodes, prediction discrepancies still existed, especially the candidates whose unique chemical environments significantly contributing to their top performance were difficult to be described. More recently, more sophisticated models using deep learning algorithms have also been utilized in materials research [155]. As reported by Lee et al., by combining Monte Carlo tree search with the recurrent neural network (RNN) algorithm, well-performing MOFs can be predicted. The performance of the designed MOFs using the algorithm was tested for CH_4 storage; hypothetical MOFs were found to have a higher deliverable capacity than existing MOFs possessing the same metal node and topology.

11.4.3 Hydrogen Storage

Hydrogen storage in MOFs has also drawn considerable attention, and employing ML techniques to facilitate the search for optimal material candidates has been of particular focus in recent years. For

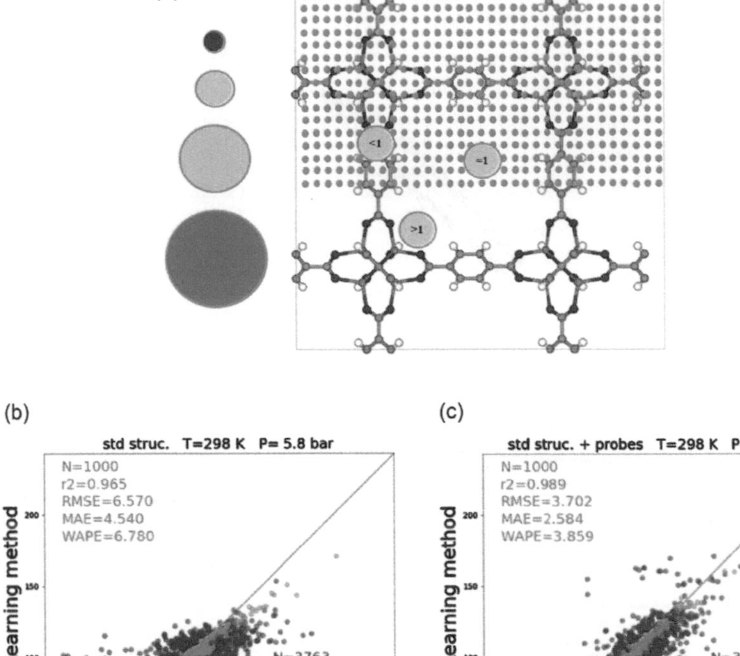

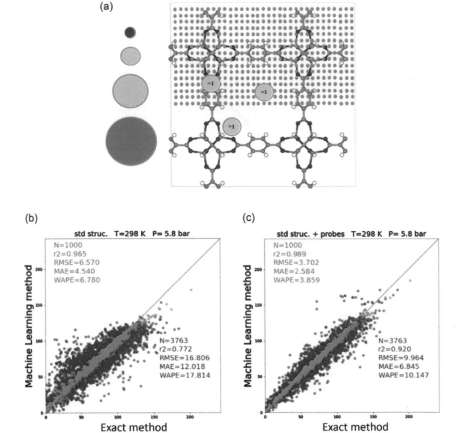

Figure 11.12 (a) Illustration of the Boltzmann factors, $e^{-\beta E}$, computed using probe particles of different sizes at each grid of a MOF structure. ML-predicted vs. GCMC-calculated CH_4 uptake for the CoRE MOFs in the training set (green symbols) and testing set (red symbols) at 5.8 bar and 298 K. In (b), only simple structural features were considered, while in (c) both the structural and the energy-based descriptors are included. *Source:* Reproduced from [151]. © 2019, American Chemical Society.

instance, Thornton et al. [156] utilized ML techniques to probe the performance limits of H_2 storage in over 850,000 nanoporous materials, including MOFs, ZIFs, COFs, zeolites, and porous polymer networks (PPNs). Using the deliverable energy of H_2 as an objective function, ML-assisted screening was employed to reduce the number of GCMC simulations required to more effectively and efficiently identify top-performance nanoporous materials. In this study, GCMC simulations and artificial neural network (ANN) training were conducted iteratively. At each iteration, the adsorption capacities obtained from GCMC calculations at 100 bar and 1 bar were used as inputs for training an ANN model to identify materials to study in the next iteration. Because of the limited set of materials studied in each iteration, prediction made by the trained ANN can be inconsistent with the GCMC-computed results and therefore there is a need to re-train the ANN model with the newly computed dataset. After a few iterations, materials having the best H_2 storage performances were identified, and some of them had never before been considered for H_2 storage. The results obtained during the iterative training are shown in Figure 11.14. This work highlights that ML techniques can be employed in an iterative manner to significantly reduce the required computational cost in a large-scale screening.

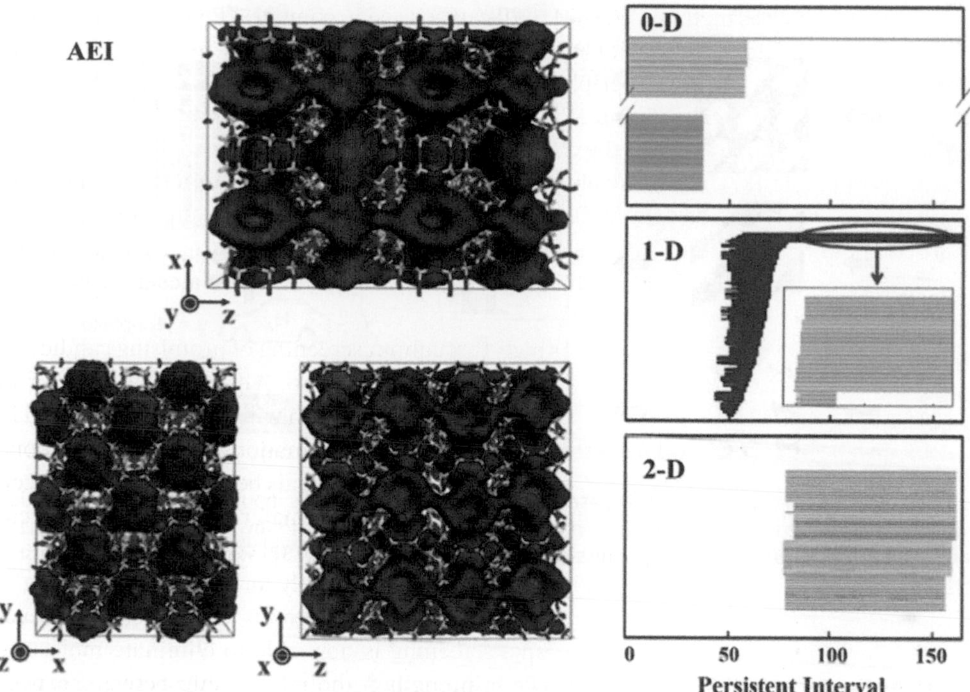

Figure 11.13 Barcodes in zeolite AEI. The left three pictures show the framework structure (ball and stick model) and the pore space (blue colored shape) of AEI in three directions. The right three panels show the 0-, 1-, and 2-dimensional barcodes. *Source:* Reproduced from [153]. © 2019, American Chemical Society.

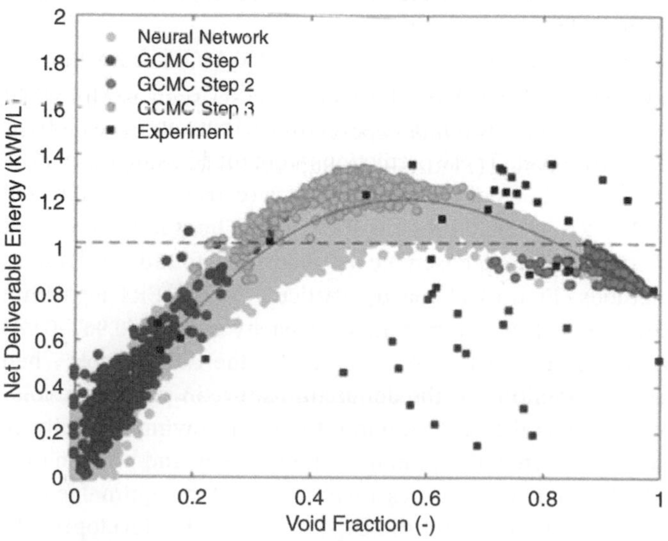

Figure 11.14 Deliverable energy of hydrogen as a function of void fraction for the computationally predicted and experimental measured data at 77 K between 100 and 1 bar. *Source:* Reproduced from [156]. © 2017, American Chemical Society.

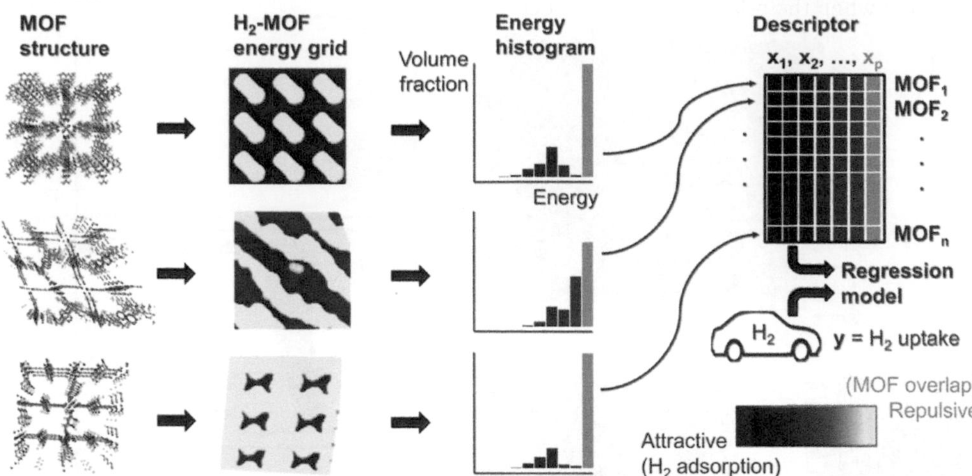

Figure 11.15 Illustration of LASSO models based on the energy histogram as input features. The interaction energy of an H_2 probe at each grid point within a MOF structure is computed, followed by mapping into a 1D histogram. *Source:* [78] Reproduced with permission from Bucior, B. J. et.al [2019] / Royal Society of Chemistry.

In a work by Bucior et al. [78], energy-based descriptors, as described in previous sections, were also used to train a LASSO model with GCMC-computed reference data to predict the H_2 delivery capacity of MOFs. The workflow of the ML model is illustrated in Figure 11.15. The LASSO regression model achieved an excellent prediction accuracy of $R^2 = 0.96$ for hMOFs. The transferability of the model was also investigated with MOFs included in the ToBaCCo and CCDC MOF database, also showing a reasonable accuracy. In this study, more than 50 promising MOFs were identified from the CCDC MOF database. MFU-4l was further synthesized with its H_2 isotherms experimentally measured, which demonstrated a deliverable capacity as large as 36 g/L at 77 K using adsorption and desorption conditions of 100 and 5 bar, respectively.

Most recently, Ahmed et al. [157] collected a diverse set of over 900,000 MOFs from 19 databases and the delivery capacity of H_2 under different conditions (i.e., an isothermal pressure swing (PS) at T = 77 K between 5 and 100 bar and a combined temperature-PS (TPS) between 77 K/100 bar and 160 K/5 bar) was predicted via ML techniques. These predictions were made using seven crystallographic features as inputs for large variations of 14 ML algorithms (extremely randomized trees (ETR), boosted decision trees (BDT), RF, gradient boosting (GB), etc.). The study represented a comprehensive and extensive effort. Tree-based ensemble methods are found superior as compared to the other examined methods (Figures 11.16a–d). Particularly, the ERT algorithm showed the best prediction performance ($R^2 > 0.997$ for gravimetric capacity and $R^2 = 0.967$–0.984 for volumetric capacity). The feature importance was also evaluated for the ERT model. When only a single feature was considered, void fraction was the dominant feature in predicting volumetric delivery capacity, while pore volume is the most important one for gravimetric delivery capacity. When only two features were considered, the combination of density and pore volume was the key. For highly accurate prediction, only five features are needed, and the optimal feature combination depends upon the operating condition and the capacity type. The developed ML models identified 8282 MOFs with performance exceeding the current state-of-the-art materials, and the top structures are illustrated in Figures 11.16e–f.

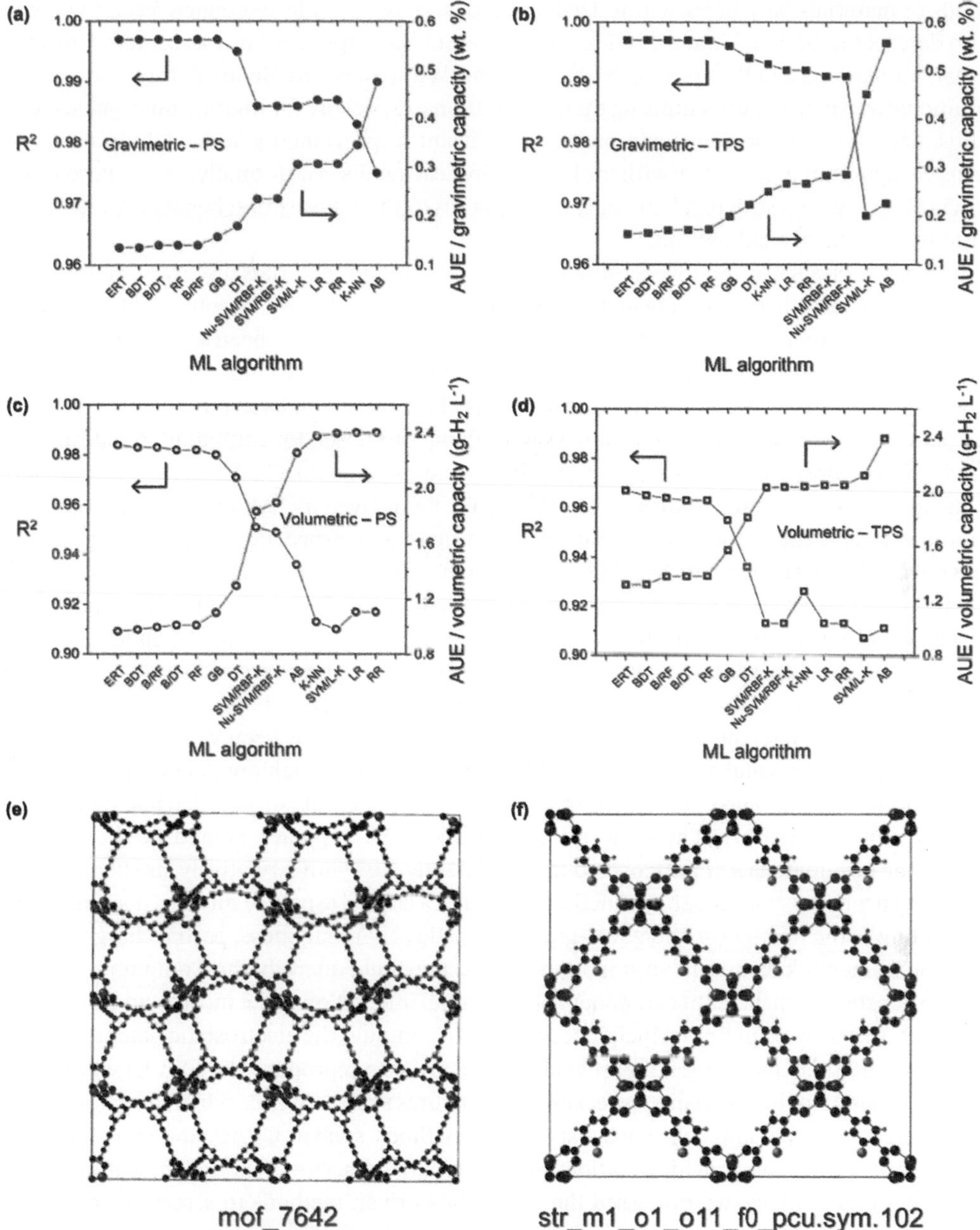

Figure 11.16 Comparison of different ML algorithms for predicting H_2 delivery capacity in MOFs. (a, c) and (b, d) show the performance at PS and TPS conditions, respectively, while (a, b) and (c, d) show the performance for predicting gravimetric and volumetric delivery capacity, respectively. (e) and (f) show the structure predicted to offer the highest capacity at the PS and TPS condition, respectively. *Source:* Reproduced from [157]. © 2021, Cell Press.

11.5 Summary and Outlook

In this chapter, we have discussed various aspects of data-driven materials discovery for adsorption applications. With dramatic improvements in materials synthesis and computational methods, thousands of materials have been synthesized and orders of magnitude more have been hypothesized. To date, computational materials discovery for adsorption applications has centered around molecular simulations and first principles calculations. While these are detailed, they can also be computationally expensive, thus limiting their applications, especially for high-throughput screening. Thus, data-driven approaches can provide significant improvements, especially in terms of balancing computational efficiency with the level of physical details. Additionally, these approaches can provide insights into structural attributes of materials to their adsorption behaviors and establish structure–property relationships.

The availability of reliable and diverse structures is important for materials discovery. To this end, hypothetical structures have been created through algorithms combining various building blocks. More recently, the experimental structures reported in the literature have been analyzed and made "computation-ready" by correcting disorder, and removing solvents, molecules, etc., for instance, in the CoRE MOF database. While many studies are based on these databases, it has recently been pointed out that care must be taken to ensure that the database chosen for computation indeed contains diverse structures for their interpretations to be valid in general [67]. Thus, the creation and use of databases for screening studies must account for their diversity. Another crucial aspect of studying materials is the features that are used to describe them, especially in structure–property relationships. In the past, aggregate features such as pore volumes, LCDs, PLDs, heats of adsorption, surface areas, etc., have been used to study materials. While these can work for applications or adsorbates when the adsorption mechanisms are relatively simple and volume dependent such as methane storage, they were shown to be inadequate for more complex cases such as water adsorption. Thus, more localized features such as cluster properties, cavity shapes, etc., can improve performance. Furthermore, the screening and discovery of materials has been viewed as an optimization problem. Experimental synthesis or molecular simulations, which are generally expensive, can provide more accurate and detailed information about the materials surveyed, while less expensive methods are less reliable. Thus, the materials exploration approach should be designed to ensure that the best materials are discovered most efficiently, i.e., with a relatively smaller number of evaluations using the more expensive methods. Besides, a critical aspect of molecular simulations is the modeling of the host–guest and guest–guest interactions. Among these, partial charge assignments to the framework atoms are an important part as they help quantify the Coulombic interactions which constitute a significant component of the overall interactions. The most accurate method to accomplish this is through first-principle techniques to compute the electrostatic potential (ESP) around atoms and then assign a set of partial charges in order to reproduce the ESP. However, this method is computationally expensive, especially for structures such as MOFs, which may have up to several hundred atoms per unit cell. Thus, data-driven methods, such as CBAC, and more recently, neural networks [121], can provide solutions with comparable accuracy and significantly lower computational time. Finally, we discussed the use of data-driven methods to screen materials for post-combustion carbon capture, methane, and hydrogen storage applications as case studies.

We believe that data-driven screening can provide useful insights to lead the discovery of nanoporous materials for adsorption applications. While studies involving uncharged molecules such as methane have dominated the literature in the past, investigating the storage and separation of charged molecules such as water, H_2S, etc., is anticipated to become significant in future. The challenges to screening materials for these applications will include their complex adsorption mechanisms dependent on both their host–guest and guest–guest interactions. Thus, appropriate features that can capture these, including more advanced and localized features, will become critically important. Additionally, the selection of datasets that are representative of the true structural space, along with the efficient exploration of the materials space, will gain prominence for these applications.

References

1 Morris, R.E. and Wheatley, P.S. (2008). Gas storage in nanoporous materials. *Angew. Chemie Int. Ed.* 47 (27): 4966–4981.

2 Hönicke, I.M., Senkovska, I., Bon, V. et al. (2018). Balancing mechanical stability and ultrahigh porosity in crystalline framework materials. *Angew. Chemie Int. Ed.* 57 (42): 13780–13783.

3 Zhang, X., Chen, Z., Liu, X. et al. (2020). Overview of the activation and porosity of metal–organic frameworks. *Chem. Soc. Rev.* 49 (20): 7406–7427.

4 Chung, Y.G., Camp, J., Haranczyk, M. et al. (2014). Computation-ready, experimental metal–organic frameworks: a tool to enable high-throughput screening of nanoporous crystals. *Chem. Mater.* 26 (21): 6185–6192.

5 Chung, Y.G., Haldoupis, E., Bucior, B.J. 1 et al. (2019). Advances, updates, and analytics for the computation-ready, experimental metal–organic framework database: CoRE MOF 2019. *J. Chem. Eng. Data* 64 (12): 5985–5998.

6 Boyd, P.G., Chidambaram, A., García-Díez, E. et al. (2019). Data-driven design of metal–organic frameworks for wet flue gas CO_2 capture. *Nature* 576 (7786): 253–256.

7 Tong, M., Lan, Y., Qin, Z. et al. (2018). Experimental covalent organic framework for methane delivery: screening and material design. *J. Phys. Chem. C* 122 (24): 13009–13016.

8 Ongari, D., Talirz, L., and Smit, B. (2020). Too many materials and too many applications: an experimental problem waiting for a computational solution. *ACS Cent. Sci.* 6 (11): 1890–1900.

9 Database of zeolite structures http://www.iza-structure.org. (accessed 9 December 2021).

10 Pophale, R., Cheeseman, P.A., and Deem, M.W. (2011). A database of new zeolite-like materials. *Phys. Chem. Chem. Phys.* 13 (27): 12407–12412.

11 Giannakoudakis, D.A. and Bandosz, T.J. (2018). *Detoxification of Chemical Warfare Agents*. Cham: Springer International Publishing.

12 Barrer, R.M. (1948). Syntheses and reactions of mordenite. *J. Chem. Soc.* 2158. https://pubs.rsc.org/en/content/articlelanding/1948/jr/jr9480002158/unauth.

13 Bennett, T.D., Coudert, F.-X., James, S.L. et al. (2021). The changing state of porous materials. *Nat. Mater.* 20 (9): 1179–1187.

14 Li, H., Eddaoudi, M., Groy, T.L. et al. (1998). Establishing microporosity in open metal–organic frameworks: gas sorption isotherms for Zn(BDC) (BDC = 1,4-Benzenedicarboxylate). *J. Am. Chem. Soc.* 120 (33): 8571–8572.

15 Li, H., Eddaoudi, M., O'Keeffe, M. et al. (1999). Design and synthesis of an exceptionally stable and highly porous metal-organic framework. *Nature* 402 (6759): 276–279.

16 Côté, A.P., Benin, A.I., Ockwig, N.W. et al. (2005). Porous, crystalline, covalent organic frameworks. *Science (80)* 310 (5751): 1166–1170.

17 El-Kaderi, H.M., Hunt, J.R., Mendoza-Cortés, J.L. et al. (2007). Designed synthesis of 3D covalent organic frameworks. *Science (8)* 316 (5822): 268–272.

18 Freund, R., Canossa, S., Cohen, S.M. et al. (2021). 25 years of reticular chemistry. *Angew. Chemie Int. Ed.* 60 (45): 23946.

19 Freund, R., Zaremba, O., Arnauts, G. et al. (2021). The current status of MOF and COF applications. *Angew. Chemie Int. Ed.* 60 (45): 23975.

20 Simon, C.M., Kim, J., Gomez-Gualdron, D.A. et al. (2015). The materials genome in action: identifying the performance limits for methane storage. *Energy Environ. Sci.* 8 (4): 1190–1199.

21 Broom, D.P. and Thomas, K.M. (2013). Gas adsorption by nanoporous materials: future applications and experimental challenges. *MRS Bull.* 38 (5): 412–421.

22 Czaja, A.U., Trukhan, N., and Müller, U. (2009). Industrial applications of metal–organic frameworks. *Chem. Soc. Rev.* 3 (5): 1284.

23 Dzubak, A.L., Lin, L.-C., Kim, J. et al. (2012). *Ab initio* carbon capture in open-site metal–organic frameworks. *Nat. Chem.* 4 (10): 810–816.

24 Lee, K., Howe, J.D., Lin, L.-C. et al. (2015). Small-molecule adsorption in open-site metal–organic frameworks: a systematic density functional theory study for rational design. *Chem. Mater.* 27 (3): 668–678.

25 Huck, J.M., Lin, L.-C., Berger, A.H. et al. (2014). Evaluating different classes of porous materials for carbon capture. *Energy Environ. Sci.* 7 (12): 4132–4146.

26 Yang, C., Kshirsagar, A.R., Eddin, A.C. et al. (2018). Tuning gas adsorption by metal node blocking in photoresponsive metal–organic frameworks. *Chem. – A Eur. J.* 24 (57): 15167–15172.

27 Trickett, C.A., Helal, A., Al-Maythalony, B.A. et al. (2017). The chemistry of metal–organic frameworks for CO_2 capture, regeneration and conversion. *Nat. Rev. Mater.* 2 (8): 17045.

28 Broom, D.P., Webb, C.J., Fanourgakis, G.S. et al. (2019). Concepts for improving hydrogen storage in nanoporous materials. *Int. J. Hydrogen Energy* 44 (15): 7768–7779.

29 Kumar, K.V., Preuss, K., Titirici, M.M. et al. (2017). Nanoporous materials for the onboard storage of natural gas. *Chem. Rev.* 117 (3): 1796–1825.

30 Mueller, U., Schubert, M., Teich, F. et al. (2006). Metal–organic frameworks: prospective industrial applications. *J. Mater. Chem.* 16 (7): 626–636.

31 Matsuda, R., Kitaura, R., Kitagawa, S. et al. (2005). Highly controlled acetylene accommodation in a metal–organic microporous material. *Nature* 436 (7048): 238–241.

32 Samsonenko, D.G., Kim, H., Sun, Y. et al. (2007). Microporous magnesium and manganese formates for acetylene storage and separation. *Chem. – An Asian J.* 2 (4): 484–488.

33 Wheatley, P.S., Butler, A.R., Crane, M.S. et al. (2006). NO-releasing zeolites and their antithrombotic properties. *J. Am. Chem. Soc.* 128 (2): 502–509.

34 McKinlay, A.C., Allan, P.K., Renouf, C.L. et al. (2014). Multirate delivery of multiple therapeutic agents from metal–organic frameworks. *APL Mater.* 2 (12): 124108.

35 Bae, M.N., Kim, Y., and Seff, K. (1998). Crystal structure of a carbon monoxide sorption complex of dehydrated fully manganese(II)-exchanged zeolite X. *Microporous Mesoporous Mater.* 26 (1–3): 101–107.

36 Sharma, A.C. and Borovik, A.S. (2000). Design, synthesis, and characterization of templated metal sites in porous organic hosts: application to reversible dioxygen binding. *J. Am. Chem. Soc.* 122 (37): 8946–8955.

37 Yong, Z., Mata, V., and Rodrigues, A. (2002). Adsorption of carbon dioxide at high temperature: a review. *Sep. Purif. Technol.* 26 (2–3): 195–205.

38 Dathe, H., Jentys, A., and Lercher, J.A. (2005). Sulfate formation on SOx trapping materials studied by Cu and S K-edge XAFS. *Phys. Chem. Chem. Phys.* 7 (6): 1283–1293.

39 Li, L. and King, D.L. (2005). High-capacity sulfur dioxide absorbents for diesel emissions control. *Ind. Eng. Chem. Res.* 44 (1): 168–177.

40 Kirik, S.D., Dubkov, A.A., Dubkova, S.A.et al. (1992). X-ray powder diffraction and t.p.d. study of SO_2 adsorption on type Y zeolite. *Zeolites* 12 (3): 292–298.

41 Liu, C.Y. and Aika, K. (2003). Ammonia adsorption on ion exchanged Y-zeolites as ammonia storage material. *J. Japan Pet. Inst.* 46 (5): 301–307.

42 Sturluson, A., Huynh, M.T., Kaija, A.R. et al. (2019). The role of molecular modelling and simulation in the discovery and deployment of metal–organic frameworks for gas storage and separation. *Mol. Simul.* 45 (14–15): 1082–1121.

43 Cho, E.H., Lyu, Q., and Lin, L.-C. (2019). Computational discovery of nanoporous materials for energy- and environment-related applications. *Mol. Simul.* 45 (14–15): 1122–1147.

44 Amit, S. (2018). *Applications of Nanoporous Materials in Gas Separation and Storage*. University of Nevada, Las Vegas.

45 Wilmer, C.E., Leaf, M., Lee, C.Y.et al. (2012). Large-scale screening of hypothetical metal–organic frameworks. *Nat. Chem.* 4 (2): 83–89.

46 Haldoupis, E., Nair, S., and Sholl, D.S. (2010). Efficient calculation of diffusion limitations in metal organic framework materials: a tool for identifying materials for kinetic separations. *J. Am. Chem. Soc.* 132 (21): 7528–7539.

47 Haldoupis, E., Nair, S., and Sholl, D.S. (2011). Pore size analysis of >250,000 hypothetical zeolites. *Phys. Chem. Chem. Phys.* 13 (11): 5053–5060.

48 Kim, J., Abouelnasr, M., Lin, L.-C. et al. (2013). Large-scale screening of zeolite structures for CO_2 membrane separations. *J. Am. Chem. Soc.* 135 (20): 7545–7552.

49 Deng, X., Yang, W., Li, S. et al. (2020). Large-scale screening and machine learning to predict the computation-ready, experimental metal-organic frameworks for CO_2 capture from air. *Appl. Sci.* 10 (2): 569.

50 Lin, L.-C., Berger, A.H., Martin, R.L. et al. (2012). *In silico* screening of carbon-capture materials. *Nat. Mater.* 11 (7): 633–641.

51 Hu, J., Gu, X., Lin, L.-C. et al. (2021). Toward sustainable metal–organic frameworks for post-combustion carbon capture by life cycle assessment and molecular simulation. *ACS Sustain. Chem. Eng.* 9 (36): 12132–12141.

52 Datar, A., Witman, M., and Lin, L.-C. (2021). Improving computational assessment of porous materials for water adsorption applications via flat histogram methods. *J. Phys. Chem. C* 125 (7): 4253–4266.

53 Datar, A., Witman, M., and Lin, L.-C. (2021). Monte Carlo simulations for water adsorption in porous materials: best practices and new insights. *AIChE J.* 67 (12): e17447.

54 James, G., Witten, D., Hastie, T. et al. (2013). *An Introduction to Statistical Learning*, vol. 103. New York: Springer Texts in Statistics; Springer.

55 Ramsey, F. and Schafer, D. (2012). *The Statistical Sleuth: A Course in Methods of Data Analysis*. Cengage Learning.

56 Hollander, M., Wolfe, D.A., and Chicken, E. (2015). *Nonparametric Statistical Methods*. Wiley Series in Probability and Statistics. Wiley. https://www.wiley.com/en-us/Nonparametric+Statistical+Methods%2C+3rd+Edition-p-9780470387375.

57 Koutsoukas, A., Monaghan, K.J., Li, X. et al. (2017). Deep-learning: investigating deep neural networks hyper-parameters and comparison of performance to shallow methods for modeling bioactivity data. *J. Cheminform.* 9 (1): 42.

58 Hall, S.R., Allen, F.H., and Brown, I.D. (1991). The crystallographic information file (CIF): A new standard archive file for crystallography. *Acta Crystallogr. Sect. A Found. Crystallogr.* 47 (6): 655–685.

59 Brown, I.D. (1996). A standard for crystallographic data interchange. *J. Res. Natl. Inst. Stand. Technol.* 101 (3): 341.

60 CIF (crystallographic information file). Crystallographic information framework https://www.iucr.org/resources/cif/documentation.

61 Meier, W.M., Olson, D.H., and Baerlocher, C. (1996). *Atlas of Zeolite Framework Types*, 4th revis. London: Elsevier.

62 Treacy, M.M.J. and Higgins, J.B. (1996). *Collection of Simulated XRD Powder Diffraction Patterns for Zeolites*, 3rd revis. London: Elsevier.

63 Earl, D.J. and Deem, M.W. (2006). Toward a database of hypothetical zeolite structures. *Ind. Eng. Chem. Res.* 45 (16): 5449–5454.

64 Moosavi, S.M., Nandy, A., Jablonka, K.M.et al. (2020). Understanding the diversity of the metal-organic framework ecosystem. *Nat. Commun.* 11 (1): 4068.

65 Gómez-Gualdrón, D.A., Colón, Y.J., Zhang, X. et al. (2016). Evaluating topologically diverse metal–organic frameworks for cryo-adsorbed hydrogen storage. *Energy Environ. Sci.* 9 (10): 3279–3289.

66 Boyd, P.G. and Woo, T.K. (2016). Method for constructing hypothetical nanoporous materials of any net topology from graph theory. *CrystEngComm* 18 (21): 3777–3792.

67 Boyd, P.G., Lee, Y., and Smit, B. (2017). Computational development of the nanoporous materials genome. *Nat. Rev. Mater.* 2 (8): 17037.

68 Groom, C.R., Bruno, I.J., Lightfoot, M.P. et al. (2016). The Cambridge Structural Database. *Acta Crystallogr. Sect. B Struct. Sci. Cryst. Eng. Mater.* 72 (2): 171–179.

69 Goldsmith, J., Wong-Foy, A.G., Cafarella, M.J. et al. (2013). Theoretical limits of hydrogen storage in metal–organic frameworks: opportunities and trade-offs. *Chem. Mater.* 25 (16): 3373–3382.

70 Sikora, B.J., Winnegar, R., Proserpio, D.M. et al. (2014). Textural properties of a large collection of computationally constructed MOFs and zeolites. *Microporous Mesoporous Mater.* 186: 207–213.

71 Anderson, R., Rodgers, J., Argueta, E. et al. (2018). Role of pore chemistry and topology in the CO_2 capture capabilities of MOFs: from molecular simulation to machine learning. *Chem. Mater.* 30 (18): 6325–6337.

72 Zhang, H. and Snurr, R.Q. (2017). Computational study of water adsorption in the hydrophobic metal–organic framework ZIF-8: adsorption mechanism and acceleration of the simulations. *J. Phys. Chem. C* 121 (43): 24000–24010.

73 Jablonka, K.M., Ongari, D., Moosavi, S.M. et al. (2020). Big-data science in porous materials: materials genomics and machine learning. *Chem. Rev.* 120 (16): 8066–8129.

74 Ongari, D., Boyd, P.G., Barthel, S. et al. (2017). Accurate characterization of the pore volume in microporous crystalline materials. *Langmuir* 33 (51): 14529–14538.

75 Willems, T.F., Rycroft, C.H., Kazi, M. et al. (2012). Algorithms and tools for high-throughput geometry-based analysis of crystalline porous materials. *Microporous Mesoporous Mater.* 149 (1): 134–141.

76 Frenkel, D. and Smit, B. (2002). *Understanding Molecular Simulation*, 2nd ed. Elsevier.

77 Dubbeldam, D., Calero, S., Ellis, D.E. et al. (2016). Molecular simulation software for adsorption and diffusion in flexible nanoporous materials. *Mol. Simul.* 42 (2): 81–101.

78 Bucior, B.J., Bobbitt, N.S., Islamoglu, T. et al. (2019). Energy-based descriptors to rapidly predict hydrogen storage in metal–organic frameworks. *Mol. Syst. Des. Eng.* 4 (1): 162–174.

79 Cho, E.H. and Lin, L.-C. (2021). Nanoporous material recognition via 3D convolutional neural networks: prediction of adsorption properties. *J. Phys. Chem. Lett.* 12 (9): 2279–2285.

80 Janet, J.P. and Kulik, H.J. (2017). Resolving transition metal chemical space: feature selection for machine learning and structure–property relationships. *J. Phys. Chem. A* 121 (46): 8939–8954.

81 Sturluson, A., Huynh, M.T., York, A.H.P. et al. (2018). Eigencages: learning a latent space of porous cage molecules. *ACS Cent. Sci.* 4 (12): 1663–1676.

82 Hasell, T. and Cooper, A.I. (2016). Porous organic cages: soluble, modular and molecular pores. *Nat. Rev. Mater.* 1 (9): 16053.

83 Deshwal, A., Simon, C., and Doppa, J.R. (2021). Bayesian optimization of nanoporous materials. *Mol. Syst. Des. Eng.* 6 (12): 1066–1086.

84 Pollice, R., Dos Passos Gomes, G., Aldeghi, M. et al. (2021). Data-driven strategies for accelerated materials design. *Acc. Chem. Res.* 54 (4): 849–860.

85 Datar, A., Chung, Y.G., and Lin, L.-C. (2020). Beyond the BET analysis: the surface area prediction of nanoporous materials using a machine learning method. *J. Phys. Chem. Lett.* 11 (14): 5412–5417.

86 Chong, S., Lee, S., Kim, B. et al. (2020). Applications of machine learning in metal-organic frameworks. *Coord. Chem. Rev.* 423: 213487.

87 Chibani, S. and Coudert, F.-X. (2020). Machine learning approaches for the prediction of materials properties. *APL Mater.* 8 (8): 080701.

88 Mukherjee, K. and Colón, Y.J. (2021). Machine learning and descriptor selection for the computational discovery of metal-organic frameworks. *Mol. Simul.* 47 (10–11): 857–877.

89 Shi, Z., Yang, W., Deng, X. et al. (2020). Computational screening of high performance metal–organic frameworks. *Mol. Syst. Des. Eng.* 5 (4): 725–742.

90 Cho, E.H., Deng, X., Zou, C. et al. (2020). Machine learning-aided computational study of metal–organic frameworks for sour gas sweetening. *J. Phys. Chem. C* 124 (50): 27580–27591.

91 Miller, B.L. and Goldberg, D.E. (1996). Genetic algorithms, selection schemes, and the varying effects of noise. *Evol. Comput.* 4 (2): 113–131.

92 Chung, Y.G., Gómez-Gualdrón, D.A., Li, P. et al. (2016). *In silico* discovery of metal-organic frameworks for precombustion CO_2 capture using a genetic algorithm. *Sci. Adv.* 2 (10): e1600909.

93 Collins, S.P., Daff, T.D., Piotrkowski, S.S. et al. (2016). Materials design by evolutionary optimization of functional groups in metal–organic frameworks. *Sci. Adv.* 2 (11): e1600954.

94 Bao, Y., Martin, R.L., Haranczyk, M. et al. (2015). *In silico* prediction of MOFs with high deliverable capacity or internal surface area. *Phys. Chem. Chem. Phys.* 17 (18): 11962–11973.

95 Moosavi, S.M., Chidambaram, A., Talirz, L. et al. (2019). Capturing chemical intuition in synthesis of metal-organic frameworks. *Nat. Commun.* 10 (1): 539.

96 Graff, D.E., Shakhnovich, E.I., and Coley, C.W. (2021). Accelerating high-throughput virtual screening through molecular pool-based active learning. *Chem. Sci.* 12 (22): 7866–7881.

97 Sturluson, A., Raza, A., McConachie, G.D. et al. (2021). Recommendation system to predict missing adsorption properties of nanoporous materials. *Chem. Mater.* 33 (18): 7203–7216.

98 Talirz, L., Kumbhar, S., Passaro, E. et al. (2020). Materials cloud, a platform for open computational science. *Sci. Data* 7 (1): 299.

99 Rappe, A.K., Casewit, C.J., Colwell, K.S. et al. (1992). UFF, a full periodic table force field for molecular mechanics and molecular dynamics simulations. *J. Am. Chem. Soc.* 114 (25): 10024–10035.

100 Mayo, S.L., Olafson, B.D., and Goddard, W.A. (1990). DREIDING: a generic force field for molecular simulations. *J. Phys. Chem.* 94 (26): 8897–8909.

101 Ongari, D., Boyd, P.G., Kadioglu, O. et al. (2019). Evaluating charge equilibration methods to generate electrostatic fields in nanoporous materials. *J. Chem. Theory Comput.* 15 (1): 382–401.

102 Mulliken, R.S. (1955). Electronic population analysis on LCAO–MO molecular wave functions. II: Overlap populations, bond orders, and covalent bond energies. *J. Chem. Phys.* 23 (10): 1841–1846.

103 Hirshfeld, F.L. (1977). Bonded-atom fragments for describing molecular charge densities. *Theor. Chim. Acta* 44 (2): 129–138.

104 Bultinck, P., Van Alsenoy, C., Ayers, P.W. et al. (2007). Critical analysis and extension of the hirshfeld atoms in molecules. *J. Chem. Phys.* 126 (14): 144111.

105 Henkelman, G., Arnaldsson, A., and Jónsson, H. (2006). A fast and robust algorithm for bader decomposition of charge density. *Comput. Mater. Sci.* 36 (3): 354–360.

106 Bayly, C.I., Cieplak, P., Cornell, W. et al. (1993). A well-behaved electrostatic potential based method using charge restraints for deriving atomic charges: the RESP model. *J. Phys. Chem.* 97 (40): 10269–10280.

107 Breneman, C.M. and Wiberg, K.B. (1990). Determining atom-centered monopoles from molecular electrostatic potentials. The need for high sampling density in formamide conformational analysis. *J. Comput. Chem.* 11 (3): 361–373.

108 Campañá, C., Mussard, B., and Woo, T.K. (2009). Electrostatic potential derived atomic charges for periodic systems using a modified error functional. *J. Chem. Theory Comput.* 5 (10): 2866–2878.

109 Manz, T.A. and Sholl, D.S. (2010). Chemically meaningful atomic charges that reproduce the electrostatic potential in periodic and nonperiodic materials. *J. Chem. Theory Comput.* 6 (8): 2455–2468.

110 Manz, T.A. and Sholl, D.S. (2012). Improved atoms-in-molecule charge partitioning functional for simultaneously reproducing the electrostatic potential and chemical states in periodic and nonperiodic materials. *J. Chem. Theory Comput.* 8 (8): 2844–2867.

111 Limas, N.G. and Manz, T.A. (2016). Introducing DDEC6 atomic population analysis. Part 2: Computed results for a wide range of periodic and nonperiodic materials. *RSC Adv.* 6 (51): 45727–45747.

112 Cho, E.H. and Lin, L.-C. (2019). Electrostatic potential optimized molecular models for molecular simulations: CO, CO_2, COS, H_2S, N_2, N_2O, and SO_2. *J. Chem. Theory Comput.* 15 (11): 6323–6332.

113 Martín-Calvo, A., Lahoz-Martín, F.D., and Calero, S. (2012). Understanding carbon monoxide capture using metal–organic frameworks. *J. Phys. Chem. C* 116 (11): 6655–6663.

114 Potoff, J.J. and Siepmann, J.I. (2001). Vapor–liquid equilibria of mixtures containing alkanes, carbon dioxide, and nitrogen. *AIChE J* 47 (7): 1676–1682.

115 Rappe, A.K. and Goddard, W.A. (1991). Charge equilibration for molecular dynamics simulations. *J. Phys. Chem.* 95 (8): 3358–3363.

116 Wilmer, C.E., Kim, K.C., and Snurr, R.Q. (2012). An extended charge equilibration method. *J. Phys. Chem. Lett.* 3 (17): 2506–2511.

117 Wilmer, C.E. and Snurr, R.Q. (2011). Towards rapid computational screening of metal–organic frameworks for carbon dioxide capture: calculation of framework charges via charge equilibration. *Chem. Eng. J.* 171 (3): 775–781.

118 Xu, Q., Zhong, C., and General, A. (2010). Approach for estimating framework charges in metal—organic frameworks. *J. Phys. Chem. C* 114 (11): 5035–5042.

119 Zou, C., Penley, D.R., Cho, E.H. et al. (2020). Efficient and accurate charge assignments via a multilayer connectivity-based atom contribution (m-CBAC) approach. *J. Phys. Chem. C* 124 (21): 11428–11437.

120 Nazarian, D., Camp, J.S., Sholl, D.S. et al. (2016). Set of high-quality point charges for simulations of metal–organic frameworks. *Chem. Mater.* 28 (3): 785–793.

121 Raza, A., Sturluson, A., Simon, C.M. et al. (2020). Message passing neural networks for partial charge assignment to metal–organic frameworks. *J. Phys. Chem. C* 124 (35): 19070–19082.

122 Wang, Y., Fass, J., Stern, C.D. et al. (2019). Graph nets for partial charge prediction. https://arxiv.org/abs/1909.07903.

123 Unke, O.T. and Meuwly, M. (2019). PhysNet: a neural network for predicting energies, forces, dipole moments, and partial charges. *J. Chem. Theory Comput.* 15 (6): 3678–3693.

124 Wells, B.A., De Bruin-Dickason, C., and Chaffee, A.L. (2015). Charge equilibration based on atomic ionization in metal–organic frameworks. *J. Phys. Chem. C* 119 (1): 456–466.

125 Annual Energy Outlook 2020 https://www.eia.gov/outlooks/aeo/pdf/AEO2020FullReport.pdf.

126 NETL DOE Post-Combustion Carbon Capture https://netl.doe.gov/coal/carbon-capture/post-combustion.

127 Bhown, A.S. and Freeman, B.C. (2011). Analysis and status of post-combustion carbon dioxide capture technologies. *Environ. Sci. Technol.* 45 (20): 8624–8632.

128 Martin, R.L., Willems, T.F., Lin, L.-C. et al. (2012). Discovery of zeolite materials for adsorption-based separations. *ChemPhysChem* 13 (16): 3595–3597.

129 Martin, R.L., Smit, B., and Haranczyk, M. (2012). Addressing challenges of identifying geometrically diverse sets of crystalline porous materials. *J. Chem. Inf. Model.* 52 (2): 308–318.

130 Fernandez, M., Boyd, P.G., Daff, T.D. et al. (2014). Rapid and accurate machine learning recognition of high performing metal organic frameworks for CO_2 capture. *J. Phys. Chem. Lett.* 5 (17): 3056–3060.

131 Krishnapriyan, A.S., Montoya, J., Haranczyk, M. et al. (2021). Machine learning with persistent homology and chemical word embeddings improves prediction accuracy and interpretability in metal–organic frameworks. *Sci. Rep.* 11 (1): 8888.

132 Edelsbrunner, H. and Harer, J. (2007). Persistent homology: a survey. *Contemp. Math.* 453: 257–282.

133 Fanourgakis, G.S., Gkagkas, K., Tylianakis, E. et al. (2020). A universal machine learning algorithm for large-scale screening of materials. *J. Am. Chem. Soc.* 142 (8): 3814–3822.

134 Chanut, N., Bourrelly, S., Kuchta, B. et al. (2017). Screening the effect of water vapour on gas adsorption performance: application to CO_2 capture from flue gas in metal-organic frameworks. *ChemSusChem* 10 (7): 1543–1553.

135 Altintas, C., Altundal, O.F., Keskin, S. et al. (2021). Machine learning meets with metal organic frameworks for gas storage and separation. *J. Chem. Inf. Model.* 61 (5): 2131–2146.

136 Nandy, A., Duan, C., and Kulik, H.J. (2021). Using machine learning and data mining to leverage community knowledge for the engineering of stable metal–organic frameworks. *J. Am. Chem. Soc.* 143 (42): 17535–17547.

137 Magnin, Y., Dirand, E., Orsikowsky, A. et al. (2022). A step in carbon capture from wet gases: understanding the effect of water on CO_2 adsorption and diffusion in UiO-66. *J. Phys. Chem. C* 126 (6): 3211–3220.

138 Kolle, J.M., Fayaz, M., and Sayari, A. (2021). Understanding the effect of water on CO_2 adsorption. *Chem. Rev.* 121 (13): 7280–7345.

139 Zhang, X., Zhang, K., Yoo, H. et al. (2021). Machine learning-driven discovery of metal–organic frameworks for efficient CO_2 capture in humid condition. *ACS Sustain. Chem. Eng.* 9 (7): 2872–2879.

140 Reardon, H., Hanlon, J.M., Hughes, R.W. et al. (2012). Emerging concepts in solid-state hydrogen storage: the role of nanomaterials design. *Energy Environ. Sci.* 5 (3): 5951.

141 Sculley, J., Yuan, D., and Zhou, H.-C. (2011). The current status of hydrogen storage in metal–organic frameworks: updated. *Energy Environ. Sci.* 4 (8): 2721.

142 He, Y., Zhou, W., Qian, G. et al. (2014). Methane storage in metal–organic frameworks. *Chem. Soc. Rev.* 43 (16): 5657–5678.

143 Makal, T.A., Li, J.-R., Lu, W. et al. (2012). Methane storage in advanced porous materials. *Chem. Soc. Rev.* 41 (23): 7761.

144 Burchell, T. and Rogers, M. (2000). Low pressure storage of natural gas for vehicular applications. *SAE Trans.* 109: 2242–2246.

145 Target explanation document: onboard hydrogen storage for light-duty fuel cell vehicles https://www.energy.gov/sites/default/files/2017/05/f34/fcto_targets_onboard_hydro_storage_explanation.pdf.

146 Fernandez, M., Woo, T.K., Wilmer, C.E.et al. (2013). Large-scale quantitative structure–property relationship (QSPR) analysis of methane storage in metal–organic frameworks. *J. Phys. Chem. C* 117 (15): 7681–7689.

147 Fernandez, M., Trefiak, N.R., and Woo, T.K. (2013). Atomic property weighted radial distribution functions descriptors of metal–organic frameworks for the prediction of gas uptake capacity. *J. Phys. Chem. C* 117 (27): 14095–14105.

148 Pardakhti, M., Moharreri, E., Wanik, D. et al. (2017). Machine learning using combined structural and chemical descriptors for prediction of methane adsorption performance of metal organic frameworks (MOFs). *ACS Comb. Sci.* 19 (10): 640–645.

149 Wu, X., Xiang, S., Su, J. et al. (2019). Understanding quantitative relationship between methane storage capacities and characteristic properties of metal–organic frameworks based on machine learning. *J. Phys. Chem. C* 123 (14): 8550–8559.

150 Tsamardinos, I., Fanourgakis, G.S., Greasidou, E. et al. (2020). An automated machine learning architecture for the accelerated prediction of metal–organic frameworks performance in energy and environmental applications. *Microporous Mesoporous Mater.* 300: 110160.

151 Fanourgakis, G.S., Gkagkas, K., Tylianakis, E. et al. (2019). A robust machine learning algorithm for the prediction of methane adsorption in nanoporous materials. *J. Phys. Chem. A* 123 (28): 6080–6087.

152 Lee, Y., Barthel, S.D., Dłotko, P. et al. (2017). Quantifying similarity of pore-geometry in nanoporous materials. *Nat. Commun.* 8 (1): 15396.

153 Zhang, X., Cui, J., Zhang, K. et al. (2019). Machine learning prediction on properties of nanoporous materials utilizing pore geometry barcodes. *J. Chem. Inf. Model.* 59 (11): 4636–4644.

154 Simon, C.M., Kim, J., Lin, L.-C. et al. (2014). Optimizing nanoporous materials for gas storage. *Phys. Chem. Chem. Phys.* 16 (12): 5499–5513.

155 Zhang, X., Zhang, K., and Lee, Y. (2020). Machine learning enabled tailor-made design of application-specific metal–organic frameworks. *ACS Appl. Mater. Interfaces* 12 (1): 734–743.

156 Thornton, A.W., Simon, C.M., Kim, J. et al. (2017). Materials genome in action: identifying the performance limits of physical hydrogen storage. *Chem. Mater.* 29 (7): 2844–2854.

157 Ahmed, A. and Siegel, D.J. (2021). Predicting hydrogen storage in MOFs via machine learning. *Patterns* 2 (7): 100291.

12

Big Data Science in Nanoporous Materials

Datasets and Descriptors

Maciej Haranczyk[1] *and Giulia Lo Dico*[1,2]

[1] *IMDEA Materials Institute, C/Eric Kandel 2, Madrid, Spain*
[2] *Tolsa Group, Carretera de Madrid a Rivas Jarama, 35, Madrid, Spain*

12.1 Introduction

Nanoporous materials contain complex networks of void channels and cages that are exploited in many industrial applications. The zeolite class of these materials is the most well-known, as they have found wide use in industry since the late 1950s, with common applications as chemical catalysts, membranes for separation, and water softeners [1]. In addition to zeolites, ordered organic or inorganic–organic polymers such as metal–organic frameworks (MOFs) [2], covalent–organic frameworks (COFs) [3], (quasi)-ordered porous polymer networks (PPNs) [4], as well as porous molecular materials, have been prototyped for applications in separation [5–8], gas storage [4,9,10], energy storage and functional electronics [11,12], sensing [13,14], drug delivery [15], catalysis [16], and others. A key requirement for the success of any nanoporous material is that the chemical composition and pore morphology must be optimal under the given conditions for a particular application. Finding the optimal material in the space of possible structures is the objective of (nanoporous) material discovery. Materials discovery makes use of available theories to predict the performance of prototype materials, identifies the high-performing structures using known structure–property relationships, and resorts to exploratory, i.e., trial and error, experimentation when no adequate theory exists (and discovers the corresponding structure–property relationships during the process).

Early, before the personal computer and pre-Internet era (i.e., before the Information Age), material discovery efforts were slow due to low throughput of nearly all steps of research and development, e.g., sequential execution of experiments, slow paper-based data storage, and limited access to data, hindering fast formulation of new research hypotheses or theories, etc. The Digital Revolution has brought widespread adoption and proliferation of digital computers and digital record-keeping. They have allowed for massive parallel property evaluation for large material sets using either high-performance computing or robot-involving high-throughput experiential screening. Similarly, widespread Internet-based connectivity has enabled immediate data-sharing among researchers worldwide. These developments have provided unprecedented opportunity to accelerate materials discovery, providing the researchers can efficiently analyze the large and complex data. Such data, as well as the required methodology, is collectively referred to as Big Data (or its various forms, such as Big Data analytics, Big Data approach, Big Data science, etc.). It broadly includes generating and capturing data, data storage, data analysis, search, sharing, transfer, visualization, querying, updating, and others. Furthermore, Big Data adopts machine learning and artificial intelligence approaches to identify trends in the data, such as those representing structure–property relationships, and even moves toward autonomous workflows in which new theories or prototype materials can be generated and tested.

AI-Guided Design and Property Prediction for Zeolites and Nanoporous Materials, First Edition. Edited by German Sastre and Frits Daeyaert.

In the context of nanoporous materials, Big Data is currently primarily associated with the multitude and diversity of crystal structures that are both known and predicted for families of nanoporous materials [17]. The unmatched versatility of these advanced crystalline polymers, such as MOFs and COFs, has its origin in their modular structure. They are formed by connecting distinct chemical building blocks (secondary building units: SBUs) into periodic networks; the basis of reticular chemistry [18]. By exchanging or modifying chemical building blocks, we can both explore the unlimited space of possible structural chemistry within an isoreticular (same topology) series, as well as achieve a wide range of alternative topologies. In recent years, significant progress has been made in identifying and classifying the underlying topologies or nets of crystal structures; a repository of ca. 3000 of them is maintained within the searchable Reticular Chemistry Structural Resource (RCSR) database [19]. These nets have been used as topological blueprints for new materials to be virtually constructed from sets of SBUs, leading to databases of nearly 1 million structures already becoming available in the late 2010s. It is in high contrast to the much smaller sets, typically not more than 20–50 materials, involved in earlier studies, and which could still be accommodated by manual workflows and analyzed by one-by-one visualization approaches. Clearly, the early enumeration and prediction of thousands of hypothetical zeolites [20] have triggered the adoption of Big Data in nanoporous materials, as the large structure databases required new tools and approaches to characterize structures, analyze trends, and make new predictions.

Large structure datasets have opened up many research avenues to material discovery, which can be explored through Big Data approaches. Typical research tasks focused on a database of nanoporous structures include:

1) *Performing searches and surveys*. Given a query structure, a researcher may look for structures with similar characteristics such as a largest cavity diameter, a value of internal surface area, or a specific pore arrangement. Similarly, the dataset may be surveyed to explore the distribution of a property of interest among structures, e.g., by analysis of a histogram representing void fractions of the structures included in the dataset.
2) *Diversity analysis, sampling, and clustering*. Given a criterium for defining (dis)similarity between structures, a researcher may choose to analyze all structures to identify the groups of the most (dis)similar structures or reduce the size of the database by selecting the most representative structures for further studies.

Once the structure database is appended with further data containing information about the corresponding material properties or application-specific performance information, the research tasks can be expanded to include:

3) *Investigation of structure–property relationships (SPRs)*. Given a choice of structural features, a researcher can investigate relationships and correlations between these structures, represented by the features and the corresponding properties or performance measures of interest. Depending on a number of factors, such as the complexity of the structures and property, and the desired level of intuitive interpretability, etc., the SPRs can assume various forms from simple multiple linear regression models operating on intuitive features to complex machine learning models operating on multi-dimensional, computer-generated structure representations. Nevertheless, the main motivation is that the extracted trends can expand the knowledge pool and/or can be employed to assess the properties, or performance, of newly considered or novel prototype materials.

Execution of the presented typical tasks requires assembly of a computational workflow involving a number of components. First, the material structure datasets need to be available in their curated form (unambiguously readable by computer codes). Second, the material structures need to be expressed in a numerical format. Typically, they are analyzed by automated, high-throughput codes to identify their characteristic features referred to as descriptors or feature vectors. Third, if the to-be-performed task involves analysis of properties and performance, the property datasets have to be collected, and usually it comes from literature analysis or designed databases performing consistent experiments or molecular simulations. Finally, we need a methodological framework to combine the above datasets into useful workflows. The following sections will highlight the typical datasets, descriptors, and methods used to implement typical Big Data tasks 1–3. Finally, we highlight more advanced and emerging aspects of Big Data that enhances the Big Data material discovery efforts.

12.2 Repositories of Nanoporous Material Structures

Large sets of material structures are the center point of the current Big Data applications to nanoporous materials discovery. A convenient representation of structures are all-atom models, which not only contain the crystal structure but also, indirectly, the information about the corresponding porosity, which is often critical to the materials' performance in an application of interest. Specifically, atomistic models of crystal structures are represented by a unit cell (repeating unit of the crystal structure), the definition of the geometry of the unit cell as well as the list of atoms present in the cell together with their coordinates. The latter can also sometimes be presented in the form of the smallest repeating fragment of the crystal together with the symmetry operations required to build the entire unit cell. A common file format for distribution of crystal structures is the crystal information file (CIF). CIF is a text file format promulgated by the International Union of Crystallography (IUCr). CIFs can also accommodate metadata such as the information on how and by who the crystal structure was obtained, etc. Furthermore, the format also allows encoding further crystal structure details such as partial occupancy of crystal sites, the presence and location of solvent molecules, and many others. Due to this flexibility to accommodate diverse information, the CIF file format has become *de facto* the standard for sharing experimentally obtained crystal structures, and it is used by crystal structure databases such as the Cambridge Structure Database (CSD) [21], the Inorganic Crystal Structure Database (ICSD) [22,23], and Crystallography Open Database (COD) [24]. Aside from CIFs, many computational structure analysis programs as well as material modelling and simulation packages use their own formats to represent the materials' periodic unit cells. These are typically reduced in content w.r.t. experimental CIFs to facilitate structure handling, e.g., without any metadata. Such file formats are CSSR-based on CSD Crystal Structure Search and Retrieval data used, for example, in codes like RASPA and native formats of simulation packages such as DL_POLY and LAMMPS. Crystal structures of nanoporous materials typically come from either crystallographic measurements or computational predictions, both of which will be outlined in the following sections.

12.2.1 Experimental Crystal Structures

Newly synthesized crystalline materials are typically reported in scientific journals. Generally, researchers attempt to obtain large crystals in order to solve the structures, i.e., obtain information on positions of atoms, using single crystal diffraction experiments. If successful, the corresponding CIF is reported in a publication outlining the synthesis of a new material (e.g., in the supporting information files accompanying it). Often, the CIF is simultaneously submitted to be included in a database of crystal structures such as the CSD. Such experimentally refined crystal structures for nanoporous materials often include solvent molecules and partially occupied or disordered atoms. Such features are crystallographically meaningful but must be removed prior to computer-based structure and porosity analysis. In the case of zeolite structures, the International Zeolite Association (IZA) has undertaken efforts to perform such curation, i.e., IZA database contains CIF files representing nearly 250 all-siliceous zeolites [25]. Curation of other families of nanoporous materials such as MOFs is a more challenging endeavor, as these are reported in much greater numbers, their structure complexity is much greater, and their disorder is more common. Such efforts have been initially conducted by isolated groups in a form of creating in-house structure repositories [26]. More recently, the researchers of the Nanoporous Materials Genome Center, and supported by the broader research community, have undertaken a more systematic effort to curate the experientially reported MOFs. It resulted in the computation-ready, experimental (CoRE) MOFs datasets (versions 2014 and 2019) [27,28]. The most recent release contains ca. 14,000 curated 3D MOFs derived from the structures deposited in the CSD, accompanying publications reported in the literature but not yet deposited in the CSD (obtained via a search based on the Web of Science literature database), those contributed by the research community and, finally, those derived through semi-automated reconstruction of disordered structures. Similar efforts have been undertaken to curate the reported structures of COF materials, i.e., CoRE COF dataset based on the CSD structures [29].

12.2.2 Predicted Crystal Structures

Crystal structure models of nanoporous materials can also be predicted computationally ahead of material synthesis. Initial efforts had been directed toward zeolite structures. Deem et al., for example, employed an algorithm to explore the space of zeolites that involved systematic search through possible geometries and symmetries of the periodic unit cell with random exploration of the configurations of Si atoms within [20], with later augmentations of the positions of oxygen atoms. The structures were refined by relaxing their positions using classical force fields [30]. The resulting database contains over 2.6 M unique structures. Roughly 15% of these are within +30 kJ/mol of α-quartz, the band in which most of the known zeolites lie. Other groups have also developed and applied methodology for the prediction of zeolite structures [31].

Advanced porous materials, such as MOFs and COFs, offer much greater diversity, and therefore potential of discovery of structures for various applications. Early work in the area of structure prediction included a variety of approaches, including chemical substitution or functionalization based on modifying known experimental structures [32]; and recursive, geometry-based assembly of materials' building units in their various combinations [33]. Modification of known structures allows for the design of new, topologically analogous (i.e., isoreticular) materials based on a reliable experimental reference structure, but does not permit easy exploration of alternative underlying topologies. Recursive geometric assembly, which resembles making construction of LEGO blocks by stitching them one after another, allows for expansive, combinatorial design of large datasets of material models; however, this strategy provides little direct control over the resulting topology, symmetry, etc., of material models. Furthermore, it requires very precise tuning of nontrivial building block geometry in order to achieve particular (especially simpler) topologies.

More recently, an alternative approach to construction of crystalline structures which allows for both high-throughput enumeration of large datasets of materials, and precise control over their underlying crystal topologies, has been implemented by a number of groups [34–36]. In this approach, information on the geometry/topology of molecular building blocks is combined with reference data describing a net. By utilizing topology-specific connectivity information and symmetry operations, building units (SBUs) can be positioned in accordance with a particular underlying net (e.g., those provided by the RCSR), exploiting the essence of reticular chemistry [18]. In the following paragraph, this approach is outlined in more detail.

The specific information of interest for structure assembly is the representation of a desired topology as a periodic graph (net), i.e., the set of vertices and edges between vertices that are unique by symmetry, and the symmetry operations which must be performed in order to complete a periodic unit cell. The RCSR database provides this information in the *Systre* file format [37]. Each new cell is named using three-letter codes, for example, **dia** is diamond topology. Expanding the basic vertices and edges using the symmetry space group provided results in the complete periodic graph illustrating the net (Figure 12.1, left). The vertices and edges of this graph encode the relative positions and connectivity of the material's building blocks. The structure assembly algorithm constructs corresponding material framework models by positioning molecular building units according to this periodic graph.

In addition to topology information, the algorithm requires molecular building blocks to be provided as input. Specifically, the molecules' connection sites (i.e., carboxylic functional groups by which organic components coordinate to metal/metal-organic components) need to be identified and indicated with temporary markers (e.g., dummy atoms), which provide a convenient connectivity-based abstraction such that the framework assembly algorithm need only consider arranging building blocks based on these markers (e.g., a copper paddlewheel is essentially abstracted into its arrangement of sites, i.e., a square: Figure 12.2). By overlaying SBUs described in this manner with the specific vertices they correspond to, a particular topology can be guaranteed in the resulting framework model. Specifically, this process involves the following steps:

1) *Interpreting the net.* The net information provided (vertices, edges, and symmetry space group) is used to construct the complete periodic graph of the net by performing the required symmetry operations on the symmetrically distinct vertices/edges provided (Figure 12.1, left).
2) *Aligning the building blocks.* Having parsed the net information, the algorithm now parses the building block information. The connectivity of the building blocks (i.e., the number of connecting positions they exhibit, and their arrangement) is analyzed, and each building block is

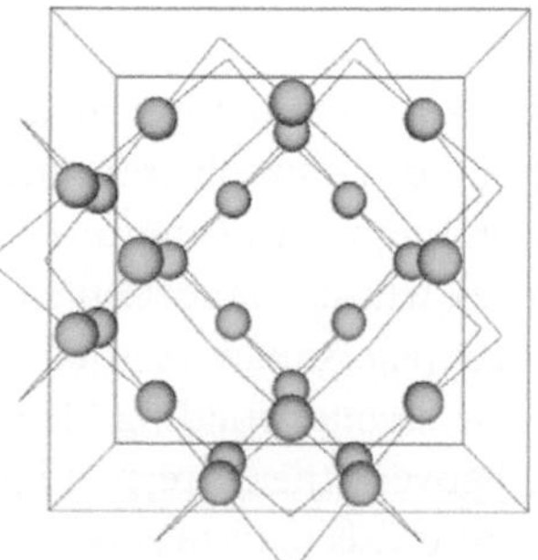
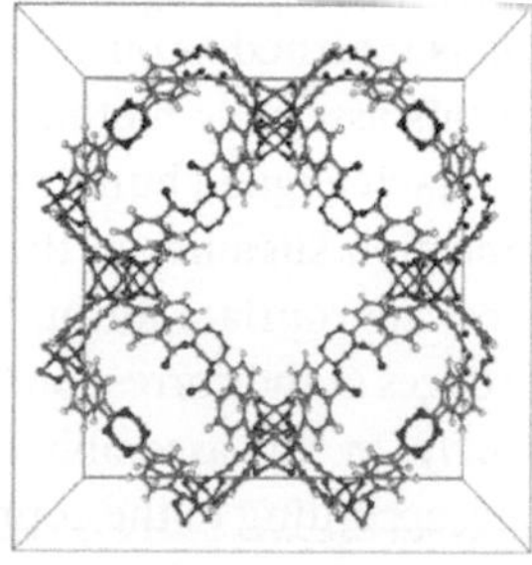

Figure 12.1 Example using **rhr** net. Left: the **rhr** net, showing the unit cell (black box), vertices (green spheres), and edges (red lines). Right: a framework assembled to exhibit the **rhr** net, comprising the building blocks illustrated in Figure 12.2: Cu_2 paddlewheels as the metal SBU (brown atoms; replacing the green spheres) and benzene 1,4-dicarboxylic acid as the organic linker (gray C atoms, red O atoms, and white H atoms; replacing the red edges).

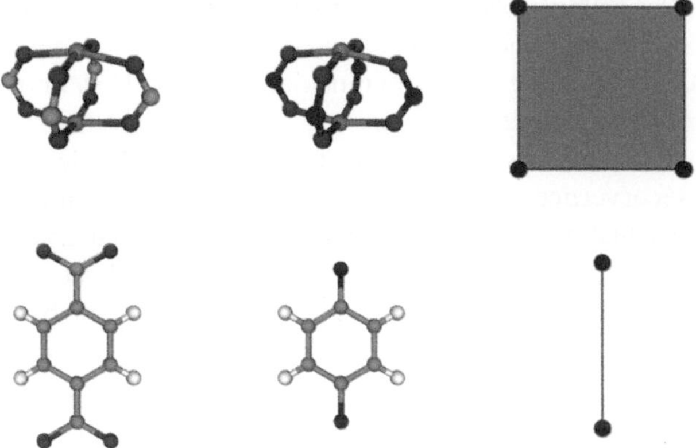

Figure 12.2 Building blocks for an example MOF. Left: a copper paddlewheel (top) and a linear dicarboxylic linker (bottom) are illustrated. These building blocks coordinate to one another by shared COO groups; these are shown in both building blocks only for clarity. Center: the markers identifying the connection sites in each SBU are shown in black. During framework construction, these markers are used to align the building blocks, to produce a framework wherein a single COO group is seen to connect both building blocks: Figure 12.1, right). Note that the COO group is preserved on the metal SBU only; this allows for the linker to be treated as a rigid entity, while implicitly allowing for torsion or other rotation about this bond, which may be required for modeling certain topologies. These markers are removed from the final framework model. Right: the geometric shape representing each SBU's arrangement of connection sites. Brown represents Cu atoms, gray C atoms, red O atoms, and white H atoms.

allocated to the appropriate symmetrically distinct vertices of the net. Furthermore, each building block is oriented in 3D space such that its site markers are aligned to the edges of the corresponding vertex, and symmetry operations are performed. This procedure results in a set of SBUs, each aligned to a vertex. In this manner, a framework can be constructed by simply translating these SBUs in 3D space, such that they are positioned in accordance with the specified net (step 3).

3) *Assembling the framework*. The complete set of aligned building blocks must now be positioned relative to one another in order to achieve the connectivity described by the net. Two approaches to framework assembly are explored, as described below. Resulting frameworks are evaluated based on the maximum distance between pairs of connectivity markers in adjacent building blocks (the ideal case being zero distance, i.e., all connections are made perfectly); the framework exhibiting the lowest deviation is returned as output.
   - *Method 1, net-based assembly*. In this assembly method, the unit cell of the net is scaled to accommodate the building blocks. Each aligned building block can then be simply translated such that the centroid of its connection sites lies at the corresponding vertex (Figure 12.1, right). This elegant method is most appropriate for building blocks whose connection sites are arranged very similarly to the edges of the corresponding vertices (e.g., square vs. square).
   - *Method 2, connection-based assembly*. In this assembly method, the aligned building blocks are connected together one by one, according to the connections between the corresponding vertices. Once all building blocks are connected, the resulting unit cell is determined. This method allows the unit cell to scale, stretch, or shear freely based on the SBU shapes provided, and is more appropriate for building blocks whose connection sites are arranged distinctly to the edges of the corresponding vertices (e.g., the irregular hexagonal arrangement of sites in hexacarboxylic acids such as those in **ntt**-net MOFs [38]).
4) *Collision detection (optional)*. This algorithm can also provide for an optional collision-detection routine, which rejects framework geometries resulting in atoms positioned too close to one

another. This feature is optional since some degree of atomic collision may be resolvable with subsequent application of framework relaxation techniques.

5) *Relaxation (optional).* The resulting assembled 3D structure models can be relaxed to identify the configuration corresponding to a minimum energy structure, though in the case of many reported structures, the initial conformations resulting from Steps 1–4 are already of satisfying quality for many Big Data tasks. Nevertheless, depending on the required accuracy, the atomic composition of the material and the available computational resources, there are many techniques available to be used in this task, from density function theory, through semiempirical methods, as well as classical force fields.

The topology-based structure assembly algorithm has been implemented in Zeo++, an open-source, high-throughput porous structure assembly, characterization, and comparison tool as well as in a number of in-house codes. These programs have been used to rapidly enumerate large datasets of isoreticular material models using the RCSR topologies. The frequency of occurrence of specific topologies experimentally has also been examined, leading to an important result in the field of reticular chemistry, namely, that the highest symmetry, minimal transitivity nets are generally most likely to form [39, 40]. Accordingly, assembling structure models based on the most prevalent net(s) for a given set of building block shapes ensures that the generated models are likely to be similar to materials that can be, or have been, achieved in experiment. Similarly, when predicting new structures, we can limit the sets of building blocks to reuse only those seen in experientially reported structures or extend these sets to include diverse structures mined from the catalogs on commercially available molecules or molecular databases such as PubChem [41]. Finally, then working with multiple datasets of structures, it is often necessary to detect and flag duplicates. MOFid and MOFkey are systematic MOF identifiers [42], which are based on algorithms that involve deconstruction of periodic MOF structures and unique representation of the building blocks and topology involved. In both cases, they are available in both open-source and Web-based tool versions.

12.3 Descriptors

Performing Big Data analysis tasks on a large set of nanoporous structures requires transforming 3D structure models into numerical representations, which emphasize their relevant structural features, and allow to query and compare various structures on equal footing. These features are referred to as descriptors or feature vectors, i.e., $x = (x_1,x_2,...,x_m)$ of m-dimensional structure presentation. The features allow formulating structure similarity metrics as well as correlating them with the material properties in the context of structure–property relationship analysis. To obtain numerical features, the structures need to be analyzed. Traditional analysis approaches heavily rely on structure visualization to obtain key measurements via a program's Graphical User Interface (GUI), and therefore suffer from too limited throughput to analyze large datasets. The emergence of initial databases of predicted structures in the late 2000s has triggered the development of algorithms and software tools focused on automatic analysis of the structures and the porosity in nanoporous materials. The currently used features can be divided into the two groups outlined in the following paragraphs.

12.3.1 Handcrafted Descriptors

This group, the handcrafted descriptors, aims to capture and numerically represent structural features that are identified by a researcher as useful to describe the structure or its properties. The development of this type of descriptors generally require domain knowledge to identify the targeted features, and is often expensive to implement. The descriptors can focus on the features representing the structure, its corresponding void space, or the surfaces separating the two (Figure 12.3A). The numerical descriptors can assume real values or be assigned binary values indicating presence of the feature (the latter are often referred to as fingerprints). Most commonly used descriptors are based on geometrical features present in the structure, i.e., representing atoms are hard spheres of chemically relevant radii such as van der Waals radii, as these can be efficiently calculated for large sets of structures. Sporadically, structural features are represented by specific value ranges or distributions. In these cases, we can choose to either simplify this information, e.g., represent it by an average or median value, or encode it "as is," i.e., as a feature vector, corresponding to important histogram bins. The commonly used descriptors are:

- crystal density (ρ), in (kg/m^3);
- pore limiting diameter (PLD), in (Å), the diameter of the largest sphere to percolate through a material;

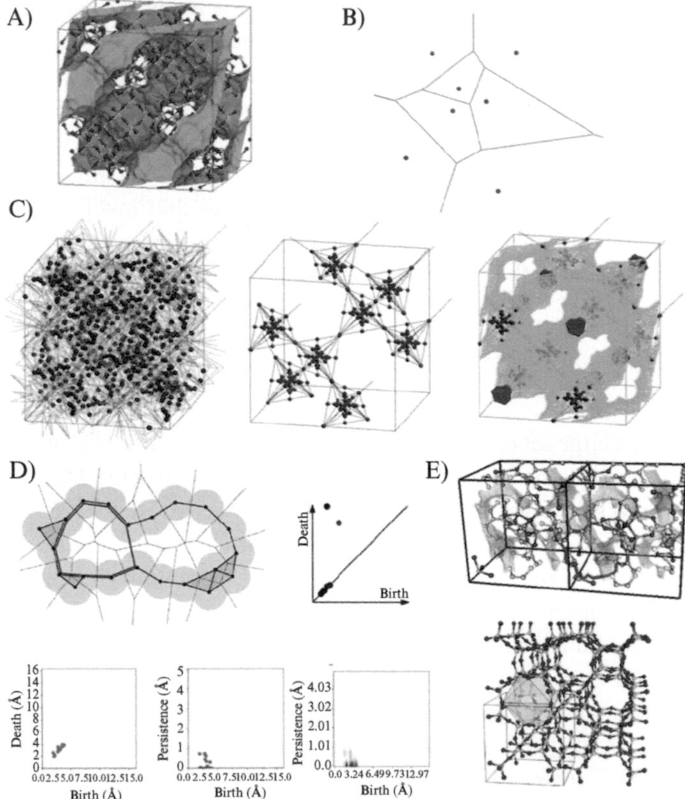

Figure 12.3 (A) An example of a nanoporous material structure (FAU zeolite) with highlighted pore surface in green; (B) An example of a Voronoi diagram constructed for 7 points in 2D; (C) An example of a workflow implemented in Zeo++ (FAU zeolite): (left) Voronoi network, (center) filtered to highlight only nodes and edges accessible to a methane probe, (right) surface of the material accessible to methane probe (green) and corresponding to inaccessible pockets (red); (D) (top) Illustration of persistence homology and (bottom) generation of persistence images; (E) Example of 1D and 2D features identified by persistent homology: (top) a cycle and (bottom) a void.

- largest cavity diameter (LCD), in (Å), the diameter of the largest sphere than can fit inside the material's pore system;
- accessible volume (AV), in (cm^3/g or cm^3/cm^3) for respectively gravimetric and volumetric AV; and
- accessible surface area (ASA), in (m^2/cm^3)

There have been a number of algorithmic frameworks for descriptor calculations. Accessible volume and accessible surface area are typically calculated using efficient Monte Carlo algorithms, such as explored by Duren et al. [43] and Do et al. [44]. They randomly sample the volume of the unit cell and the surfaces of its atoms to identify the fraction of total, volume, and surface that is not occupied and overlapped by atoms. AV and ASA are then obtained by multiplication of this fraction by the total values, which can be calculated analytically. First et al. developed ZEOMICS and MOFOMICS tools that fragment and approximate the void space inside porous materials as a set of geometrical blocks such as cylinders and spheres, the geometrical parameters of which can approximate cavity sizes [45,46]. The space contained in a material's unit cell can also be discretized in the form of a 3D grid, points of which can be assigned information such as being flagged if they correspond to the void space or the distance to the material surface. The resulting 3D grids can be further analyzed to predict guest molecule accessibility and geometrical parameters describing pores. For example, Haldoupis et al. have adopted this approach to calculate PLD and LCD [47].

A number of approaches have successfully employed a computational geometry technique, the Voronoi decomposition (or Delaunay triangulation). In the Voronoi decomposition, the space surrounding n atoms is divided into n polyhedral cells such that each cell face is a plane equidistant from the two atoms sharing that face (Figure 12.3B). Edges of such cells overlap with lines equidistant to three neighboring atoms, whereas vertices of cells, the Voronoi nodes, are equidistant from four neighboring atoms (in a general asymmetric case). The Voronoi network, a three-dimensional (3D) graph comprising such nodes and edges, maps the void space surrounding the atoms. Analysis of such a network can provide parameters such as the diameters of the largest included and free spheres as well as detailed information about void space geometry and topology [48]. For example, void space regions inaccessible to a given probe can be identified, and this information can be utilized in the calculation of ASA, AV, and others (Figure 12.3C). The Voronoi network also serves as a starting point for the development of novel structural descriptors which encode: i) the entire shape of a material's void space: the Voronoi hologram [49], which is a histogram representation of the guest-accessible portion of the Voronoi network; and ii) the stochastic ray approach [50], which can capture surface texture information. Zeo++ is an example of an open-source package that can perform multifacet characterization of material porosity [48,51] relying on the Voronoi tessellation with sub-0.1 Å accuracy [52] and high speed of C++ code. Other authors have also used Voronoi/Delaunay techniques, including Foster et al., who prototyped calculation of the diameter of the largest included and the largest free spheres in zeolites [53], with Estela and Blatov using it to derive other descriptors of crystal structures [54,55], and Simon et al., who used it to identify guest-molecule binding pockets and used their estimated binding energy as the structure descriptor [56].

In summary, the handcrafted descriptors are very useful to capture and communicate important features but are typically costly to develop. In fact, only a handful of descriptors are implemented in common software tools such as Zeo++ [51] or POREBLAZER [57]. Furthermore, when selecting the descriptors, specific domain knowledge is required. Each application may require careful selection of features that an expert recognizes as important for the specific task. For example, in the case of gas adsorption at high pressure, guest molecules tend to occupy the entire void space of the material, so the void fraction can be a good indicator of total adsorption. In contrast, for gas adsorption at low pressures, guest molecules tend to aggregate in the strongly binding regions of the material's pore, and therefore descriptors reflecting the presence of such binding sites are good indicators for adsorption. For these reasons, there is an increasing interest in automating descriptor development and creating feature vectors to systematically capture all the length scales of features present in the structures.

12.3.2 Toward Automatically Generated and Multi-Scale Descriptors

Due to limitations of singular descriptors, it is often desirable to involve representations that address a broader spectrum of structural features. One of the initial representations of such multi-scale descriptors is the pore size distribution (PSD) histogram. PSD histograms indicate the fraction of the void space volume that corresponds to certain pore sizes. However, they lack information about pore connectivity. Algorithms for PSD and their implementations were discussed by Do et al. [57] and Sarkisov et al. [58], and involved computational geometry and grid-based approaches. Similarly, Zeo++ employs the Voronoi tessellation technique to calculate PSD histograms [50]. Zeo++ also offers additional related approaches, namely the Voronoi holograms [49] and stochastic rays [50], which are histograms summarizing, respectively, the node-edge-node triplets of the Voronoi network representing the probe-accessible space and the width of pores measured in random directions. Both later representations aim to include some information about the pore shape but are much more difficult to interpret, and may be more suitable for applications involving machine learning.

More recently, researchers have begun developing more diverse approaches allowing to simultaneously capture structural features across various length scales as well as dedicating less effort to laborious manual feature implementation or curation. Specifically, persistent homology from the topological data analysis field can be adapted to describe the geometry of nanoporous materials at various scales [59,60]. In this approach, the structure's atoms are represented as a union of hard spheres centered on its atoms. While increasing the radii of these spheres, the changes in the topology of their union is monitored. The changes come in two types: a topological feature, such as a loop or a void, which either appears or disappears. An important consequence of the algebraic formulation of this process is that these events can be paired uniquely, resulting in a set of birth–death pairs of radii, called a persistence diagram (Figure 12.3D). There are two persistence diagrams relevant to the analysis of nanoporous materials: a diagram that tracks births and deaths of loops that correspond to cross-sections of channels (these are called one-dimensional (1D) features), and a diagram that tracks voids that correspond to pockets (two-dimensional (2D) features) (Figure 12.3E). The difference in birth–death values is called the persistence of the pair. Pairs of larger persistence capture more prominent pores in nanoporous structures. The analysis aims to identify all the pairs present in a structure. In order to normalize this analysis across a repository of materials of various unit cell sizes, a computational cell of equal size, e.g., $(100\ \text{Å})^3$ cell, is constructed for each structure while ignoring periodic boundary conditions. The size of this cell is chosen to be large enough to capture the statistics of the distribution of the topological features in every structure. Finally, persistence diagrams are translated into more practical feature vectors suitable for Big Data analysis tasks via a persistence images approach, introduced by Adams et al. [61]. The birth–death pairs (b, d) are transformed into birth–persistence pairs $(b, d - b)$. They are then convolved with Gaussians and discretized onto a grid of a fixed size, e.g., 50×50, by integrating the resulting mixture of Gaussians in the cells of the grid. The latter step introduces "fuzziness" of the representation facilitating identification of similar structures. A similar approach involving topological data analysis was also proposed by Lee et al., and involved persistent barcodes to represent morphology of the pore surfaces formed by nanoporous materials' voids [62].

Another approach that was designed to encompass features across length scales is the energy histogram implemented by Snurr et al. [63]. In this approach, a probe atom is chosen, e.g., argon or carbon. Then the energy value, for this probe species on a 3D grid of equally-spaced points placed throughout the unit cell of a given structure, is calculated using the Lennard–Jones potential:

$$V=4\epsilon\left[\left(\frac{\sigma}{r}\right)^{12}-\left(\frac{\sigma}{r}\right)^{6}\right] \tag{12.1}$$

where ϵ and σ are the LJ well depth and size, and r is the distance between the LJ sites. The Lorent–Berthelot (LJ) mixing rules were used to obtain the LJ parameters between unlike sites. The LJ parameters for the framework atoms were taken from commonly used force fields such as the Universal Force Field (UFF) [64]. Finally, a 1D histogram from the 3D energy grid data is obtained, and can be optionally simplified to reduce the number of poorly populated bins such as those corresponding to points located inside large pores, for which the energy values are close to zero. The advantage of this approach is that it directly incorporates information about the chemistry of the system through ϵ and σ.

Besides geometry and topology, chemical makeup of the internal surfaces is an important descriptor of nanoporous structures. Thus far, approaches have constructed chemical descriptors by incorporating information from building blocks of a material such as MOFs and their functional groups [65,66]. A recent report indicated the possibility of adopting chemical descriptors developed in the context of inorganic materials. Specifically, word embeddings of the chemical elements may be employed to represent a given material's stoichiometric formula, and by doing so incorporate the relevant chemical information. The chosen embeddings are constructed from a large corpus of abstracts with the word2vec algorithm [67]. While the use of word embeddings to featurize composition do represent an implicit knowledge that the chemical elements are distinct, they use no explicit element-specific properties. Instead they carry information on common word associations that an element encounters in the scientific literature derived from an unsupervised learning procedure on raw text. A recent study on MOFs exploited matminer open-source tool [68] to generate feature vectors that correspond to 200 embedding dimensions, with the minimum, maximum, range, mean, and standard deviation for each dimension, for a total of 1000 values. It has been demonstrated that addition of such descriptors to geometry-based and topology-based descriptors, such as those described in the earlier sections, can significantly improve the machine learning models [59].

12.4 Properties

In addition to the structure datasets and their representations, investigation of the structure–property relationships requires datasets of properties collected for structures present in the dataset. Collecting such data experimentally is very challenging due to experimental costs, variations in material samples (e.g., presence of defects), and their preparation (e.g., activation protocols). The NIST/ARPA-E Database of Novel and Emerging Adsorbent Materials is a notable example. Fortunately, material modeling techniques can be effectively utilized to predict properties of nanoporous materials, given their all-atom structure models. The electronic structure methods, dominantly the density functional theory (DFT), are employed in structure relaxation as well as investigation of electronic and optical properties [69]. Classical mechanics, first-principle molecular simulations, and DFT calculations are used to shed light on thermal (e.g., negative thermal expansion [70], heat capacity [71]) and mechanical (e.g. elastic constants [72], and stiffness tensor [73]) properties. Similarly, Grand Canonical Monte Carlo simulations are employed in the prediction of guest–molecule adsorption [36]. These techniques have been the subject of recent reviews [74–77]. More importantly, the increasing computational resources have allowed applications of many of these approaches to sets of thousands of structures [56,33,78–81], with the resulting datasets being available by researchers through dedicated websites [82,83].

Various approaches have been developed and implemented to collect data from materials science literature, e.g., by automatic data extraction using natural language processing (NPL) techniques. For example, Jensen et al. presented efficient tools to automatically extract zeolite synthesis and topology data from journal articles and these have been used to predict synthesis pathways for new zeolite structures [84]. Kim et al. employed NPL tools to extract the values of MOF surface areas for scientific journals with high accuracy [85]. Without any doubt, the near future will bring much wider implementation of NPL to extract even more complex properties from the bulk of scientific literature, especially in relation to difficult-to-simulate phenomena related with synthesis, scale-up and defects.

12.5 Data Analysis

Once the sets of materials structures and their descriptors are available, we can implement basic data analysis functions. Specifically, we can assess the characteristics of materials by comparing the distribution of descriptor values. We can also filter and search structures by comparing the descriptor values of the structures in the dataset with given query values. Often, the descriptor space is of high dimensionality (e.g., more than 20), and the descriptors are somewhat related with each other. In these cases, data analysis tools such as principal component analysis (PCA) can be employed as a dimensionality reduction technique. In PCA, the original vector space, with n features, is transformed to new k components (k = n). The latter are created by projecting each pristine dimension, contributing with different weights, into the novel components. Each k vector contributes to the total variance with a certain score, thus the space may be reduced by removing of the least important k component, while preserving information of the data.

12.5.1 Material Similarity and Distance Measures

More complex analysis of materials often requires definition of similarity, or distance, between two structures. The motivation is to employ the Similarity Principle [86], which states that the materials similar in terms of their structure do exhibit similar properties. The challenge is how to implement the definition of similarity, as the complexity of material structures provides many facets that can be considered. Some domain knowledge is necessary to ensure that the similarity measures consider the features which are important for the property or the application of interest. The process requires measure of similarity or distance in a multi-dimensional descriptor space. For example, for binary vectors, where each bit indicates the presence ("1") or absence ("0") of a particular feature, the Tanimoto (or Jaccard) similarity coefficient has been widely utilized. For features represented by real vectors, Euclidean or Manhattan distance are commonly used. Table 12.1 features similarity and distance measures used in chemical applications. The selection of a similarity, or distance, measure typically requires either domain knowledge or some trail-and-error experimentation on smaller sets of examples, i.e., to verify if the considered measure produces results in par with expectations before it is deployed to study large datasets.

Table 12.1 The most commonly used similarity and distance measures with the corresponding definition.

Similarity or Distance method	Equation	Definition
Tanimoto (or Jaccard) similarity coefficient	$S_{AB} = c/(a+b-c)$	The value of the coefficient ranges from zero to one. An S value of one indicates that the materials have identical descriptor representations and a value of zero indicates that there is no similarity (i.e., there are no bits in common between the two materials).
Dice similarity coefficient (Hodgkin index)	$S_{AB} = 2c/(a+b)$	
Cosine similarity (Carbó index)	$S_{AB} = c/\sqrt{(a+b)}$	
Manhattan distance	$d_{x,y} = \sum_{i=1}^{n} \lvert x_i - y_i \rvert$	The distance between the two material (x,y) in *n*-dimensional space is defined as the sum of the distances in each dimension.
Euclidean distance	$d_{x,y} = \sqrt{\sum_{i=1}^{n} (x_i - y_i)^2}$	The distance of a line segment connecting two points into the vector space.
Weighted Euclidean distance	$d_{x,y} = \sqrt{\sum_{i=1}^{n} w_i (x_i - y_i)}^{2}$	Euclidean distance where each *i*th dimension is weighted by w_i. It is useful to solve problems when the descriptors contribute with no-equal scores affecting the targets.
Minkowski distance	$d_{x,y} = \left(\sum_{i=1}^{n} \lvert x_i - y_i \rvert^p\right)^{1/p}$	Generalization of Manhattan and Euclidean measures. When $p = 1$, Minkowski score is reduced to Manhattan definition, while for $p = 2$, Minkowski score is equal to Euclidean distance.

The measures operating on binary representations use a, b, c, values, where a is the number of bits set to "1" in material A; b is the number of bits set to "1" in material B; and c is the number "1" bits common to both A and B. For the measures operating on real values, x_i and y_i are the values of the *i*th property assigned to *x*th and *y*th material in the vector space.

12.5.1.1 Diversity Selection

Working with large sets of materials, it is often required to reduce the size of the sets to facilitate handling. Random sampling of structures to form smaller subsets of the original set is commonly performed. However, when the original set is not homogenous in terms of coverage of the material space, the randomly sampled subset will carry the same characteristics. Instead, we may choose to select subsets that maximize the diversity of the sampled set to better capture the variety present in the original set.

MaxMin maximum-dissimilarity-based selection is a useful technique to perform this task. Additionally, it does not require calculation of pairwise distances for all the structures in the original set. The MaxMin approach [87] proceeds as follows:

1) Select a starting structure, e.g., using random selection.
2) Calculate the distance of the remaining structures against every structure which has been selected, storing the minimum observed distance, d.
3) Add the structure which exhibits the highest d to the set of selected structures.
4) If the specified end criteria have not been met, go back to step 2.

12.5.1.2 Cluster Analysis

An investigated set of materials can be analyzed to identify the groups of similar structures using an unsupervised machine learning technique such as agglomerative hierarchical clustering. The algorithm is unsupervised because it has access to input values (descriptors) and infer patterns without target values (i.e., properties). It is a bottom-up approach that can be outlined in the following procedure:

1) Compute the distance matrix between the objects contained in the dataset.
2) Each object creates its own cluster.
3) The pair of clusters identified as the most similar is merged to form a new cluster. The similarity between clusters can be defined in various ways (Table 12.1).
4) The hierarchy is moved up by repeating point 3 until every object is contained in the same cluster

An alternative is the top-down approach, i.e., divisive hierarchical clustering, can also be formulated. The linkage between different clusters reflects their distance, and it can be defined in various ways (Table 12.2). One of the advantages of hierarchical clustering is that it can be depicted with a dendrogram in which on one of the sides is the cluster containing all objects, while the opposite

Table 12.2 List of possible linkage methods for connecting diverse clusters and building the dendrogram.

Linkage method	Equation	Definition
Single linkage	$d_{u,v} = \min\left(\mathrm{dist}\left(u[i], v[j]\right)\right)$	Nearest Point Algorithm. Minimizes the distance for all objects i in the cluster u and j in the cluster v.
Complete linkage	$d_{u,v} = \min\left(\mathrm{dist}\left(u[i], v[j]\right)\right)$	Farthest Point Algorithm or Voor Hees Algorithm. Maximizes the distance for all objects i in the cluster u and j in the cluster v.
Average linkage	$d_{u,v} = \sum_{i,j} \frac{d\left(u[i], v[j]\right)}{\left(\lvert u\rvert * \lvert v\rvert\right)}$	Mean distance between all objects *i* and *j* contained in the *u* and *v* clusters, in which cardinalities are expressed by $\lvert u\rvert$ and $\lvert v\rvert$, respectively.
Weighted linkage	$d_{u,v} = (\mathrm{dist}(s,v) + \mathrm{dist}(t,v))/2$	Arithmetic mean of the average distances between u and v, where u was formed with cluster s and t, and v is a remaining cluster.
Centroid linkage	$\mathrm{dist}(s,t) = \lVert c_s - c_t \rVert_2$	Euclidean distance between the centroid of cluster (s) and the centroid of the remaining cluster (t).
Ward's linkage	$d(u,v) = \sqrt{\frac{\lvert v\rvert + \lvert s\rvert}{T} d(v,s)^2 + \frac{\lvert v\rvert + \lvert t\rvert}{T} d(v,t)^2 - \frac{\lvert v\rvert}{T} d(s,t)^2}$	Useful for minimization of the variance. When *u* is a cluster created by merging *s* and *t*, *v* is a remaining cluster, and *T* is the sum of the cardinality $\lvert v\rvert + \lvert s\rvert + \lvert t\rvert$.

side of this plot shows all clusters containing single elements (singletons). The analysis of the dendrogram can help to identify the appropriate number of clusters based on the distance between clusters. The main disadvantage of hierarchical clustering is that it requires a precalculation of pairwise distances which can be computationally expansive for large sets of materials. In such cases, alternative approaches such as k-means clustering can be employed [88].

12.6 Machine Learning Models of Structure–Property Relationships

Given a set of material structures, and their descriptors as well as the corresponding set of properties, it is possible to construct a mathematical model of the relationships between descriptors and properties. Statistical machine learning provides numerous frameworks in which the structure–property functions can be learnt from example data via supervised learning. Specifically, these frameworks seek to identify

$$P = f(x_1, x_2, x_3, \ldots, x_n) \tag{12.2}$$

where P is the property of interest, and f is the function that calculates this property from the descriptors $x_1 \ldots x_n$. The form of $f(x)$ could be linear, multilinear, or nonlinear. More complex machine learning algorithms employ definitions of f based on decision trees, neural networks. or other frameworks, specification of which is expressed by hyperparameters. The latter are usually optimized to achieve the highest-quality models given a specific framework. Meanwhile, the selection of frameworks is done by assessment of the quality of the resulting models [89]. The models are usually trained and validated using different subsets of the available data (e.g., by splitting the available 80/20). Some of the most commonly used frameworks are multilinear regression, random forests, neural networks, and support vector machines. They are implemented in statistical analysis package R as well as numerous Python libraries (Sklearn, Scikit-learn, Keras, and TensorFlow). In the following, we will outline the decision tree-based frameworks in more detail, as they have a proven track record in nanoporous materials-related applications.

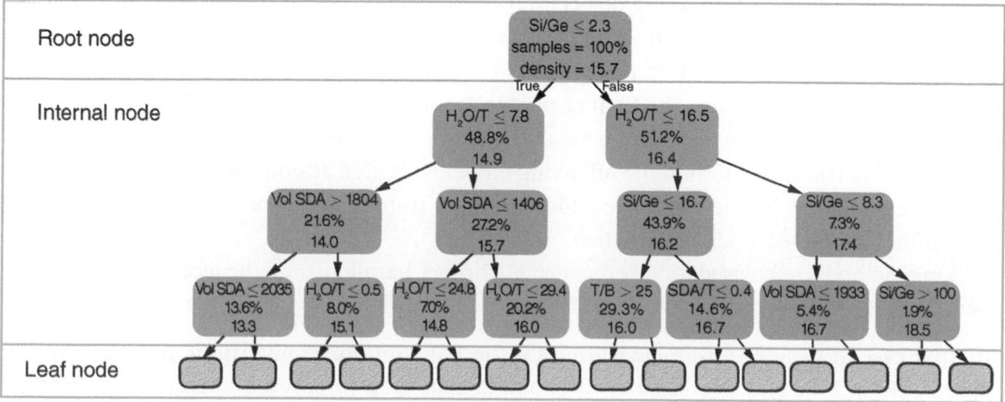

Figure 12.4 A single decision-tree regression model trained to predict zeolite framework density from synthesis conditions. *Source:* From [91].

The Decision Tree algorithm, introduced by Leo Breiman in 1984, is one of the most common models used in data mining due to its easy interpretability and versatility [90]. Figure 12.4 depicts an example of a typical workflow of a decision-tree classifier which predicts zeolite framework density from synthesis conditions [91]. The tree is essentially a series of top-down questions which build decision rules (branch) starting from the root node, passing to internal node (decision node) representing the features, and finally ending at the leaf nodes that correspond to the target classes.

The core algorithm employs Gini index (G_i) or Gini impurity (Equation 12.3) measuring the degree or probability (p_i) of a particular variable being wrongly classified when randomly chosen.

$$G_i = 1 - \sum_{i=1}^{n} p_i^2 \tag{12.3}$$

In the same way, the Entropy (E_i) is estimator of homogeneity (Equation 12.4), achieving zero value for a homogeneous sample and a value of one if the sample is equally divided. The decision path proceeds to reduce the level of entropy (Information Gain) selecting the feature/attribute which gives the maximum information about the class and aims. The latter also provides visualization of the most important features extracting useful information.

$$E_i = -\sum_{i}^{n} p_i \log_2 \tag{12.4}$$

Despite easy interpretability and fast execution times, decision trees are prone to overfitting, especially when trees are particularly deep.

Alternatively, ensemble learning combines multiple imperfect models (referred to as “weak learners”) to obtain more accurate and robust predictions. The final outcomes depend upon the proper architecture of the weak learner, and on the method employed to combine them. When a single learning algorithm is used (“homogeneous weak learners”), we can implement: i) bagging, e.g., Random Forest; or (ii) boosting, e.g., Extreme Gradient Boosting, XGBoost, while if we want to combine diverse types of weak learners (“heterogeneous models”) we may consider; iii) “stacking” approaches. The latter are explained in the following:

i) The bagging method is also called **b**ootstrap **agg**regat**ing** (bagging). The bootstrap creates a subset by randomly picking objects and duplicates them until the same size as the original dataset is reached. The remaining objects represent the out-of-bag subset, used to assess the model. Once the data is partitioned, m models are trained on m bootstrap samples and their outcomes are then aggregated by averaging.
ii) The boosting method creates the final ensemble model by sequentially building learners in a adaptive way, i.e., the new model depends on the previous ones. The flow starts from a model trained on the training set, then the following model is created attempting to correct the errors of the previous one. The learners are added until the training set is perfectly predicted or the maximum number of models is reached.
iii) Stacking methods, often used for heterogeneous learners, parallely train the models and combines them in a meta-model to compute a prediction based on different weak models predictions.

A particular case of bagging ensemble learning is the Random Forest algorithm, proposed for the first time by Ho in 1995 and referred to by Breiman in 2001 [90]. It represents an ensemble method in which the final prediction is made by a bunch of imperfect predictors (i.e., the wisdom of a crowd). The idea of Random Forest architecture is to make predictions by averaging multiple deep decision trees. The latter are parallely trained by a random subset of the training set (bootstrapped), hence reducing the variance. The Random Forest algorithm applies the general technique of bagging to increase the predictive power, returning an average of *n* individual predictions (regression case,) or the most voted class (classification case). A typical workflow can be summarized as follows:

1) Select *n* random subsets from the training set.
2) Build *n* decision trees splitting, each tree by a random subset of features.
3) Collect the prediction of the untouched test set from each individual decision tree.
4) Take the major voted (classifier) or average the decision tree's outcomes (regressor).

The learning process can be controlled by tuning hyperparameters. The most common approach to assess the models is k-fold cross-validation. The data is partitioned into k groups, where k–1 of them is used to train the model which is validated on the remaining fold. The algorithm repeats the process iteratively k times. The extreme case is for k = n, which is named the leave-one-out cross-validation (LOOCV). The latter configuration ensures each row of data is given an opportunity to represent the entirety of the testing set and it is particularly useful for small amounts of available data. The most important hyperparameters in Random Forest are:

n_estimators: Number of trees in the forest, used to build the model before taking the maximum voting or averages of predictions. Normally, a high number of trees gives accurate performance but makes the coding slower.
max_features: Number of features kept in each individual tree. By increasing this parameter, generally, the model improves its predictability due to the many options considered at each node. However, too many features could lead overfitting.
min_sample_leaf: Minimum number of data points allowed in a leaf node. Smaller leaf helps to capture noise in training data.
min_samples_split: Minimum number of data points placed in a node before the splitting of the node itself.
max_depth: Maximum number of levels which are considered in each decision tree.
bootstrap: Method for sampling data points (with or without replacement).
random_state: Fixing the reproducibility of model predictions. By setting this parameter, the model always returns the same prediction for a particular dataset.

A common challenge related with building structure–property models are inhomogeneous and/or small datasets. In these cases, the selection of the splitting method into training and testing set become crucial to build reasonable predictive models. For example, we can image in our dataset or we have a reduced amount of data exploring a certain descriptor. If that particular information is not given to train our trees, the model would be unable to process an accurate prediction. On the contrary, if the information is exclusively used to train the model, we lose the possibility to evaluate its predictability for that specific data. In these cases, the implementation of out-of-bag error (oob_score) may be an efficient solution for machine learning models, which utilize bootstrap aggregating. When the oob_score is applied, the dataset is not partitioned into training and testing sets, while the sampling with replacement performed by bagging process, leaves, iteratively, one data to assess the model performance. This approach allows to use information about all the available data without leakages.

12.7 Current and Future Applications

The development of structure databases, descriptors, and material property datasets have enabled implementation of Big Data approaches in the discovery of nanoporous materials for various applications. Statistical machine learning models are routinely employed to predict adsorption of guest species in the context of hydrogen or methane storage, and separation of noble gases, CO_2 mixtures, and hydrocarbons [47,56,33,75,80,89,92–94]. They have also been used to predict mechanical and optical properties [95,96].

Alongside the discovery applications that follow the workflow presented in this chapter, there are also other emerging implementations of Big Data, which focus on generation of new material structures. Specifically, generative neural network algorithms can generate new structures with specified properties practically indistinguishable from the known structures provided as a training set, and they have been used to design zeolites with specific heats of adsorption of methane as well as MOFs for CO_2 separations [97,98]. Another exciting new direction of Big Data application to nanoporous materials is data-driven experimentation, which allows for building structure–property models based on experimentally accessible descriptors and experimentally measured properties. In these cases, the structures are represented by a set of numbers that encode information on how a structure was obtained, e.g., synthesis temperature, concentration of substrates, type of solvent used, etc. Such approaches can be utilized in cases where exact crystal structure representations are not available or not applicable, and have been used to employ models representing the quality of crystalline materials [99] and catalytic properties of chemically-modified structures [100]. Finally, there are also "auxiliary" applications of Big Data to nanoporous materials. An example of such is generation of force field parameters such as practical atomic charges for new systems based on a dataset of charges obtained from density functional theory [101].

References

1 Li, Y. and Yu, J. (2021). Emerging applications of zeolites in catalysis, separation and host–guest assembly. *Nat. Rev. Mater.* 6: 1–19. Available from: https://www.nature.com/articles/s41578-021-00347-3(accessed 25 October 2021).

2 Zhou, H.-C., Long, J.R., and Yaghi, O.M. (2012). Introduction to metal–organic frameworks. *Chem. Rev.* 112 (2): 673–674. Available from: https://pubs.acs.org/doi/full/10.1021/cr300014x (accessed 21 October 2021).

3 Feng, X., Ding, X., and Jiang, D. (2012). Covalent organic frameworks. *Chem. Soc. Rev.* 41 (18): 6010–6022. Available from: https://pubs.rsc.org/en/content/articlehtml/2012/cs/c2cs35157a (accessed 21 October 2021).

4 Makal, T.A., Li, J.-R., Lu, W. et al. (2012). Methane storage in advanced porous materials. *Chem. Soc. Rev.* 41 (23): 7761–7779. Available from: https://pubs.rsc.org/en/content/articlehtml/2012/cs/c2cs35251f (accessed 21 October 2021).

5 Li, J.-R., Sculley, J., and Zhou, H.-C. (2011). Metal–organic frameworks for separations. *Chem. Rev.* 112 (2): 869–932. Available from: https://pubs.acs.org/doi/abs/10.1021/cr200190s (accessed 21 October 2021).

6 Sumida, K., Rogow, D.L., Mason, J.A. et al. (2011). Carbon dioxide capture in metal–organic frameworks. *Chem. Rev.* 112 (2): 724–781. Available from: https://pubs.acs.org/doi/abs/10.1021/cr2003272 (accessed 21 October 2021).

7 Herm, Z.R., Wiers, B.M., Mason, J.A. et al. (2013). Separation of hexane isomers in a metal-organic framework with triangular channels. *Science* 340 (6135): 960–964. Available from: https://www.science.org/doi/abs/10.1126/science.1234071 (accessed 21 October 2021).

8 Cychosz, K.A., Ahmad, R., and Matzger, A.J. (2010). Liquid phase separations by crystalline microporous coordination polymers. *Chem. Sci.* 1 (3): 293–302. Available from: https://pubs.rsc.org/en/content/articlehtml/2010/sc/c0sc00144a (accessed 21 October 2021).

9 Düren, T., Sarkisov, L., Yaghi, O.M. (2004). Design of new materials for methane storage. *Langmuir* 20 (7): 2683–2689.

10 Yuan, D., Zhao, D., Sun, D. (2010). An isoreticular series of metal–organic frameworks with dendritic hexacarboxylate ligands and exceptionally high gas-uptake capacity. *Angew. Chem. Int. Ed.* 49 (31): 5357–5361. Available from: https://onlinelibrary.wiley.com/doi/full/10.1002/anie.201001009 (accessed 21 October 2021)

11 Narayan, T.C., Miyakai, T., Seki, S. (2012). Mobility in a Tetrathiafulvalene-based microporous metal–organic framework. *J. Am. Chem. Soc.* 134 (31): 12932–12935. Available from: https://pubs.acs.org/doi/abs/10.1021/ja3059827 (accessed 21 October 2021).

12 Saravanan, K., Nagarathinam, M., Balaya, P. (2010). Lithium storage in a metal organic framework with diamondoid topology: a case study on metal formates. *J. Mater. Chem.* 20 (38): 8329–8335. Available from: https://pubs.rsc.org/en/content/articlehtml/2010/jm/c0jm01671c (accessed 21 October 2021).

13 Kreno, L.E., Leong, K., Farha, O.K. (2011). Metal–organic framework materials as chemical sensors. *Chem. Rev.* 112 (2): 1105–1125. Available from: https://pubs.acs.org/doi/abs/10.1021/cr200324t (accessed 21 October 2021).

14 Sarkisov, L. (2012). Toward rational design of metal–organic frameworks for sensing applications: efficient calculation of adsorption characteristics in zero loading regime. *J. Phys. Chem. C* 116 (4): 3025–3033. Available from: https://pubs.acs.org/doi/abs/10.1021/jp210633w (accessed 21 October 2021).

15 Horcajada, P., Gref, R., Baati, T. et al. (2011). Metal–organic frameworks in biomedicine. *Chem. Rev.* 112 (2): 1232–1268. Available from: https://pubs.acs.org/doi/abs/10.1021/cr200256v (accessed 21 October 2021).

16 Lee, J., Farha, O.K., Roberts, J. et al. (2009). Metal–organic framework materials as catalysts. *Chem. Soc. Rev.* 38 (5): 1450–1459. Available from: https://pubs.rsc.org/en/content/articlehtml/2009/cs/b807080 (accessed 21 October 2021).

17 Jablonka, K.M., Ongari, D., Moosavi, S.M. (2020). Big-data science in porous materials: materials genomics and machine learning. *Chem. Rev.* 120 (16): 8066–8129. Available from: https://pubs.acs.org/doi/full/10.1021/acs.chemrev.0c00004 (accessed 25 October 2021).

18 Yaghi, O.M., O'Keeffe, M., Ockwig, N.W. et al. (2003). Reticular synthesis and the design of new materials. *Nature* 423 (6941): 705–714. Available from: https://www.nature.com/articles/nature01650 (accessed 21 October 2021).

19 O'Keeffe, M., Peskov, M.A., Ramsden, S.J. et al. (2008). The Reticular Chemistry Structure Resource (RCSR) database of, and symbols for, crystal nets. *Acc. Chem. Res.* 41 (12): 1782–1789. Available from: https://pubs.acs.org/doi/abs/10.1021/ar800124u (accessed 21 October 2021).

20 Earl, D.J. and Deem, M.W. (2006). Toward a database of hypothetical zeolite structures. *Ind. Eng. Chem. Res.* 45 (16): 5449–5454.

21 Groom, C.R., Bruno, I.J., Lightfoot, M.P. et al. (2016). The Cambridge Structural Database. *Acta Crystallogr. Sec. B: Struct. Sci. Cryst. Eng. Mater.* 72 (2): 171–179.

22 Moghadam, P.Z., Li, A., Wiggin, S.B. et al. (2017). Development of a Cambridge Structural Database subset: a collection of metal–organic frameworks for past, present, and future. *Chem. Mater.* 29 (7): 2618–2625.

23 Technology NI of S and NIST Inorganic Crystal Structure Database.

24 Graulis, S., Chateigner, D., Downs, R.T. et al. (2009). Crystallography open database: an open-access collection of crystal structures. *J. Appl. Crystallogr.* 42 (4): 726–729.

25 International Zeolite Association. (2018). Database of Zeolite structures.

26 Goldsmith, J., Wong-Foy, A.G., Cafarella, M.J. et al. (2013). Theoretical limits of hydrogen storage in metal–organic frameworks: opportunities and trade-offs. *Chem. Mater.* 25 (16): 3373–3382. Available from: https://pubs.acs.org/doi/abs/10.1021/cm401978e (accessed 21 October 2021).

27 Chung, Y.G., Camp, J., Haranczyk, M. et al. (2014). Computation-ready, experimental metal-organic frameworks: a tool to enable high-throughput screening of nanoporous crystals. *Chem. Mater.* 26 (21): 6185–6192.

28 Chung, Y. (2014 Computation-ready, experimental metal–organic frameworks: a tool to enable high-throughput screening of nanoporous crystals. *Chem. Mater.* 26 (21): 6185–6192.

29 Tong, M., Lan, Y., Yang, Q. et al. (2017). Exploring the structure-property relationships of covalent organic frameworks for noble gas separations. *Chem. Eng. Sci.* 168: 456–464.

30 Pophale, R., Cheeseman, P.A., and Deem, M.W. (2011). A database of new zeolite-like materials. *Phys. Chem. Chem. Phys.* 13 (27): 12407–12412.

31 Treacy, M.M.J., Rivin, I., Balkovsky, E. et al. (2004). Enumeration of periodic tetrahedral frameworks. II: Polynodal graphs. *Micropor. Mesopor. Mater.* 74 (1–3): 121–132.

32 Lewis, D.W., Ruiz-Salvador, A.R., Gómez, A. et al. (2009). Zeolitic imidazole frameworks: structural and energetics trends compared with their zeolite analogues. *CrystEngComm* 11 (11): 2272–2276. Available from: https://pubs.rsc.org/en/content/articlehtml/2009/ce/b912997a (accessed 21 October 2021).

33 Wilmer, C.E., Leaf, M., Lee, C.Y. et al. (2012). Large-scale screening of hypothetical metal-organic frameworks. *Nat. Chem.* 4 (2): 83–89.

34 Colón, Y.J., Gómez-Gualdrón, D.A., Snurr, R.Q. et al. (2017). Automated construction of metal–organic frameworks and their evaluation for energy-related applications. *Cryst. Growth Des.* 17 (11): 5801–5810. Available from: https://pubs.acs.org/doi/abs/10.1021/acs.cgd.7b00848 (accessed 21 October 2021).

35 Addicoat, M.A., Coupry, D.E., and Heine, T. (2014). AuToGraFS: automatic topological generator for framework structures. *J. Phys. Chem. A* 118 (40): 9607–9614.

36 Martin, R.L. and Haranczyk, M. (2014). Construction and characterization of structure models of crystalline porous polymers. *Cryst. Growth Des.* 14 (5): 2431–2440. Available from: https://pubs.acs.org/doi/abs/10.1021/cg500158c (accessed 21 October 2021).

37 Delgado-Friedrichs, O. and O'Keeffe, M. (2003). Identification of and symmetry computation for crystal nets. *Acta Crystallogr. Sec. A* 59 (4): 351–360. Available from: https://onlinelibrary.wiley.com/doi/full/10.1107/S0108767303012017 (accessed 21 October 2021).

38 Farha, O.K., Eryazici, I., Jeong, N.C. et al. (2012). Metal–organic framework materials with ultrahigh surface areas: is the sky the limit? *J. Am. Chem. Soc.* 134 (36): 15016–15021. Available from: https://pubs.acs.org/doi/abs/10.1021/ja3055639 (accessed 21 October 2021).

39 Li, M., Li, D., O'Keeffe, M., and Yaghi, O.M. (2013). Topological analysis of metal–organic frameworks with polytopic linkers and/or multiple building units and the minimal transitivity principle. *Chem. Rev.* 114 (2): 1343–1370. Available from: https://pubs.acs.org/doi/abs/10.1021/cr400392k (accessed 25 October 2021).

40 Ockwig, N.W., Delgado-Friedrich, O., O'Keeffe, M. et al. (2005). Reticular chemistry: occurrence and taxonomy of nets and grammar for the design of frameworks. *Acc. Chem. Res.* 38 (3): 176–182. Available from: https://pubs.acs.org/doi/abs/10.1021/ar020022l (accessed 25 October 2021).

41 Kim, S., Chen, J., Cheng, T. et al. (2019). PubChem 2019 update: improved access to chemical data. *Nucleic Acids Res.* 47 (D1): D1102–D1109.

42 Bucior, B.J., Rosen, A.S., Haranczyk, M. et al. (2019). Identification schemes for metal–organic frameworks to enable rapid search and cheminformatics analysis. *Cryst. Growth Des.* 19 (11): 6682–6697. Available from: https://pubs.acs.org/doi/full/10.1021/acs.cgd.9b01050 (accessed 25 October 2021).

43 Düren, T., Millange, F., Férey, G. et al. (2007). Calculating geometric surface areas as a characterization tool for metal–organic frameworks. *J. Phys. Chem. C* 111 (42): 15350–15356.

44 Do, D.D., Herrera, L.F., and Nicholson, D. (2011). A method for the determination of accessible surface area, pore volume, pore size and its volume distribution for homogeneous pores of different shapes. *Adsorption* 17 (2): 325–335. Available from: https://link.springer.com/article/10.1007/s10450-010-9314-2 (accessed 21 October 2021).

45 First, E.L. and Floudas, C.A. (2013). MOFomics: computational pore characterization of metal-organic frameworks. *Micropor. Mesopor. Mater.* 165: 32–39.

46 First, E.L., Gounaris, C.E., Wei, J. et al. (2011). Computational characterization of zeolite porous networks: an automated approach. *Phy. Chem. Chem. Phys.* 13 (38): 17339–17358.

47 Haldoupis, E., Nair, S., and Sholl, D.S. (2010). Efficient calculation of diffusion limitations in metal organic framework materials: a tool for identifying materials for kinetic separations. *J. Am. Chem. Soc.* 132 (21): 7528–7539. Available from: https://pubs.acs.org/doi/abs/10.1021/ja1023699 (accessed 21 October 2021).

48 Willems, T.F., Rycroft, C.H., Kazi, M. et al. (2012). Algorithms and tools for high-throughput geometry-based analysis of crystalline porous materials. *Microporous Mesoporous Mater.* 149 (1): 134–141.

49 Martin, R.L., Smit, B., and Haranczyk, M. (2011). Addressing challenges of identifying geometrically diverse sets of crystalline porous materials. *J. Chem. Inf. Model.* 52 (2): 308–318. Available from: https://pubs.acs.org/doi/abs/10.1021/ci200386x (accessed 21 October 2021).

50 Pinheiro, M., Martin, R.L., Rycroft, C.H. et al. (2013). Characterization and comparison of pore landscapes in crystalline porous materials. *J. Mol. Graph. Model.* 44: 208–219. Available from: http://dx.doi.org/10.1016/j.jmgm.2013.05.007 (accessed 21 October 2021).

51 Zeo++: high throughput analysis of crystalline porous materials. Available from: http://www.zeoplusplus.org (accessed 22 October 2021).

52 Pinheiro, M., Martin, R.L., Rycroft, C.H. et al. (2013). High accuracy geometric analysis of crystalline porous materials. *CrystEngComm* 15 (37): 7531–7538. Available from: https://pubs.rsc.org/en/content/articlehtml/2013/ce/c3ce41057a (accessed 21 October 2021).

53 Foster, M.D., Rivin, I., Treacy, M.M.J. et al. (2006). A geometric solution to the largest-free-sphere problem in zeolite frameworks. *Micropor. Mesopor. Mater.* 90 (1–3 Spec. Iss.): 32–38.

54 Carr, D.A., Lach-hab, M., Yang, S. et al. (2009). Machine learning approach for structure-based zeolite classification. *Micropor. Mesopor. Mater.* 117 (1–2): 339–349.

55 Blatov, V.A. and Shevchenko, A.P. (2003) A Analysis of voids in crystal structures: the methods of "dual" crystal chemistry. *Acta Crystallographica Section A* 59(1): 34–44.

56 Simon, C.M., Mercado, R., Schnell, S.K. et al. (2015). What are the best materials to separate a Xenon/Krypton mixture? *Chem. Mater.* 27 (12): 4459–4475. Available from: https://pubs.acs.org/doi/abs/10.1021/acs.chemmater.5b01475 (accessed 21 October 2021).

57 Sarkisov, L. and Harrison, A. (2011). Computational structure characterisation tools in application to ordered and disordered porous materials. *Mol. Simul.* 37 (15): 1248–1257. http://dx.doi.org/101080/08927022011592832 Available from: https://www.tandfonline.com/doi/abs/10.1080/08927022.2011.592832(accessed 21 October 2021).

58 Do, D.D. and Herrera, L.F. (2008). A new method to determine pore size and its volume distribution of porous solids having known atomistic configuration. *J. Colloid Interface Sci.* 328 (1): 110–119. Available from: https://pubmed.ncbi.nlm.nih.gov/18834598 (accessed 21 October 2021).

59 Krishnapriyan, A.S., Montoya, J., Haranczyk, M. et al. (2021). Machine learning with persistent homology and chemical word embeddings improves prediction accuracy and interpretability in metal–organic frameworks. *Sci. Rep.* 11 (1): 1–11. Available from: https://www.nature.com/articles/s41598-021-88027-8 (accessed 21 October 2021).

60 Krishnapriyan, A.S., Montoya, J., Haranczyk, M. et al. (2020). Topological descriptors help predict guest adsorption in nanoporous materials. *J. Phys. Chem. C* 124 (17): 9360–9368.

61 Adams, H., Emerson, T., Kirby, M. et al. (2017). Persistence images: a stable vector representation of persistent homology. *J. Mach. Learn. Res.* 18: 1–35. Available from: http://jmlr.org/papers/v18/16-337.html (accessed 22 October 2021).

62 Lee, Y., Barthel, S.D., Dłotko, P. et al. (2017). Quantifying similarity of pore-geometry in nanoporous materials. *Nat. Commun.* 8 (1): 1–8. Available from: https://www.nature.com/articles/ncomms15396 (accessed 22 October 2021).

63 Bucior, B.J., Bobbitt, N.S., Islamoglu, T.T. et al. (2019). Energy-based descriptors to rapidly predict hydrogen storage in metal–organic frameworks. *Mol. Syst. Des. Eng.* 4 (1): 162–174. Available from: https://pubs.rsc.org/en/content/articlehtml/2019/me/c8me00050f (accessed 25 October 2021).

64 Rappe, A.K., Casewit, C.J., Colwell, K.S. et al. (2002). UFF, a full periodic table force field for molecular mechanics and molecular dynamics simulations. *J. Am. Chem. Soc.* 114 (25): 10024–10035. Available from: https://pubs.acs.org/doi/abs/10.1021/ja00051a040 (accessed 21 October 2021).

65 Borboudakis, G., Stergiannakos, T., Frysali, M. et al. (2017). Chemically intuited, large-scale screening of MOFs by machine learning techniques. *NPJ Comput. Mater.* 3 (1): 1–7. Available from: https://www.nature.com/articles/s41524-017-0045-8 (accessed 22 October 2021).

66 Anderson, R., Rodgers, J., Argueta, E. et al. (2018). Role of pore chemistry and topology in the CO_2 capture capabilities of MOFs: from molecular simulation to machine learning. *Chem. Mater.* 30 (18): 6325–6337. Available from: https://pubs.acs.org/doi/abs/10.1021/acs.chemmater.8b02257 (accessed 21 October 2021).

67 Tshitoyan, V, Dagdelen, J., Weston, L. et al. (2019). Unsupervised word embeddings capture latent knowledge from materials science literature. *Nature* 571 (7763): 95–98. Available from: https://pubmed.ncbi.nlm.nih.gov/31270483 (accessed 22 October 2021)

68 matminer (Materials Data Mining): matminer 0.7.4 documentation. Available from: https://hackingmaterials.lbl.gov/matminer (accessed 22 October 2021).

69 Marmier, A., Lethbridge, Z.A.D., Walton, R.I. et al. (2010). ElAM: a computer program for the analysis and representation of anisotropic elastic properties. *Comput. Phys. Commun.* 181 (12): 2102–2115.

70 Han, S.S. and Goddard, W.A. (2007). Metal–Organic Frameworks Provide Large Negative Thermal Expansion Behavior. *J. Phys. Chem. C.* 111 (42): 15185–15191.

71 Tan, J.C. and Cheetham, A.K. (2011). Mechanical properties of hybrid inorganic–organic framework materials: establishing fundamental structure–property relationships. *Chem. Soc. Rev.* 40 (2): 1059–1080. Available from: https://pubs.rsc.org/en/content/articlehtml/2011/cs/c0cs00163e (accessed 21 October 2021).

72 Bristow, J.K., Tiana, D., and Walsh, A. (2014). Transferable force field for metal–organic frameworks from first-principles: BTW-FF. *J. Chem. Theory Comput.* 10 (10): 4644–4652. Available from: https://pubs.acs.org/doi/abs/10.1021/ct500515h (accessed 21 October 2021).

73 Butler, K.T., Hendon, C.H., and Walsh, A. (2014). Electronic chemical potentials of porous metal–organic frameworks. *J. Am. Chem. Soc.* 136 (7): 2703–2706. Available from: https://pubs.acs.org/doi/abs/10.1021/ja4110073 (accessed 21 October 2021).

74 Cho, E.H., Lyu, Q., and Lin, L.-C. (2019). Computational discovery of nanoporous materials for energy- and environment-related applications. *Mol. Simul.* 45 (14–15): 1122–1147. Available from: https://www.tandfonline.com/doi/abs/10.1080/08927022.2019.1626990 (accessed 22 October 2021).

75 Boyd, P.G., Lee, Y., and Smit, B. (2017). Computational development of the nanoporous materials genome. *Nat. Rev. Mater.* 2 (8): 1–15. Available from: https://www.nature.com/articles/natrevmats201737 (accessed 22 October 2021).

76 Colón, Y.J. and Snurr, R.Q. (2014). High-throughput computational screening of metal–organic frameworks. *Chem. Soc. Rev.* 43 (16): 5735–5749. Available from: https://pubs.rsc.org/en/content/articlehtml/2014/cs/c4cs00070f (accessed 22 October 2021).

77 Evans, J.D., Fraux, G., Gaillac, R. et al. (2016). Computational chemistry methods for nanoporous materials. *Chem. Mater.* 29 (1): 199–212. Available from: https://pubs.acs.org/doi/full/10.1021/acs.chemmater.6b02994 (accessed 22 October 2021).

78 Sikora, B.J., Wilmer, C.E., Greenfield, M.L. et al. (2012). Thermodynamic analysis of Xe/Kr selectivity in over 137,000 hypothetical metal-organic frameworks. *Chem. Sci.* 3 (7): 2217–2223.

79 Colón, Y.J., Gómez-Gualdrón, D.A., Snurr, R.Q. et al. (2017). Automated construction of metal-organic frameworks and their evaluation for energy-related applications. *Cryst. Growth Des.* 17 (11): 5801–5810.

80 Bobbitt, N.S., Chen, J., and Snurr, R.Q. (2016). High-throughput screening of metal-organic frameworks for hydrogen storage at cryogenic temperature. *J. Phys. Chem. C* 120 (48): 27328–27341.

81 Wilmer, C.E., Farha, O.K., Bae, Y.S. et al. (2012). Structure-property relationships of porous materials for carbon dioxide separation and capture. *Energy Environ. Sci.* 5 (12): 9849–9856.

82 Metal organic framework database | Mofs. Available from: https://mof.tech.northwestern.edu/databases

83 Available from: https://nanoporousmaterials.org/databases (accessed 21 October 2021).

84 Jensen, Z., Kim, E., Kwon, S. et al. (2019). A machine learning approach to zeolite synthesis enabled by automatic literature data extraction. *ACS Cent. Sci.* 22: 27. Available from: https://pubs.acs.org/doi/full/10.1021/acscentsci.9b00193 (accessed 22 October 2021).

85 Park, S., Kim, B., Choi, S. et al. (2018). Text mining metal–organic framework papers. *J. Chem. Inf. Model.* 58 (2): 244–251. Available from: https://pubs.acs.org/doi/full/10.1021/acs.jcim.7b00608 (accessed 25 October 2021).

86 Johnson G.M.M. (1990). *Concepts and Applications of Molecular Similarity*. New York: John Wiley & Sons.

87 Snarey, M., Terrett, N.K, and Willett, P. (1997). Comparison of algorithms for dissimilarity-based compound selection. *J. Mol. Graph. Model.* 15 (6): 372–385. Available from: https://pubmed.ncbi.nlm.nih.gov/9704300 (accessed 25 October 2021).

88 Li, Y. and Wu, H. (2012). A clustering method based on K-means algorithm. *Phys. Procedia* 25: 1104–1109.

89 Pardakhti, M., Moharreri, E., Wanik, D. et al. (2017). Machine learning using combined structural and chemical descriptors for prediction of methane adsorption performance of Metal–Organic Frameworks (MOFs). *ACS Comb. Sci.* 19 (10): 640–645.

90 Breiman, L. (2001). Random forests. *Mach. Learn.* 45: 5–32.

91 Jensen, Z., Kim, E., Kwon, S. et al. (2019). A machine learning approach to zeolite synthesis enabled by automatic literature data extraction. *ACS Cent. Sci.* 5: 892–899.

92 Wilmer, C.E. and Snurr, R.Q. (2011 Jul 15). Towards rapid computational screening of metal-organic frameworks for carbon dioxide capture: calculation of framework charges via charge equilibration. *Chem. Eng. J.* 171 (3): 775–781.

93 Peng, L., Zhu, Q., Wu, P. et al. (2019). High-throughput computational screening of metal–organic frameworks with topological diversity for hexane isomer separations. *Phys. Chem. Chem. Phys.* 21 (16): 8508–8516. Available from: https://pubs.rsc.org/en/content/articlehtml/2019/cp/c8cp07527a (accessed 25 October 2021).

94 Shah, M.S., Fetisov, E.O., Tsapatsis, M. et al. (2018). C_2 adsorption in zeolites: *in silico* screening and sensitivity to molecular models. *Mol. Sys. Des. Eng.* 3 (4): 619–626. Available from: https://pubs.rsc.org/en/content/articlehtml/2018/me/c8me00004b (accessed 25 October 2021).

95 Evans, J.D. and Coudert, F.-X. (2017). Predicting the mechanical properties of zeolite frameworks by machine learning. *Chem. Mater.* 29 (18): 7833–7839. Available from: https://pubs.acs.org/doi/full/10.1021/acs.chemmater.7b02532 (accessed 25 October 2021).

96 Moghadam, P.Z., Rogge, S.M.J., Li, A. et al. (2019). Structure-mechanical stability relations of metal-organic frameworks via machine learning. 1(1): 219–234. Available from: https://doi.org/10.1016/j.matt.2019.03.002 (accessed 25 October 2021).

97 Kim, B., Lee, S., and Kim, J. (2020). Inverse design of porous materials using artificial neural networks. *Sci. Adv.* 6 (1): essa9324. Available from: https://www.science.org/doi/abs/10.1126/sciadv.aax9324 (accessed 25 October 2021).

98 Yao, Z., Sánchez-Lengeling, B., Bobbitt, N.S. et al. (2021). Inverse design of nanoporous crystalline reticular materials with deep generative models. *Nat. Mach. Intell.* 3 (1): 76–86. Available from: https://www.nature.com/articles/s42256-020-00271-1 (accessed 25 October 2021).

99 Moosavi, S.M., Chidambaram, A., Talirz, L. et al. (2019). Capturing chemical intuition in synthesis of metal-organic frameworks. *Nat. Commun.* 10 (1): 1–7. Available from: https://www.nature.com/articles/s41467-019-08483-9 (accessed 25 October 2021).

100 Dico, G.L., Nuñez, Á.P., Carcelén, V. et al. (2021). Machine-learning-accelerated multimodal characterization and multiobjective design optimization of natural porous materials. *Chem. Sci.* 12 (27): 9309–9317. Available from: https://pubs.rsc.org/en/content/articlehtml/2021/sc/d1sc00816a (accessed 25 October 2021).

101 Kancharlapalli, S., Gopalan, A., Haranczyk, M. et al. (2021). Fast and accurate machine learning strategy for calculating partial atomic charges in metal–organic frameworks. *J. Chem. Theory Comput.* 17 (5): 3052–3064. Available from: https://pubs.acs.org/doi/abs/10.1021/acs.jctc.0c01229 (accessed 25 October 2021).

13

Efficient Data Utilization in Training Machine Learning Models for Nanoporous Materials Screening

Diego A. Gómez-Gualdrón[1], Cory M. Simon[2], and Yamil J. Colón[3]

[1] *Chemical and Biological Engineering, Colorado School of Mines, Colorado, USA*
[2] *School of Chemical, Biological, and Environmental Engineering, Oregon State University, Oregon, USA*
[3] *Department of Chemical and Biomolecular Engineering, University of Notre Dame, Notre Dame, Indiana, USA*

Machine learning is becoming a key tool in the study of nanoporous materials and promises to play a continuous crucial role in the discovery of new materials in the foreseeable future. It is important to keep in mind that machine learning is a "data hungry" approach, whose success in any field is predicated on the ability of the pertinent research community to generate and use data as efficiently as possible. In the past ten years, the nanoporous materials community has seen an explosion in the availability of data, mostly due to the application of molecular simulation to calculate adsorption properties in nanoporous materials databases. Thus, the prediction of adsorption properties has served as a natural "testbed" for the application of machine learning approaches to nanoporous materials discovery. However, it is important to put things into perspective and note that while "big data" in other areas (e.g., social media) refers to billions of datapoints, in the nanoporous materials community, data generation (even for adsorption) has rarely hit the million datapoints. The latter holds true despite research groups worldwide pushing their computational resources to the limit. From this perspective, machine learning applications in the field of nanoporous materials have been explored within a context of "data scarcity." This chapter uses select machine learning efforts to predict adsorption properties from the past eight years (going through them somewhat chronologically) as a point of reference to discuss different topics pertinent to the various decisions that need to be made when attempting to efficiently train a machine learning model to predict a nanoporous material property. An attempt is made throughout this chapter to consistently bridge these different decisions to how they could affect the efficacy with which data is used in model training. The first half of the chapter focuses on the most basic decisions we face when developing a machine learning model to predict material properties. For instance, how to represent the material (i.e., descriptor selection), which materials to use to train the model, and what kind of model to train to make the predictions. The second half of the chapter focuses on more advanced strategies adopted in recent years, seeking to more directly address the data scarcity issue. These strategies include but are not limited to, transfer learning and active learning. The lessons learned during the past eight years are starting to come together, to the point where a single machine learning model can predict adsorption in nanoporous materials as distinct as zeolites, metal–organic frameworks (MOFs), and hyper-cross-linked polymers [1]. But most of these lessons have been learned through the study of MOFs, which is why this chapter primarily focuses on these materials. Finally, we hope that while the discussion of machine learning approaches in this chapter is "anchored" to the examples of adsorption property predictions, our attempt to present the rationale behind different model training aspects or approaches stripped down to their basics can make the insights provided in this chapter somewhat application-agnostic and useful for the reader interested in the prediction of nanomaterial properties other than adsorption.

AI-Guided Design and Property Prediction for Zeolites and Nanoporous Materials, First Edition. Edited by German Sastre and Frits Daeyaert.

13.1 Descriptor Selection

Large datasets enable simpler features that can be used as input for a model. The reason is that a large dataset enables models that are better at "filling the knowledge gap" that establishes the relationship between an output and a given set of inputs. Conversely, the use of more elaborate, advanced features reduces the burden on data generation. To illustrate this idea, consider the relationship between y and x in Equation 13.1:

$$y = 5 \log x \tag{13.1}$$

There could be a machine learning model that could first figure out that the relationship between the output y and the input x is logarithmic in nature (i.e., $y \propto \log x$), and subsequently that the proportionality constant between y and $\log x$ is "5." The caveat is that the model needs to see a lot of data to learn this. On the other hand, if by using "domain knowledge" about the problem we "engineer" the feature "$\log x$" as input to a machine learning model, from a much smaller number of datapoints, the model could figure out that output y and the input $\log x$ are related by a proportionality constant of value "5."

However, both approaches have pros and cons. In the "data-driven" approach, the burden is primarily on data generation, but does not require as deep a priori understanding of the relationship between x and y. By contrast, in the "domain knowledge-driven" approach, the burden is primarily on deep a priori understanding of the relationship between x and y, but allows us to get away with relatively scarce data generation.

13.1.1 Engineering of Advanced Features

Expectedly, the first attempts to train machine learning models to predict adsorption in nanoporous materials had the idea of "helping the model" to use the data as efficiently as possible. Thus, these attempts leaned heavily toward the domain knowledge-driven approach, putting a lot of effort into feature engineering. To examine some thoughtful examples of advanced feature engineering, let us consider first the work by Woo et al. [2], who were among the first to attempt predicting adsorption loadings in nanoporous materials using machine learning. In 2013, these authors aimed to predict CO_2 adsorption in MOFs, for which they extracted advanced features from so-called "atomic property weighted radial distribution functions (AP-RDFs)" and used them as inputs for support vector regression (SVR) models. So let us inspect a general AP-RDF, which is defined in Equation 13.2 as:

$$AP - RDF\,(R) = \sum_{i,j}^{all\ i,j\ pairs} P_i P_j e^{-B(r_{ij}-R)^2} \tag{13.2}$$

In this equation, r_{ij} is the distance between atoms i and j, P_i and P_j are the respective values of a property P of interest (e.g., electronegativity) for these atoms, R is the general distance variable, and B is a tunable parameter. Figure 13.1 illustrates what an AP-RDF looks like for the well-known material IRMOF-1, when P is the electronegativity of a given MOF atom. The interpretation of AP-RDFs is perhaps less intuitive than for characterization properties routinely obtained by experimentalists such as density, pore size, and specific surface area (sometimes called first-order descriptors). Nonetheless, interpretation can be attempted.

The AP-RDF for a MOF can be interpreted as a probability distribution that aims to simultaneously inform a machine learning model about both how atoms are spatially distributed within the MOF unit cell and how electronegative these atoms are. For instance, the AP-RDF in Figure 13.1 reveals that the most electronegative atoms in the IRMOF-1 unit cell are typically separated by

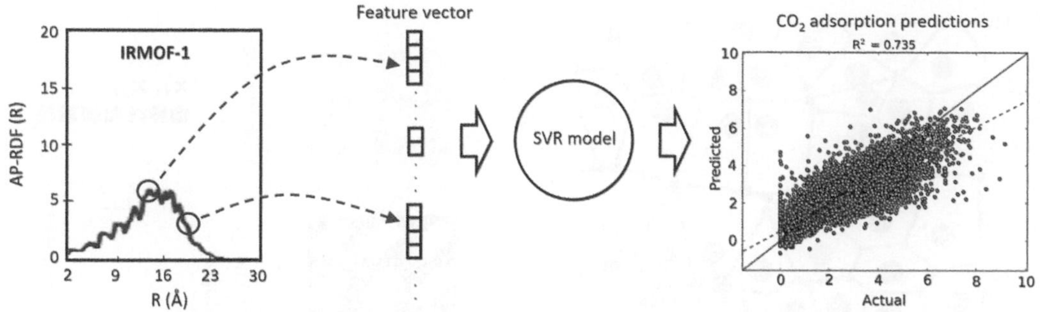

Figure 13.1 Example of an AP-RDF calculated in the well-known MOF IRMOF-1 when the atom property *P* considered is electronegativity and *B* is 10. The values of AP-RDF at different values of *R* can be used as components of the feature vector representing the material. The rightmost parity plot illustrates the accuracy of an SVR model predicting CO_2 adsorption in around 20,000 MOFs from a 90% CH_4 10% CO_2 mixture at 5 bar. The feature vectors in this example were made of 113 AP-RDF values at 113 different R values for electronegativity-, polarizability-, and van der Waals size-weighted AP-RDFs. *Source:* Adapted from [2]. © 2013, The American Chemical Society.

around 16 Å. We would expect a machine learning model to find this information somewhat useful to predict CO_2 adsorption, because more electronegative atoms tend to create higher bond dipole moments in the MOFs to which CO_2 could be attracted.

Woo et al. [2] trained their SVR model using molecular simulation adsorption data for around 20,000 MOFs, with each MOF represented by a 339-component feature vector made out of the values of three different AP-RDFs (P_i = electronegativity, polarizability, and van der Waals radius, respectively) at 113 different *R* values. The parity plot in Figure 13.1 illustrates the level of accuracy of the SVR model when trained with molecular simulation data for CO_2 adsorption from a 90% CH_4 and 10% CO_2 mixture at 5 bar, which can be summarized by a Pearson correlation factor (R^2) of 0.735 (where $R^2 = 1.00$ would represent error-free predictions).

As a second example, let us consider work in 2015 by Simon et al. [3] to predict Xe and Kr adsorption from a 20% Xe, 80% Kr mixture at 1 bar and 298 K in nanoporous crystals (i.e., MOFs, COFs, zeolites, etc.). These authors represented each material as a 7-component feature vector, which was used as input for a random forest regression (RFR) model. Six components of the feature vector were routine material characterization properties (void fraction, volumetric and surface densities, specific surface area, largest pore diameters, and diffusion-limiting pore diameter), but the seventh component was an average "Voronoi energy," E_v, proposed by the authors to be calculated as in Equation 13.3:

$$E_v = \frac{1}{N}\sum_{i=0}^{N} E(v_i) \tag{13.3}$$

where *N* is the number of accessible Voronoi sites and $E(v_i)$ is the interaction energy for an Xe atom placed at a Voronoi site, with the relevant energy calculated using molecular simulation.

Figure 13.2a illustrates the Voronoi sites for a "toy" two-dimensional (2D) nanoporous material, whose atoms are represented by red spheres. The Voronoi sites are the vertices in the illustrated "wireframe" built around the atoms and represent locations in space that are equidistant to at least three atoms. As with AP-RDFs, we can also attempt to interpret the meaning of E_v. To that end, we can think of accessible Voronoi sites as roughly equivalent to adsorption sites, in which case E_v is a sort of average adsorption energy for Xe in a given material. Ultimately, Simon et al. [3] used adsorption data for 15,000 materials to train their RFR model, whose performance was partly evaluated by the model ability to predict Xe/Kr adsorption selectivity. The root mean squared error (RMSE) for selectivity predictions was 1.25, which

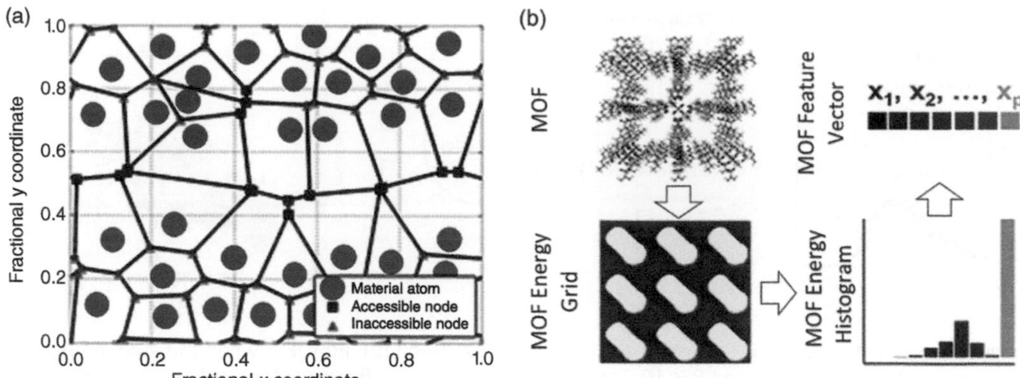

Figure 13.2 (a) Example Voronoi lattice for a "toy" 2D nanoporous material whose atoms are represented in red. Accessible and inaccessible Voronoi sites indicated by (blue) squares and (green) triangles, respectively. (b) Steps to build a MOF feature vector based on hydrogen adsorption energetics. First, the unit cell of a MOF is tessellated into a grid with 1 Å spacing and the interaction energy of hydrogen is determined at each grid point. Second, a 12-bin histogram is built for the interaction energies. Finally, the probability density associated with each histogram bin becomes one of the components of the feature vector. *Source:* Adapted from [4].

was promising considering that the highest selectivities to be predicted (which are the ones of interest) reached values up to 35.

Note that descriptors such as the E_v from Equation 13.3 make intuitive sense (at least for molecular simulation specialists), since what drives adsorption is the strength of interaction between adsorbed species and the adsorption sites of the material. This can be quickly realized if we recall traditional adsorption models such as the Langmuir model depicted in Equation 13.4:

$$N = N_o \frac{KP}{1 + KP} \tag{13.4}$$

where the adsorbed amount N depends on the number of adsorption sites N_o, the pressure P, and the equilibrium constant K, where the latter depends on the free energy of interaction between the adsorbed species and the adsorption sites of the material, ΔG_i, according to Equation 13.5:

$$K = e^{\Delta G_i / RT} \tag{13.5}$$

Thus, perhaps reflecting of the importance given to adsorption site energetics by the molecular simulation community, in 2019, Snurr et al. [4] proposed a 12-component material feature vector that provided somewhat more granular information on the energetics of adsorption sites than E_v. Their goal was to predict hydrogen adsorption. Thus, to build the feature vector of a material, these authors tessellated the material unit cell into a regularly spaced grid and determined the hydrogen interaction energy at each grid point, ultimately generating a histogram of interaction energies (Figure 13.2b). The 12 components of the feature vector corresponded to the probability associated with a specific interaction energy range in the histogram.

Then Snurr et al. [4] used the described probability density feature vector as input to a least absolute shrinkage and selection operator (LASSO) regression model with the goal of predicting hydrogen deliverable capacity between 100 bar and 2 bar at 77 K. These authors had hydrogen adsorption data from simulations in 54,776 MOFs but used a cross-validation procedure where the LASSO model was trained only on data for 1000 MOFs and tested on a separate set of 1250 MOFs, all at a given time. The RMSE of these predictions was determined as 2.3 g/L, which was promising considering that the highest capacities to be predicted reached values up of to 50 g/L.

13.1.2 Engineering of Simpler Features

While the features described in Section 13.1.1 make intuitive sense to molecular simulation specialists, the "know-how" to calculate these advanced features can constitute a significant barrier for the utilization of machine learning models by researchers not in the molecular simulation field. Additionally, these features may hinder leveraging the model to build "chemical intuition" (which is part of getting the most out of a given generated dataset) that can be more easily translated into more tangible design rules for synthesis of optimal materials. Accordingly, there have been efforts to train models using simpler, more easily interpretable features.

For instance, in 2017, Srivastava et al. [5] proposed a 26-component feature vector as input to four different machine learning models that were trained to predict methane adsorption in MOFs. Seven components of the feature vector corresponded to traditional characterization properties (e.g., void fraction, specific surface area, largest pore size, etc.), while the remaining 19 components corresponded to chemical information about the MOFs. The chemical information was simple, including but not limited to counts of elements (e.g., N, C, O, H, F, etc.) in the unit cell, ratios of atoms (e.g., metallic percentage = number of metal atoms divided by number of C atoms, ratio of electronegative atoms = number of F, Cl, Br, I atoms divided by total atoms, etc.), and the identity of the metal node (e.g., Zn, Cu, V, Zr).

The above-described chemical features can be readily calculated by most researchers (even if not highly specialized in molecular simulation), given that they can be calculated with simple codes. The interpretation of these features also plays well with chemical intuition. For instance, the metallic percentage is indicative of how many metal nodes appear in the MOF unit cell, which could be informative to the models, because nodes are generally strong adsorption sites, either due to the presence of undercoordinated metal sites or due to cooperative interaction of the MOF linkers that converge to the node. For a second instance, the identity of the metal node is somewhat indicative of MOF topology, given that a node of a given metal tends to be associated with a given coordination pattern. This is especially true for the MOFs in the study in question, which were taken from a database featuring only six topologies.

Ultimately, Srivastava et al. [5] trained their models with data for methane adsorption at 35 bar and 298 K in 10,433 MOFs, achieving prediction accuracies of 8.75% according to the mean absolute percent error (MAPE) when using RFR models. Interestingly, these authors also trained an SVR model, allowing us to make a somewhat direct comparison with earlier SVR models by Woo et al. [2], hence enabling us to compare the performance of simpler chemical and structural features versus more elaborate (advanced) features such as AP-RDFs. Note that the comparison is not completely direct because the SVR model by Woo et al. [2] was trained with adsorption data at lower pressure than the model by Srivastava et al. [5]. However, both models were trained with methane adsorption data from tens of thousands of MOFs from the same database (the hMOF database [6]). Intriguingly, the SVR model with simpler features had somewhat higher R^2 (0.92 vs. 0.88) than the SVR model with advanced features.

Indeed, simple features are not incompatible with efficient data usage as long as they are well thought out. For instance, in 2018, Gomez-Gualdron et al. [7] trained various machine learning models to predict CO_2 adsorption from CO_2 mixtures in MOFs using molecular simulation data for over 400 materials. These authors proposed to describe the MOF with a 13-component feature vector catenating traditional structural features (void fraction, density, specific surface area,

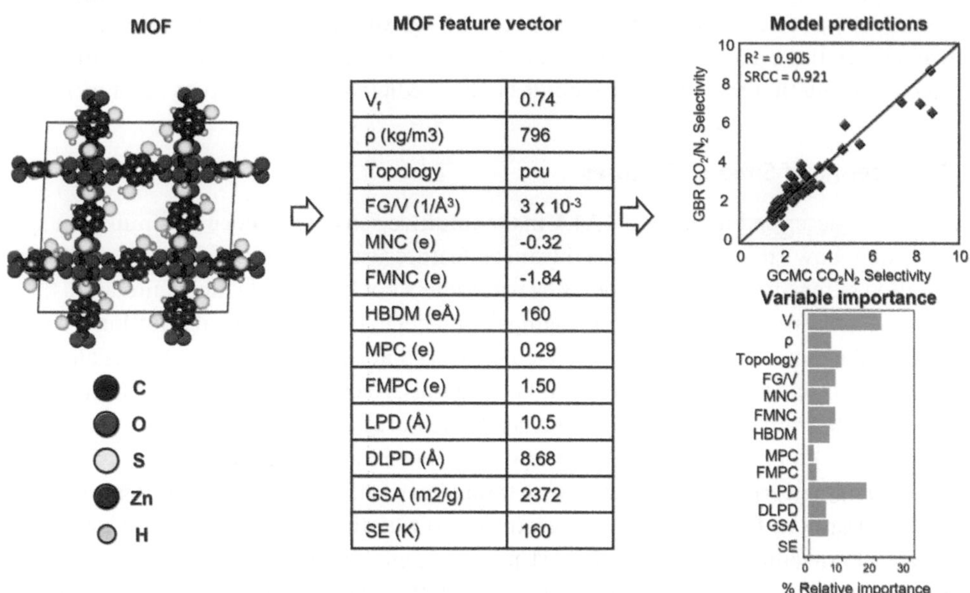

Figure 13.3 Left: a thiol-functionalized MOF analog to IRMOF-1; Center: Representation of MOF in left panel by a 13-component feature vector based on properties traditionally used for MOF characterization (void fraction, V_f; density, ρ; largest pore diameter, LPD; diffusion-limiting pore diameter (DLPD), gravimetric surface area (GSA), and topology), counts of functional groups per unit cell volume (FG/V), and physical–chemical MOF properties indicative of CO_2 attraction by the MOF (functional group most negative (MNC) and most positive charge (MPC), highest bond dipole moment (HBDM), and sum of epsilons (SE); along with framework most negative (FMNC) and most positive charge (FMPC)). Right: Parity plot comparing CO_2/N_2 selectivity predictions by a GBR model vs. values by (GCMC) simulations, along with corresponding variable importance. *Source:* Adapted from [7].

largest pore diameter, and diffusion-limiting pore diameter), with the addition of framework topology (as a qualitative feature mapped into an integer value), and simple physical-chemical descriptors (Figure 13.3). The latter includes simple counts of functional groups (per volume) and characterization of these groups according to their most positive and most negative charges (MPC and MNC, respectively), their highest bond dipole moment (HBDM), and the sum of epsilons (SE) of the group constituent atoms. The underlying porous framework is also characterized by the most positive and most negative framework charge (FMPC and FMNC, respectively).

The selection of physical-chemical features by Gomez-Gualdron et al. [7] is rationalized by the expectation that CO_2 adsorption is primarily driven by electrostatic interactions. For example, MPC, MNC, and HBDM inform the models to what extent a particular functional group may electrostatically attract CO_2. The count of functional groups, on the other hand, informs how many sites with that particular affinity to CO_2 are distributed around the material (per unit volume). The addition of topology to the feature vector is justified by the idea that indicating the topology of a MOF implicitly informs a machine learning model about the number of MOF structural details such as node connectivity, pore shape, pore polydispersity, and perhaps others.

Ultimately, Gomez-Gualdron et al. [7] trained five of the most common types of models trained for adsorption prediction at the time (i.e., multivariable linear regression (MLR), decision trees (DTs), SVR, RFR, artificial neural networks (ANNs), in addition to gradient boosted regressors (GBR). The goal was to predict six different CO_2 adsorption properties at 313 K, including CO_2 loading at 1 bar, CO_2 selectivities and working capacities for 15/85 CO_2/N_2 mixtures at 1 bar, and 20/80 CO_2/H_2 mixtures at 20 bar. The accuracy of the predictions varied both with the type of property predicted and the type of model used, but the predictions of the GBR models were notably good, especially given the small dataset used in this work. For instance, predictions of CO_2/N_2 selectivity by the GBR model were characterized by an $R^2 = 0.91$ (Figure 13.3).

GBR models, as for any model based on decision trees, readily provide information of variable importance, which can be helpful to build (or refine) chemical intuition. For instance, Figure 13.3 presents variable relative importance for the GBR model predicting CO_2/N_2 selectivity, revealing that void fraction, largest pore diameter, and topology are more important than more chemistry-dependent features on modulating selectivity. This informs chemists aiming to synthesize more CO_2-selective MOFs of the importance of first conceiving an underlying framework with suitable pore structure, which then can be imparted with the right chemistry to further boost selectivity. Feature importance also reveals that to boost CO_2/N_2 selectivity, more attention is needed on modulating the negative charge in the functional group than the positive one. This type of analysis is facilitated because of the simplicity and interpretability of the components of the feature vector.

13.2 Material Selection

The strategy to select materials on which to train machine learning models also impacts the burden on data generation. To better understand this, let us return to the relationship given by Equation 13.1. This relationship ($y = 5 \log x$) is illustrated by the continuous line in Figure 13.4. Imagine a training set that includes the blue points (Figure 13.4a) versus one that includes the red points (Figure 13.4b). Although both training sets have the same number of points, it is apparent that a machine learning model would learn more about the relationship between y and x from the blue set than from the red set, making training with the blue set more efficient. In other words, if we were to test a model trained on the "red set" to predict one of the blue points, we would probably encounter larger prediction errors than if the situation was reversed.

An additional notion from the above discussion is that adding the white points (Figure 13.4c) to either training set may not improve the model. If added to the red set, the absence of points at larger x values would probably still mean that the model remains fundamentally incapable of learning the relationship between y and x to its full extent. If added to the red set, the effects would probably be less significant, but could make the data less efficient as the model would not learn much new information. On the other hand, if we were to add too many "white-like" points to the training set, the latter could become "imbalanced," and the model could focus on learning the relationship at small x values but not at large x values.

Translating the above ideas into a strategy for material selection could be summarized by "diverse is best." In 2020, Kulik et al. [8] examined the diversity of material sets that have been used to train machine learning models to predict adsorption in MOFs. Although in some instances these material sets are generated for specific work using "MOF generation" codes such as ToBaCCo (or others), in most instances the material sets are extracted from existing MOF databases such as the hMOF [6], ToBaCCo [9], BW [10], and CoRE MOF [11] databases. Kulik et al. [8] presented a comparison of diversity for the latter three databases through visualization and clustering approaches.

Figure 13.5a is constructed using a t-distribution stochastic neighbor embedding (t-SNE), which is a statistical visualization method to project high-dimensional data onto low-dimensional space. To apply the method, Kulik et al. [8] represented MOFs with 165-component feature vectors, with 9 components corresponding to traditional structural features (i.e., density, specific surface area, largest pore size, etc.), and the remaining 156 components corresponding to values of revised autocorrelation (RAC) functions calculated as in Equation 13.6:

$$RAC(d) = \sum_i^{start} \sum_j^{scope} f(P_i P_j) \delta(d_{ij}, d) \tag{13.6}$$

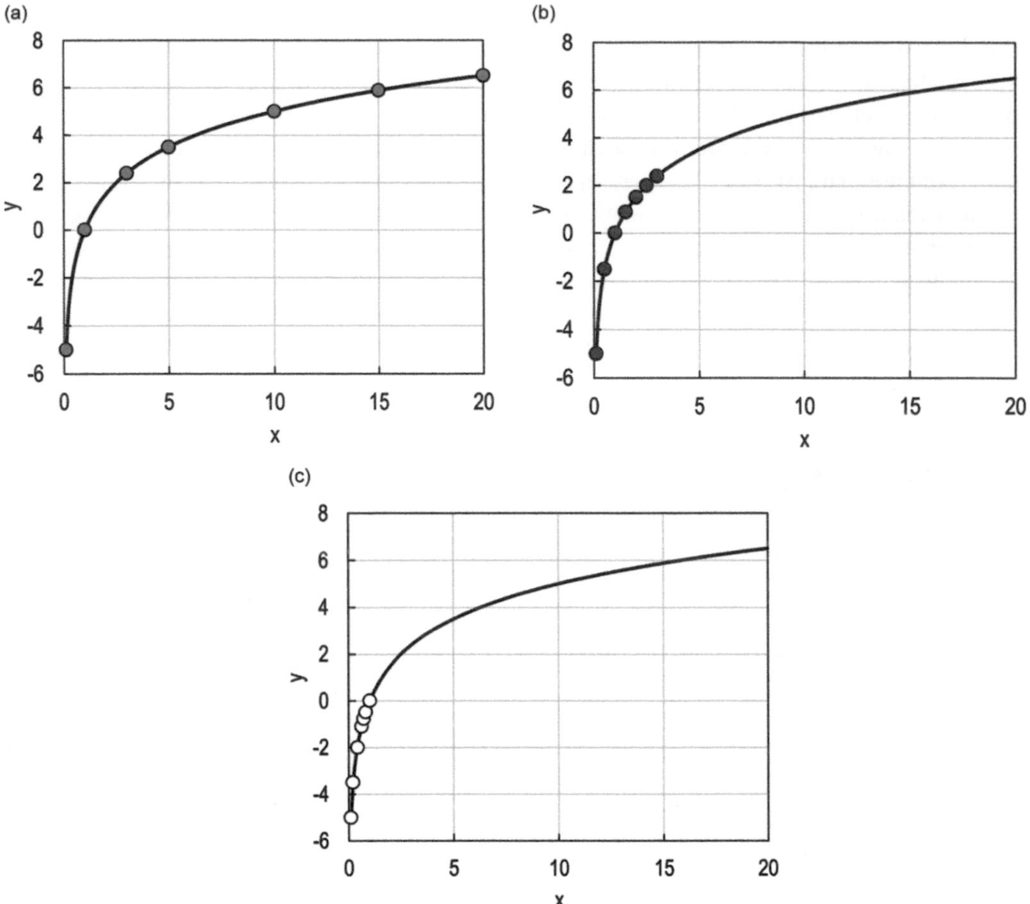

Figure 13.4 Various distributions of training points to learn the y = 5 log x relationship. (a) Blue points represent a balanced and diverse training dataset. (b) Red points represent a biased (less diverse) training dataset. (c) White points represent potential additional training data points. However, their addition to points in (b) do not add much extra information, while their addition to points in (a) may imbalance the training dataset. *Source:* Diego Gómez-Gualdrón.

which can be thought of as discrete counterparts of the AP-RDFs calculated with Equation 13.2, particularly when $f(P_i, P_j)$ is taken as the product of properties P_i and P_j of atoms i and j. However, the RACs considering $f(P_i, P_j)$ to be the difference between Pi and Pj were also calculated by these authors. The properties considered by these authors relate to the chemistry of the material (i.e., atom identity, connectivity, electronegativity, covalent radii, nuclear charge, and polarizability), thus their RACs can be considered as chemical features. The "discretization" arises from the delta Dirac function $\delta(d_{ij}, d)$, which is zero except when d_{ij} (an integer corresponding to the shortest number of bonds atoms i and j are away from each other) is equal to the integer d.

The t-SNE is such that the closeness between two points in the projection is indicative of the similarity of the corresponding high-dimensional feature vectors, and hence the corresponding MOFs. This is partly reflected by large-pore and small-pore MOFs congregating in separate regions, respectively. Using pore size as reference, Figure 13.5a illustrates the idea that none of the current MOF databases covers the whole "material space" perfectly, hence not being perfectly diverse. The imperfect diversity of the CoRE, BW-DB, and ToBaCCo databases is concisely summarized in the spider plots in Figure 13.5b, with respect to pore geometry, metal chemistry, linker chemistry, and functional groups.

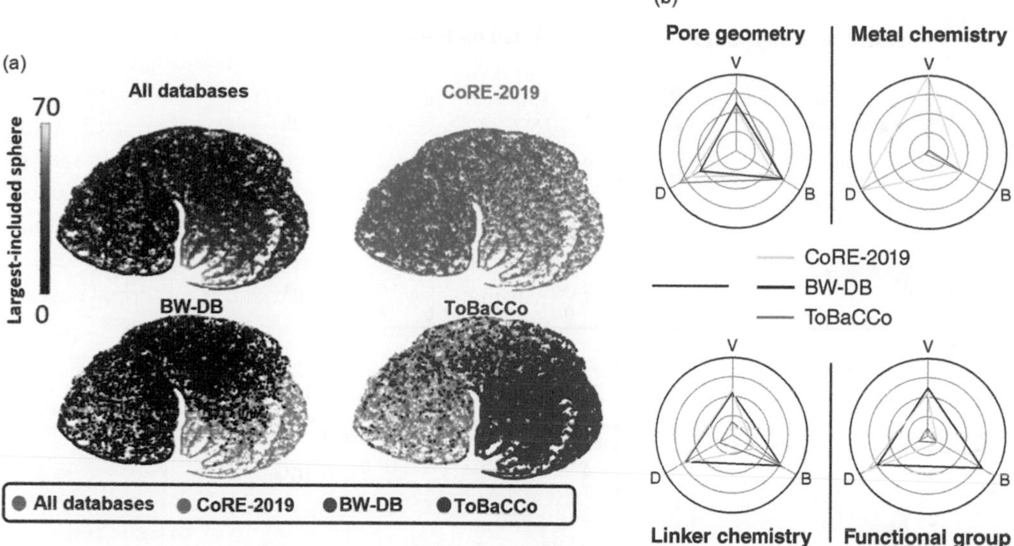

Figure 13.5 (a) Example of t-SNE plots illustrating the different kinds of pore sizes favored by three common MOF databases (CoRE-2019, BW-DB, and ToBaCCo). Each point represents a MOF. Points in the top left plot are colored by pore size and in the remaining plots are colored according to the database the MOFs belong to. (b) Spider plots illustrating the diversity (D), bias (B), and variety (V) of the MOFs in the three databases considered in panel (a). *Source:* From [8] © 2020, Springer Nature.

The practical importance of diversity is illustrated in Figure 13.6a, in which Kulik et al. [8] showed that an RFR model trained on (a training set from) the BW-DB database to predict CO_2 adsorption at 0.15 bar dramatically loses accuracy when tested on (a test set from) the CoRE MOF database instead of (a test set from) the BW-DB database. By contrast, an RFR model trained on equally sized but more diverse training sets made up by MOFs from both databases presents similar accuracy when tested on either database. Importantly, the learning curves in Figure 13.6b show that constraints on model accuracy stemming from imperfect diversity are unlikely to be solved by increasing the training set size skip after a critical set size. For instance, the model trained on the BW-DB database plateaus at an accuracy inferior to that for the diverse training set. These plots are also indicative that there is an optimal training set size, which is something we would like to know to generate (and use) data more efficiently.

13.3 Model Selection

Perhaps most apparent when datasets are small, the model selection may also affect prediction accuracy. For instance, in the earlier discussed work by Gomez-Gualdron et al. [7], which used a dataset of around 400 MOFs, differences in model accuracy could be observed despite using the same datasets and components of the feature vector. The relative performance of the models trained by these authors somewhat changed with the property to be predicted, which echoes the idea that there is not a universally better or worse type of machine learning model that performs best for all types of problems (something conveyed by the so-called "no free lunch" theorem). Rather, we need to find the right model for the right problem. For instance, we do not want too complex a model when the relationship between output and features is simple, or conversely, we do not want too simple a model when the relationship is rather complex. Note that when inspecting the literature,

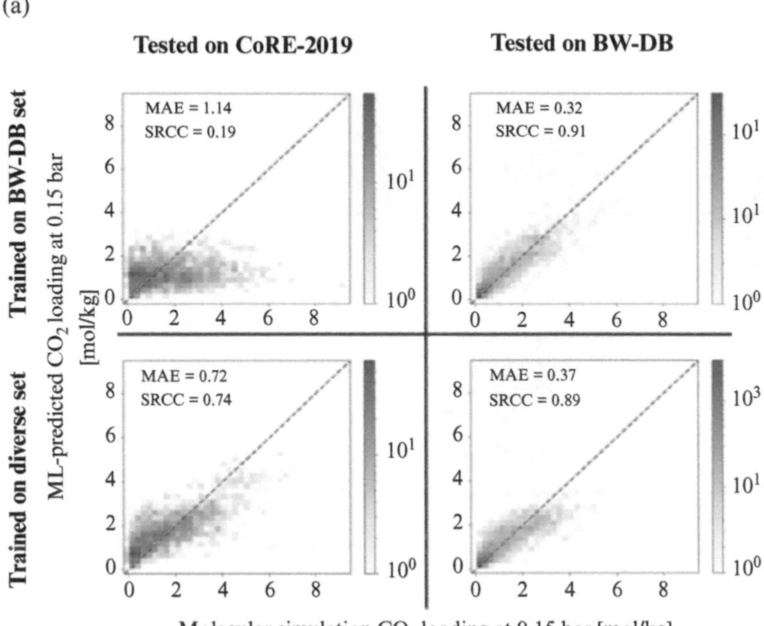

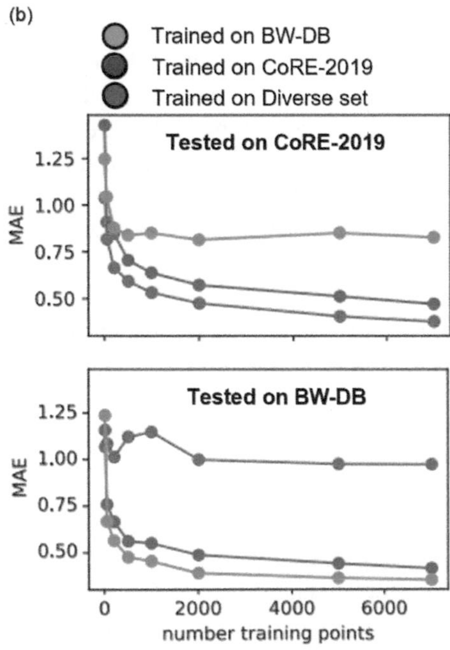

Figure 13.6 (a) Parity plots indicating the accuracy of RFR models for CO_2 predictions on the CoRE-2019 and BW-DB databases, depending on the nature of the training set (either from the BW-DB database or a "diverse set" sharing MOFs from different databases). *Source:* [8] 2020, Springer Nature. (b) Learning curves indicating the dependence of mean absolute error (MAE) on the number of training points during the training of RFR models to predict CO_2. *Source:* Adapted from [8].

the rationale for model selection in a given work is not always clear or directly stated by the authors. In some cases, factors such as the availability of machine learning tools, as well as ongoing model selection trends at the time a prediction problem is tackled, may play a role. However, given the role the model type may play on prediction accuracy and efficient data use, a cursory inspection on the basics of some of the most common models now follows, which we hope informs the reader on the basic features (and pros and cons) that these models bring to the table.

13.3.1 Linear Regression

Multivariable linear regression (MLR) is one of the simplest types of models to train. An MLR model assumes that the relationship between the output y and the n components of the feature vector $\boldsymbol{x}$ ($x_1, x_2, \ldots x_n$) is linear, as given by Equation 13.7:

$$y = b + \sum_{i=1}^{i=n} m_i x_i \tag{13.7}$$

where the m_i and b are the regression coefficients. In analogy to simple linear regression, where $y = b + mx$, b can be thought as a "y-intercept" and each m_i as the "slope" of the relationship between y and the corresponding x_i feature. The way an MLR model is trained is to find a set of regression parameters that minimize:

$$J = \frac{1}{N} \sum_{j=1}^{N} (y_{pred_j} - y_j)^2 \tag{13.8}$$

where y_{predj} and y_j are the predicted and actual adsorption property for material j, N is the number of materials in the training (and/or validation) set, and J is a so-called "cost function," which in the case of Equation 13.8 is the RMSE. The success of training the MLR model depends on there actually being a (nearly) linear relationship between the output and the inputs (components of the feature vector). It is not uncommon for authors working on adsorption prediction in nanoporous materials to train an MLR model, first to establish a baseline for prediction accuracy, and then train more complex models to determine whether they make better predictions.

Arguably, the appeal of MLR models (especially in earlier efforts) comes from: i) the perceived success of empirical linear relationships such as Chahine's rule [12], which is widely used by experimentalist to relate hydrogen adsorption capacity with the specific surface area of a material; and ii) the simple interpretability of the model, e.g., the strength of the relationship between a given feature x_i and the adsorption property y is indicated by the magnitude of m_i, while the nature of the relationship is indicated by the sign of m_i.

Incidentally, MLR is also ideal to introduce one of the tools that can be used to reduce the tendency to overfit a model: regularization. Regularization essentially means that some sort of control is introduced to the model parameters. In the case of MLR models, we typically want unimportant features x_i to simply have values for the corresponding m_i equal to zero. One way to accomplish this goal is by altering the cost function in Equation 13.8 to become:

$$J = \frac{1}{N} \sum_{j=1}^{N} (y_{pred_j} - y_j)^2 + \lambda \sum_{j=1}^{N} |m_i| \tag{13.9}$$

which essentially adds a "cost" to having large magnitudes for m_i according to a proportionality (regularization) constant λ, which itself needs to be optimized. The modification to the cost function presented in Equation 13.9 corresponds to the so-called "LASSO models," such as the model by Snurr et al. [4] discussed earlier in this chapter to predict hydrogen adsorption.

13.3.2 Supported Vector Regressors

Concisely stated, SVR finds the "hyperplane" that best approximates the relationship between the target property y (e.g., CO_2 loading) and the components of a feature vector x (e.g., AP-RDF values at different R values). If the material feature vector only had two components, then the hyperplane would become a plane, or if the material feature vector somehow only had one component, then the hyperplane would become a line. Let us use the latter (unlikely) scenario, which is illustrated in Figure 13.7a, to provide an intuitive description of SVR. In Figure 13.7a, ideally we would like to find a line (hyperplane) that contains all the observations of y in N materials. Namely:

$$y = mx + b. \tag{13.10}$$

Since this is impossible for the data in Figure 13.7a, the next best thing is to have the line (hyperplane) from Equation 13.10 define a region that contains all the observations of y in N materials, where the region would be defined in relation to this line (hyperplane) according to Equations 13.11 and 13.12:

$$y = mx + b + \varepsilon \tag{13.11}$$

$$y = mx + b - \varepsilon \tag{13.12}$$

where ε is the maximum error that we feel comfortable with to deem a prediction as correct.

The issue is that even finding a region bound by parallel lines (hyperplanes) that contains all observations of y is rarely possible, it must be accepted that a number of observations will unavoidably be left out (Figure 13.7a). In this scenario, if we use a "hard margin" approach, then we would settle for finding the line (hyperplane) that defines the region that contains the most observations of y. Notice once again that the definition of optimal line (hyperplane) for an SVR model is different than the definition of optimal line for a linear regression model.

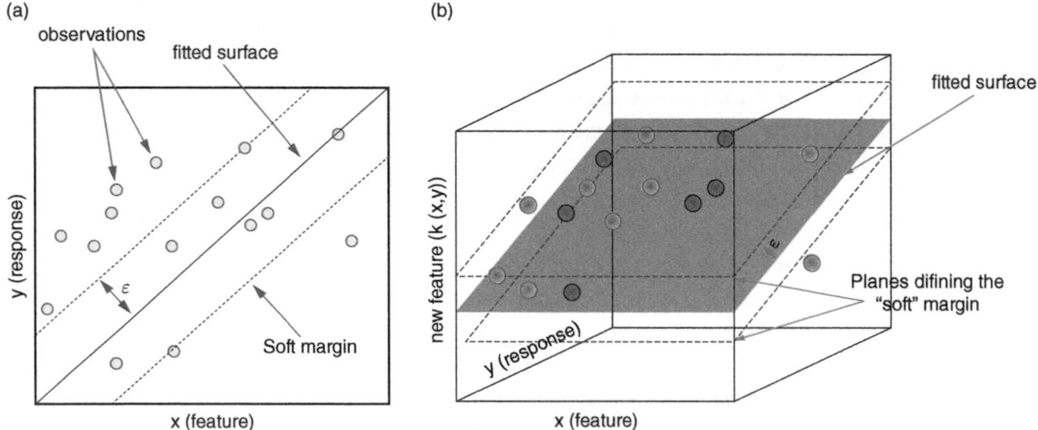

Figure 13.7 (a) Visualization of how an SVR model (with soft margin) works for the simple case of a single-component feature vector, where the fitted hyperplane becomes a line. The values of observation y that fall within the two soft margins (distanced by ε from the fitted surface) do not count toward the error model. SVR models typically try to balance maximizing the number of observations within the two soft margins and minimizing the distance of points outside the margins to the fitted surface. *Source:* Adapted from [7]. Reproduced with permission of American Chemical Society. © 2018 American Chemical Society. (b) Schematization of the "kernel trick," which would take the data in panel (a) and create a third dimension, where a fitted hyperplane (which now becomes a regular plane instead of line) may represent the data better. *Source:* Adapted from [7].

On the other hand, if we use a "soft margin" approach, then we would settle for a region that contains as many y observations as possible while minimizing the distances, ξ, of the left-out observations $y_{leftout}$ to the region boundaries. This is done by adding the term in Equation 13.13:

$$C\sum_{K=1}^{M}|\xi_k| \tag{13.13}$$

to the cost function, where k indicates a left-out observation and the summation is the total M of left-out observations. The importance of minimizing ξ versus maximizing the number of observations of y in the region is determined by a proportionality (regularization) constant, C.

The described procedure thus far, for an output y and a single material feature x, suggests that the trained SVR models would be expected to work better the more the relationship between y and x looks linear. However, SVR offers the possibility of improving predictions by mapping the data into a space with higher dimensionality using the so-called "kernel trick." Without really diving into the underlying mathematics, a practical application of the kernel trick could be, for instance, to transform the 2D data in Figure 13.7a into three-dimensional (3D) data, where the observations could be better described by a plane (Figure 13.7b).

As for the hyperparameter ε, note that it depends on the type of property to be predicted and the context of the application. For instance, if we intend to use an SVR model to predict hydrogen adsorption at 77 K and 100 bar to discover a new MOF for hydrogen storage, perhaps an ε of 3 g/L would be acceptable, given that hydrogen loadings (which would be y) at that condition have been seen to vary between 0 and 60 g/L. From the practical aspect of training SVR models, note that SVR models are perhaps more suitable for datasets including up to tens of thousands of data entries, at which point other models may be easier to train.

13.3.3 Decision Tree-based Regressors

The structure of a simple decision tree regressor (DTR) is schematized in Figure 13.8a. Thus, a DTR starts with a "root" node that may sequentially split into additional nodes. Only two additional nodes can be created with each splitting, and when a node does not split into additional nodes, it is known as a "leaf" node. With the exception of leaf nodes, each node represents a "condition" that determines to which of the subsequent nodes a given data entry should go, with each data entry eventually falling into one of the leaf nodes. Ideally, this would create a scenario where all data entries in a given leaf node have the observed values of y. More practically, all data entries in a given leaf node have very similar observed values of y, as indicated by the variance, σ:

$$\sigma = \frac{\sum_{k=1}^{M}\left(y_k - y_{average}\right)^2}{M} \tag{13.14}$$

where y_k are the observations of y in the node, and $y_{average}$ is the average value of y among the M observations that end up in the node.

The variance is key in training DTRs, as training consists in finding for each node the condition that leads to a split of the data that minimizes the average of the variances of the two "child" nodes (weighted by the number of observations ending in each child node) relative to the variance of the "parent" node. The condition at each node is based on the value of feature x_i in the feature vector $\boldsymbol{x}$. For instance, we could place a data entry in the first child node if the value of the third feature in the feature vector is larger than 0.5 (i.e., $x_3 \geq 0.5$), or else in the second child node.

Upon training, the DTR can make quantitative property predictions for a new data entry (in the context of this chapter, typically a new material) by taking the feature vector and move across the

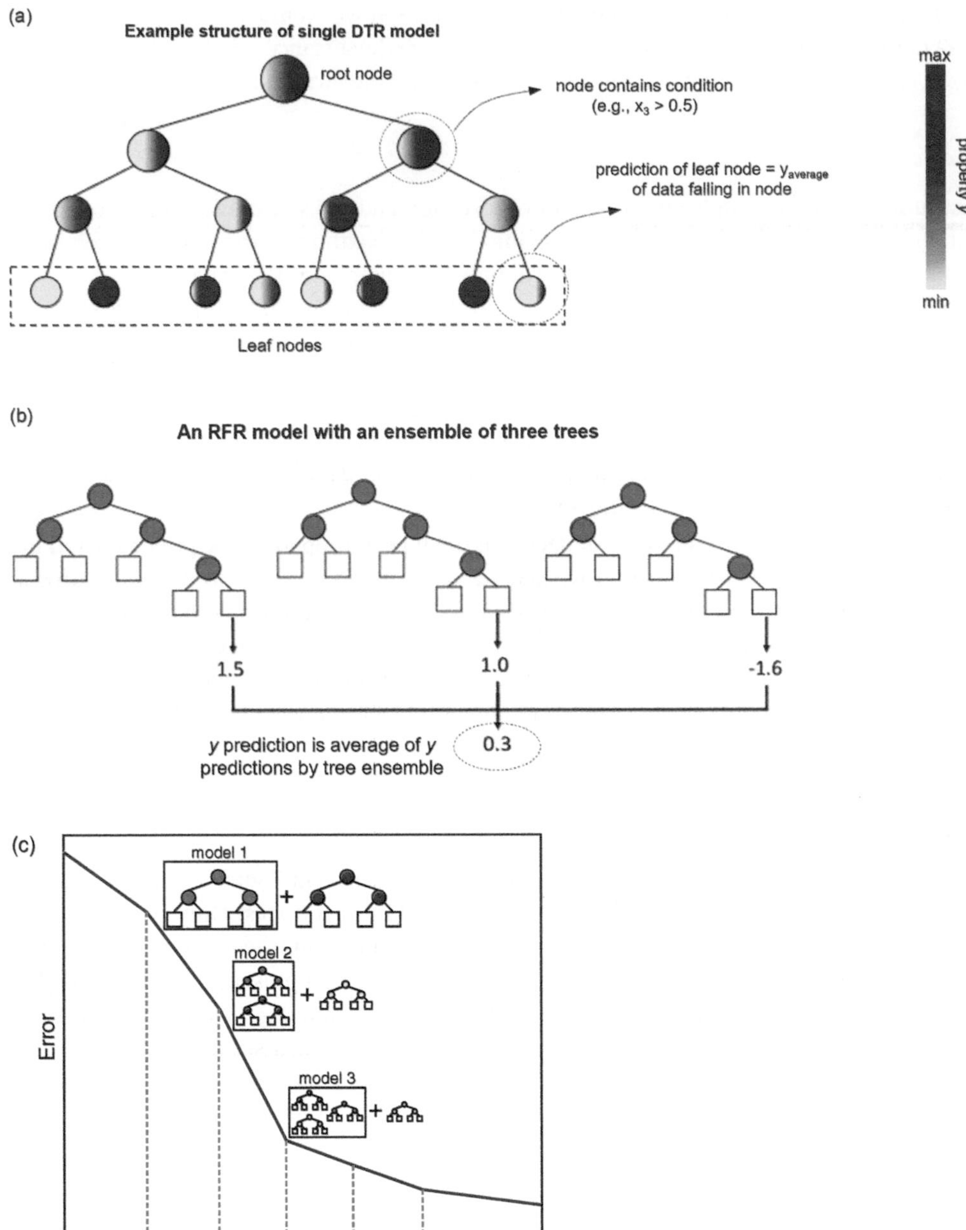

Figure 13.8 (a) Schematics of the structure of a single DTR model. Each of the nodes in the tree are colored to indicate the distribution of the values of the observation of y associated with the node. Data splitting in each node is conditioned on a threshold value of a feature *Source:* Adapted from [7]. (b) Illustration of voting to determine the predicted value of y in an RFR mode. (c) Evolution of the prediction error as trees are sequentially added to a GBR model. *Source:* (b and c) Adapted from [7]. Reproduced with permission of American Chemical Society. © 2018 American Chemical Society.

"tree" until the new data entry is placed into one of the leaf nodes. At this point, the predicted value of y for the new data entry is said to be equal to the $y_{average}$ associated with the corresponding leaf node. Given that the number of leaf nodes in a DTR is finite, the predictions of y values are effectively quantized which, depending on the problem, may hinder prediction accuracy. For instance, revisiting the work by Gomez-Gualdron et al. [7] on CO_2 adsorption predictions, DTR models (along with MLR models) tended to underperform compared with other machine learning models.

One way to overcome potential limitations of the single DTR model is to train an ensemble of "trees," where each tree make its own prediction for the value of y, with the final predicted value of y corresponding to the average of the y prediction of each tree in the ensemble. The schematic of a small parallel ensemble of trees is shown in Figure 13.8b, where such type of model configuration refers to the earlier discussed RFR models trained by Simon et al. [3] to predict Xe/Kr adsorption selectivity. The random character of the "forest of trees" comes both from: i) each individual tree being trained on a random subset of the training data; and ii) each node in each tree only considering a random subset of components of the feature vector to set the splitting condition. The random subsets to train each tree can even include some duplicated data entries in what is referred to as "bootstrapping." All of the above acts as a workaround to the particular sensitivity of a given tree to the specifics of the training data, resulting in more robust prediction models.

Alternative to the parallel ensemble characteristic of RFR models, another possibility is to create an ensemble of trees is in sequential fashion (Figure 13.8c) in a type of configuration known as gradient boosting regressors (GBRs). One of the differences with RFR models is that each tree in the sequential GBR ensemble is trained not to predict the actual value of observation y, but the error (referred as "residual") of the prediction of y made by the preceding sequence of trees in the ensemble. If we consider that a sequence of trees in the GBR model has M trees and that the residual predicted by each tree is E_m, then the prediction of the ensemble up to tree M is given by Equation 13.15:

$$y = y_{average} + \lambda \sum_{m=1}^{M} E_m \tag{13.15}$$

where $y_{average}$ is the average value of y in the whole dataset, and λ is a proportionality constant (and hyperparameter) referred to as the "learning rate." The residuals that the very first tree aims to learn are simply the differences between the actual values of y and the $y_{average}$, while for subsequent trees the residuals are the differences between the actual values of y and the predicted values. As the number of trees in the ensemble increases, the error of the predictions is expected to decrease (Figure 13.8c).

Decision tree-based regressors models, most notably RFR models, are perhaps one of the most commonly found type of models in the literature of machine learning prediction of adsorption properties in nanoporous materials. Perhaps the most appealing aspect of these types of models is that information such as variable importance naturally emerges from training, facilitating the building of chemical intuition (particularly if the components of the feature vectors correspond to chemically intuitive properties). Additionally, it is possible to see the "thought process" of the model as it figures out what prediction to make for property y, enhancing the interpretability of the model to build chemical intuition.

13.3.4 Artificial Neural Networks

Artificial Neural Networks (ANN) are inspired by the biological neural networks that make up animal brains, where neurons (nodes) receive signals from other neurons and, upon processing, send their own signals to other neurons. One of the most common architectures of ANN is a multilayer

(a)

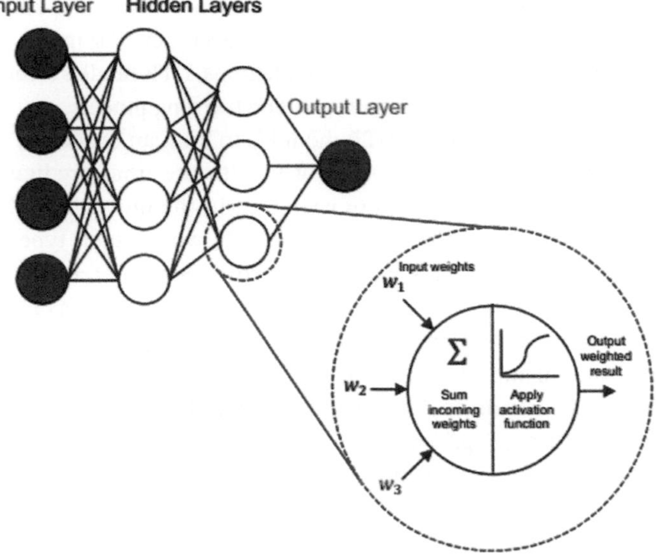

(b)

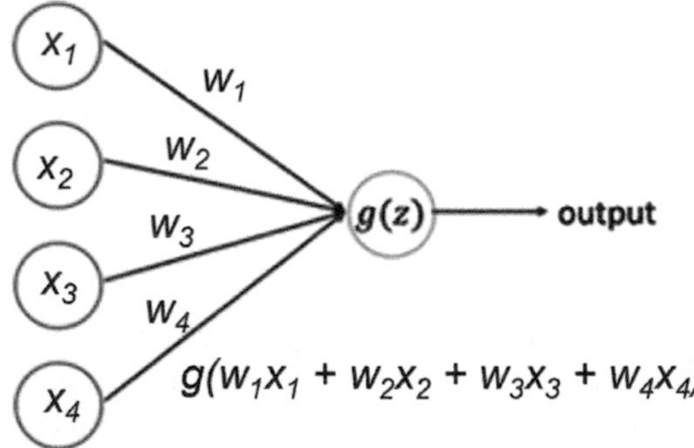

(c)

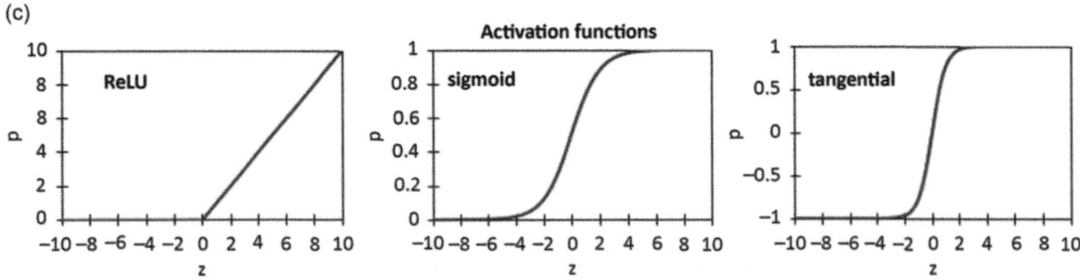

Figure 13.9 (a) Schematic of common type of ANN model: a perceptron. (b) Schematic of the deconstruction of a neuron. Alternatively, this can also be interpreted as the simplest ANN model there can be. No hidden layers. Just the input nodes feeding their signals (multiplied by a corresponding weight) to the output node, which then applies an activation function to produce the prediction of observation *y*. (c) Examples of common activation functions. *Source:* Adapted from [7].

perceptron, one example of which is schematized in Figure 13.9a. In a multilayer perceptron, all nodes in a given intermediate (hidden) layer process "signals" coming from all the nodes in the preceding layer, and in turn send signals to all the nodes in the subsequent layer. The first and last layers are not "hidden" and are known as the input and output layer, respectively.

The input layer has as many nodes as components of the feature vector $\boldsymbol{x}$ (four in Figure 13.9a), because each node in the input layer needs to receive the value of one of the components of the feature vector (e.g., the first node receives the value of x_1, the second node receives the value of x_2, etc.), so that these values can be converted into signals that are propagated through the intermediate hidden layers, and eventually to the output layer. So let us now discuss how these signals are processed and transmitted.

The signal coming from an input node i to a node in the first hidden layer node j is simply the value of x_i multiplied by a weight factor w_{ij}. The weight factor can be thought of as modulating the intensity of the signal, such as the same input layer node can send signals of different intensity to different nodes in the first hidden layer. A given node j in the first hidden layer takes all the signals coming from the N nodes i in the preceding layer and adds these to obtain z_j, as in Equation 13.16:

$$z_j = \sum_{i-1}^{N} w_{ij} x_i \tag{13.16}$$

The value of z_j is then fed to a so-called "activation function," $g(z)$, which produces a value p_j. Common types of activation functions are illustrated in Figure 13.9c, which include: i) the rectified linear unit (ReLU):

$$p_j = \max(0, z_j) \tag{13.17}$$

which dictates that p_j is equal to z_j unless z_j is negative, in which case p_j becomes zero; ii) the logistic or sigmoid activation function:

$$p_j = \frac{1}{1+ e^{-z_j}} \tag{13.18}$$

which essentially makes p_j equal to one if z_j is too positive, equal to zero if z_j is too negative, and between zero and one for intermediate values of z_j; and iii) the hyperbolic tangent activation function:

$$p_j = \tanh(z_j) = \frac{2}{1+ e^{-2z_j}} - 1 \tag{13.19}$$

which essentially makes p_j equal to one if z_j is too positive, equal to minus one if z_j is too negative, and between minus one and one for intermediate values of z_j. The node j in the first hidden layer then sends its output p_j to all the nodes k in the subsequent layer but multiplied by a weight factor w_{jk}, so that node k takes all the signals from the M nodes in the preceding layer and adds them to obtain z_k:

$$z_k = \sum_{j=1}^{M} w_{jk} p_j \tag{13.20}$$

Then, analogous to what happens to nodes in the first hidden layer, the sum is used as input for the node activation function, whose output is subsequently transmitted to nodes in subsequent layers.

Eventually, nodes in the output layer are reached, which, through similar operations to those just described, yield the actual predictions. Notice that the output layer must have as many nodes as (types of) predictions that are to be made. For instance, if only one property y is to be predicted, then the output layer must have only one node.

If we have decided on the architecture of the ANN (i.e., the number of layers and numbers of nodes in each layer) and the type of activation functions that go into each node, then the training of an ANN model consists of the optimization of all the weights by which all the "signals" that travel through the nodes in the network are multiplied. This optimization is done using a so-called "backpropagation algorithm." Briefly, to train the model, we start by "initializing" all the weights in the ANN with random values. Then, a first prediction of the observation y is made for all the data entries in the training set, with a cost function (e.g., an RMSE as in Equation 13.8) being calculated to determine the overall goodness of the ANN predictions. Then, we update the weights sequentially starting from the output layer and moving back toward the input layer. For each update, the value of the weight of interest is changed in the direction that is expected (based on derivative calculations) to reduce the cost function. When all weights have been updated once, the update sequence resumes at the output layer and starts to "backpropagate" again.

If there were no hidden layers (essentially the diagram in Figure 13.9b), the prediction of y by the output (node) would correspond to a linear combination of the components of the feature vector with some subsequent non-linear scaling by the node activation function. This makes it apparent that the ability of ANNs to model non-linear relationships between y and the components of a feature vector x stems from the non-linear character of the activation functions. It is also apparent that as more layers (and more nodes) are used in an ANN, the way the model yields a prediction becomes more notably convoluted than for other previously discussed models in this chapter. However, this degree of convolution hinders interpretability, which has led many to describe ANNs as "black box" models. Importantly, however, ANN models can be powerful predictive tools, becoming increasingly effective as the size of the dataset used for training increases.

13.4 Data Usage Strategies

A source of inefficient data usage in the application of machine learning to adsorption predictions in nanoporous materials perhaps arises from the early "zeitgeist" for the intended scope of machine learning models. Until recently, the dominant approach in machine learning efforts has been to generate (as large as possible) training datasets running molecular simulations for materials in the database of choice, and then train models to predict specific adsorption properties (e.g., deliverable capacity, adsorption selectivity) of specific adsorbates (e.g., methane or hydrogen) or adsorbate mixtures (e.g., CO_2/N_2, Xe/Kr) under specific conditions (e.g., temperature, pressure, composition). The catch is that since extensive molecular simulation data was already generated for the specific adsorption property in the database of interest, the possible practical uses for the trained machine model become limited.

In principle, if the model was trained with data for only a part of the database, the model could be used to complete the exploration of the remaining materials in the database, or someone could use the model to explore other databases. In practice, however, this rarely happens. And because the "as is" model cannot be used to explore adsorption of the same adsorbate or mixture at different conditions than originally trained, much less to explore the adsorption of other adsorbates, the payoff for data generation and model training is close to nill. But what if the data or models used to predict the adsorption properties for an adsorbate or mixture at a thermodynamic condition could be leveraged to predict adsorption at a different condition instead? Or even for a different adsorbate or mixture?

In such a case the payoff could be tremendous given that there is probably an abundance of nanoporous materials that could have a significant impact on numerous adsorption-based applications beyond those that capture the most attention from the research community (e.g., CO_2 capture, hydrogen adsorption, etc.), and these applications probably involve adsorbates for which large datasets with their adsorption properties have not been generated (at least, not yet); whereas fast adsorption predictions at different thermodynamic conditions would open the door to coupling material and process optimization. In pursuing these goals, the research community has been pursuing different model training strategies, which we now proceed to discuss with some examples.

13.4.1 Transfer Learning

The concept of transfer learning is one that is often brought up in machine learning applications where we have to deal with data scarcity. To discuss the central idea behind transfer learning, consider again the relationship between y and x given in Equation 13.1. When using this relationship to discuss feature engineering at the beginning of this chapter, we noted that we could use data more efficiently if we engineered a "log x" feature, as with relatively few data points a machine learning model could figure the proportionality constant (i.e., 5) relating "y" and "log x." Now, let us say that there is another problem for which the variables u and v followed the relationship in Equation 13.21:

$$u = 2 \log v \tag{13.21}$$

Perhaps from "domain knowledge" we could intuit that the relationship between u and v shares similarities with the relationship between y and x. In that case, we would probably be inclined to engineer the feature "log v" and use relatively few data points to train a new model that would rapidly figure the proportionality constant between u and $\log v$ (i.e., 2). If collecting numerous data points for u at a given v is difficult and costly, then being able to figure out the relationship in Equation 13.21 with just a few points would be a highly welcome scenario. At this point, let us note that we would probably use the same domain knowledge that informed us to design the $\log v$ feature to impose a linear model to describe the relationship between u and $\log v$.

Now imagine a scenario where the feature "log x" was not something that we designed, but rather a feature that was implicitly learned by a model that we trained using x as input (instead of $\log x$) and y as output. Imagine that we borrow the "old" model, "keep" the pieces of the model that represents the learning of the "log x" feature, and only modify (retrain) the pieces of the model that represents the proportionality between y and $\log x$, but now with data for u and v. Conceivably, with just a few data points, the new model would then figure out the new proportionality constant relating u and $\log v$ (i.e., 2). As abstract as it sounds, the above scenario describes transfer learning: the idea that something learned by an old model can be transferred and retained by a new model.

Fortunately, the layered structure of ANNs confers them with the modularity necessary to materialize the idea of "keeping" one part of a model while "retraining" another, making them ideal for transfer learning (further boosting the increase popularity of ANN models in recent years). The suitability of ANNs for transfer learning is further boosted by the understanding that ANNs learn in a "piecemeal" fashion; an example usually brought up to illustrate this, is convolutional ANNs in image recognition. Say that there is an ANN trained to recognize human faces. It is said that the first layers in an ANN trained for this purpose first learn to recognize "edges" within an image, while subsequent layers learn how these edges spatially relate to each other to form more complex features (say, intermediate layers learn to recognize "mouths," while the final layers learn to recognize "faces"). Relevant to our transfer learning discussion, we could think that an ANN model to recognize dogs would still need to first detect "edges," although later it would need to learn how these edges spatially relate to each other to form, say, "snouts." Accordingly, if we were to train an ANN to recognize dogs, we may as well borrow the "edge-recognizing" first layers of the ANN that recognizes human faces, and only retrain subsequent layers so that they can learn how these edges associate into dog-specific features.

To ground this discussion to the world of nanoporous materials, consider that numerous applications of these materials, while involving different molecules (and mixtures thereof) and operating conditions, all fall under the umbrella of "adsorption applications." This means that the properties we desire to predict are governed by similar fundamentals (e.g., attraction between adsorbates and material atoms). Accordingly, perhaps some layers of an ANN trained to predict, say, methane adsorption, could have learned something useful to predict, say hydrogen.

Study Case 1: Transfer learning in the prediction of CH_4, H_2, and Xe/Kr adsorption Colón et al. [13] recently explored to what extent transfer learning was possible when training models to predict methane, hydrogen, and Xe and Kr adsorption. These authors leveraged publicly available adsorption data for 13,506 MOFs in over 40 topologies in the previously mentioned ToBaCCo database [9], which consisted of molecular simulation results for adsorption of hydrogen at 100 bar at four temperatures, adsorption of methane at 298 K at two pressures, and for co-adsorption of an 20/80 Xe/Kr mixture at 298 K and three pressures. Rationalizing that the adsorption of methane, hydrogen, Xe, and Kr in the studied MOFs was primarily driven by dispersion forces, in all their work these authors used a short 5-component feature vector based on MOF textural properties: volumetric and gravimetric surface areas, void fraction, diffusion-limiting pore diameter, and largest pore diameter (Figures 13.10a–b).

Colón et al. [13] then chose the prediction of hydrogen adsorption at 243 K and 100 bar as the "source task," which was done by an ANN model trained in leveraging the full extent of the ToBaCCo dataset (i.e., the 13,506 observations) randomly partitioned into training, validation, and test sets with 78 to 12 to 10 proportions, respectively. The ANN model trained on the source task consisted of two hidden layers stacked between the 5-node input layer and the 1-node output layer (node). These authors then sought to transfer the learning acquired by the ANN model trained on the source task to the new ANN models each trained to achieve three new "target tasks:" i) hydrogen adsorption under different conditions (100 bar and 130 K); ii) methane adsorption at 100 bar and 298 K; and iii) Xe/Kr adsorption at 5 bar and 298 K. Since the tasks are different, but the domain (MOF database) on which to make the prediction is the same, this is considered an example of inductive transfer learning.

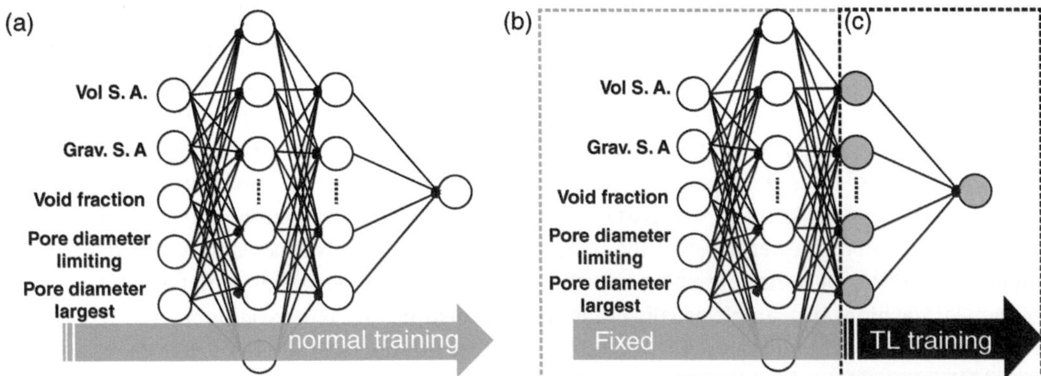

Figure 13.10 Illustration of transfer learning for gas adsorption in MOFs. (a) Schematics of the original ANN trained on the source task. (b) Schematics of the new ANN trained on the target task, indicating the "freezing" of the first hidden layer (i.e., keeping the same weights as in the original ANN). (c) Schematics of 1000 random draws to form 100-datapoint training set to evaluate transfer learning versus direct learning. *Source:* From [13]. Reproduced with permission of American Chemical Society. 2020 American Chemical Society.

Because the idea behind transfer learning is to use data more efficiently to train the new ANN models, these authors used (randomly drawn) datasets about 100-fold smaller than the dataset used to train the ANN for the source task. The transfer learning procedure starts by taking the ANN trained on the source task and having it predict, say, hydrogen adsorption at 100 bar and 130 K. Upon the first evaluation of the cost function (e.g., a RMSE as in Equation 13.8), the back-propagation algorithm starts to update the weights in the ANN. To preserve some of the knowledge gained by the original ANN, Colón et al. [13] followed the standard convention of freezing the layers closer to the input layer (the idea being that these first layers "learn" the features that describe the materials). Specifically, these authors made sure that only the weights of the hidden layer closest to the output layer (the last hidden layer) were allowed to change, so that the weights of the first hidden layer of the new ANN are identical to those of the ANN trained on the source task.

The effectiveness of transfer learning was then evaluated by comparing R^2 for ANNs trained using transfer learning against those trained directly on the same data. To make sure statistically significant conclusions were made about the effectiveness of transfer learning, these authors made the comparison for 1000 transfer learning (TL) ANNs and 1000 "direct learning" (DL) ANNs. Each of these ANNs was trained on one of 1000 distinct, randomly drawn batches of 100 data points (Figure 13.10c). For training on the new target task of hydrogen adsorption under different conditions (100 bar and 130 K), on average, Colón et al. [13] saw generally good predictions by both DL- and TL-ANNs. However, the accuracy of TL-ANNs was more consistent as indicated by a narrower range of R^2 among the 1000 TL-ANNs than in the 1000 DL-ANNs, which means that transfer learning can limit the possibility of obtaining a bad model despite the limited amount of data. For example, one case where the TL-ANN obtained an R^2 value of 0.96, the corresponding DL-ANN showed an R^2 value of −0.259!

These authors obtained similarly encouraging results for transfer learning when training on the new target task of methane adsorption at 100 bar and 298 K, which is a new adsorbate under different adsorption conditions, with R^2 values for TL-ANNs consistently above 0.90. On the other hand, these authors did not observe either good transfer learning or direct learning performance when trying to predict Xe/Kr mixture adsorption. Colón et al. [13] attributed these results to the inadequacy of the features in the feature vector to describe Xe/Kr mixture adsorption, given that not even direct learning was capable of producing models with sufficient accuracy. The latter is a useful reminder of the importance of feature selection for efficient data use.

Study Case 2: Transfer learning in the prediction of alkanediol-solvent desorption: Another successful example of transfer learning during ANN training to predict adsorption in nanoporous materials was presented recently by Siepmann et al. [14]. These authors demonstrated transfer learning during the training of ANNs to predict equilibrium loadings of mixtures of alkanediol with a solvent (either water or alcohol) in zeolites. The versatility of transfer learning is showcased by some interesting differences in the work by these authors with respect to the example just discussed. One difference arises from the feature vectors used in each work. For instance, in the work by Colón et al. [13] the feature vector was a representation of the material, with material textural properties used as inputs to predict the adsorption loading of one molecule. On the other hand, the work by Siepmann et al. [14], to predict desorption loadings, used as input to the ANN a feature vector that represents the state variables that characterize Gibbs ensemble Monte Carlo (GEMC) simulations (i.e., N_1, N_2, V, and T: Figure 13.11a), which are the types of simulations these authors used to model the desorptive drying of the above-mentioned alkanediol–solvent mixtures.

Another difference arises from the need to predict two compositions (y_1 and y_2 of the alkanediol and solvent components, respectively) instead of a single adsorption loading. Accordingly, the ANN by Siepmann et al. had a "multitask" structure where the ANN starts with a number of

(a)

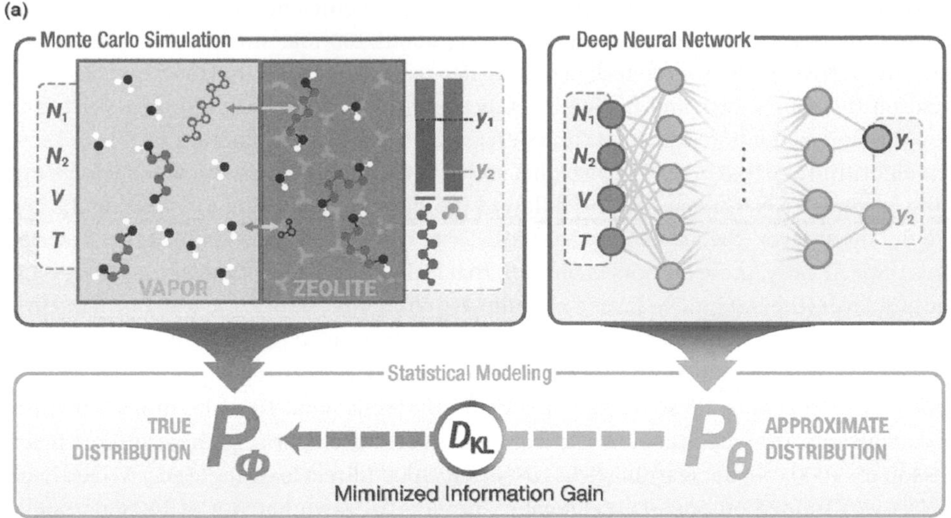

(b)

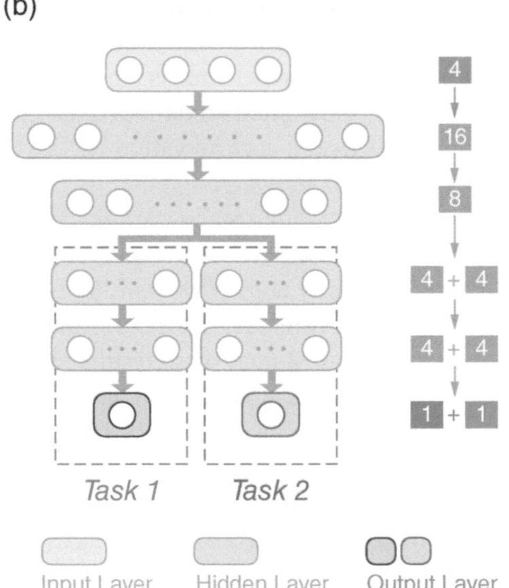

Figure 13.11 (a) Methods to predict desorption of alkanediol-solvent mixtures from zeolites. Left: Gibb ensemble Monte Carlo (GEMC) simulations, which start with N_1 and N_2 alkanediol and solvent molecules at temperature T within a zeolite unit cell of volume V. Right: Multitask artificial neural network (ANN), which uses N_1, N_2, V, and T as input to predict the final compositions y_1 and y_2. (b) Details of the ANN architecture and illustration of "frozen" layers (i.e., layers where weights were maintained at the original values inherited from the ANN trained on the source task) during transfer learning procedure. *Source:* From [14] The Royal Society of Chemistry/CC BY-3.0.

regular hidden layers (in this case two hidden layers) but eventually "branches" into different sets of hidden layers (in this case two branches, each with two hidden layers), where each branch ends in a different output node (Figure 13.11b).

The above differences are indicative of the different types of prediction scenarios that could be tackled with machine learning, which may lead to subtle changes on how the transfer learning procedure works. For instance, one of the most intriguing results by Siepmann et al. [14] is that these authors demonstrated "good" transfer learning (as indicated by the mean squared error (MSE) of y_1

and y_2 predictions) by keeping the weights of the layers closer to the outputs in the original ANN (which was trained to predict compositions for the pentanediol-water system in zeolite MFI), while fine-tuning the weights of the layers closer to the input layer to yield the new ANN, which runs against the "typical" expectation in transfer learning of keeping the earlier layers of the original ANN.

To rationalize these results, let us point out first that the target tasks by each new ANN were predicting compositions for the following systems: i) butanediol-water system in zeolite MFI; ii) pentanediol-ethanol system in zeolite MFI; and iii) pentanediol-water system in zeolite LTA, which involve related, but somewhat different types of interactions. Second, Siepmann et al. [14] noted that it is likely that the earlier, non-branched part of the original ANN is the part that "learns" information about adsorbate–adsorbent and adsorbate–adsorbate interactions, and that the later branched parts of the original ANN are the parts that interpret the learned interaction information and translates into composition predictions. Accordingly, Siepmann et al. [14] noted that it was more appropriate in their case to retain the knowledge that translates interactions into composition predictions (which should be based on the same principles across different systems) and instead retrain the earlier part that learns the interactions of a system (which changes as material and/or adsorbates are changed in the target tasks).

13.4.2 Multipurpose Models

Instead of waiting for a new application to "come up" to then attempt transfer learning, ideally we could build a model that is inherently capable of handling any adsorption scenario. This goes in the spirit of the development of early 20th century adsorption models, who had the idea of "universality" in mind. For instance, while the Langmuir model in Equation 13.4 is limited due to strict underlying assumptions, it had universality in mind in that it is "material" and "adsorbate" agnostic. That is, not really developed for a specific system, but rather for a generic system with a given number of sites and a given interaction between adsorbate and adsorption sites.

In principle, a "universal" machine learning model to predict adsorption could be trained by adding adsorbate properties as part of the feature vector (in addition to thermodynamic conditions and material properties). Thus, the feature vector would represent adsorption "scenarios" instead of "adsorbents." Then, when a new adsorption scenario emerges, we would only need to input the feature values representing the new adsorption scenario. As a preliminary demonstration of the potential of such universal models, in work published in 2020, Gomez-Gualdron et al. [15] demonstrated that the same ANN model was capable of predicting full (room temperature) isotherms for the following adsorbates: He, Ar, Xe, Kr, N_2, H_2, CH_4, C_2H_6, in MOFs.

For the adsorption scenario feature vector, these authors used a 28-component feature vector that contained: i) one feature to represent the thermodynamic condition for adsorption, which in this case was fugacity; ii) 23 features to represent each MOF, with six features corresponding to traditional MOF characterization properties (i.e., void fraction, specific surface area, density, largest pore diameter, diffusion-limiting pore diameter, and the pore size standard deviation), and 17 features corresponding to simple counts of chemical "patterns" found in the MOFs (Figure 13.12); and iii) four features to represent each adsorbate, with these features being inspired by the interaction parameters we use in molecular simulations to model these adsorbates. One consideration is that although different adsorbates may have a different number of atoms (hence a different number of interaction parameters) each adsorbate must still be represented by four features. Accordingly, the features corresponded to an effective attraction parameter ($\varepsilon_{effective}$), an effective size parameter ($\sigma_{effective}$), a largest charge parameter (q_{max}) and the bond length associated with q_{max}.

Notice that the representation of a given adsorbate used by these authors was not a direct representation of its "chemistry," but rather of its "physics," given its representation by parameters that

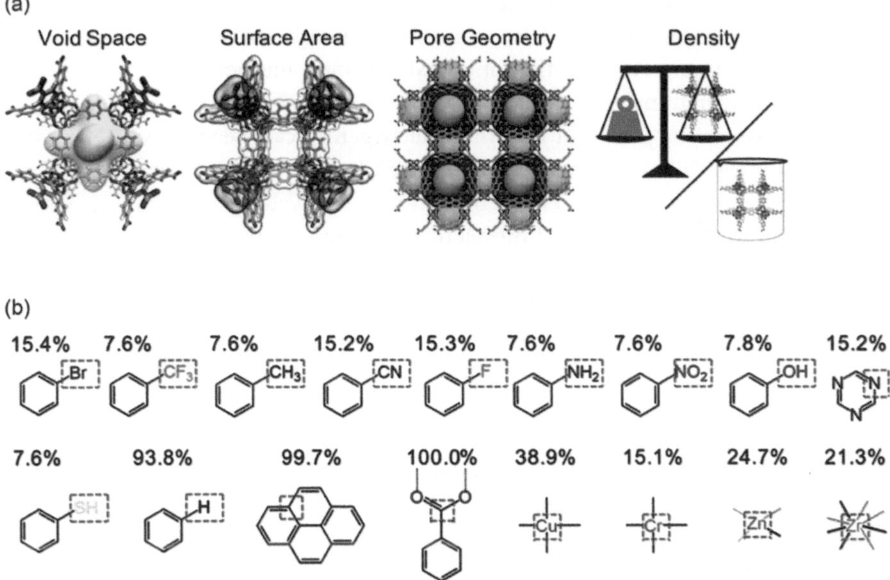

Figure 13.12 Illustration of 23 properties of a 28-component feature vector used to represent adsorption scenarios to train a "multipurpose" model to predict adsorption of various adsorbates (remaining 5 features correspond to 4 adsorbate properties and fugacity). (a) Illustration of MOF textural properties included in the feature vector. (b) Illustration of MOF chemical properties as indicated by simple counts of the illustrated chemical patterns [15]. *Source*: Adapted from [15]. Reproduced with permission of The American Chemical Society. © 2020 American Chemical Society.

are indicative of the forces the adsorbate would experience during adsorption (e.g., the dispersion forces are indicated by $\varepsilon_{effective}$) and of to what extent a nanoporous material can be saturated with the said adsorbate (e.g., $\sigma_{effective}$ is indicative of how much pore volume an individual adsorbate occupies during adsorption). This situation allowed these authors to train the ANN model with "hypothetical" adsorption scenarios engendered by so-called "alchemical" adsorbates: hypothetical molecules with interaction parameters that were "physically reasonable" but not necessarily associated with an actual existing molecule.

To generate a training and validation dataset, Gomez-Gualdron et al. [15] created 400 of these alchemical adsorbates and simulated their adsorption loadings at 6 fugacities (between 1 and 100 bar) in 2000 MOFs that were "custom-built" for this work using the ToBaCCo code. No MOF, alchemical adsorbate, or fugacity used in one dataset appears in the other. The test set, on the other hand, was comprised of an additional set of 400 custom-built MOFs for which adsorption loadings for the real molecules of interest were simulated. Upon model training, the optimal ANN architecture was determined to comprise four hidden layers with 50, 20, 20, and 20 nodes, respectively.

The success of the training procedure can be illustrated in two ways. First, by a broad picture provided by parity plots such as those in Figure 13.13a, which compare the adsorption loadings predicted by the ANN model for Ar, CH_4, Kr, Xe, N_2, and C_2H_6 at all studied fugacities in the 400 MOFs in the test set against corresponding values from molecular simulation. These plots indicate unprecedented excellence in prediction accuracy, with associated R^2 values consistently above 0.99 and mean absolute percent errors (MAPE) below 5%. Second, by a more focused picture, where the comparison is based on full isotherm predictions, such as in the representative comparisons for the C_2H_6 case shown in Figure 13.13b. The accuracy of the prediction of an isotherm takes account of different isotherm points and is assessed based on the isotherm mean percentage error (IMPE). Based on this metric, these authors classified their isotherm prediction into five

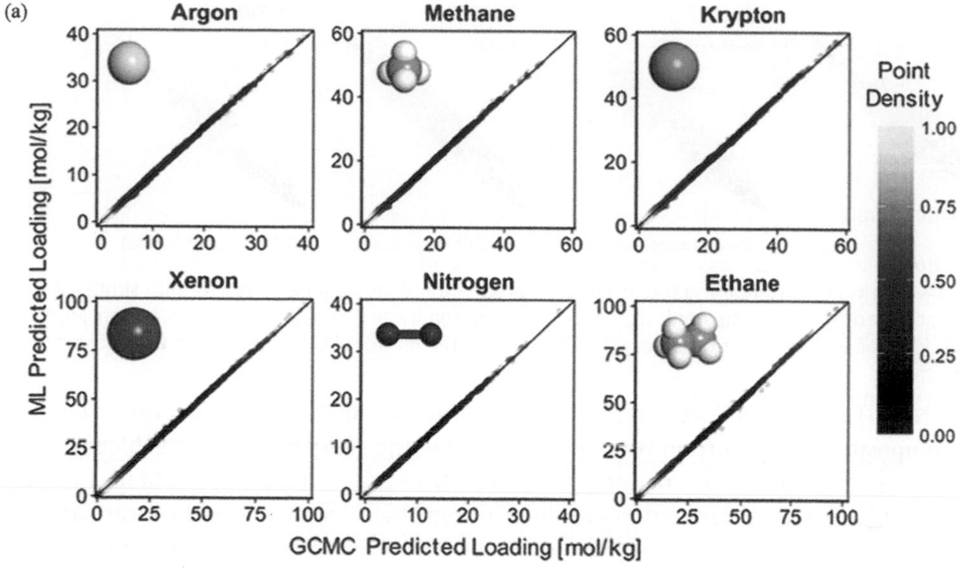

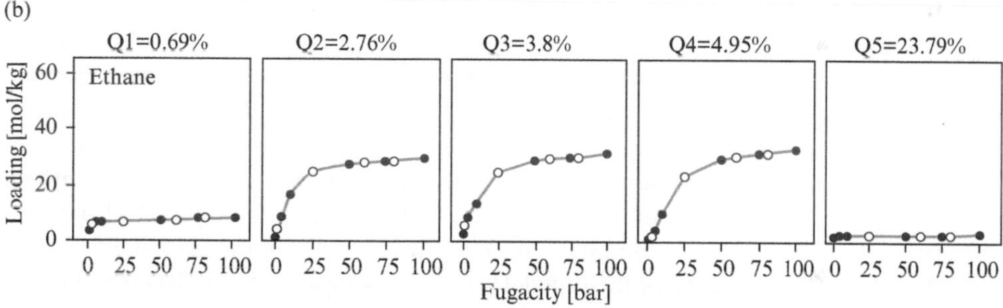

Figure 13.13 (a) Parity plots illustrating the adsorption loading predictions for different adsorbates using the same ANN model against adsorption loadings from simulations. (b) Examples of full isotherm predictions by ANN model (here for ethane) versus isotherm points obtained from simulations. Examples are for the different quantiles based on prediction accuracy, with the isotherm mean percentage error at the top of each plot [15]. *Source*: Adapted from [15]. Reproduced with permission of The American Chemical Society. © 2020 American Chemical Society.

quantiles, with the first quantile (Q1) corresponding to the top 20% best predictions, the second quantile (Q2) to the next 20% best prediction, and so forth, with predictions in the Q1 to Q4 quantiles presenting IMPE below 5%.

Illustrative of the advantage of having a model that uses thermodynamic conditions as input, in subsequent work, Anderson and Gomez-Gualdron [16] grew the training dataset by adding simulation data at fugacities below 1 bar, and retrained the ANN model, which was then capable of predicting full isotherms down to nearly zero pressure. It is apparent that we could add data for new adsorbates and then retrain the model (probably adding components to the feature vector) to expand its prediction capabilities to more "sophisticated" adsorbates. Nonetheless, the current capabilities of the new ANN model already allowed these authors to use its predictions for single-component adsorption as input to ideal adsorption solution theory (IAST), to (in an example of maximizing the payoff for training data) generate mixture adsorption data for binary mixtures involving the adsorbates in Figure 13.13a. An illustration of the accuracy of these predictions is presented in Figure 13.14 for the Ar/Kr, Kr/Xe, CH_4/

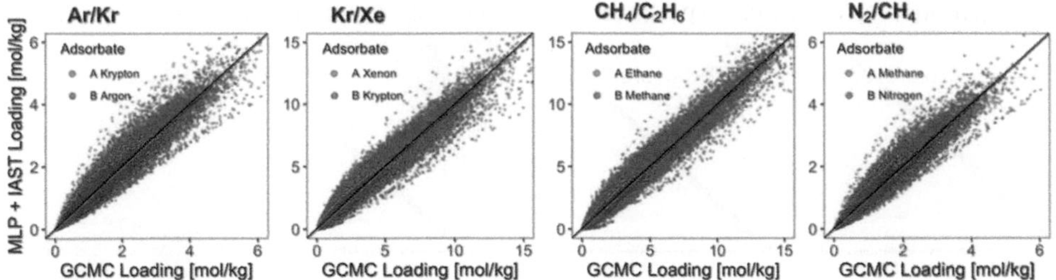

Figure 13.14 Parity plots comparing the mixture adsorption loadings obtained from (GCMC) simulations with those obtained by combining single-component adsorption by an ANN (multilayer perceptron, MLP) model with ideal adsorption solution theory (IAST). *Source:* Adapted from [16]: Reproduced with permission of AIP Publishing.

C_2H_6, and N_2/CH_4 binary mixtures cases, with each parity plot including all adsorption data for mixture composition ranging from 20% to 80% for one of the components. The achieved accuracy was high enough that in a screening using the ANN+IAST mixture predictions, the 90th percentile MOFs (based on separation capability) from the screening would capture around 90% of the 95th percentile MOFs that would be obtained from a traditional screening using molecular simulation.

In another instance of multipurpose models, Sholl et al. [17] have also been exploring the development of a multipurpose "algebraic" adsorption equation that would predict the adsorption in numerous $C_xH_yO_uN_vS_w$ molecules. To derive the adsorption equation, these authors used multiple regression programming (MRGP), with the adsorption equation using 28 descriptors as input to the model, including 3 adsorbate descriptors (the parameters for the Peng Robinson equation of state: critical pressure, critical temperature, and acentric factor), 5 MOF geometric descriptors (e.g., void fraction), and 20 chemical descriptors corresponding to counts of chemical patterns (e.g., number of 6-membered rings in the MOF), once again highlighting the usefulness of relatively simple descriptors. The training of their model was done using adsorption data for 24 molecules in 460 MOFs taken from the CoRE-2019 database.

Notably, the algebraic model by Sholl et al. [17] predicted adsorption loadings within a promising 1 mol/kg error threshold for around 70% of adsorption scenarios (where a "scenario" corresponded to a MOF + adsorbate + pressure combination). On the other hand, in around 7% of adsorption scenarios, the error prediction by the model was larger than 1000%, which motivated these authors to use the same adsorption data and the MRGP technique to train an additional classification model (with output equal to 0 or 1), which would indicate if an adsorption scenario fell or not under the "application domain" of their algebraic adsorption equation. In another example of efficient data usage, these authors ultimately leveraged their generated molecular simulation adsorption data (via their algebraic model) to screen MOFs with the potential to separate near-azeotrope binary mixtures, exploring more than 4.9×10^{10} adsorption scenarios involving molecules other than those included in the training dataset!

13.4.3 Material Recommendation Systems

An interesting concept based in the same way as using data for one application based on one adsorbate, to "discover" materials for another application based on a different adsorbate, is the recommendation system proposed recently by Simon et al. [18]. These authors noted that if they had a complete material–property "matrix," containing measurements of every adsorption property of interest (columns) of every candidate nanoporous material (rows), matching materials with applications would be a trivial look-up problem. The issue is that, in practice, many (nanoporous material, gas adsorption property) measurements are missing. Indeed, whether in the laboratory or in a simulation: i) only a small proportion of the properties of each material have been measured; and ii) only a small proportion of materials have been investigated for each application.

Curiously, a similar situation arises in commercial settings, which has an abundance of items that could be rated highly by many different customers. A complete item–customer matrix, whose entries contain ratings of items by customers, would convert item recommendations to customers into a look-up problem. However, in practice: i) each customer has rated only a small proportion of the items; and ii) each item has been rated by only a small proportion of the customers. The idea behind a commercial recommendation system is to take an incomplete item–customer matrix (containing missing values) and leverage the observed entries to, in a principled way, "fill in" the missing entries of the matrix in order to recommend items to customers.

Along the same lines, a material recommendation system [18] aims to predict the missing entries in a material-property matrix through a machine learning strategy, using the observed entries (Figure 13.15). One implementation of a material recommendation system is based on a low rank matrix model of the material–property matrix $\boldsymbol{A}$, whose entry A_{mp} represents property p of material m. The low rank model assumes that: i) the materials and properties can each be represented by low-dimensional, latent vectors $\{\boldsymbol{m}_m\}$ and $\{\boldsymbol{p}_p\}$; and ii) the affinity of material m for property p is expressed by the inner product of the latent vectors $\boldsymbol{m}_m$ and $\boldsymbol{p}_p$. Specifically, a low rank model specifies entry (m, p) of $\mathbf{A}$ as:

$$\mathrm{A}_{\mathrm{mp}} = \boldsymbol{m}_m^T \boldsymbol{p}_p + \mu_m, \tag{13.22}$$

with μ_m being a material bias reflecting the tendency for a material to exhibit high/low values of the properties, independent of material–property interactions. Notably, we can "machine-learn" the latent vectors $\{\boldsymbol{m}_m\}$ and $\{\boldsymbol{p}_p\}$ and material biases $\{\mu_m\}$ using only the observed entries; after predictions for the unobserved entries followed from Equation 13.21. Moreover, the latent vectors $\{\boldsymbol{m}_m\}$ and $\{\boldsymbol{p}_p\}$ themselves are useful for drawing a low-dimensional map of materials, wherein materials with similar properties congregate.

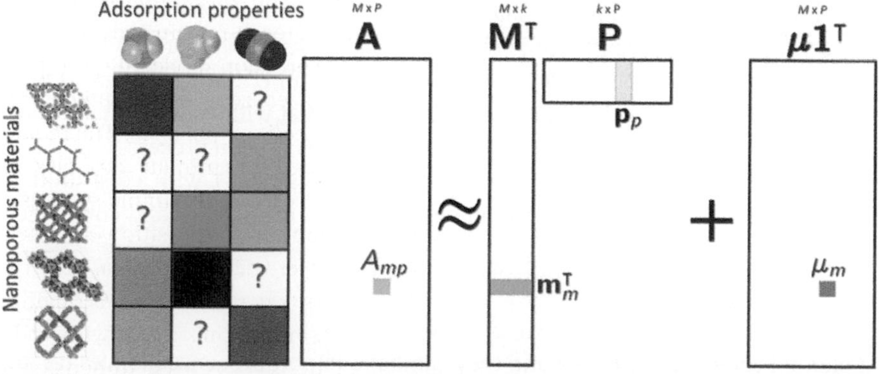

Figure 13.15 In a material recommendation system, we are tasked with "filling out" the missing values of an incomplete material–property matrix. Here, nanoporous materials are represented in rows and gas adsorption properties are represented in columns; question marks denote missing entries. One implementation of a material recommendation system is based on a low rank matrix model, where we machine-learn the vector representations of the materials and properties and material biases from the observed entries; predictions for the unobserved entries then follow [18]. *Source*: Adapted from [18]. Reproduced with permission of The American Chemical Society. © 2021 American Chemical Society.

13.4.4 Active Learning

In generating data for machine learning, simulation experts often generate as many datapoints as possible by straining computational resources. However, techniques such as active learning can help balance the accuracy of a predictive model with the number of data points required. Indeed, active learning can be an important tool to efficiently generate and use datasets to train machine learning models to predict adsorption in nanoporous materials. Consider a scenario where accessing a larger number of structure is inexpensive (maybe drawing from some of the databases mentioned so far in this chapter, e.g., the hMOF, ToBaCCo, BW, or CoRE databases), but where adsorption data is relative hard to collect from simulation. Perhaps, it was possible with some effort to run simulations for some of the materials in the database, to the extent that enough data points to attempt training a machine learning model were collected. Now say that we train a machine learning model to predict adsorptions only to find that its accuracy, while encouraging, still needs improvement. Broadly speaking, an active learning framework would help decide for which additional materials to run simulations (to grow the original dataset), so that the accuracy of the subsequently trained model is more likely to dramatically improve.

Active learning can be particularly attractive in situations where the "feature space" is very large. Notice that we refer to the "feature space" instead of "material space," because as first mentioned in our discussion on transfer learning, we have now seen that the feature vector can also include thermodynamic variables such as temperature and pressure. Accordingly, scenarios where we train models applicable to multiple materials but also to multiple adsorption conditions are particularly attractive for active learning, even if the individual cost of the relevant simulation is deceptively low. A scenario where the simulations (or experiments) to produce the training data are individually already resource and time intensive are obviously suited for active learning. Indeed, active learning approaches are increasing in popularity with molecular simulation researchers [19]. For instance, in the context of nanoporous materials, Santos et al. [20] presented an example in 2020, where a dataset was built by performing simulations where new data was chosen based on the model uncertainty via the "query-by-committee" approach. As a result, these authors estimated a reduction of one order of magnitude on the number of simulations needed to build a dataset that would facilitate a machine learning model prediction (a convolutional ANN) of methane density profiles within random pore shapes.

In another instance, Colón et al. recently implemented an active learning workflow to make predictions for full methane and carbon dioxide adsorption isotherms in HKUST-1. The workflow intelligently selects the next adsorption simulation to be performed that will result in an accurate machine learning model. The model of choice by these authors was a Gaussian process (GP). A GP is a non-parametric machine learning model that describes a real process, $f(x)$, with a distribution over functions that have a joint Gaussian distribution described by a mean $\mu(x)$ and covariance $K(x,x')$ function:

$$f(x) \sim \mathcal{N}(\mu(x), K(x,x')) \tag{13.23}$$

There are many potential choices for $K(x,x')$ and they can have important consequences on the resulting model. A rational quadratic kernel (Equation 13.24) has been used before to describe adsorption loading in MOFs:

$$K(x,x') = \left(1 + \frac{d(x,x')^2}{2\alpha l^2}\right)^{-\alpha} \tag{13.24}$$

where $d(x,x')$ is the Euclidean distance between x and x', l is the length scale of the kernel, and α is the scale mixture parameter. The parameters of the kernel, l and α, are found by maximizing the log-marginal likelihood. Importantly, the GP provides the predicted mean value of the functions and their uncertainty.

An important aspect of active learning approaches is the "acquisition function," which provides the means of choosing "the next" simulation to be performed. Following the active learning premise, Colón et al. [21] chose to explore the space by selecting simulations that were deemed to be the most informative to the model. So, for iteration n+1, they chose conditions (e.g., temperature and pressure) x_{n+1} where the variance σ_n^2 from the GP regression is a maximum at iteration n: $x_{n+1} = \arg\max_{(x \in X)} \sigma_n^2(x)$. This is known as active learning MacKay, which was originally proposed in the context of ANNs. Seo et al. [22] implemented the idea for GPs. Once the new simulation is performed, the GP is refit including the new data and the procedure is repeated until uncertainty is below some threshold. At the beginning of the procedure, a GP prior is fit using data spread in x, using Latin hypercube sampling (LHS). Because GPs are non-parametric, the data are part of the resulting model. So, using the resulting model requires the data used to develop it. This makes it easy for the model to accommodate new additions to the database and new users can directly see the data used in the model.

Based on the ideas above, Colón et al. [21] implemented an active learning workflow for adsorption in simulations in porous materials. The workflow starts by fitting a prior set of adsorption data to a GP and then it automatically chooses the next simulation condition. The result from the simulation is added to the dataset, the GP is retrained, and the uncertainties are estimated again. Once the relative predicted uncertainty from the GP is below 2%, the iterative process ends, and the GP fit is taken as the final model. This was implemented into an active procedure to do molecular simulations of adsorption of pure components CH_4 and CO_2 in the MOF HKUST-1. The resulting GP takes temperature and pressure as an input and outputs the amount of gas adsorbed in the MOF.

Experimentally, an adsorption isotherm measurement can consist of 10–15 pressure points. If temperature is also considered, intervals of 5 K, each with its own isotherm, are reasonable. So, to obtain that resolution in a pressure and temperature range of 10^{-6}-300 bar (10^{-6}-100 bar for CO_2) and 100–300 K, respectively, it would require between 200 and 600 measurements or simulations. Active learning can produce an accurate model in those same conditions using less than a hundred simulations. For methane and carbon dioxide adsorption, only a total of 83 simulations were needed to achieve a relative error of less than 3%, which is comparable to the uncertainty of the molecular simulations themselves (Figure 13.16). This represents an order of magnitude saving in the number of simulations required to obtain an accurate picture of the adsorption landscape.

13.4.5 Machine Learning to Speed Up Data Generation

In some cases, machine learning could be exploited to speed up the generation of adsorption data itself by addressing key bottlenecks during the molecular simulation pipeline. A well-known bottleneck in the generation of adsorption data in MOFs is the need to provide partial point charges to MOF atoms to perform simulations in adsorption scenarios where electrostatic interactions are important (e.g., CO_2 adsorption). These charges must reproduce the electrostatic potential (ESP) inside the MOF pores, with the "gold standard" for calculating MOF charges requiring "days-long" periodic density functional theory (DFT) calculations. These calculations yield the ESP within the MOF pores, which is used by methods such as the Repeating Electrostatic Potential Extracted Atomic (REPEAT) method [23] or the Density-Derived Electrostatic and Chemical (DDEC) method [24] to fit charges that reproduce the ESP.

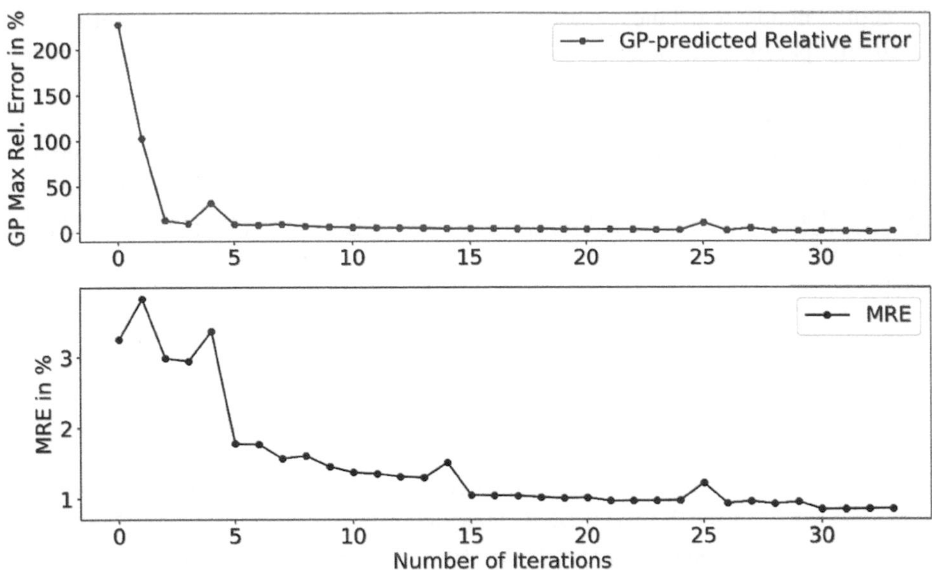

Figure 13.16 Evolution of GP to predict methane adsorption while navigating temperature (100–300 K) and pressure conditions (10^{-6} to 3×10^{-2} bar). Increasing iterations improves the predicted GP maximum relative error and the process ends when it reaches 2%. The mean relative error of the resulting model with the GCMC simulations is simultaneously tracked and it shows clear improvement with increasing iterations. In this case, the final error of the resulting model is below 1% [21].

Alternatives to assign MOF point charges, including methods based on "classical" theories such as charge equilibration [25], have been often considered, but the accuracy of the assigned point charges is arguably low. But interestingly, in recent years there have been a few studies aiming to develop machine learning methods to assign partial charges to MOFs with both high accuracy and low computational cost. Each of these studies has leveraged the existence of a database of around 3000 (experimental) MOFs in the CoRE MOF database, for which Sholl et al. calculated DDEC charges [26]. A machine learning model trained on this data would then be able to quickly generate charges for hundreds of thousands of MOFs, opening the door for accurate, large-scale data generation.

The success of training machine learning models to predict charges is predicated on the hypothesis that the partial point charge assigned to an atom is dictated by its identity (atomic species) and local (e.g., bonding) environment. For example, Snurr et al. [27] designed a feature vector that described an atom and its local bonding environment, and then used the training example to teach a model to map each feature vector to a charge. The feature vector of an atom is comprised by electronegativity and ionization potential (characterizing the atom itself), and the coordination number, the average distance from the atom to a bonded atom, the average electronegativity and ionization potential of the bonded atoms, and the average electronegativity of atoms two hops from the atom (characterizing the bonding environment). A RFR model was the trained to yield predicted charges.

In an interesting approach proposed by Simon et al. [28], instead of hand designing a feature vector of the local bonding environment of an atom, message passing neural networks (MPNNs) can automatically learn charge-prediction-specific vector representations of each node within its local bonding environment. MPNNs take as input a node-labeled graph representation of a MOF (where nodes = atoms, edges = bonds, labels = chemical species) and, in an equivariant manner, output a list of charges on each node of the MOF. MPNNs learn this mapping from node-labeled graph and charge-labeled nodes examples.

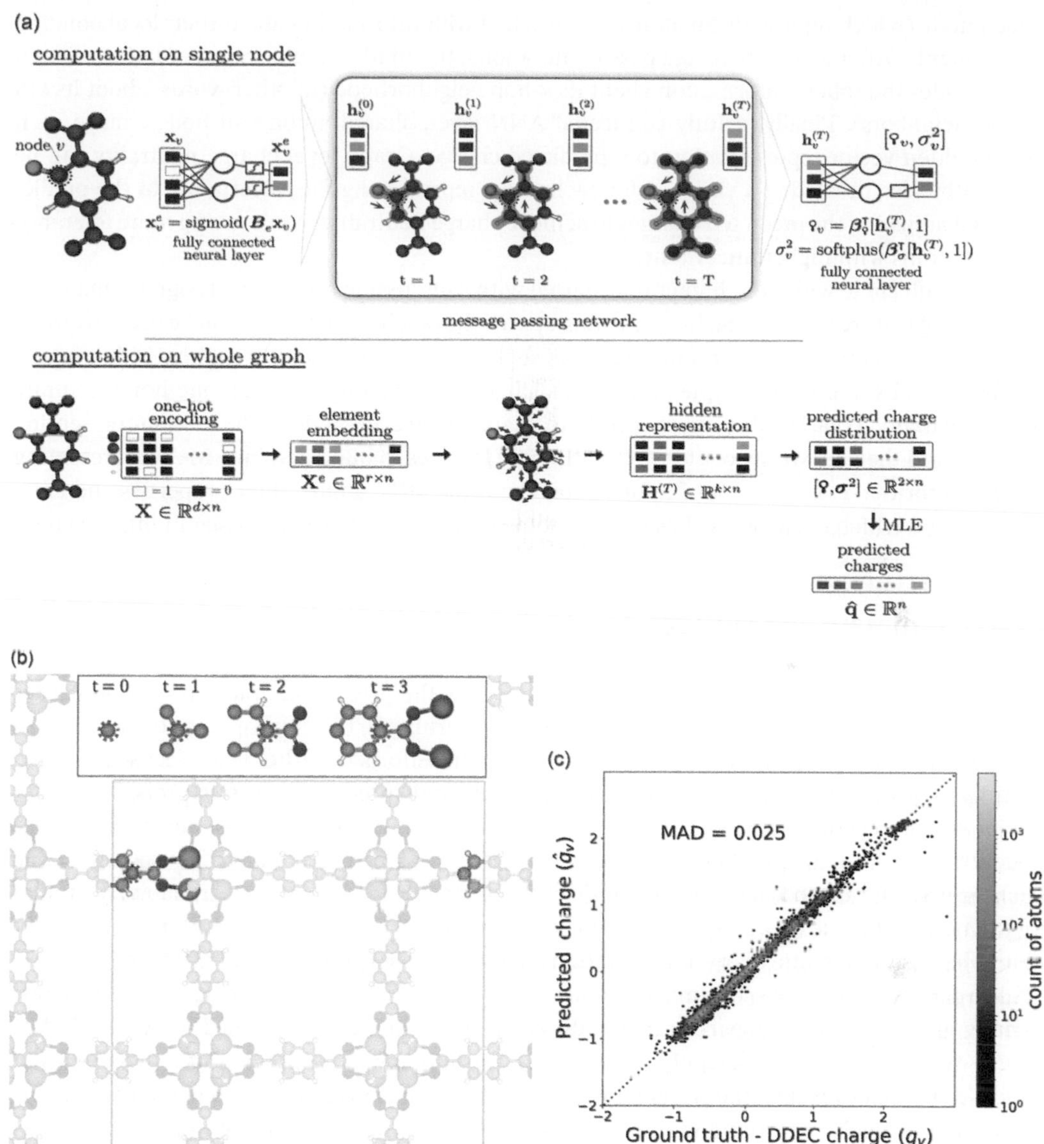

Figure 13.17 Message passing neural networks (MPNN) for partial charge assignment. (a) The architecture of a MPNN for charge prediction, broken into node-level and graph-level computations. (b) After t iterations of message passing, the hidden representations of the nodes $\boldsymbol{h}_v^{(t)}$ encode information about its t-hop neighborhood. (c) A parity plot of the MPNN: predicted vs. actual charges on test MOFs with DDEC charges [28]. *Source:* Adapted from [28]. Reproduced with permission of The American Chemical Society. © 2020 American Chemical Society.

We now describe how the node-labeled graph propagates through the MPNN until charges are assigned, which is a process illustrated in Figure 13.17. First, each node is assigned a "hidden" vector representation, initialized using an embedding layer that maps the one-hot encoding of its atom type into a dense, low-dimensional vector representation of the atom type. Then, each node iteratively: i) collects the hidden representations of its neighbors; ii) collates them into a message vector; and iii) updates its hidden representation based on the message. In layperson terms, by allowing neighboring nodes to iteratively exchange information, the hidden vector representations of each node (which represents an atom) are enriched with information about their local bonding environments. After k such message passing iterations, the hidden vector representation of each node encodes the salient information about its k-hop neighborhood (in other words, about its kth nearest neighbors). Finally, a fully connected ANN layer, shared among all nodes, maps each node's hidden vector representation to a predicted charge. Graph-level charge neutrality can be enforced by also predicting a variance for each node, representing the uncertainty in the prediction, then adjusting the predicted charges to achieve charge neutrality, while giving more tolerance to adjust atoms with higher uncertainty.

Coming full circle with this chapter that started with a discussion on feature design for material-representing feature vectors, machine learning approaches such as MPNNs could be used to design features for prediction of adsorption properties such as those discussed throughout this chapter. The user could still provide simple properties such as atom identity (through one-hot encoding), electronegativity, polarizability, and so forth, to characterize each node of the graph representing the MOF, and then an MPNN could work with these features to build a unique MOF-representing feature vector that encodes a lot of information about the MOF atoms, their properties and environment, and which could be used as input to some of the model types discussed in this chapter.

13.5 Summary and Outlook

The discussion presented in this chapter illustrates how the prediction of adsorption properties has served as a testbed for the application of machine learning to the development of nanoporous materials. The evolution of these approaches is apparent. Efforts in earlier years focused on (successfully) demonstrating that machine learning models could predict adsorption properties under specific operating conditions. Realizing that, even for adsorption applications, machine learning model training lives in a world of "data scarcity," efforts in recent years have focused on testing machine learning approaches that more efficiently generate and/or use data. Arguably, the unifying theme about these efforts is the idea of taking advantage of adsorption data for a given set of materials, adsorbates, and operating conditions to predict adsorption for a different set.

Adsorption will probably continue as an ideal testbed for exploring the application of machine learning approaches to nanoporous materials. Particularly, the lessons learned to deal with data scarcity will be very useful to apply the machine learning model to predict other properties or aspects of nanoporous material development for which data is even harder to come by. Exciting areas to apply machine learning in nanoporous materials include, but are not limited to, the prediction of thermal and charge transport [29,30] properties, mechanical and chemical stability [31,32], optical and electronic properties [33], as well as synthesis-related predictions [34,35].

Recognizing how difficult it is for a single lab to generate large enough datasets for the properties above (and even for adsorption properties), there is an increased awareness in the field for the need to "pool" experimental (and simulation) data from different labs around the world. For instance, Norquist et al. [36] set up a publicly accessible online repository to collect synthesis data deposited by different labs, which was used to train a model to predict whether a set of synthesis conditions would successfully yield a desired crystal compound. Finally, there is potential to exploit all the data that has already been deposited in the literature through the years. In this case, the application of recent advances in natural language processing (NLP) that can automatedly scout the literature and extract relevant data will be powerful to assist in the creation of new datasets or complement existing ones. As an example of this emerging approach, consider the work by Olivetti et al. [37], where NLP was used to collect data to train a model that predicts zeolite structure based on synthesis conditions.

References

1 Sun, Y., DeJaco, R.F., Li, Z. et al. (2021). Fingerprinting diverse nanoporous materials for optimal hydrogen storage conditions using meta-learning. *Sci. Adv.* 7 (30): eabg3983.

2 Fernandez, M., Trefiak, N.R., and Woo, T.K. (2013). Atomic property weighted radial distribution functions descriptors of metal–organic frameworks for the prediction of gas uptake capacity. *The J. Phys. Chem. C* 117 (27): 14095–14105.

3 Simon, C.M., Mercado, R., Schnell, S.K. et al. (2015). What are the best materials to separate a Xenon/Krypton mixture? *Chem. Mater.* 27 (12): 4459–4475.

4 Bucior, B.J., Bobbitt, N.S., Islamoglu, T. et al. (2019). Energy-based descriptors to rapidly predict hydrogen storage in metal–organic frameworks. *Mol. Sys. Des. Eng.* 4 (1): 162–174.

5 Pardakhti, M., Moharreri, E., Wanik, D. et al. (2017). Machine learning using combined structural and chemical descriptors for prediction of methane adsorption performance of metal organic frameworks (MOFs). *ACS Comb. Sci.* 19 (10): 640–645.

6 Wilmer, C.E., Leaf, M., Lee, C.Y. et al. (2011). Large-scale screening of hypothetical metal–organic frameworks. *Nat. Chem.* 4: 83.

7 Anderson, R., Rodgers, J., Argueta, E. et al. (2018). Role of pore chemistry and topology in the CO_2 capture capabilitics of MOFs: from molecular simulation to machine learning. *Chem. Mater.* 30 (18): 6325–6337.

8 Moosavi, S.M., Nandy, A., Jablonka, K.M. et al. (2020). Understanding the diversity of the metal-organic framework ecosystem. *Nat. Commun.* 11 (1): 4068.

9 Colón, Y.J., Gómez-Gualdrón, D.A., and Snurr, R.Q. (2017). Topologically guided, automated construction of metal–organic frameworks and their evaluation for energy-related applications. *Cryst. Growth Des.* 17 (11): 5801–5810.

10 Boyd, P.G., Chidambaram, A., García-Díez, E. et al. (2019). Data-driven design of metal–organic frameworks for wet flue gas CO_2 capture. *Nature* 576 (7786): 253–256.

11 Chung, Y.G., Haldoupis, E., Bucior, B.J. et al. (2019). Advances, updates, and analytics for the computation-ready, experimental metal–organic framework database: CoRE MOF. *J. Chem. Eng. Data* 64 (12): 5985–5998.

12 Poirier, E., Chahine, R., and Bose, T.K. (2001). Hydrogen adsorption in carbon nanostructures. *Int. J. Hydrog. Energy* 26 (8): 831–835.

13 Ma, R., Colón, Y.J., and Luo, T. (2020). Transfer learning study of gas adsorption in metal–organic frameworks. *ACS Appl. Mater. Interfaces* 12 (30): 34041–34048.

14 Sun, Y., DeJaco, R.F., and Siepmann, J.I. (2019). Deep neural network learning of complex binary sorption equilibria from molecular simulation data. *Chem. Sci.* 10 (16): 4377–4388.

15 Anderson, R., Biong, A., and Gómez-Gualdrón, D.A. (2020). Adsorption isotherm predictions for multiple molecules in MOFs using the same deep learning model. *J. Chem. Theory Comput.* 16 (2): 1271–1283.

16 Anderson, R. and Gómez-Gualdrón, D.A. (2021). Deep learning combined with IAST to screen thermodynamically feasible MOFs for adsorption-based separation of multiple binary mixtures. *The J. Chem. Phys.* 154 (23): 234102.

17 Gharagheizi, F., Tang, D., and Sholl, D.S. (2020). Selecting adsorbents to separate diverse Near-Azeotropic chemicals. *The J. Phys.Chem. C* 124 (6): 3664–3670.

18 Sturluson, A., Raza, A., McConachie, G.D. et al. (2021). Recommendation system to predict missing adsorption properties of nanoporous materials. *Chem. Mater.* 33 (18): 7203–7216.

19 Graff, D.E., Shakhnovich, E.I., and Coley, C.W. (2021). Accelerating high-throughput virtual screening through molecular pool-based active learning. *Chem. Sci.* 12 (22): 7866–7881.

20 Santos, J.E., Mehana, M., Wu, H. et al. (2020). Modeling nanoconfinement effects using active learning. *The J. Phys. Chem. C* 124 (40): 22200–22211.

21 Mukherjee, K., Dowling, A.W., and Colón, Y.J. (2022). Sequential design of adsorption simulations in metal–organic frameworks. *Mol. Syst. Des. Eng.* 7 (3): 248–259.

22 Seo, S., Wallat, M., Graepel, T. et al. (2000). Gaussian process regression: active data selection and test point rejection. In: *Mustererkennung* (ed. G. Sommer, N. Krüger, and C. Perwass), 27–34. Berlin, Heidelberg: Springer.

23 Campañá, C., Mussard, B., and Woo, T.K. (2009). Electrostatic potential derived atomic charges for periodic systems using a modified error functional. *J. Chem. Theory Comput.* 5 (10): 2866–2878.

24 Manz, T.A. and Sholl, D.S. (2010). Chemically meaningful atomic charges that reproduce the electrostatic potential in periodic and nonperiodic materials. *J. Chem. Theory Comput.* 6 (8): 2455–2468.

25 Ongari, D., Boyd, P.G., Kadioglu, O. et al. (2019). Evaluating charge equilibration methods to generate electrostatic fields in nanoporous materials. *J. Chem. Theory Comput.* 15 (1): 382–401.

26 Nazarian, D., Camp, J.S., and Sholl, D.S. (2016). A comprehensive set of high-quality point charges for simulations of metal–organic frameworks. *Chem. Mater.* 28 (3): 785–793.

27 Kancharlapalli, S., Gopalan, A., Haranczyk, M. et al. (2021). Fast and accurate machine learning strategy for calculating partial atomic charges in metal–organic frameworks. *J. Chem. Theory Comput.* 17 (5): 3052–3064.

28 Raza, A., Sturluson, A., Simon, C.M. et al. (2020). Message passing neural networks for partial charge assignment to metal–organic frameworks. *The J. Phys. Chem. C* 124 (35): 19070–19082.

29 Zanca, F., Glasby, L.T., Chong, S. et al. (2021). Computational techniques for characterisation of electrically conductive MOFs: quantum calculations and machine learning approaches. *J. Mater. Chem. C* 9: 13584–13599.

30 Babaei, H., McGaughey, A.J., and Wilmer, C.E. (2017). Effect of pore size and shape on the thermal conductivity of metal–organic frameworks. *Chem. Sci.* 8 (1): 583–589.

31 Moghadam, P.Z., Rogge, S.M., Li, A. et al. (2019). Structure-mechanical stability relations of metal-organic frameworks via machine learning. *Matter* 1 (1): 219–234.

32 Batra, R., Chen, C., Evans, T.G. et al. (2020). Prediction of water stability of metal–organic frameworks using machine learning. *Nat. Mach. Intell.* 2 (11): 704–710.

33 Rosen, A.S., Iyer, S.M., Ray, D. et al. (2021). Machine learning the quantum-chemical properties of metal–organic frameworks for accelerated materials discovery. *Matter* 4 (5): 1578–1597.

34 Anderson, R. and Gómez-Gualdrón, D.A. (2020). Large-scale free energy calculations on a computational metal–organic frameworks database: toward synthetic likelihood predictions. *Chem. Mater.* 32 (19): 8106–8119.

35 Daeyaert, F., Ye, F., and Deem, M.W. (2019). Machine-learning approach to the design of OSDAs for zeolite beta. *Proc. Natl. Acad. Sci.* 116 (9): 3413–3418.

36 Raccuglia, P., Elbert, K.C., Adler, P.D. et al. (2016). Machine-learning-assisted materials discovery using failed experiments. *Nature* 533 (7601): 73–76.

37 Jensen, Z., Kim, E., Kwon, S. et al. (2019). A machine learning approach to zeolite synthesis enabled by automatic literature data extraction. *ACS Cent. Sci.* 5 (5): 892–899.

14

Machine Learning and Digital Manufacturing Approaches for Solid-State Materials Development

Lawson T. Glasby[1], Emily H. Whaites[1], and Peyman Z. Moghadam[1,2]

[1] *Department of Chemical and Biological Engineering, University of Sheffield, Sheffield, United Kingdom*
[2] *Department of Chemical Engineering, University College London (UCL), London, United Kingdom*

14.1 Introduction

Solid state chemistry, often referred to as materials chemistry, is a field of chemistry concerned with studying the synthesis, structure, and properties of materials in the solid phase. These solids are often classified as crystalline, amorphous, organic, inorganic, or nano-materials, depending on the type, and the arrangements of their constituent atoms. Some notable examples include zeolites, covalent organic frameworks (COFs), metal–organic cages (MOCs), and metal–organic nanosheets (MONs).

One intensely studied class of solid state materials, and the primary example used throughout this chapter, are metal–organic frameworks (MOFs), crystalline structures synthesized from organic and inorganic building blocks to form an extended framework material. The building-block approach creates the opportunity for the synthesis of tens of thousands of combinations where they can be tailored to achieve particular properties for a multitude of applications, and since the start of the 1990s, thousands of MOF materials have been synthesized at laboratory scale [1–6]. However, despite their great promise for a wide range of applications, only a handful have been successfully commercialized [7].

In general, the production of MOFs is largely a manual process and because of the complex multi-dimensional nature of their synthesis, the development process can be time-consuming and inefficient when exploring the entire MOF synthesis space. To tackle these challenges, there is a clear need for the adoption of technologies that can expedite the way MOF materials are designed and developed with optimum properties. One way to address these challenges involves the deployment of state-of-the-art computer simulations and digital technologies. This approach includes a wide range of techniques from database generation, to high-throughput screening, machine learning (ML), and the use of novel digitalization tools to overcome "data-poor" processes that characterize the complex chemistry of MOFs. Figure 14.1 demonstrates gas adsorption applications in MOFs.

To fully understand the current size of the MOF space, Moghadam et al. reported nearly 70,000 experimentally synthesized MOF structures in the Cambridge Structural Database (CSD) upon the creation of the CSD MOF subset [8]. Moving forward to the present day, based on data published in the latest CSD release (version 5.3, 2021), there are now 114,373 experimentally synthesized MOFs. There has been a rapid increase in the quantity of submitted MOF structures within the CSD with an estimated 10% of all deposited structures meeting the criteria of a MOF, and an incredible 15,998 experimentally synthesized materials were deposited and approved during the last quarter (June 2022). This large, rapidly expanding chemical space continually adds to the difficulty faced by chemists in identifying useful MOFs for their chosen applications. Many research groups

AI-Guided Design and Property Prediction for Zeolites and Nanoporous Materials, First Edition. Edited by German Sastre and Frits Daeyaert.

Figure 14.1 Schematic showing the applications of gas separation where CO_2 is captured and methane is separated (left), and the storage of gaseous methane (right) in MOFs. The structures here are represented in a general form where a typical metal-oxo cluster is seen as a metal node, and the organic linkers are drawn as straight connecting bars.

within the materials field have begun to quickly adopt ML techniques in the quest to enable rapid and reliable materials discovery and synthesis. The development of ML tools should be seen as a means to ease the process at every stage, from the identification of useful MOFs, to reducing the cost of synthesis for novel structures.

To maximize the potential for MOF discovery, it is essential to develop robust data collection techniques to obtain crucial information from the large volume of published work on MOFs and their properties. Moghadam et al. used keyword searching to produce the CSD MOF subset, an automatically updating database, with specific criteria for material selection from the Cambridge Crystallographic Data Centre's 1,000,000+ structures [8]. This database, alongside the Computation-Ready Experimental (CoRE) MOF database developed by Chung et al. are two of the largest and highly curated collections of experimental MOF materials, and both allow for high-throughput computational screening of MOFs for a wide range of applications [9]. Due to the nature of these porous structures, particularly those that are constructed in 3D, there are many applications in gas storage and separation but many other studies have focused on sensing, catalysis, drug delivery, and chemical removal [10–15].

Often, large-scale, high-throughput screening is not thorough enough to filter large databases into small subsets of interesting structures without significant input from more computationally expensive studies. The development of ML tools for property prediction has commonly been touted as the next step toward a rapid and inexpensive computational approach. ML usage has seen a significant increase in all fields of scientific discovery and data development over recent years, in part due to the popularity of user-friendly tool kits such as RDKit [16]. ML models can be created and developed to make highly accurate predictions, although the success of such models is dependent on a sufficiently large amount of data taken from materials databases for use as training sets.

Structure–property data is often used as a foundation for investigating the synthesis requirements of novel structures. One of the most promising applications of new ML approaches is to predict optimal reaction synthesis conditions for stable structures with sufficient crystallinity. This approach is not limited to MOFs and can, and has been, applied to a range of different materials, examples of which are discussed in this chapter.

Typically, the time required to obtain optimal MOF synthesis is long and expensive due to the highly diverse synthesis conditions required in the creation of these materials. This diversity means there is a lack of general synthesis recipes, and for many cases predicted synthesis conditions are non-transferable. Key parameters for synthesis include solvents, reaction temperature, and time [17]. Due to the novelty of using ML for synthesis of MOFs, only a handful of studies have been reported to date. The use of Natural Language Processing (NLP), a sub-field of ML and artificial intelligence (AI), has been investigated to augment chemists' expertise when approaching experimental design. Luo et al. [18] and Zhang et al. [19] have utilized NLP to study the MOF databases, CoRE MOF, and CSD MOF subset, respectively, in an effort to train models that are able to predict the optimal synthesis conditions for new MOF structures.

If we consider the use of a digital manufacturing approach when investigating MOF synthesis, a considerable bottleneck in synthesis improvement is the availability of synthesis data. Digital manufacturing techniques, such as digital twins, rely on the constant availability of high-quality data to feed back to the synthesis process. A digital twin is a virtual representation of a physical object or process, in this case a synthesis process, used for the continual improvement of engineering activities. When developing an initial synthesis path to create, for example, a MOF in a laboratory, a similar ML approach can be used, despite the requirement for greater flexibility in experimental design. By implementing ML models for small quantity experiments to predict synthesis conditions, the number of real experimental tests can be reduced, as only the most viable reactions are chosen for experimental trial, and although this process is perhaps more suited to the improvement of larger-scale materials synthesis, it is still applicable at lab scale.

A shift from batch production to a closed loop process is a significant driver toward the introduction of digital manufacturing. This approach, when combined with ML prediction techniques described previously, can be used to redevelop existing larger-scale synthesis techniques. By increasing the available information about the synthesis of a particular material, including accurate ML predicted data such as approximate cost, reaction time, and required reagents, the implementation of a digital twin becomes easier. Considering the savings made due to the reduction in labor and time when attempting to optimize the synthesis process, focus can be directed toward the development of other structures and their use in desired applications. The use of ML should be seen as a means to achieve the goal of automated synthesis, and as a tool to drive continuous improvement.

Improving the manufacturing pathway of novel or complex materials is often seen as a large hurdle for many chemists. Accessing the relevant data and hosting the technology for synthesis development can be expensive and time-consuming. Digital manufacturing opens the doors for significant changes in the development of material production. With digital infrastructure such as synthesis servers and synthesis databases, high-cost equipment can be accessed remotely, making the research process much less cost intensive, expanding the research field further to groups without a wide array of computational or experimental resources. New tools and databases designed for digitization can increase collaboration of computational researchers and experimentalists to aid in their development of novel material manufacturing techniques for new and existing solid state materials.

14.2 The Development of MOF Databases

The progress of ML would begin to stagnate without the access to sufficient training data. However, the field of solid state chemistry sees thousands of materials synthesized each year, with tens of thousands published in the past decade. However, due to the substantial amount of text accompanying the interesting data in each subsequent publication, obtaining large and reliable databases of ML training material has the potential to become a significant bottleneck to development without continuous addition of new information. The number of MOFs in the CSD MOF subset alone has increased by 30,000 structures since 2017, and additional publications with information about synthesis of new structures are published regularly [8]. The latest CSD update contained over 16,000 additions to the database with an estimated 10% of structures being MOF or MOF-like. To allow for chemists to efficiently process information, data must be presented and submitted in a fashion that is accurate, organized, and machine readable.

In 2014, Chung et al., as part of the Materials Genome Initiative, developed the first MOF database CoRE MOF, containing 4700 porous structures [9]. The data originated from the CSD, and included some additional properties to ensure these entries were suitable for molecular simulations. The conditions for entry into the database included only 3D structures with pore sizes larger than 2.4 Å. These criteria were chosen specifically to allow for the screening of MOF structures for use in gas storage, separation, and catalysis. In 2019, the CoRE MOF database was manually updated to include over 14,000 porous 3D structures that have been reported in published literature [20]. Value was also added to the database with the introduction of reconstructed disordered structures, and new pore analytics and physical property data.

Following the release of the CoRE MOF database, the CSD MOF subset was developed in 2017. Moghadam et al. expanded the criteria used to identify MOFs by allowing for the inclusion of materials of other dimensionalities and pore sizes. One notable advantage of the CSD MOF subset over the CoRE MOF database is that the CSD MOF subset has been designed to be automatically updated, adding new materials quarterly to ensure all structural data is current. The CSD MOF subset includes 1D, 2D, and 3D structures, and can be organized into these specific categories at the click of a button. The CSD MOF subset is primarily accessed as part of the CCDC's software suite and is compatible with their structure search tool ConQuest, and the CSD PythonAPI.

The CSD MOF subset has been used in a significant number of studies, and has formed the primary data source for the development of DigiMOF, a text-mined structure information database created by Gubsch et al. [21]. The aim of the DigiMOF database is to provide MOF synthesis data alongside other key properties which are lacking within previous databases. Synthesis data, when collated within a database, will allow for synthesis routes to be compared and assessed, leading to the minimization of failed or inefficient experiments, and increased likelihood of scalability and profitability when compared with current trial and error synthesis development. Due to the diverse range of applications for MOF structures, several other significant contributions have been developed and published which focus on specific applications due to certain properties. Notable examples include the CoRE MOF 2014-DDEC database by Nazarian et al. [22], in which DFT-derived partial atomic charges were determined, and the quantum (QMOF) database by Rosen et al. [23]. The QMOF database enables the searching of structures based on properties such as charges, bond orders, or band gaps for use in photo-catalysis, and other similar applications. As part of their works carried out with the use of ML for MOF synthesis, Luo et al. [18] and Nandy et al. [24] have also created small publicly accessible databases to investigate synthesis conditions of MOF structures and solvent extraction data, respectively. Aside from databases containing "already-synthesized" structures, there are several examples of collections of hypothetical structures such as the hMOF database by Wilmer et al. [25], containing 137,953 hypothetical MOFs, the 13,512 MOF structures created in specific topologies using a material generation algorithm called ToBaCCo by Colón et al. [26], and more recently by Majumdar et al. [27], with ca. 20,000 hypothetical MOFs designed specifically for a diverse chemical design space. The use of ML in synthesis prediction is a step toward finding feasible pathways for the eventual creation of many of these hypothesized materials without the need for expensive trial and error experimentation.

14.3 Natural Language Processing

One of the biggest limitations of ML is the requirement for data availability, accurate deep generative models typically requiring training datasets with a size of the order of 10^6 [28]. In terms of MOF structures, properties, and applications, an abundance of data is collated within currently available databases; however, they rarely contain information regarding the synthesis conditions and parameters required to prepare them. Alternatives to creating new or real-time data include NLP algorithms that are frequently being used to extract synthesis data from published scientific literature. NLP can be structured into a series of four steps: i) article retrieval; ii) conversion and paragraph classification; iii) word tokenization; and iv) extraction and manual verification.

Datasets for ML training can be generated through NLP of new experimentation or by accessing existing data held within experimental logbooks. One example where NLP would have been useful is from Xie et al., who collected the synthesis parameters of 486 reactions from archived experimental notebooks of both successful and failed experiments [29]. This data was used to train models for synthesis condition prediction of metal–organic nanocapsules (MONCs). An eXtreme gradient boost (XGBoost) algorithm topped the table of prediction accuracy, using 17 descriptors, at 91% with solvents, modulators (molar mass and mole), and cations emerging as the dominant factors in the formation of single-crystal MONCs.

Additionally, it is worth noting here that a recent perspective contribution by Jablonka et al. has sought to address the issue of inaccessible, non-digital, and non-reported experimental practices [30]. As paper-based lab records are often still the norm in many institutions, the authors suggested that the development of a modular open science platform would benefit not only the data mining studies highlighted in this chapter, but also beyond that. In recent years the introduction of electronic lab notebooks (ELNs) aimed to address these concerns surrounding data management, and increase the reusability of experimentally gathered data. However, although the argument that the technology is already available to begin the collection of this work, the adoption of these ELNs must be suitable for the synthetic work performed by chemists and materials scientists, and developed as an easily accessible and open source resource that demonstrates fair principles and practices. Only, once an acceptable and agreed format is chosen, would such a tool become useful for the gathering of data for the application of ML.

The use of NLP can be found throughout the field, with a heavy focus on using ML for the prediction of synthesis parameters. Kim et al. [31] used a cross-reference application programming interface (API) to retrieve a list of articles focused on the synthesis of titanium nanotubes. From approximately 100 different journal articles, several hundred paragraphs were manually labeled as either synthesis paragraphs, or other, before being fed into a logistic regression classifier. Paragraphs underwent a word embedding approach such that they could be represented as real-value vectors, followed by binary labels, with "1" indicating synthesis information present, and "0" as unrelated. The post logistic regression data had an overall accuracy of 95% on unseen test data.

The next stage saw relevant synthesis paragraphs undergo transformation into dependency parse trees using ChemDataExtractor and SpaCy parsers [32]. Word tokenization and speech tagging are performed to split sentences into constituent words, and grammatical labels are added to each word token. Synthesis verbs of interest are detected by a neural network (NN) approach upon the traversal of these dependency parse trees, which are then iterated along to find operating parameters. Nouns are then scanned and matched against the PubChem database and validated against the ChemDataExtractor model to confirm meaningfulness [33]. Trained on ~5000 human annotated words, this NN approach yielded an overall accuracy of ~86% as measured against a set of 100 human-annotated synthesis articles.

In recent years, NLP has seen increasing use in the field of MOF synthesis data extraction. Luo et al. used NLP to extract data for use in predicting MOF synthesis conditions, with the structure selection based on those found exclusively in the CoRE MOF database [18]. The NLP successfully extracted synthesis conditions for 983 MOF structures. The parameters have been collected and made available within an open source synthesis database; however, due to the lack of standardization when reporting reaction conditions, some key parameters are missed due to ambiguity.

Therefore, accurate and reliable NLP currently requires parallel manual extraction. A total of six relevant parameters were extracted from the CoRE MOF into the new automatically created SynMOF-A database, alongside metal and linker information taken directly from the associated crystallographic information files (CIFs) of each material. Additional manual versions, SynMOF-M, and SynMOF-ME, were also created to ensure accurate data sources before being used to train ML models to discover similarity patterns in the synthesis conditions.

Park et al. [34] extracted data for 46,071 MOF synthesis reactions from 28,565 papers using a newly-developed data extraction code. This study categorized paragraphs in papers much in the same way as Kim et al. [33] with a binary approach to synthesis information, but with a notably larger test set of 180 papers. However, this study extracted synthesis conditions from solid state materials found only within the CSD MOF subset database. The NLP method here used named entity recognition (NER) to extract chemical names and then categorized them using neural networks, a 100-dimensional bi-directional-LSTM which is able to consider the forward and backward context, alongside a conditional random field (CRF) layer used to predict each label of sequence data. Here the NLP was used to extract MOF names, precursor, and solvents, with a high precision (~98%) but significantly varying recall. The data collected by Park et al. [34] focused on single-step reactions, as their NLP algorithm was unable to differentiate between multiple steps within MOF synthesis.

After a review of current text-mining and MOF synthesis literature, Gubsch et al. chose to develop parsers that would extract the information on the solvents used, the inorganic and organic precursors, and their synthesis methods [21]. The parser training technique used in this study is visualized in Figure 14.2.

DigiMOF database is the first within the MOF field to utilize the tool ChemDataExtractor, in conjunction with the CSD PythonAPI, to produce a property database using text mining software. MOF topologies were also extracted for further synthesis route analysis, and compared with the building blocks to investigate potential trends. A blacklist was employed to filter frequently found misidentifications within the published text. The DigiMOF project is made open source to encourage collaboration for further improvements in the gathering of useful synthesis information. The current version of the DigiMOF database consists of 43,281 unique MOF properties with a precision of 77%. This work, among other attempts such as that of Luo et al. [18], is the foundation for

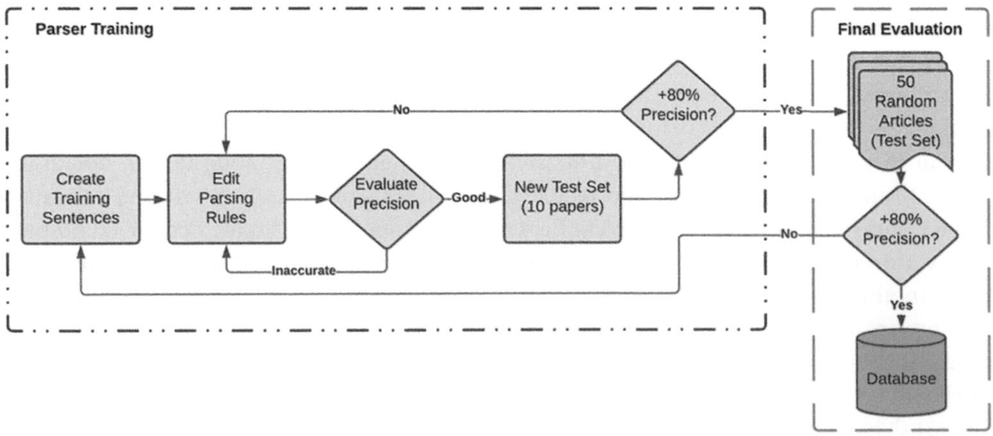

Figure 14.2 A flow diagram which shows the process of developing suitably precise parsers for data extraction by text mining [21], licensed under CC BY NC 4.0.

the future of digital manufacturing, hoping to provide a searchable database for key MOF properties that can allow for assessment of viable reactions.

Despite these promising studies into the utilization of NLP to create useful synthesis datasets, a bottleneck associated with extracting data from published journals is the bias toward successful reactions. Without proper weighting or failed reaction data, training sets for ML models will include bias. However, the data extracted in these studies still has multiple uses, from collation into a searchable database for material evaluation, to use as training data for further development of ML models. The availability of more and more extracted data helps to shift ML development toward overcoming frustrating "data-poor" situations that have previously been a hindrance for the development of digital synthesis.

14.4 An Overview of Machine Learning Models

ML models are typically chosen based on the available data and the desired result. Certain models are able to handle higher levels of complexity and often have variable levels of interpretability. ML tools have become increasingly accessible to the general researcher in most scientific fields, with the emergence of downloadable software packages including Scikit-learn and TensorFlow, combined with an abundance of online help, ML training courses, and video guides [35, 36]. These packages provide access to a multitude of models from the simplest to increasingly intricate.

Arguably, the simplest model is linear regression, a linear approach used to model the relationship between a scalar response and one variable. Conroy et al. [37] used linear regression as part of their work to identify key property descriptors required to predict synthesis routes for zeolite LTA synthesis, and predict the quantitative output of synthesis routes. Linear regression works on the basis that the output is linearly relative to the feature inputs, and the predicted outcome is a sum of the weighted features [38]. The assumption of linearity limits both the potential uses and results obtained from this method, although due to the simplicity of the model, the results are often transparent and can be easily interpreted.

Non-linear models include random forest regression, a supervized learning algorithm based on selecting decision trees after being trained on a dataset and averaging the results. This technique makes predictions based on the outputs of merging multiple decision trees, combining multiple predictions to make a more accurate prediction than a single model. The results of this model are typically easily interpreted, with easy to visualize results due to the simple nature and easy visualization of a decision tree. A tree with a depth of five, for example, would be easy for most people to follow after a brief explanation. However, limitations include overfitting and slow prediction speed when using a large number of trees, and the combination with random forest regression reduces the interpretability, particularly as the maximum number of trees can begin to exceed several hundreds.

Jensen et al. used random forest regression to predict the densities of zeolite based on their synthesis conditions, following the use of NLP and text markup parsing tools to automatically extract information from 70,000 zeolite journal articles [39]. They trained a random forest regression model using sci-kit learn, across 100 decision trees with splits determined by mean squared error. The model was cross-validated on syntheses that resulted in a pure phase zeolite, including 898 synthesis routes. Support vector regression, simple neural network, and Gaussian process regression models were also used and compared to random forest, with the random forest model exhibiting the highest accuracy with the added benefit of human interpretability.

Support vector machines (SVM) are a popular example of kernel models, a class of algorithms used for pattern analysis. The general task of kernel models is to compare new data with the data found in training datasets to make predictions. An SVM training algorithm builds a model that assigns examples to one category or another in a non-probabilistic binary classification. The SVM maps training examples to points in space, creating a gap between two categories, then new examples are mapped into the same space and predicted to belong to either side, depending on which side of the gap they fall.

Raccuglia et al. employed SVM when exploring chemical space, focusing particularly on inorganic–organic hybrid materials, based on failed experimental data found in archived lab notebooks [40]. The vast majority of unreported failed reactions are archived in notebooks that are typically inaccessible. To guide future efforts toward successful synthesis, a web-accessible public database was created to enable initial data entry from existing notebooks and current experimental data. The dataset, 3955 unique and complete reactions, was split into groups, 1/3 as test and 2/3 as training. A single SVM model was used to predict the likelihood of crystallization based on synthesis parameters, with an accuracy of 78% in describing all reaction types, and 79% when considering only vanadium-selenite reactions.

When dealing with larger datasets, it is more common to see neural networks being used as the primary ML models. Neural networks are designed to mimic the process through which the human brain operates, with a combination of hidden layers with input and output layers. The node connection between each layer forms a network where each node has an associated weight. The capability of a neural network is highly dependent on the quality and size of the dataset used, but these models can adapt to a changing input so that the network can generate the best possible result without requiring any redefinition of the output criteria.

Park et al. [34] trained their artificial neural network (ANN) model using a positive-unlabeled learning (PU learning) algorithm. This model was trained to predict the synthesizability of MOFs based on given input synthesis parameters, and was able to differentiate between amorphous and crystalline forms of the same MOF material. This research has been some of the first in the field of MOFs to use "big data" to achieve meaningful insights into ideal synthesis conditions.

XGBoost is an open source software library composed of gradient boosting ML algorithms. The aim of gradient boosting is to find patterns within the data and make predictions based on these relationships. It gives a prediction model in the form of ensemble learning, where multiple learning algorithms are used to obtain better predictive performance, typically a fixed set of alternative models which allows some flexibility. The gradient boosting prediction model most often consists of decision trees, and where a decision tree is the weak learner, the result is a gradient-boosted trees algorithm which usually outperforms random forest. While this boosting can increase the accuracy of linear regression or a decision tree, it may sacrifice intelligibility and interpretability. To recover performance and interpretability, some model compression techniques exist which allow for the transformation of an XGBoost into a single decision tree that can approximate the original decision function [41].

For prediction of the likelihood of crystallization of MONCs, Xie et al. utilized a multi-model method to allow for cross-validation and comparison of results [29]. The models included linear regression, Gaussian Naïve Bayes (GNB), k-nearest neighbors (KNN), SVM, decision tree, random forest, XGBoost, and multilayer perceptron (MLP). All of the tested models achieved an accuracy of 82% or higher, and an F1 score which exceeded 81%. The result of training and evaluation of these nine ML models found XGBoost to have the highest accuracy.

It is important to note that all ML models offer individual advantages and limitations, and when choosing an ML model, it is essential to consider the importance of high accuracy and interpretability as well as ease of operation. When choosing a model that best fits the input dataset, the following factors must be taken into consideration. First, the runtime of models such as random tree will become progressively longer as the number of trees increases, and the output will be achieved much more slowly. This can be impractical when using large and complex datasets. Some models also lack interpretability, making it difficult to know exactly what the results of the study are, and the user may be unable to interpret the factors that have caused the predicted output. In a synthesis context, the relationships within the data provide the essential information required to aid process and material development. Without interpretability of the data's relationships, it becomes progressively difficult, as models increase in complexity, to further understand low accuracies and slow runtimes.

Evaluation is a key stage in ensuring the return of accurate ML model results, and evaluation metrics allow chemists to quantify the performance of models on data that has yet to be seen. Accuracy, the ratio between correct predictions and total predictions, is most useful with balanced data. For data with imbalances, F1 scores are used to evaluate the outcome, combining the precision of the results with recall. Additionally, R^2 represents the proportion of variance from the original dataset, and is best used with regression models such as the random forest regression.

Another approach, Bayesian optimization, is becoming increasingly popular in the development of synthesis digitization [42–44]. This algorithm allows for continued optimization of a closed loop process when used in conjunction with automated physical systems, helping to optimize products and synthesis conditions, and increase the feasibility of reactions. Using input data to take an initial guess about a chosen function, it continues to refine this first prediction as data is added with each iteration, and builds a probable model of the objective function that is being explored. It is composed of two models, a surrogate model and an acquisition function. The surrogate model defines the probability distributions over the chosen function, built using the

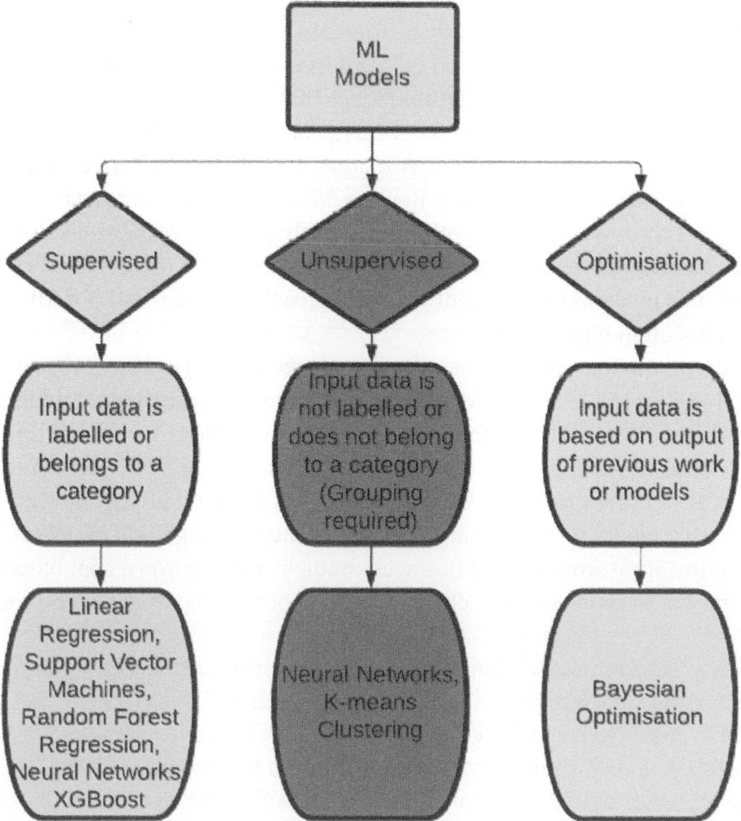

Figure 14.3 A flow diagram demonstrating the classification of some basic, and some more complex, machine learning model types.

sample data provided. The acquisition function selects the next samples from the search space and these new sample points are used to update the surrogate model, increasing the accuracy. The cycle of update and optimization continues to produce an accurate model of the chosen objective function. The surrogate model most used in Bayesian optimization is the Gaussian process, due to its flexibility to fit wide ranges of data, and the construction of a Gaussian distribution. Other models for the surrogate include the tree-structured parzen estimator (TPE), a sequential model based the optimization approach which constructs models to approximate the performance of parameters based on historical measurements, and chooses new parameters to test based on this model.

A simple, flow-style overview of the different approaches to ML models that have been discussed in this section can be seen in Figure 14.3.

14.5 Machine Learning for Synthesis and Investigation of Solid State Materials

The application of ML in the realm of solid state chemistry is becoming commonplace for predicting realizable structures as well as their synthesis and intrinsic properties. In addition to benefiting from lower research and production costs as the sector moves away from trial and error experimentation, improvements to the field include smaller starting material requirements, fewer failed structures, and less reliance on the intuition of chemists.

The availability of pre-generated data that can currently be found in journals and databases has made it easier to train models for property predictions. This, combined with user-friendly models that can be used by chemists with limited coding experience, means that new developments in ML for material synthesis are being made every day. The year 2022 has so far seen a significant increase in publications which employ ML for material synthesis predictions, with an expectation that this trend will continue throughout the decade. The Thomson Reuters' Web of Science offers researchers the ability to search for key words and subjects, a useful trend analysis tool which can be used to confirm the increase in popularity of certain topics (found here https://www.webofscience.com/wos/woscc/basic-search). The tool was used to confirm the increasing trend in ML for MOF synthesis, and the results can be seen in Figure 14.4.

ML tools began with property prediction, enhancement, and analysis of the structure–property relationships. Kennicutt et al. [45] used SVM, a supervized machine learning algorithm, to predict the adsorption rate of activated carbon, specifically choosing SVM to reduce the risk of overfitting and for ease of use. The model was trained on 95 compounds with 23 structure descriptors, and reached a final training dataset R^2 accuracy of 0.932. Although there is a high R^2 value, it may not necessarily be a suitable predictive model for compounds outside of the training domain. Efforts were focused on data for Calgon Filtrasorb 400 (F400), a well-studied, microporous, coal-based adsorbent and so descriptors for carbon surface chemistry and pore properties were not considered.

More recently, Dico et al. used several independent ML models to assess natural nano-porous clays for their use in adsorption and catalysis [46]. An extremely randomized trees (Extra Trees) regression algorithm was used to characterize the raw clay characteristics, additive characteristics, and processing conditions assessed against simple decision tree, random forest, and MLP models. The final R^2 for the extra tree regressor of 0.77 is reasonable for a dataset of this size, based on 41 feature representations, with a variation in R^2 values when reduced to 20 descriptors. The data has

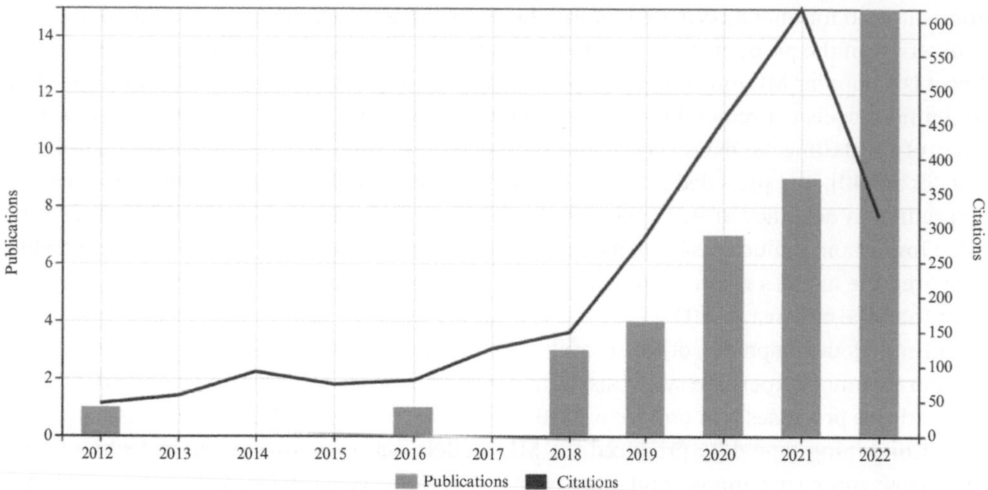

Figure 14.4 A Web of Science search for trends in publications using the key words "metal–organic framework," "synthesis," and "machine learning," as of July 2022 (https://www.webofscience.com/wos/woscc/citation-report/c0f28728-bd2b-4392-834e-7bc24ac6334b-474cb7be).

sufficient accuracy for its predictive ability of most parameters, which will enable its use as a tool for further assessment of the viability for parameters required in the processing of natural porous materials, particularly porous clays. The analysis of the models sheds light on the features of raw minerals which significantly affect the internal and external surface area, and pore volume.

Following the rise in models aimed at finding the relationships between property and structure, chemists began to develop models that would look to find and confirm synthesis–structure relationships. These models work to extract significant synthesis descriptors from substantial amounts of data and realize patterns unable to be seen by humans. For example, Jensen et al. used ML models to investigate less intuitive relationships between zeolite synthesis parameters and the final structure [39]. They trained a single decision tree regression model on 898 pure phase zeolite systems specifically to demonstrate ML synthesis intuition, rather than predictive ability. This model was able to reproduce the framework density of training data with an impressive R^2 value of 0.97.

Raccuglia et al. developed an algorithm focused on exploring the chemical space to predict a range of reaction conditions that will result in crystallization of template vanadium selenites [40]. Information gathered from "dark" reactions, failed or unsuccessful hydrothermal syntheses, trained an ML model to predict reaction success. The model successfully predicted conditions for new products with a final success rate of 89%. The authors concluded that the model may be used to reveal further relationships between reaction conditions and product formation.

Muraoka et al. also investigated the relationships between reaction conditions and products using ML synthesis prediction techniques [47]. The dataset was formed using NLP from zeolite synthesis records, alongside a similarity network of the crystal structure and synthesis descriptors, producing a dataset of 686 synthesis conditions. Multiple ML models including random forest and XGBoost were used to predict the synthesis results from these custom descriptors. The XGBoost was found to have the highest accuracy at 75–80%, and while not within the scope of the work, they envisioned that the XGBoost model could be used to provide further informatics into the likelihood of formation of specific zeolites. The combination of similarity networks with synthesis prediction allowed for unexplored areas of the chemical space to be found and populated, increasing the diversity of the products formed, a major challenge with zeolites and MOFs alike.

XGBoost is a popular ML model used for predicting synthesis conditions for reactions. Xie et al. [29] used nine models to predict the synthesis conditions of MONCs and successfully synthesized a new set of crystalline MONCs. The data originated from archived lab notebooks, similar to Raccuglia et al. [40], and provided a dataset of 486 reactions. The XGBoost was found to have the highest prediction accuracy at 91% and was able to quantify "chemical intuition." By providing quantified importance values, reaction parameters were ranked for future reactions. These results also compared the model's synthesis prediction against a chemist, achieving a prediction accuracy 5% higher than the chemist, at 80%.

The continuous development of ML models lays the foundation for new techniques that can enable accurate and inexpensive synthesis paths for solid state materials. The reaction data gathered for synthesis processes will provide a starting point for further analysis by chemists and data scientists. Combining new data produced via ML models, with improvements in technique for design of faster, more sustainable, and more economic synthesis pathways, will see the field shift more quickly to full digitization for a wide range of material requirements.

14.6 Machine Learning in Design and Discovery of MOFs

Developments in the field of ML for structurally similar materials, such as zeolites and organic porous structures, have also inspired MOF scientists to perform synthesis prediction studies. Due to the large volumes of data produced, ML is quickly becoming a necessity for efficient exploration of the MOF material space. For the large library of existing MOF structures and considering its continued growth, high throughput screening techniques on their own are no longer fast enough to identify promising materials for synthesis. It is very difficult to computationally screen the vast material space for a single application, restricting the likelihood of the most effective structure being identified. The ML models which are discussed in this section have contributed to recent advances in the shift toward a new data-driven and digitalized paradigm to design and discover new MOFs.

Inverse design is a tool that is seeing a growth in popularity for chemists with advanced ML tools, particularly deep learning, being developed to aid this process. Zhang et al. developed an inverse design algorithm to directly create novel MOFs for use in carbon capture [19]. In this contribution, the authors combined Monte Carlo tree search with recurrent neural networks (RNN) based on input from 10 different combinations of metal nodes and topologies from previously reported experimental MOFs, and these can be seen in Figure 14.5. Using the criteria obtained by the ML study, the algorithm hypothesized a novel MOF based on IRMOF-15, which was expected to perform in carbon capture applications with a CO_2 adsorption capacity up to 543% higher than the input structure. Other examples include a 165% increase on the current capability of MOF-118, and 11% increase for MOF-119. A small increase in CO_2 adsorption of MOF-119 is expected considering the already high loading capacity of 8.18 mmol/g, particularly compared with the increase in loading from 1.29 to 8.30 mmol/g hypothesized for IRMOF-15.

Hardian et al. developed a new ML module that combines a design of experiments (DoE), an SVM, an evolutionary algorithm, and a desirability function to predict the optimal conditions for sustainable ZIF-8 synthesis [48]. DoE is an alternative systematic approach to achieve good balance between a reduced number of experiments and efficiency, allowing different factors to be varied and investigated simultaneously, in the hope of accelerating the process of discovery and optimization.

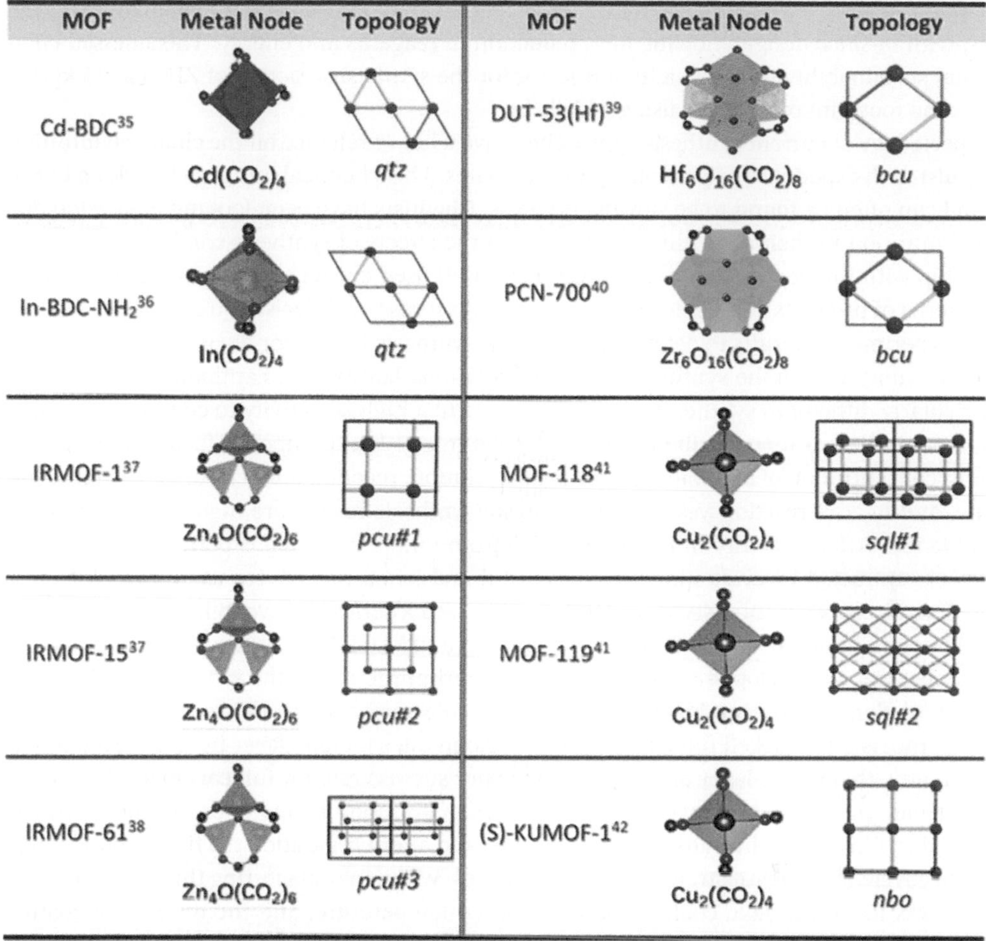

Figure 14.5 Input topologies of novel experimental MOFs for use in an inverse design algorithm targeting structures for top performance in carbon capture applications. *Source:* Reproduced with permission from [19]. © 2020, American Chemical Society.

The data was first obtained by experimental face-centered central composite design, completing 27 runs. The structures' crystallinity was analyzed using X-ray diffraction (XRD), and after confirmation of sharp XRD peaks which well matched the calculated XRD pattern, AI 1 was used with the experimental data output from the DoE as input for the SVM algorithm, followed by a grid search to generate 456,976 virtual data points. The results found that highest product quality was achieved at high voltage, long reaction time, low electrolyte concentration, and high linker concentration. AI 2 followed a similar method to AI 1, followed by 50 random virtual data generations as the initial population for the evolutionary algorithm, followed by the implementation of a desirability function for the last optimization step. Two optimization steps were considered, first to maximize the product quality, and second to maximize product quality and process sustainability.

DoE results were insufficient to identify optimal conditions, AI 1 created an unnecessarily large dataset which increased the computational cost, and AI 2 with the incorporation of the evolutionary algorithm made it possible to screen only the best data. These predicted conditions were used to synthesize ZIF-8, and the resulting structure had 100% purity, 88% yield with 86% crystallinity.

A major argument toward using ML for material synthesis is to improve the sustainability of the process, with *in-silico* design allowing for a reduction in reagents and energy. The assessed environmental sustainability provided a final E-factor for the synthesis reaction of ZIF-8 at 11 kg/kg, and a carbon footprint of 27 kg/kg using 7 kWh/kg.

As expected, most current synthesis approaches have a heavy reliance on the chemical intuition of a chemist with experience of previous synthesis routes. This chemical intuition is lacking in ML tools, and can often be found to be slowing progress. Chemists have been looking to develop this chemical intuition in their algorithms by evaluating the effects of synthesis conditions and their relationship with the final structure. Kitamura et al. [49] used cluster analysis to classify powder XRD patterns of products and determine experimental success, and decision tree analysis to visualize the experimental results to determine dominant synthesis factors for the production of KGF-3, a building unit used in the synthesis of MOFs containing lanthanide. Lanthanide-based MOFs are particularly difficult to synthesize and predict, due to a high sensitivity to condition changes, leading to overall poor reproducibility. Data collected from 108 experiments, focusing on lanthanide ions, concentration of metal ion and/or ligand solution, reaction temperature and time, cooling time, and type of reaction vessel found that the synthesis results are highly affected by the lanthanide ion. After difficulty in isolating KGF-3 from initial screening experiments, dominant factors were extracted by evaluating both successful and failed procedures using ML. With this information, the experimentalists successfully synthesized a series of novel pillar-layered lanthanide MOFs containing the double-layer-based building units KGF-3.

Huelsenbeck et al. developed an active learning algorithm to aid in the synthesis of HKUST-1 thin film [50]. Some MOFs can be grown on multiple substrates using a roll-to-roll process; however, this process often lacked full coverage. Other techniques include layer-by-layer growth, solvothermal growth, and gel-layer growth, with a greater success rate for full coverage. Drawbacks for these techniques include slow crystallization, lack of orientation, and poor thickness control. HKUST-1 thin film is used in transistors and sensors, and after preparation the final product must have a full coverage of the substrate with no void spaces. When manufacturing thin films, the synthesis process must also take coating speed, substrate temperature, and the number of coating passes into account. A pool-based active learning (PAL) and regression method was used to efficiently guide the solution-shearing synthesis. Each iteration chose 18 diverse and representative solution-shearing process parameters for validation and feedback; these were composed of samples created using parameters determined by a generalized subset design (GSD). Each sample was replicated three times and characterized using optical microscopy to label the samples as "fully covered" or "not fully covered," where the substrate is visible between HKUST-1 particles. The results showed that 22% of the 18 initial experimentally synthesized samples were fully covered. An ensemble SVM model was trained to classify the coverage into the same two tiers, which was then used to predict coverage for over 11 million parameter combinations in the unexplored dataset. A second set of experiments were performed to obtain thin films based on the 18 parameter sets, with optical micrographs showing 67% of conditions having full coverage. These were used to retrain the SVM ensemble, resulting in a final virtual parameter space predicting 13% of parameter combinations to have full coverage. The use of active learning combined with a solution-shearing process resulted in a final product with a large full coverage and a minimum thickness of 2.2 μm. This work not only uses ML to predict synthesis conditions, but also the results from each experimental stage to optimize performance, allowing for an increased pace of material development; a key step for automating and digitizing manufacturing.

Luo et al. collected data based on the CoRE MOF database, with a training dataset composed of MOFs extracted from published and predominantly successful experimental synthesis data, focusing on six relevant parameters [18]. Random forest and neural network were the best performers used to predict the synthesis time and temperature of MOF structures. The results were compared, showing the random forest approach had the higher accuracy across all predicted parameters; however, the neural network predictions will become more accurate as datasets grow, and additionally may be able to exploit correlations between different synthesis parameters. Therefore, the neural network has greater potential as the field develops where more complex models are expected to outperform random forest in the not too distant future. A comparison between the expert trial and error approach, and an ML model implementation is visualized in Figure 14.6.

Prediction of MOF synthesis conditions is a difficult task as for many cases there is not one true solution, and a whole range of conditions can lead to a successful synthesis. Some reactions may be optimized for yield, while others may be chosen for environmental or financial cost. Despite what can be described as low R^2 values for reaction temperature prediction at 0.286, and predicted reaction time at 0.076, comparisons between the predictions of 11 human MOF synthesis experts revealed that the ML model out-predicted all experts who had an R^2 value much closer to zero, even after averaging across all human estimates. Even small correlations learned and exploited by ML modeling can help to better estimate synthesis conditions without the availability of big data.

Park et al. created a database of 46,701 structures from 28,565 published papers found in the CSD MOF subset, and a PU learning algorithm was chosen to reduce the positive bias found in the dataset [34]. PU learning algorithms are typically used in cases where the proportion of positive to negative data is heavily skewed, and this approach has been used previously to predict synthesis information in inorganic materials [51]. Extracted synthesis conditions were used as positive data, and to be regarded as positive data the entry must include composition or temperature, plus one other parameter, and this is shown in Figure 14.7. In total, 3748 pieces of extracted MOF synthesis information were considered as the positive data, with 2998 positive data points used in the training set and 750 in the test set. *In-silico* data generated by randomly sampling parameters for the extracted data was classified as unlabeled data. Randomly generated unlabeled data totaled 1 million pieces, with 900,000 used in the training set and 100,000 used in the test set. Often, ML methods such as decision tree or SVM are used as the binary classifier for PU learning, but in this case

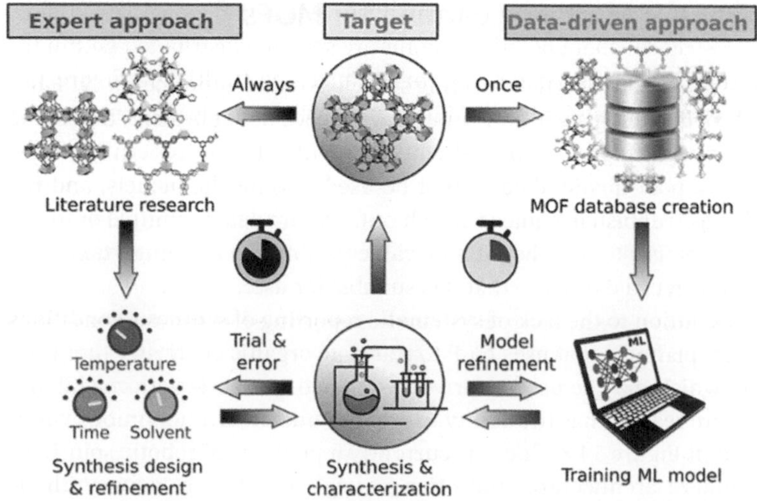

Figure 14.6 A comparison of the trial and error approach vs. training machine learning models to predict synthesis conditions. *Source:* From [18] Luo et al. (2022), John Wiley & Sons, CC BY 4.0.

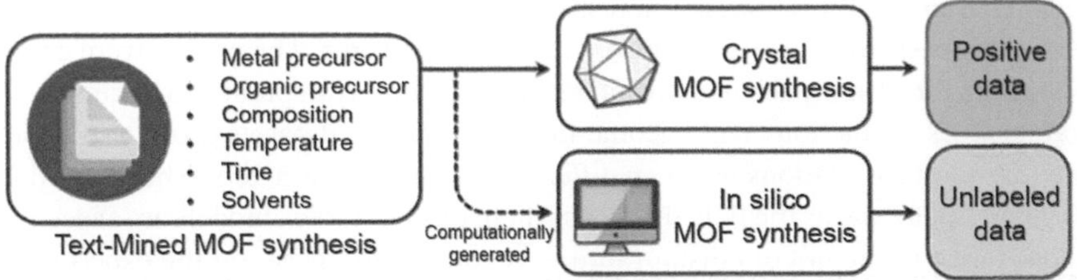

Figure 14.7 A schematic showing the criteria used to differentiate the positive (P) and unlabeled (U) data. *Source:* Reproduced with permission from [34]. © 2022, American Chemical Society.

that approach was not appropriate and a neural network with simple dense layers was used instead. At each iteration the binary classifier is trained, with positive and unlabeled data in a ratio of 1:10, using the Adam optimizer implemented using Tensorflow [52].

The model used the input synthesis conditions to predict the output crystallinity of a structure. Final output crystallinity scores, with "1" representing conditions for high crystalline structure and "0" indicating conditions for low crystalline structures, were determined by averaging the prediction scores of the binary classifier over the total number of iterations. The final score is denoted as the "crystal score" and crystallinity is classified based on the output of the PU learning algorithm exceeding 0.5. A positive-negative (PN learning) learning model was implemented as an alternative method to evaluate the performance of the PU learning model. For the test set, the recalled scores were PU learning at 83.1% and PN learning at 50.3%. False negatives are also low for PU learning, but higher for PN learning.

ML models are used in conjunction with MOF synthesis for all stages of material design, discovery, and manufacture. These approaches highlighted in this section are concerned with not only discovering new materials, but with discovering new synthesis techniques for existing materials, and the outcomes of new techniques such that the yield and crystallinity can be determined before experimental synthesis even begins to take place. However, despite the ongoing adoption of ML, there are still many limitations to the process of applying ML to MOFs at every stage.

14.7 Current Limitations of Machine Learning for MOFs

A major factor in the chance of success for all ML is meeting sufficient data requirements. NLP has been used to access thousands of documents within the published space but at present, due to the lack of systematic labeling when reporting experimental synthesis, NLP often misses critical information. With all ML algorithms, poor-quality data cannot be used to train the models, and the process can often be referred to as "rubbish in equals rubbish out." Where data is omitted or incorrectly parsed, it might require chemists to find the missing values within the literature, taking up valuable time and resources to correct a dataset so that it is suitable for use.

Mehr et al. have proposed a solution to the lack of systematic reporting of synthesis conditions [53]. They developed a software platform that uses NLP to translate organic chemistry literature directly into an editable code which may be used to drive automated synthesis within a laboratory setting. Automatic literature reading has the ability to create a universal autonomous workflow, and this is demonstrated in Figure 14.8. There is currently a plethora of robotic solutions, but they lack a common standard architecture, and often struggle to adapt to new synthetic methods. The standard of recording and reporting of new chemical compound synthesis varies greatly and procedures are typically explained in ambiguous and incomplete passages of text, relying on expert chemical intuition to bridge the gaps. The quality of reaction database data

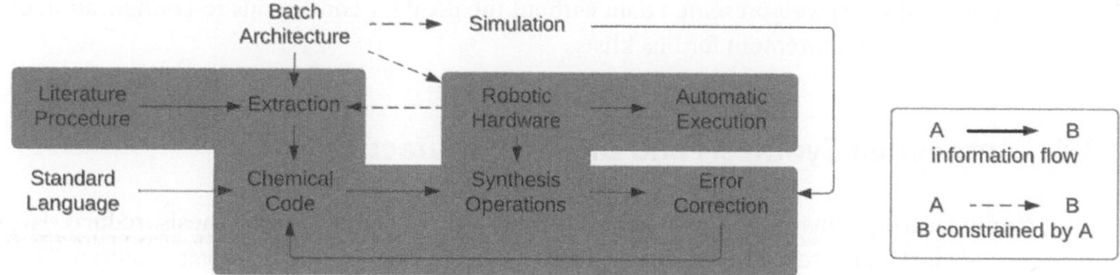

Figure 14.8 A universal system for the automatic execution of chemical synthesis from literature. Extraction of the procedure is followed by an algorithmic process for producing the code that conforms to a standard hardware and software architecture. Manual error correction and simulated execution ensure reliability and safety. *Source:* From [53] / American Association for the Advancement of Science.

therefore can be sporadic, and this presents many problems in terms of reproducibility and for the development of reliable digital methods that aim to predict synthesis properties for new materials and functionalization.

This leads the discussion to the importance of ensuring that there is an availability of high-quality training data to give an ML model the best chance of making reliable and accurate predictions. Most published work reports only successful experiment and synthesis data, particularly in the field of MOFs; however, zeolite studies are noted as being particularly good for reporting both successful and failed experiments [39]. In cases where most failed procedures are confined to personal notes and lab books, the use of purely successful data found in published literature and databases may lead to bias within training sets.

Moosavi et al. investigated a robotic synthesis approach guided by ML for HKUST-1 synthesis optimization [54]. This approach enabled the variation of multiple synthesis conditions such as solvent composition, temperature, and method (e.g., conventional heating, microwave, electrochemistry) and enabled the recording of both successful and failed data. This regenerated experimental data was used to optimize the ML models with consideration toward the removal of potential bias. Xie et al. and Raccuglia et al. took advantage of archived lab notebooks holding all experimental results in an attempt to overcome bias; however, this approach used real-time failed experimental output data [29, 40]. These techniques worked well on a small scale, but for larger datasets, alternatives must be proposed. Park et al. worked to overcome the lack of negative data available, with regards to the crystallinity of MOFs, by using PU learning algorithms [34]. Although these are intuitive solutions to maximize ML potential, new publications should endeavor to ensure the suitability of data for digitization and chemists should begin to realize the importance of all experimental data and begin to publish all results, including those of failed synthesis. Moosavi et al. produced a public webpage on the MaterialsCloud as part of the work on HKUST-1 [54]. The aim of this web application is to involve a large number of groups involved in MOF synthesis to collectively report failed or partially successful experiments, offering the potential to change the way the community approaches synthetic chemistry. Chemists can document all synthesis reactions, with public access to all reaction data, increasing the amount of negative data available for training.

Lastly, the development of automated systems that may run reactions and create molecules are often hindered because of a lack of a machine readable standard within experimental publications. Gubsch et al. also suggested that publications introduce new standardized submission templates for material synthesis articles so that the data is presented in a suitable manner for reliable and accurate text mining and parsing tools [21]. Taking this approach would significantly simplify the parsing techniques for all NLP approaches, and simple methods can be used to extract key parameters from tables of well presented data without the need for continuous re-configuration of parsers or the current requirement for blacklists.

14.8 Automated Synthesis and Digital Manufacturing

Digital manufacturing aims to use previously collected data to streamline synthesis, reduce risk, and locate the most viable reaction pathways and cost-effective materials for a given material. This involves the use of a database in conjunction with the material property data, NLP for accessing published reactions, and ML for property and synthesis prediction. King et al. were one of the first to acknowledge the influence of computation in the scientific process, and their development of Robot Scientist "Adam" was used to generate genomic hypotheses about yeast, which were confirmed with manual experiment [55]. In the years following, there has been a significant influx in the number of publications focused on automated synthesis, which include: Burger et al. [56], who created a mobile robot that ran autonomously for eight days, driven by a Bayesian search algorithm to search for improved photocatalysis for hydrogen production from water; Sun et al. [57], who developed a meta-learning model to predict the adsorption loading of materials over a range of temperatures and pressures; and Domingues et al. [58], who chose genetic algorithms to obtain conditions that provide excellent crystallinity and yield for the microwave-based high-throughput robotic synthesis of Al-PMO.

Pyzer-Knapp et al. used Bayesian optimization, a branch of ML, in combination with an energy structure function map (ESF) to aid in the discovery of porous crystals for methane capture [59]. The molecules T2, P2, and T2E, as seen in Figure 14.9, were chosen and screened for methane deliverable capacity, as they have been predicted to have stable crystal structures. ESF maps are very computationally expensive for this particular application due to the large energy range of predicted crystal structures, plus the effect of solvent stabilization and methane adsorption calculations. It took around 800,000 CPU hours to compute an ESF map for a single molecule (T2E) in this study.

Bayesian optimization was used to selectively acquire energy and property data to generate the same levels of insight as ESF at a fraction of the computational cost. Without Bayesian optimization, the generation of the energy structure maps are highly computationally expensive and computational cost increases with complexity, a particular drawback for porous materials where the energy range across the crystal structure is extended by solvent templating. Bayesian optimization techniques, including Thompson sampling for parallel optimization, and greedy sampling, were

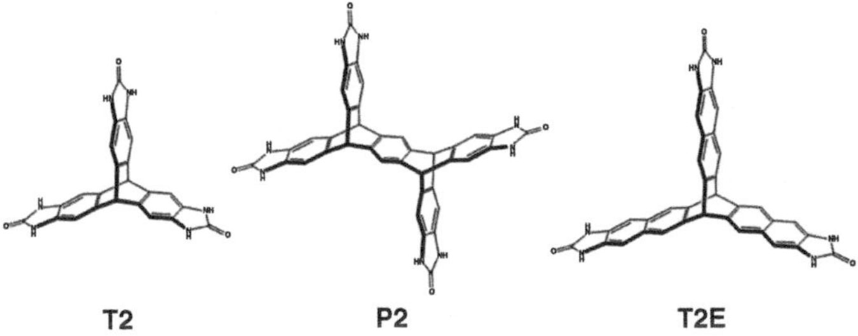

Figure 14.9 Skeletal structures of T2, P2, and T2E. These specific structures are used by Pyzer-Knapp et al. as example materials in their study to accelerate the computational discovery of porous solids through improved navigation of ESF maps. *Source:* From [59] / American Association for the Advancement of Science / Licensed under CC BY-4.0.

compared across the three systems, with a clear preference for the Bayesian approach in the T2E and T2 systems. Using this technique, an enormous 544,955 hours of computational time were saved. In cases where density functional theory (DFT) calculations would be required for lattice energy rankings, then the savings would become even greater. During this time saved, many more candidate molecules can be screened, increasing the likelihood of finding better candidates for methane uptake in the same duration, although it is important to remember that as with all computational accelerations using an ML approach, there may not be the same completeness to the study compared with using the ESF mapping approach, as some parameters may not have been calculated by ML models.

The application of ML in synthesis and digitization of materials production are relatively new and exciting fields of research. One particular area of interest is focused on automated synthesis improvement, also referred to as flow chemistry, where the design of synthesis systems is continually improved using synthesis steps extracted from literature as inputs, and an ML approach repeatedly alters the inputs based on the output data. This continual improvement approach centered around ML data allows for minimal physical input from chemists during reactions, and it is hoped that this technique can outperform the reaction modifications made by chemists from their intuition alone. This powerful new approach has the added bonus of freeing up time for chemists to continue research, instead of being required to complete laborious and repetitive experiments.

Granda et al. noticed the progress in automated chemistry, online analytics, and real-time optimization, suggesting it was possible to construct robots which can autonomously explore chemical reactivity [60]. They designed, built, and programmed an organic synthesis robot to autonomously perform reactions based on the Suzuki–Miyaura reaction, comprising of inline spectroscopy, real-time data analysis, and feedback mechanisms. In this experimental set-up, the robot was configured to perform up to 6 experiments in parallel, producing up to 36 sets of successful and failed experimental data each day for use with ML. A schematic of the feedback loop which was used in conjunction with the experimental set-up can be seen in Figure 14.10, showing the use of ML to aid in the process of continual reaction improvement. For almost all experimental techniques investigating chemical reactivity, generating data is time-consuming and cost intensive, and so employing ML to make more educated guesses at each iteration is a significant step to better discovering new pathways.

Moosavi et al. created an ML methodology focused on capturing chemical intuition from a set of partially failed MOF synthesis attempts, to find the optimal synthesis conditions for yielding the highest surface area HKUST-1 product [54]. Synthesis data was gathered using a genetic algorithm (GA), a robust global optimization algorithm for searching complex space, and the optimal conditions were synthesized, including the largest BET area HKUST-1 to date. After 120 failed and partly successful experiments which did not achieve the largest BET area, a random decision forest was used to assess the relative importance of synthesis variables to determine their impact on crystallinity and phase purity. The results found that temperature change had up to three times the impact when compared to adjusting the reactant ratio. This is considered to be "chemical intuition" and ML allows for the transfer of this knowledge into subsequent experiments.

Weighting of the 9 model parameters using the previously determined chemical knowledge shrinks the chemical space of HKUST-1, and can be transferred to a new synthesis. Conditions for synthesis of Zn-HKUST-1 were predicted across a weighed set of 20 diverse conditions, and 2 methods of synthesis for Zn-HKUST-1 that resulted in crystals were revealed. These 20 intuition-based samples would need to be replaced by an estimated 5000 random samples to maintain the same sampling accuracy. While the work was limited to a small subset of MOFs (HKUST-1), the quantification of synthesis variables can be applied to other future synthesis, particularly in the case where the chemistry is too specific for exact conditions be transferable.

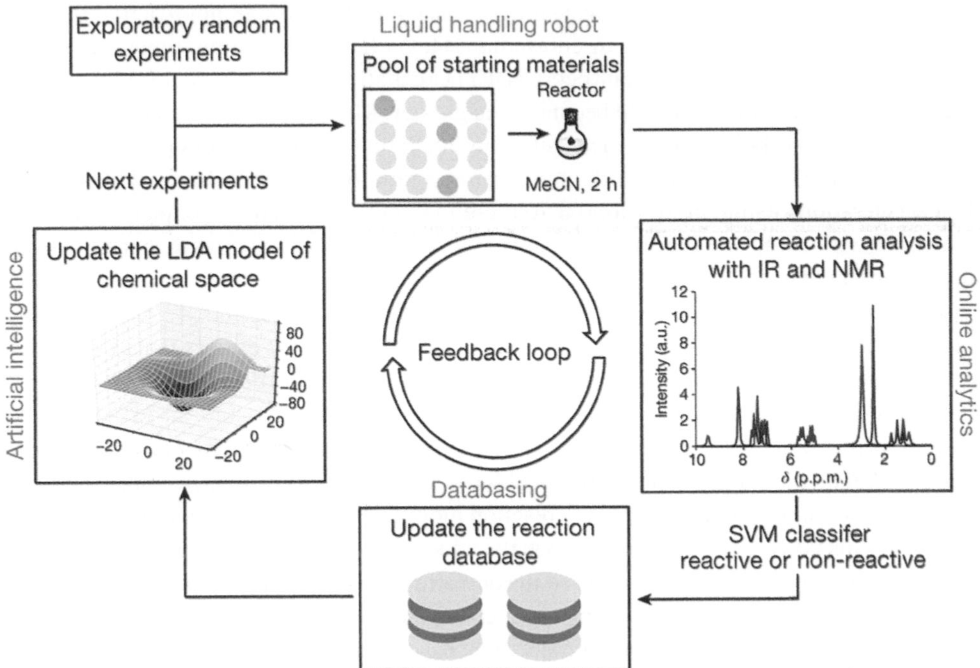

Figure 14.10 Schematic of the feedback loop for data generation for the Suzuki-Miyaura reaction. *Source:* From [60] / Springer Nature.

Perhaps sensibly, Wilbraham et al. [61] suggest that the digitization of chemistry is not simply about implementing ML or AI to process chemical data, nor is it about the development of increasingly capable automation hardware, but that it should be focused on unambiguous development of a chemical state machine that uses ontology to connect precise instruction sets with hardware performing chemical transformations. Setting a universal standard should result in an increase in collaboration, reproducibility, and safety while decreasing the labor required to make new compounds and broaden chemical space. Similar to the proposal by Mehr et al., the authors seek to create a universal programming language that is machine readable, with the ability to be exported and executed on robotic platforms, and it should facilitate the unambiguous dissemination of these procedures [53, 61]. A shift from fixed-configuration synthesis machines to a robotic platform is required to enable the processing of reactions while collecting real-time data. In fact, the most recent developments within automation have even reached beyond flow chemistry with a focus on bespoke workflows no longer exclusive to synthesis procedures. As technology in this field develops, digital chemical robot systems will require feedback from simple sensors as well real-time online analytics to navigate process space autonomously and enable efficient synthesis optimization and novel reaction discovery.

Mehr et al. [53] also investigated a system for autonomous workflow combined with NLP. To use batch synthesis for digitization of chemistry, the robot's hardware must be connected to practical synthesis by an executable hardware-independent programming language, as this allows for the execution of laboratory synthetic procedures without manual adaptation or modification. The developed system was designed to be accessible to all, with the goal that instructions should be able to be translated between chemistry and robot without loss of information. This is achieved by allowing users to directly execute procedures imported from literature on an automated synthesis platform such as the Chemputer, a potentially crucial step for large-scale digitization of solid state material production. The authors devized a new chemical programming language, Chemical Description Language (XDL), allowing users to encode procedures without ambiguity, and which represents syntheses as sequences of processes. The system includes a chemical integrated development environment (ChemIDE) which enables the importation of procedures from literature using a NLP called SynthReader, and although this is useful for mining vast literature datasets, a machine-readable representation of procedures with unambiguous details is required, including strict tagging of chemical entities, locations of reagents, implicit process details, and an environment in which the user can manually edit the output. Currently, this system is constrained by the capability of SynthReader; however, it has successfully tagged relevant text entries, converted them to a list of actions, added process information in an XDL format, and synthesized target modules upon execution of the aforementioned XDL file on an automated platform.

The Crystputer, in a similar vein of digitization to the Chemputer, is a cyber-physical system developed by Zhao et al. that has been developed to enable the digital manufacture of nanocrystals via convergence between digital and physical systems [62]. The synthesis process explored in this work is the creation of colloidal Au nanocrystals. The system combines the physical modular set-up containing pipette modules and a 6-axis robotic arm, with an ultra-sensitive camera, performing over 2300 experiments autonomously to develop an Au nanocrystals genome. The process begins with NLP, and a python-based algorithm scans literature to design the experiments based on reported parameters and conditions. These are then exported to the robotic arm and the parsed experiments are conducted on the autonomous physical system. Data is collected and used to train an ML model to pinpoint the relationships between the original synthesis conditions and the product's properties, which can then be used to aid in the retrosynthesis and scale up of targeted Au nanorods. The Crystputer is yet another development which contributes further to the advancement into automation of production, facilitating the shift of data-driven materials innovation to intelligent manufacturing.

Salley et al. developed an automated robotic platform for the synthesis of gold nanoparticles based on a Darwinian approach [63]. Genetically-inspired optimization has already been used in a range of applications, such as catalysis and in light emitting materials, although not for autonomous synthesis. Here, a genetic algorithm approach was used to mimic natural material evolution for a robotic platform in an attempt to optimize the production of gold nanoparticles over many cycles by discovering new synthesis conditions for known nanoparticle shapes. Over three independent cycles of material evolution the system produced spherical nanoparticles, rods, and octahedral nanoparticles by using optimized rods as seeds.

The system begins with an established spectral target for spherical nanoparticles. Synthesis conditions are extracted from published literature, and these spherical particles are synthesized to obtain a target for the automated system. These spheres were analyzed using in-line UV spectroscopy and the platform was given the next set of reagents (which are estimated and optimized by the genetic algorithm) to synthesize nanorods, alongside a new spectral target. These materials can be used as seeds for further cycles, and in this study the authors set their own targets for this stage rather than choosing a literature value. This automated closed loop approach has created reliable known materials without bias, and can be used to discover complex nano-constructs using desired spectroscopic responses. This methodology offers many benefits compared with a manual approach including automation, speed, safety, and reproducibility (via the use of a digital code in an automatic platform), in addition to providing researchers with a new tool to aid in the understanding of nanoparticle formation and in the development of new application areas.

Epps et al. integrated ML with flow chemistry to automate the synthesis of inorganic perovskite quantum dots (QDs) [64]. By digitizing the process, the self-driving Artificial Chemist is able to create made-to-measure inorganic QDs from 11 precision tailored QD synthesis compositions that were obtained without prior knowledge, within 30 hours, and using less than 210 mL of QD starting solution. Artificial Chemist was pre-trained to use new precursors to further accelerate the synthetic discovery of QD compositions without user selection of experiments, and further enhance the optoelectronic properties of the in-flow synthesized QDs. This fully autonomous closed loop experiment selection method expedited the tedious process of synthetic path discovery at a fraction of time and material cost when compared with user dependent experiment selection; the full process is shown in Figure 14.11.

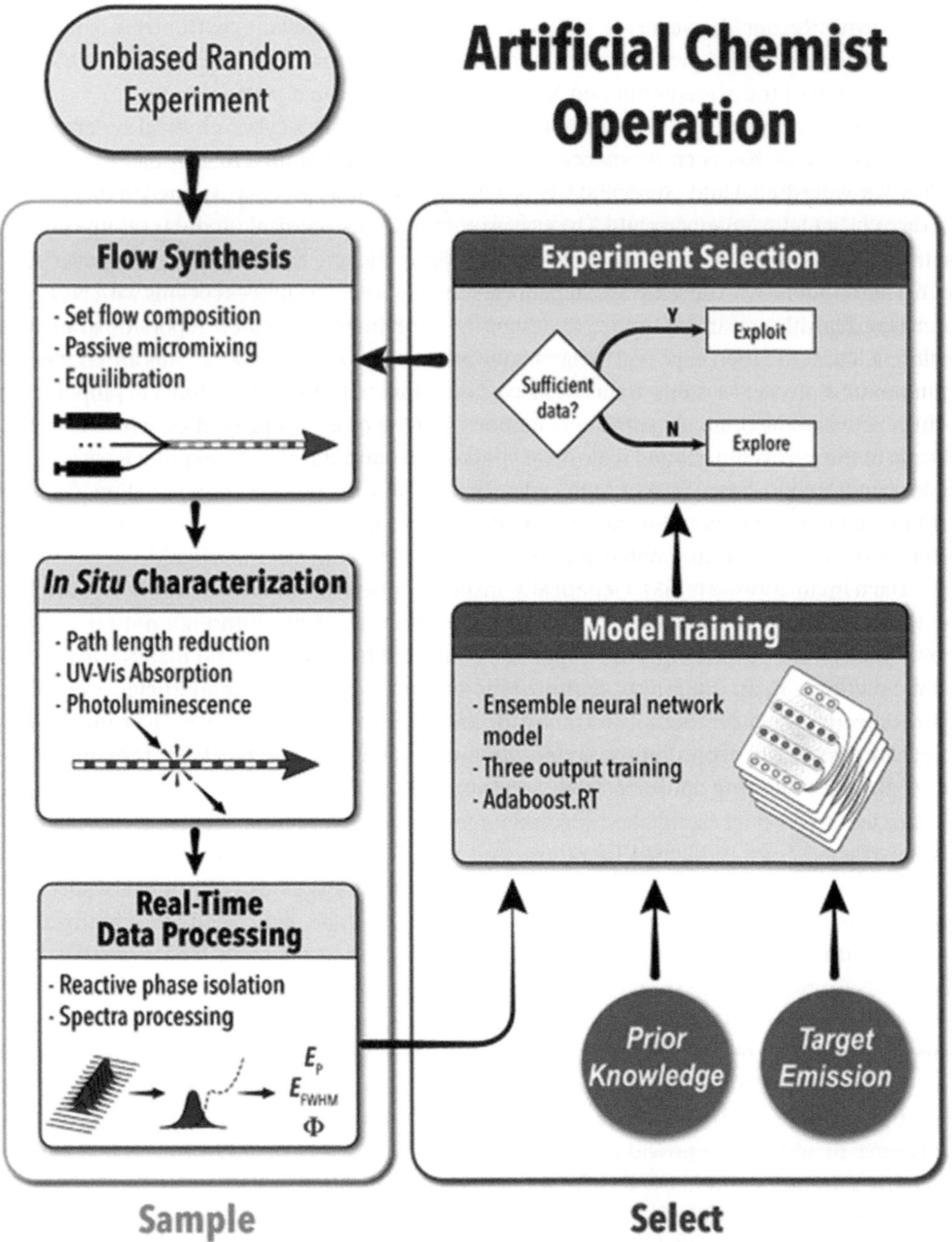

Figure 14.11 Process flow diagram for the automatic synthesis of quantum dots from initial random experimentation, using flow synthesis and real-time data processing, to new experimental selection. *Source:* From [64] / John Wiley & Sons.

The Artificial Chemist uses plug-and-play fluidic micro-reactors, which are capable of autonomous synthesis across multiple target parameters simultaneously, to explore the chemical space of colloidal QDs, learn synthesis pathways, identify composition and relevant routes, transfer knowledge to experiments, and continually synthesize rapidly optimized QDs on demand. This QD technology utilizes UV absorption and photoluminescence monitoring alongside a real-time ML-based Bayesian optimization approach. Artificial Chemist studied over 1400 reactions with 11 target values across 8 different optimization algorithms. The final technique comprises a central control system which responds to a constantly updated ensemble neural network-based Bayesian optimization algorithm with intelligent decision-making.

Abdel-Latif et al. continued the focus on digital and autonomous manufacturing, with experimentation into nanocrystal lead halide perovskite (LHP) QD synthesis [65]. They identified the optimal formulation of emerging inorganic LHP QDs, which with their vast colloidal synthesis universe and multiple synthesis/post synthesis processing parameters, was previously a challenging undertaking for material- and time-intensive batch synthesis strategies. A modular microfluidic synthesis strategy was introduced, integrated with an AI-guided decision-making agent for intelligent navigation through the complex synthesis universe of LHP QDs with 10 individually controlled synthesis parameters and an accessible parameter space exceeding 2×10^7. The developed autonomous microfluidic experimentation strategy rapidly identified the optimal formulation of LHP QDs through a two-step colloidal synthesis and post-synthesis halide exchange reaction. In this study, the use of two in-series microfluidic reactors enabled continuous band gap engineering of LHP QDs via in-line halide exchange reactions, and using an inert gas within a three-phase flow format resulted in accelerated closed-loop formulation optimization and end-to-end continuous manufacturing of LHP QDs. These QD crystals have similar applications to certain MOFs, with fields including optoelectronics and photovoltaic devices, as a result of their high photoluminescence yield. An example of the reactor design used to produce these LHP QDs in closed-loop formulation is shown in Figure 14.12.

This project improved yield by varying the starting concentration, volumetric injection ratio, halide salt concentrations, and compositions to improve peak emission energy. Current production of the crystals is achieved via a batch flask colloidal synthesis process which can lead to slow production as well as a lack of consistency within final products. Large-scale production of nanocrystals requires large reactor designs, synthesis modifications, and oversight for prevention of defects at nanoscale. By integrating AI with modular microfluidic reactors for synthesis of the crystals, including 3 precursor models, 2 in-series microfluidic reactors, and an online spectral characterization model, the system was designed to have an autonomous run time of 24 hours. To allow for the closedloop manufacturing, optical properties were monitored in real time, and fed forward to an ensemble neural network. After multiple runs were conducted, 10 optimized products were identified within a total of 250 experiments, demonstrating end-to-end manufacturing.

Figure 14.12 Schematic of the reactor design for the continuous synthesis of lead halide perovskite quantum dots. *Source:* From [65] Abdel-Latif et al. (2021), John Wiley & Sons, CC BY 4.0.

Chang et al. also understood the necessity of closed-loop systems for digitization, and they developed a method using a system called the autonomous research system (ARES) in combination with Bayesian optimization to improve the growth rate of carbon nanotubes (CNT) [66]. Two comparable Bayesian optimization models, to allow for full evaluation, were used to validate the use of ML in CNT growth rates. In the initial stage, seed experiments were manually conducted and analyzed, which consisted of a series of input and output variables: total system pressure, flow rate of ethylene, flow rate of hydrogen, total water vapor, and growth temperature. Following this stage, more meaningful variables were calculated for the algorithm including partial pressures of ethylene and partial pressures of hydrogen, followed by a final critical output of maximum growth rate. Model BO-1 was seeded using 25 experiments which were manually selected and had been confirmed to produce successful growth. In comparison, model BO-2 was randomly seeded from a random selection of 48 unbiased growth conditions. After receiving seed data, the ARES conducted the experiment and the BO-1 and BO-2 models then suggested new growth conditions. ARES executed the new experiment and updated the dataset, performing autonomous improvement in a closed-loop fashion. Overall, BO-1 and BO-2 met the goal of improved growth with both converging within 100 experiments, with a growth rate increase of up to a factor of eight.

Moving on from the conversion of synthesis processes from batch approaches to continuous manufacturing, it must be noted that modular systems allow for the easy conversion to adaptable industrial and commercial-scale production. However, scaling up a process, if ill-designed, may lead to inefficient production and a larger than desirable proportion of defective materials. Published literature contains thousands of synthesis processes for ZnO, although very few are able to be scaled up to industrial scale without high costs or a compromise in quality. These two key parameters, cost and performance, are a necessity when looking at the feasibility of manufacturing materials.

Jose et al. considered cost and performance when designing their annular micro-reactor synthesis (AMS) system for the large-scale (kilograms per day) production of ZnO [67]. The Thompson sampling efficient multi-objective (TSEMO) algorithm was used to increase the quality of the product and process using a simultaneous optimization approach with limited experimental evaluations. Mechanistic insights were determined following the characterization of post-optimization materials, which was assessed by comparing development time, safety, complexity, and scalability to known continuous and batch processes. This algorithm used the data obtained from 25 papers, which were surveyed for wet-chemical precipitation methods compatible with the AMS approach, to determine the synthesis variables. These variables were screened to reduce the number of redundant variables and establish conditions for optimization, producing a total of 26 different conditions for synthesis. Three iterations of TSEMO were performed and 20 experimental conditions were generated. TSEMO then fitted a Gaussian surrogate model for each objective and the next set of experimental conditions were computed to maximize the objective, then these conditions were repeated until the maximum number of iterations had been reached. If optimal conditions had not been reached, then the previous steps must be repeated. TSEMO required 6 experimental steps per iteration, a total of 18 experimental conditions. The molar concentration reached an optimal condition after only one iteration, which indicated that high concentration can produce high performance and yield, a fact not previously realized in the literature.

It is clear that the trend of combining real-time data analysis, ML, and AI into synthesis processes has spread through many areas of materials science. Chemists in all fields from QD manufacture to MOFs are utilizing the power of closed-loop processes and implementing systems of continual improvement combined with product feedback data.

14.9 Digital Manufacturing of MOFs

The previous studies worked to improve the real-time data collection for a variety of different materials, and this is seen to have provided a foundation for progression in the digitization of MOF production. For many chemists looking to produce MOFs autonomously, the first step is often a conversion from the batch process to a continuous one for use in flow chemistry. Batch processing may be difficult to scale up, due to problems with hazards and increased costs and further possibilities of batch-to-batch error and low reproducibility.

In a contribution focused on the scalability of MOF nanosheets, Jose et al. suggested that for MOF synthesis to be scalable, while maintaining sufficient precision, improvements to continuous reactors must be made [68]. Current techniques are not scalable nor precise enough to use at industry scale and the characterization of 2D MOF nanostructures is problematic due to post-processing methods. In this work, copper benzene dicarboxylic acid (CuBDC) nanosheets were synthesized using an annular flow micro-reactor with accelerated precipitation kinetics. Previous methods use liquid-phase, hydrothermal conditions in batch reactors requiring long reaction times and elevated temperatures, a method which is often hazardous, expensive, and imprecise above kilogram scales. Efficient synthesis is challenging due to difficulty in mixing, the fast kinetics of particle growth, and anisotropic growth. Micro-reactors can rigidly control mixing conditions and provide fast and continuous mixing, although at high saturation reactor clogging can become a problem. This study overcame the challenges of ton-scale MOF nanosheet synthesis and developed a more scalable technique for CuBDC, using trimethylamine at ambient temperature and pressure, by utilizing a continuous approach in recently developed AMS which enabled rapid mixing and uniform shear. The synthesized monodispersed CuBDC nanosheets were analyzed using XRD and infrared spectroscopy to determine particle size distributions. The final process showed an improved efficiency of up to 10^5 times the previous batch production techniques.

In another MOF synthesis contribution, Shukre et al. studied the crystallinity, yield, and precipitation of 45 sample variations of HKUST-1 with the aim to convert the synthesis method from a batch process to a continuous one [69]. The batch process was initially used to pinpoint the conditions for optimal synthesis of HKUST-1, and an optimized reaction was found which could be used in flow conditions with a millifluidic droplet-based reactor. Based on a detailed comparison of samples using both batch and flow techniques, the primary conditions identified for optimization were the residence time, temperature, and the diameter of the inner tubing. The flow process was able to continuously synthesize HKUST-1, with high-quality crystals that were comparable to the output of a traditional batch synthesis process. The novel configuration, a variation on the millifluidic droplet reactor, was able to synthesize HKUST-1 within a few minutes of residence time, and has shown there is great potential in the scale-up synthesis of this MOF alongside the opportunity to investigate the process of other MOFs in a similar reactor set-up. Figure 14.13 shows the experimental set-up of the millifluidic reactor and the additional equipment required to set up the new synthesis technique for HKUST-1.

Although the previous studies are not strictly focused on the digitization of the MOF synthesis process, the conversion from batch to flow chemistry opens the door for future studies to implement a digital approach more easily. To shift MOF synthesis toward digitization of production, availability of data is paramount. A review of MOF sustainability by Julien et al. praised the work completed by various chemists on their adaptation of ML for data prediction, particularly using the predictive power when approaching the development of sustainable manufacturing procedures that are conscious of environmental impact [70]. While MOFs may often be seen as the key to a sustainable future, particularly in the field of carbon capture, the synthesis of these

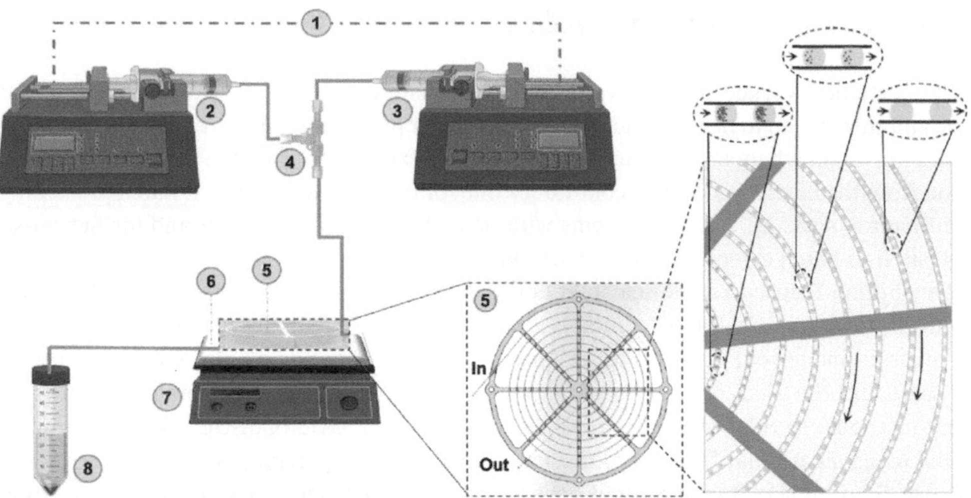

Figure 14.13 Set-up for continuous synthesis of HKUST-1 using the millifluidic reactor. The equipment as shown above: 1. Syringe pump; 2. Silicone oil in continuous phase; 3. Reactant solution in dispersed phase; 4. ETFE Tee; 5. 3D printed anchor; 6. Grooved aluminium block; 7. Hot plate; and 8. Product collection vial. *Source:* From [69] / with permission of Elsevier.

structures is not free from environmental issues. As the demand for MOFs for use in "green" applications increases, these environmental factors will become amplified as there are several stages of synthesis that have the potential to cause difficulty in scale-up for commercialization. High-energy inputs, use of water as a reaction media, unsafe building blocks, cost of raw materials, and a requirement for bulk petrochemically-derived solvents all have the potential to dismount any MOF scale-up process. The ongoing desire to commercialize MOFs requires urgent address of these challenges, introduction of ML, and the digitization of processes for non-MOF materials has already proved that a synthesis pathway can be monitored for product yield alongside developing sustainable synthesis properties, and this should be extended to MOFs.

In one of currently very few explicit MOF digitization approaches, Xie et al. combined robotic synthesis with a Bayesian optimization algorithm to accelerate the synthesis of ZIF-67 using direct laser writing apparatus, precursor injecting, and Joule-heating components [71]. The MOF synthesis reaction was automated upon the feeding of Bayesian recommended reaction parameters without prior knowledge, and the platform continually improved the crystallinity of ZIF-67 within limited iterations. Figure 14.14 shows the approach pathway of semi-automated robotic synthesis within this study. The dependencies showed that molar ratio, precursor volume, and DC voltage were much more significant factors in improving crystallinity than the duration of reaction. This study resulted in the creation of a robotic platform that enabled semi-autonomous synthesis, with lower reagent consumption and time, and involved minimal human intervention during production. The process was not completely autonomous due to manual composition measurement, although this could be overcome through the development of autonomous X-ray analysis and real-time feedback sensors throughout the platform. Other limitations of this approach included the batch-to-batch manufacturing technique, and a non-closed loop optimization which restricts the development of the robotic platform. To transfer these techniques to the future of automation requires the incorporation of a roll-to-roll approach, accompanied by *in-situ* characterization, automatic data analysis, and self-optimization.

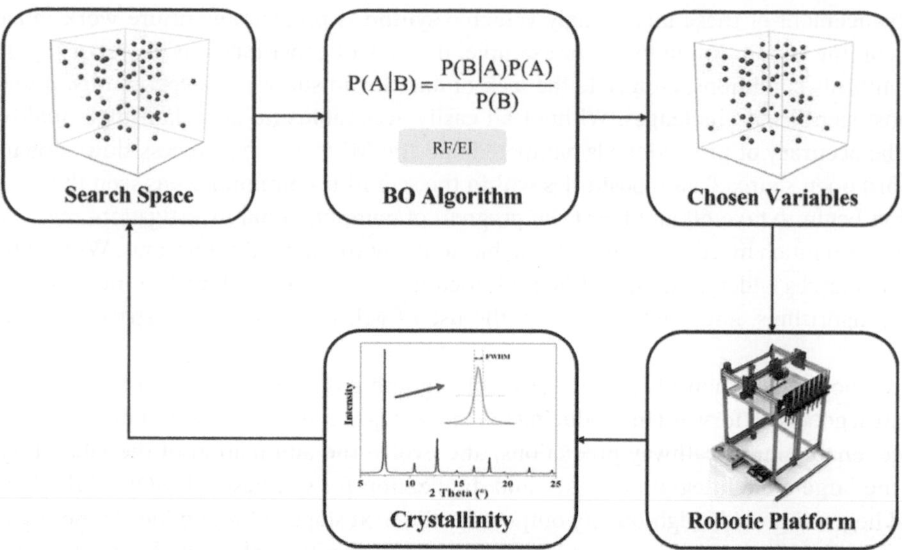

Figure 14.14 Process flow diagram for the automated synthesis of ZIF-67, showing the use of Bayesian optimization (BO) to make continual variation to the chosen variables after the output analysis is performed. *Source:* From [71] Xie et al. (2021), Reproduced with permission from American Chemical Society.

In another MOF-based study, a variation of ZIF-8, after mineralization with poly(ethylene glycol) (PEG) in the presence of bio-macromolecules, was synthesized through an autonomous computer controlled system by Wu et al. [72]. An automatic synthesis system prepared PEG mineralized ZIF-8 composite particles based on flow chemistry with microfluidic chips produced using femtosecond laser micro-machining. Ideally designed for use by non-specialists, who will be able to obtain target ZIF composites by selecting an input parameter, the process is able to monitor and regulate itself. After inputting the target size, the system calculates required concentration of reagents, and pumps are instructed to mix the set reactants. The synthesis reaction is monitored automatically by an *in-situ* UV-visual spectrometer to check the size of synthesized ZIF-8 particles and when the calculated results are equal to the experimental results, the reaction is ended and the crystals can be harvested as final product. The authors also assessed the possibility of remote control of the system, with successful control from a building 20 km away. Remote control of manufacturing is an advantage of production digitization, opening up the possibility of manufacturing server rooms that would allow small, lower budget research groups without access to expensive equipment to take advantage of digital manufacturing in their research.

14.10 The Future of Digital Manufacturing

Although ML is now a well-established field in most areas of data science and computing, with many commercial uses that people around the world benefit from every day, its use within the realm of material discovery and chemicals manufacturing is relatively new. Integration of ML within the solid state nanomaterials field has seen some significant progress, particularly since the beginning of the decade, with multiple articles in high impact publications reporting the successful prediction of novel, sustainable, or economically improved synthesis conditions. To see continued advancement of these increasingly valuable synthesis predictions, future work must focus on overcoming certain limitations. For example, the lack of consistency in the reporting of results and synthesis conditions, alongside the lack of data for unsuccessful experiments, poses one of the most significant challenges. Without an easily accessible and abundant high-quality data source, the accuracy of predictions is hampered and the ML discovery process slows down. The shift toward open source data repositories within the scientific community has seen the idea of collaboration begin to take off and boost the progress of computational investigation, making this area of research much more accessible, affordable, and environmentally conscious. With websites, GitHub resources, video guides, and helpful documentation all developed to increase the access to data, algorithms, and prediction tools, the use of ML within chemical space is set to accelerate.

Additionally, the introduction of knowledge graphs, graph-based reaction optimization, and digital twins are a good step forward in producing the knowledge required to discover new or missing knowledge, enable rapid pathway predictions, and evolve the automation of the laboratory [73–75]. For the large-scale integration of ML and digitization for synthesis of MOFs, ML algorithms should begin to provide high-quality outputs for the next stage of production. These high-quality and reliable algorithm outputs are key to provide sufficient data for continuous manufacturing and commercialization of otherwise inaccessible materials, offering essential information to maximize the potential of the input resources, synthesis process, and final product. The predicted synthesis conditions will allow manufacturers to compare the resources required for each structure from input to output, helping to ease the monetary and environmental cost of production. Automated analysis of structure properties, crystallinity, cost, and environmental effects will play a key role in the sustainable manufacturing of these highly functional materials.

It should be noted that as novel manufacturing techniques are explored with an increased use of digital processes, not all MOFs will find suitable pathways for large-scale synthesis. Due to the complexity of the nanocrystal structures of a subsection of MOFs, large-scale synthesis is unfeasible and unnecessary. Additionally, the rarity of certain transition metals on Earth limits the scale at which certain materials could be produced. These MOFs experience significant costs for materials alone, even at kilogramme levels, with certain physical constraints limiting the potential of more complex combinations. Recent world events have highlighted many problems regarding the adaptability of production processes. When facing fluctuations in demand and unreliable supply chains, it is imperative to assess the vulnerability of the manufacturing pathway such that essential materials are not restricted so that when digitizing the synthesis processes it is possible to increase resilience and efficiency. However, the full potential of digitization cannot be reached until sufficient, high-quality data is accessible, although large-scale global events, such as COVID-19 and the subsequent disruption to global supply lines, cannot be predicted.

References

1 Qiu, S. and Zhu, G. (2009). Molecular engineering for synthesizing novel structures of metal–organic frameworks with multifunctional properties. *Coord. Chem. Rev.* 253 (23): 2891–2911. doi: 10.1016/j.ccr.2009.07.020.

2 Stock, N. and Biswas, S. (2012). Synthesis of metal–organic frameworks (MOFs): routes to various MOF topologies, morphologies, and composites. *Chem. Rev.* 112 (2): 933–969. doi: 10.1021/cr200304e.

3 Tahmasian, A. and Morsali, A. (2012). Ultrasonic synthesis of a 3D Ni(II) metal–organic framework at ambient temperature and pressure: new precursor for synthesis of nickel(II) oxide nano-particles. *Inorg. Chim. Acta* 387: 327–331. doi: 10.1016/j.ica.2012.02.017.

4 Cohen, S.M. (2012). Postsynthetic methods for the functionalization of metal–organic frameworks. *Chem. Rev.* 112 (2): 970–1000. doi: 10.1021/cr200179u.

5 Miller, S.E., Teplensky, M.H., Moghadam, P.Z. et al. (2016). Metal–organic frameworks as biosensors for luminescence-based detection and imaging. *Interf. Foc.* 6 (4): 20160027. doi: 10.1098/rsfs.2016.0027.

6 Konnerth, H., Matsagar, B.M., Chen, S.S. et al. (2020). Metal–organic framework (MOF)-derived catalysts for fine chemical production. *Coord. Chem. Rev.* 416: 213319. doi: 10.1016/j.ccr.2020.213319.

7 Chen, Z., Wasson, M.C., Drout, R.J. et al. (2021). The state of the field: from inception to commercialization of metal–organic frameworks. *Faraday Discuss.* 225 (0): 9–69. doi: 10.1039/D0FD00103A.

8 Moghadam, P.Z., Li, S., Wiggin, S. et al. (2017). Development of a Cambridge Structural Database subset: a collection of metal–organic frameworks for past, present, and future. *Chem. Mater.* 29 (7): 2618–2625. doi: 10.1021/acs.chemmater.7b00441.

9 Chung, Y.G., Camp, J., Haranczyjk, M. et al. (2014). Computation-ready, experimental (CoRE) metal–organic frameworks: a tool to enable high-throughput screening of nanoporous crystals. *Chem. Mater.* 1 (69): 5–24.

10 Gascon, J., Hernández-Alonso, M.D., Almeida, A.R. et al. (2008). Isoreticular MOFs as efficient photocatalysts with tunable band gap: an operando FTIR study of the photoinduced oxidation of propylene. *ChemSusChem* 1 (12): 981–983. doi: https://doi.org/10.1002/cssc.200800203.

11 Lopez, N., Zhao, H., Ota, A. et al. (2010). Unprecedented binary semiconductors based on TCNQ: single-crystal X-ray studies and physical properties of Cu($TCNQX_2$) X=Cl, Br. *Adv. Mater.* 22 (9): 986–989. doi: https://doi.org/10.1002/adma.200903217.

12 Coudert, F.-X. (2020). Water adsorption in soft and heterogeneous nanopores. *Acc. Chem. Res.* 53 (7): 1342–1350. doi: 10.1021/acs.accounts.0c00215.

13 Islamoglu, T., Chen, Z., Wasson, M.C. et al. (2020). Metal–organic frameworks against toxic chemicals. *Chem. Rev.* 120 (16): 8130–8160. doi: 10.1021/acs.chemrev.9b00828.

14 Lawson, H.D., Walton, S.P., and Chan, C. (2021). Metal–organic frameworks for drug delivery: a design perspective. *ACS Appl. Mater. Interf.* 13 (6): 7004–7020. doi: 10.1021/acsami.1c01089.

15 Zanca, F., Glasby, L.T., Chong, S. et al. (2021). Computational techniques for characterisation of electrically conductive MOFs: quantum calculations and machine learning approaches. *J. Mater. Chem. C* 9 (39): 13584–13599. https://doi.org/10.1039/D1TC02543K (accessed 12 September 2022).

16 Landrum, G., Tosco, P., Kelley, B. et al. (2022). rdkit/rdkit: 2022_03_4 (Q1 2022) release. Zenodo, 5 July 2022. doi: 10.5281/zenodo.6798971.

17 Moosavi, S.M., Nandy, A., Jablonka, K.M. et al. (2020). Understanding the diversity of the metal–organic framework ecosystem. *Nat. Commun.* 11 (1): Art. no. 1. https://doi.org/10.1038/s41467-020-17755-8 (accessed 12 September 2022).

18 Luo, Y., Bag, S., Zaremba, O. et al. (2022). MOF synthesis prediction enabled by automatic data mining and machine learning**. *Angewandte Chemie, International Edition* 61 (19): e202200242. doi: 10.1002/anie.202200242.

19 Zhang, X., Zhang, K., and Lee, Y. (2020). Machine learning enabled tailor-made design of application-specific metal–organic frameworks. *ACS Appl. Mater. Interfaces* 12 (1): 734–743. https://doi.oth/10.1021/acsami.9b17867 (accessed 12 September 2022).

20 Chung, Y.G., Haldoupis, E., Bucior, B.J. et al. (2019). Advances, updates, and analytics for the computation-ready, experimental metal–organic framework database: CoRE MOF 2019. *J. Chem. Eng. Data* 64 (12): 5985–5998. doi: 10.1021/acs.jced.9b00835.

21 Gubsch, K., Bence, R., Glasby, L.T. et al. (2022). DigiMOF. A database of MOF synthesis information generated via text mining. *Chem R.* xiv: 1–26. https://doi.org/10.26434/chemrxiv-2022-41t70.

22 Nazarian, D., Camp, J.S., and Sholl, D.S. (2016). A comprehensive set of high-quality point charges for simulations of metal–organic frameworks. *Chem. Mater.* 28 (3): 785–793. doi: 10.1021/acs.chemmater.5b03836.

23 Rosen, A.S., Iyer, S.M., Ray, D. et al. (2021). Machine learning the quantum-chemical properties of metal–organic frameworks for accelerated materials discovery. *Matter* 4 (5): 1578–1597. doi: 10.1016/j.matt.2021.02.015.

24 Nandy, A., Terrones, G., Arunachalam, N. et al. (2022). MOFSimplify, machine learning models with extracted stability data of three thousand metal–organic frameworks. *Sci. Data* 9 (1): 1–11. doi: 10.1038/s41597-022-01181-0.

25 Wilmer, C.E, Leaf, M., Lee, C.Y. et al. (2012). Large-scale screening of hypothetical metal–organic frameworks. *Nat. Chem.* 4 (2): Art. no. 2. doi: 10.1038/nchem.1192.

26 Colón, Y.J., Gómez-Gualdrón, D.A., and Snurr, R.Q. (2017). Topologically guided, automated construction of metal–organic frameworks and their evaluation for energy-related applications. *Cryst. Growth Des.* 17 (11): 5801–5810. doi: 10.1021/acs.cgd.7b00848.

27 Majumdar, S., Moosavi, S.M., Jablonka, K.M. et al. (2021). Diversifying databases of metal–organic frameworks for high-throughput computational screening. *ACS Appl. Mater. Interf.* 13 (51): 61004–61014. doi: 10.1021/acsami.1c16220.

28 Yao, Z., Sanchez-Lengeling, N.S., Bobbitt, N.S. et al. (2021). Inverse design of nanoporous crystalline reticular materials with deep generative models. *Nat. Mach. Intell.* 3 (1): 76–86. https://doi.org/10.1038/s42256-020-00271-1.

29 Xie, Y., Zhang, C., Hu, X. et al. (2020). Machine learning assisted synthesis of metal–organic nanocapsules. *J. Am. Chem. Soc.* 142 (3): 1475–1481. doi: 10.1021/jacs.9b11569.

30 Jablonka, K.M., Patiny, L., and Smit, B. (2022). Making the collective knowledge of chemistry open and machine actionable. *Nat. Chem.* 14 (4): Art. no. 4. doi: 10.1038/s41557-022-00910-7.

31 Kim, E., Huang, K., Saunders, A. et al. (2017). Materials synthesis insights from scientific literature via text extraction and machine learning. *Chem. Mater.* 29 (21): 9436–9444. doi: 10.1021/acs.chemmater.7b03500.

32 Swain, M.C. and Cole, J.M. (2016). ChemDataExtractor: a toolkit for automated extraction of chemical information from the scientific literature. *J. Chem. Inf. Model.* 56 (10): 1894–1904. doi: 10.1021/acs.jcim.6b00207.

33 Kim, S., Cheng, T., et al. (2021). PubChem in 2021: new data content and improved web interfaces. *Nucl. Acids Res.* 49 (D1): D1388–D1395. doi: 10.1093/nar/gkaa971.

34 Park, H., Kang, Y., Choe, W. et al. (2022). Mining insights on metal–organic framework synthesis from scientific literature texts. *J. Chem. Inform. Model.* 62 (5): 1190–1198. doi: 10.1021/acs.jcim.1c01297.

35 Pedregosa, F., Varoquaux, A., Gramfort, A. et al. (2011). Scikit-learn: machine learning in Python. *J. Mach. Learn. Res.* 12 (85): 2825–2830.

36 Abadi, M., Agarwal, A., Barham, P. et al. (2016). TensorFlow: large-scale machine learning on heterogeneous distributed systems. In: *Computer Science: Distributed, Parallel, and Cluster Computing*. arXiv, 16 March 2016. https://doi.org/10.48550/arXiv.1603.04467.

37 Conroy, B., Nayak, R., Hidalgo, A.L.R. et al. (2022). Evaluation and application of machine learning principles to Zeolite LTA synthesis. *Micropor. Mesopor. Mater.* 335 (February): 111802. doi: 10.1016/j.micromeso.2022.111802.

38 Molnar, C. (2022). Interpretable Machine Learning: A Guide For Making Black Box Models Explainable. [Online]. Available: christophm.github.io/interpretable-ml-book.

39 Jensen, Z., Kim, E., Kwon, S. et al. (2019). A machine learning approach to Zeolite synthesis enabled by automatic literature data extraction. *ACS Cent. Sci.* 5 (5): 892–899. doi: 10.1021/acscentsci.9b00193.

40 Raccuglia, P., Elbert, K.C., Adler, P.D.F. et al. (2016). Machine-learning-assisted materials discovery using failed experiments. *Nature* 533 (7601): 73–76. doi: 10.1038/nature17439.

41 Sagi, O. and Rokach, L. (2021). Approximating XGBoost with an interpretable decision tree. *Inform. Sci.* 572: 522–542. doi: 10.1016/j.ins.2021.05.055.

42 Mockus, J., Tiesis, V., and Zilinskas, A. (1978). The application of Bayesian methods for seeking the extremum. *Towards Glob. Optim.* 2 (2): 117–129.

43 Snoek, J., Larochelle, H., and Adams, R.P. (2012). Practical Bayesian optimization of machine learning algorithms. *Stat. Mach. Learning*. arXiv: 29 August 2012. https//doi.org/10.48550/arXiv.1206.2944.

44 Nogueira, F. (2014). Bayesian optimization: open source constrained global optimization tool for Python. https://github.com/fmfn/BayesianOptimization.

45 Kennicutt, A.R., Morkowchuk, L., Krein, M. et al. (2016). A quantitative structure–activity relationship to predict efficacy of granular activated carbon adsorption to control emerging contaminants. *SAR QSAR Environ. Res.* 27 (8): 653–676. doi: 10.1080/1062936X.2016.1216465.

46 Dico, G.L., Nuñez, Á.P., Carcelén, V. et al. (2021). Machine-learning-accelerated multimodal characterization and multiobjective design optimization of natural porous materials. *Chem. Sci.* 12 (27): 9309–9317. doi: 10.1039/D1SC00816A.

47 Muraoka, K., Sada, Y., Miyazaki, D. et al. (2019). Linking synthesis and structure descriptors from a large collection of synthetic records of zeolite materials. *Nat. Commun.* 10 (1): 1–11. doi: 10.1038/s41467-019-12394-0.

48 Hardian, R., Liang, Z., Zhang, X. et al. (2020). Artifical Intelligence: the silver bullet for sustaimable materials development. *Green Chem.* 22: 7521–7528.

49 Kitamura, Y., Terado, E., Zhang, Z. et al. (2021). Failure-experiment-supported optimization of poorly reproducible synthetic conditions for novel lanthanide metal–organic frameworks with two-dimensional secondary building units. *Chem. Eur. J.* 27 (66): 16274–16274. doi: 10.1002/chem.202104014.

50 Huelsenbeck, L., Jung, S., Valle, R.H.D. et al. (2021). Accelerated HKUST-1 thin-film property optimization using active learning. *ACS Appl. Mater. Interfaces* 13 (51): 61827–61837. doi: 10.1021/acsami.1c20788.

51 Jang, J., Gu, G.H., Noh, J. et al. (2020). Structure-based synthesizability prediction of crystals using partially supervised learning. *J. Am. Chem. Soc.* 142 (44): 18836–18843. doi: 10.1021/jacs.0c07384.

52 Kingma, D.P. and Ba, J. (2017). Adam: a method for stochastic optimization. In: *Computer Science, Machine Learning*. arXiv, 29 January 2017. https://doi.org/10.48550/arXiv.1412.6980.

53 Mehr, S.H.M., Craven, M., Leonov, A.I. et al. (2020). A universal system for digitization and automatic execution of the chemical synthesis literature. *Science* 370 (6512): 101–108. doi: 10.1126/science.abc2986.

54 Moosavi, S.M., Chidambaram, A., Talirz, L. et al. (2019). Capturing chemical intuition in synthesis of metal–organic frameworks. *Nat. Commun.* 10 (1): 539. doi: 10.1038/s41467-019-08483-9.

55 King, R.D., Rowland, J., Oliver, S.G. et al. (2009). The automation of science. *Science* 324 (5923): 85–89. doi: 10.1126/science.1165620.

56 Burger, B., Maffettone, P.M., Gusev, V.V. et al. (2020). A mobile robotic chemist. *Nature* 583 (7815): Art. no. 7815. doi: 10.1038/s41586-020-2442-2.

57 Sun, Y., Dejaco, R.F., Li, Z. et al. (2021). Fingerprinting diverse nanoporous materials for optimal hydrogen storage conditions using meta-learning. *Sci. Adv.* 7 (30): eabg3983. doi: 10.1126/sciadv.abg3983.

58 Domingues, N.P., Moosavi, S.M., Talirz, L. et al. (2022). Using genetic algorithms to systematically improve the synthesis conditions of Al-PMOF. June 2022. *Chem R.* xiv. https://doi.org/10.26434/chemrxiv-2022-cgnf5.

59 Pyzer-Knapp, E.O., Chen, L. et al. (2021). Accelerating computational discovery of porous solids through improved navigation of energy-structure-function maps. *Sci. Adv.* 7 (33): e4763. doi: 10.1126/sciadv.abi4763.

60 Granda, J.M., Donina, L., Dragone, V. et al. (2018). Controlling an organic synthesis robot with machine learning to search for new reactivity. *Nature* 559 (7714): 377–381. doi: 10.1038/s41586-018-0307-8.

61 Wilbraham, L., Mehr, S.H.M., and Cronin, L. (2020). Digitizing chemistry using the chemical processing unit: from synthesis to discovery. *Acc. Chem. Res.* 54 (2): 253–262. doi: 10.1021/acs.accounts.0c00674.

62 Zhao, H., Chen, W., Wang, Z. et al. (2021). *Cyber-Physical System Enabled Digital Manufacturing of Nanocrystals: A Crystputer.* Rochester, NY. 3 August 2021. doi: 10.2139/ssrn.3898779.

63 Salley, D., Keenan, G., Grizou, J. et al. (2020). A nanomaterials discovery robot for the Darwinian evolution of shape programmable gold nanoparticles. *Nat. Commun.* 11 (1): Art. no. 1. doi: 10.1038/s41467-020-16501-4.

64 Epps, R.W., Bowen, M.S., Volk, A.A. et al. (2020). Artificial chemist: an autonomous quantum dot synthesis bot. *Adv. Mater.* 32 (30): 2001626. doi: 10.1002/adma.202001626.

65 Abdel-Latif, K., Epps, R.W., Bateni, F. et al. (2021). Self-driven multistep quantum dot synthesis enabled by autonomous robotic experimentation in flow. *Adv. Intell. Syst.* 3 (2): 2000245. doi: 10.1002/aisy.202000245.

66 Chang, J., Nikolaev, P., Carpena-Nunesz, J. et al. (2020). Efficient closed-loop maximization of carbon nanotube growth rate using Bayesian optimization. *Sci. Rep.* 10 (1): 1–9. doi: 10.1038/s41598-020-64397-3.

67 Jose, N.A., Kovalev, M., Bradford, E. et al. (2021). Pushing nanomaterials up to the kilogram scale: an accelerated approach for synthesizing antimicrobial ZnO with high shear reactors, machine learning and high-throughput analysis. *Chem. Eng. J.* 426 (July): 131345. doi: 10.1016/j.cej.2021.131345.

68 Jose, N.A., Zeng, H.C., and Lapkin, A.A. (2020). Scalable and precise synthesis of two-dimensional metal–organic framework nanosheets in a high shear annular microreactor. *Chem. Eng. J.* 388 (January): 124133. doi: 10.1016/j.cej.2020.124133.

69 Shukre, R., Ericson, T.E., Unruh, D.K. et al. (2022). Batch-screening guided continuous flow synthesis of the metal–organic framework HKUST-1 in a millifluidic droplet reactor. *Micropor. Mesopor. Mater.* 339: 112005. doi: 10.1016/j.micromeso.2022.112005.

70 Julien, P.A., Mottillo, C., and Friščić, T. (2017). Metal–organic frameworks meet scalable and sustainable synthesis. *Green Chem.* 19 (12): 2729–2747. doi: 10.1039/c7gc01078h.

71 Xie, Y., Zhang, C., Deng, H. et al. (2021). Accelerate synthesis of metal–organic frameworks by a robotic platform and Bayesian optimization. *ACS Appl. Mater. Interfaces* 13 (45): 53485–53491. doi: 10.1021/acsami.1c16506.

72 Wu, M., Xia, L., Li, Y. et al. (2022). Automated and remote synthesis of poly(ethylene glycol)-mineralized ZIF-8 composite particles via a synthesizer assisted by femtosecond laser micromachining. *Chinese Chem. Lett.* 33 (1): 497–500. doi: 10.1016/j.cclet.2021.07.004.

73 McDermott, M.J., Dwaraknath, S.S., and Persson, K.A. (2021). A graph-based network for predicting chemical reaction pathways in solid-state materials synthesis. *Nat. Commun.* 12 (1): Art. no. 1. doi: 10.1038/s41467-021-23339-x.

74 Bai, J., Cao, L., Mosbach, S., Akroyd, J. et al. (2022). From platform to knowledge graph: evolution of laboratory automation. *JACS Au.* 2 (2): 292–309. doi: 10.1021/jacsau.1c00438.

75 An, Y., Greenberg, J., Zhao, X. et al. (2022). Building open knowledge graph for metal–organic frameworks (MOF-KG): challenges and case studies. In: *Computer Science and Artificial Intelligence*. arXiv, 10 July (2022). (accessed 9 August 2022). [Online]. Available https://doi.org/10.48550/arXiv.2207.04502.

15

Overview of AI in the Understanding and Design of Nanoporous Materials

Seyed Mohamad Moosavi[1], *Frits Daeyaert*[2], *Michael W. Deem*[3], *and German Sastre*[4]

[1] *Department of Mathematics and Computer Science/Mathematics, Artificial Intelligence for the Sciences, Freie Universität Berlin, Berlin, Germany*
[2] *Synopsisdenovodesign, Beerse, Belgium*
[3] *Certus LLC, Houston, Texas, USA*
[4] *Instituto de Tecnología Química UPV-CSIC, Universidad Politécnica de Valencia, Valencia, Spain*

15.1 Introduction

What would have happened if Fritz Haber and Carl Bosch had approached the optimization of their catalyst [1] for ammonia synthesis using machine learning? Certainly history cannot be rewritten, but we thought this could be a case that illustrates one of the ways artificial intelligence (AI) could substitute systematic work using parametric methods. Industrial application is a realm in which AI should find almost unanimous appraisal, as "reaching a solution" is more important than "understanding the problem." However, others may argue that AI can also guide us in learning new chemistry and rationalizing trends in vast amounts of data. This does in fact play a central role in our view and in the design of many future applications of AI [2].

Nowadays, materials research is routinely generating (big-)data, be it experimental or computational [3]. AI and machine learning algorithms can extract patterns, group similarities, and make predictions from these data. This is the so-called fourth paradigm of science [4]. Following thousands of years of empirical observations of natural phenomena, theoretical science was born with Galileo Galilei. Recently the advent of computers allowed calculation of models in a much faster way and even to design robots for high-throughput experimentation, resulting in a third paradigm. It is of common opinion that rationalization stands as the highest conquest of mankind in terms of knowledge, and thus the other three paradigms must always be subordinate to the second. Remarkably, machine learning can enable tasks that otherwise would remain infeasible; in particular, when the data and interactions within the data are too complex and intractable for human understanding and conceptualization [5].

Machine learning has been successfully applied in materials science to make predictions of steel fatigue strength, physical and mechanical properties of alloys, electronic band gaps of perovskite materials, and catalytic activities. ML has lead to the identification of promising porous materials, polymer dielectrics, mixed oxide catalysts, organic light-emitting diodes (OLED), superconductors, and photovoltaic materials [6]. This book has provided an overview of this burgeoning field applied to nanoporous materials.

15.2 Databases

A foundation for most machine learning studies is identification of the right dataset, from which solutions and patterns are to be inferred. Large amounts of data are accessible from databases of microporous materials and their properties. Despite considerable efforts and progress,

AI-Guided Design and Property Prediction for Zeolites and Nanoporous Materials, First Edition. Edited by German Sastre and Frits Daeyaert.

development and maintenance of a sustainable data ecosystem requires extra attention from the community and will be crucial for the rapid growth of the field. In particular, development of standards and protocols for reporting chemical data, collecting and curating the existing data, and ensuring reproducibility are vital. Communication and collaboration between computational and experimental researchers for data reporting leads to development of comprehensive databases, allowing for the design and discovery of nanoporous materials with enhanced performance for various applications.

15.2.1 Structural Databases

Experimental structures of MOF materials are reported in the Cambridge Structural Database (CSD) [7]. We can access these structures using the ConQuest software, which now filters over 100,000 structures from the CSD. However, these structures are not ready to be used for computational atomistic modeling as they contain solvents, partial occupancies, and other extra atoms. Therefore, Chung et al. have developed a protocol for cleaning these structures, which has led to the development of the Computational Ready, Experimental (CoRE-)MOF database [8]. Similarly, for COFs, the CURATED COF database has been developed [9]. Moreover, the collection of experimentally realized zeolites is available from the IZA database [10]. In addition to these experimental structures, multiple algorithms have been developed to computationally predict MOF structures starting from structural building blocks (SBUs) and the underlying topology networks [11], as well as generating predicted zeolite structures using Monte Carlo algorithms [12] and combinatorial methods [13].

Despite these significant developments, there are still major outstanding issues with these databases that need to be resolved. In particular, an important indicator of the value of a database is its diversity. It has been shown that lack of diversity leads to poor transferability of machine learning models, incorrect chemical insights and conclusions, and even hinders discovery. Furthermore, multiple studies have indicated issues with structures in the MOF databases, including duplicates, wrong charge state, and atomic overlap. Therefore, further manual curation of structures and automation of these procedures will be necessary in the future.

15.2.2 Databases of Material Properties

Function-led design and discovery of materials requires linking structures to their properties. While we have structural databases, there has been very little systematic effort for gathering data and developing databases of materials properties. A remarkable exception is the QMOF database [14], which provides access to the band gaps of over 15,000 MOFs computed using density functional theory (DFT). This database is a good starting point for exploration of MOFs for applications in electronics and catalysis. The Deem database of predicted zeolites contains energy, density, and ring distributions, as well as the static and high-frequency dielectric constant tensors [15].

Additionally, for gas separation and storage, which has been the main focus in this book, the experimental isotherms reported in the literature are gathered in the NIST/ARPA-E Database of Novel and Emerging Adsorbent Materials (NIST-ISODB) [16]. However, a recent study has shown [17] that it is hard to link the structures in the CSD to their corresponding isotherms in the NIST database. As we have the tools for systematic development of databases, especially in the case of computed properties, developing a large database of adsorption properties will help the field to grow faster in the future.

15.2.3 Databases of Synthesis Protocols

Autonomous synthesis and self-driving labs are major components of the revolution of matter discovery using machine learning and artificial intelligence. To enable these technologies, databases of synthesis protocols and algorithms for design of experiments are under development [18, 19] and are considered in Chapter 14 [20]. These topics have received attention only recently in the field of porous materials. As an early attempt, databases of synthesis protocols for MOFs have been developed by natural language processing of reported data in the literature. The machine learning models trained on these data can predict the synthesis conditions for new MOFs. However, it has been shown that reporting failed experiments is equally important as reporting the successful ones [21, 22]. Just as an expert chemist develops an intuition for how to synthesize new materials based on a series of failed and successful experiments, so too machine learning can be used to capture chemical intuition if it is exposed to such data. Therefore, by integrating electronic lab notebooks (ELNs) and FAIRification of chemical data management, chemical synthesis will be ready for explosive growth in the future [23, 24].

15.3 Big-Data Science for Nanoporous Materials Design and Discovery

That we have access to so many materials, which are potentially promising for a wide range and ever-growing number of applications, presents a challenge of how to identify the best performing materials for a given application. Clearly, conventional approaches fall short when it comes to exploring these vast chemical spaces. In this book, we aimed to cover the development of novel big-data science methods, which are helping us to approach this design and discovery challenge for porous materials in new ways. The authors have discussed how artificial intelligence and machine learning can be used for predicting materials properties, designing experiments, learning from failed experiments, or developing self-driving labs.

From a machine learning point of view, these tasks can be divided into three main categories of supervised, unsupervised, and reinforcement learning. In supervised learning, we aim to map a feature set to predefined labels, for instance, predicting the CO_2 uptake of a MOF from its pore geometry features. On the other hand, unsupervised learning deals with datasets without labels.

Examples of unsupervised learning include clustering the databases or understanding their diversity. The third category of tasks is those that involve sequential, on-the-fly decision-making to optimize a final goal. For example, reinforcement learning can be used to plan a sequence of modification of a molecule to maximize its reactivity. Here, we briefly overview the components and methodologies for material design and discovery.

15.3.1 Representations of Chemical Data

Converting chemical structures to numbers is the first step of a machine learning approach. We might even argue that the choice of chemical representation is the most important factor for the success of the machine learning models in achieving their corresponding tasks. A meaningful representation can be achieved explicitly from physics or chemistry motivated descriptors, or it can be learned through a deep learning framework. In the case of porous materials, encoding both the chemistry and the pore geometry is crucial. As a result, significant efforts have been devoted to develop descriptors for both aspects of the materials.

Characterizing porosity and quantifying the geometric similarities of two porous structures are among the most important tasks to enable machine learning studies of nanoporous materials. Simple, yet intuitive, geometric features, such as largest cavity diameter or pore volume, can be used to describe the pore geometry. These descriptors can efficiently be computed using several available codes, e.g. Zeo^{++} [25] or PoreBlazer [26]. Despite their simplicity, in many cases these simple descriptors have shown to be effective, for example, in predicting gas adsorption properties. Additionally, more sophisticated approaches have been explored for this aim. For example, persistent homology, a tool from topological data analysis, has been used to quantify pore shape similarities of zeolites [27, 28], and shown to be effective for both supervised and unsupervised tasks of studying porous molecular materials [29]. Beside these hand-crafted descriptors, we can use neural networks to "learn" geometric features. For example, a convolutional neural network on three-dimensional (3D) grids of atomic information was used to predict gas adsorption properties successfully [30].

Encoding material chemistry can be as important as encoding the pore geometry for many applications. Several methods and descriptors have been proposed to represent the chemistry of porous materials. This includes simple heuristics, such as the statistics of elemental composition (e.g., Stoichiometry 120) or nearest neighbors, methods to represent building blocks and fragments of a MOF (e.g., RACs), and the full representation of relations using a graph representation of crystals. An overview of these methods can be found in these extensive reviews [31, 32]. While the later methods, including crystal graph-based representation, are more powerful and expressive, eventually the data size and application determine which representation suits the machine learning model. Applying representation learning to perform the task of finding the appropriate input features is a very powerful machinery (Chapter 7 [33]). However, representation learning often requires lots of data. In the case of small datasets, a priori encoding of chemistry is more effective.

Further progress is still required to better represent porous materials. A particular challenge is the definition of porosity and its description. For example, in contrast to crystalline materials, the definition of porosity for porous molecular materials and porous organic cages is not straightforward, as they often include intercrystalline porosity and packing dependent porosity.

Even for MOFs and zeolites, the definition of porosity is probe dependent. For example, using the largest included sphere to describe the pore size of a material might be too simplistic for applications related to long-chain hydrocarbons. In addition, while a crystal graph-based description of the chemistry of materials is shown to be powerful for many applications, this approach can miss the spatial information which can be crucial when we aim to study catalytic and adsorptive properties of materials. Development of descriptors and algorithms that follow physical invariances and equivariances may be important developments in the future of this field.

15.3.2 Learning Algorithms

The foundation of big-data methods is rooted in statistical learning theory. The subtle difference between machine learning and curve fitting is that the aim in machine learning is "prediction" in contrast to "fitting." In machine learning, we aim to develop models that capture the true behavior of a complex system for which we do not know the underlying laws, but instead about which we have many observations. Hence, we use expressive models that can fit the observations, yet we have to regularize the space of models to prevent overfitting the noise in the data. This principle is called the bias-variance tradeoff. An underfit model (high bias) is not expressive enough to capture the true complexity of the system. On the other hand, an overfit model (high variance) is so complex that it encodes the noise in the data. Choosing the right model and fine tuning its (hyper-) parameters are crucial steps to avoid under- and overfitting.

Various considerations need to be taken into account when choosing the learning algorithm. For supervised tasks on tabular data, for example, predefined descriptors of pore geometry, a first consideration is the size and dimension of the training data. For learning complex functions, we need more data and higher model complexity. However, if we have only small amounts of data, simpler models, such as linear or logistic regression, often work better. Furthermore, extra attention needs to be paid to the dimensionality of the data, as high dimensions can interrupt the learning process. In particular, some algorithms, including kernel based-methods and nearest neighbor methods, are more sensitive to the data dimensionality. In these cases, careful feature selection and engineering is required to ensure maximum performance. On the other hand, artificial neural networks and decision trees can internally do feature selection and suffer less from the curse of dimensionality.

In many cases in the modern application of AI in chemical sciences, we aim to infer solutions starting from raw chemical data, for example, from a string representation of a chemical structure. In these cases, we aim to use a neural network machinery for representation learning. Depending on the type of our raw data, e.g., tabular, point clouds, or text, we can choose the right method for the problem of interest. Examples of neural network classes include convolutional neural networks (CNN), recurrent neural networks (RNN), and graph neural networks (GNN). Interestingly, using neural networks opens up many fascinating scientific approaches, for example, in transfer learning, one-shot learning, and generative modeling. These methods have only recently been explored in the field of porous materials, and further exciting developments are expected in the near future.

15.4 Applications

Emerging fields in the application of big-data science in nanoporous materials design are the development of generative algorithms for the design of new materials [34, 35], of data-driven synthesis approaches [36], and of data-derived methods for molecular simulations (Chapter 12 [37]). Generative neural networks have the capability of designing novel molecules that have similar specified properties to the molecules provided in a training set. In data-driven experimentation, structure–property models are built from experimental synthesis and property descriptors and used to accelerate the discovery of new and better performing materials. Data-derived methods for molecular simulation offer promise to replace or complement the time-consuming atomistic modeling methods for molecular property prediction.

Nanoporous materials are used in a wide range of applications, including areas related to energy and environment, electronics, catalysis, and sensing [38–40]. Computational prediction and discovery of materials for gas adsorption and separation have long been centered around computationally intensive atomistic simulations based upon first principles and force field calculations. Naturally, the development of nanoporous materials structure and property databases and appropriate molecular descriptors has enabled the use of Big Data Science approaches in the discovery of nanoporous materials for these applications. Data-driven approaches provide significant improvements in terms of balancing efficiency and accuracy in this field. Emerging fields are the development of generative algorithms for the design of new materials, data-driven synthesis approaches, and data-derived methods for prediction of thermodynamic properties [41] and molecular simulations [42].

Computational prediction and discovery of materials for gas adsorption and separation have long been centered around computationally intensive atomistic simulations based upon first principles and force field calculations. Data-driven approaches provide significant improvements in terms of balancing efficiency and accuracy in this field.

Reactive neural network potentials with DFT accuracy have been recently developed, extending the sampling quality by several orders of magnitude with respect to previous attempts, while retaining DFT accuracy [43]. This has been applied to study the dynamic behavior of small platinum clusters occluded in zeolites. Using a combination of global structure prediction, microkinetic modeling, and enhanced free energy surface exploration, it was possible to identify mechanisms of Pt atom-trapping in high-silica zeolites, the effect of the presence of reducing adsorbates (CO, H_2) on Pt migration, and the energetics of cluster growth [44].

The adsorption of CO_2 mixtures in MOFs has been predicted with multivariable linear regression (MLR), decision trees (DTs), artificial neural networks (ANNs), and gradient boosted regressors (GBR) [45]. A key feature to this study was to describe the MOF with a 13-component feature vector containing structural features (void fraction, density, specific surface area, largest pore diameter, and diffusion-limiting pore diameter), plus the addition of framework topology (as a qualitative feature mapped into an integer value) and simple physico-chemical descriptors related to the polarity of the MOF constituents, since this is a crucial aspect of the interaction with CO_2.

This approach suggests the possibility of having universal models in which the feature vector represents adsorption scenarios instead of adsorbents. For new adsorbents, the only change in the model is the feature vector. Gomez-Gualdron et al. [46] demonstrated that the same ANN model was capable of predicting full (room temperature) isotherms for the following adsorbates in MOFs: He, Ar, Xe, Kr, N_2, H_2, CH_4, and C_2H_6.

Important studies have recently compared how conclusions on adsorption properties may be biased by the quality or extent of the MOF database. Moosavi et al. [47]. studied the adsorption of CO_2 at low pressure using three databases: CoRE MOF 2019, BW-20K, and ARABG-DB, using trained random forest regression models with geometric and chemical descriptors. The metal chemistry appeared as the most important factor for the CoRE MOF database, while pore geometry was the most important for BW-DB, hence showing inconsistencies arising from the different diversities of the databases. Similarly, Daglar et al. [48] observed that the choice of the database leads to large variations in calculated ideal selectivities of MOFs at low pressure, although not at high pressure. Mixture selectivities and APS (adsorbent performance score) of MOFs dramatically changed depending on the database used in molecular simulations, leading to contradictory rankings of MOFs based on these metrics, with identification of top materials being strongly biased by the database chosen. In addition to improving the diversity [49], the quality of the databases must be continuously monitored since errors have been detected [50].

MOFs have been an active area of research since the late 1990s, when some outstanding structures were presented. A few year laters, the search of MOFs for hydrogen storage started [51] and it has continued without interruption. Initially, gravimetric and volumetric targets were difficult to meet simultaneously, although when DOE requirements loosened and new materials were synthesized, targets seemed within reach. A recent study has gathered a large part of the body of knowledge, joining 19 MOF databases to collect 918,734 structures [52]. Using 14 ML methods and 7 structural features, 8282 MOFs (most of them hypothetical) were selected, whose predicted behavior would be close to top-performing MOFs in pressure swing (usable gravimetric and volumetric uptakes of >8.5% and >39.5 g/L respectively), having low densities (<0.31 g/cm^3) in combination with high surface areas (>5300 m^2/g), void fractions (0.90), and pore volumes (>3.3 cm^3/g). The extremely randomized trees (ERT) algorithm was identified as the most accurate among the ML models tested for predicting H_2 uptake, using calculations of 24,674 MOFs as training data. The ML models are freely available, allowing fast and accurate predictions with a small amount of data required as input. The use of AI methods has led to gathering scattered data, facilitating a unified computational analysis with coherent conclusions, and bringing hydrogen storage on MOFs closer to industrial application.

Crystal structure prediction (CSP) of molecular cages or PMMs using computational chemistry requires high-throughput calculations that can hardly be simplified using AI algorithms, since it is not easy to define descriptors related to the different ways of packing. By careful design it is possible to identify molecules that can pack together in the most inefficient way, against the default trend of attractive intermolecular interactions, so as to increase porosity as much as possible. In this way, surface areas of 3700 m^2/g and densities as low as 0.417 g/cm^3 have been achieved [53].

Computational predictions do usually come up with a large number of relative minima, whilst often there are only a few experimentally described stable polymorphs. AI algorithms have been used to facilitate an adequate and efficient sampling of the potential energy surface landscape at a reduced computational cost, as well as to accelerate state-of-the-art quantum chemical calculations through the analysis of results at a lower level of theory [54].

Applications of ML methods to catalysis remain difficult due to the inherent complexity of catalysis as a time-dependent dynamic event. The main obstacles to using data-driven methods for catalyst design are the lack of universal datasets for catalytic activities and selectivities and the scarcity of general and simple descriptors. Some of these descriptors of catalytic activity are surface properties such as work function, d-band center, coordination number, surface energy, and different shapes of strain in the geometry of the active site. The aim is to establish scaling relationships between the adsorption energies of reactants and intermediates, and between activation and reaction energies through the so-called Brønsted–Evans–Polanyi relationships [55]. Oxygen vacancy formation in ABO_3 perovskites has been found to be a useful descriptor of electrocatalytic activity as calculated from DFT. Chen and Ciucci explored systematically the oxygen vacancy formation energies, and its relation to stability and conductivity, when A and B are substituted by Na, K, Ca, Sr, Pb (A site) and Cu, Ni, Zn, Ag (B site) in $BaFeO_3$. The results showed that the oxygen vacancy formation energy is linearly correlated with the occupied O p-band center, which is in turn related to the basicity of the oxygen atoms [56]. Similarly, d-band center and other descriptors such as coordination number and adsorption energies have been employed as descriptors to generate an ML model to predict CO_2 electrocatalytic activity of a systematic series of copper-based intermetallic compounds [57].

Combination of ML methods with high-throughput experimentation has the potential to lead to autonomous discovery in the field of zeolites, MOFs, and PMMs. Chemical reactions, conditions, and precursors needed for synthesis can be explored. The synthetic feasibility of computationally generated structures and the optimization of specific structures for a given target application can be predicted [58].

Some remaining challenges include increasing the accuracy of modeling host–guest and guest–guest interactions for molecules with more complex adsorption mechanisms and in materials containing transition metals, and more efficient exploration of nanoporous materials space.

15.5 Zeolite Synthesis and OSDAs

Organic structure directing agents (OSDAs) were introduced in zeolite synthesis media by Richard M. Barrer in 1960, drawing parallels with the concept of molecule-micropore match which is based upon the lock and key model of protein-ligand binding. More than 60 years later, it is still not easy to make a clear map pairing thousands of OSDA ←→ zeolite successful combinations due in part to the important and simultaneous role played by other synthesis conditions such as pH, water/silica, gel composition (in particular Si, SiAl, SiGe, AlP, and SiAlP), time, temperature, mineralizing agent, and presence of alkaline cations. Within the most important and successful subset of zeolites [59], those containing Si and Al, with the general composition $OSDA^{m+}{}_{z}M^{n+}{}_{y}Al_{x}Si_{1-x}O_2$ (x=n·y+m·z, with M and OSDA being inorganic and organic cations), still defy, after decades of research, our current characterization methods to elucidate the Al framework distribution [60], with only two (electrostatically based) rules to guide our insights. Al-O-Al links are forbidden (Lowenstein's 1956 rule), and Al-(M,OSDA) tend to be close. The third item in our list of zeolite challenges is the presence (number, location, and type) of framework defects [61] when the synthesis is carried out in the usual hydroxide media, in contrast to the case in fluoride media leading to almost perfect crystals. Finally, a topic particularly suited to apply AI algorithms is the generation and application of databases of predicted zeolite structures, which has been covered in Chapter 6 of this book [62].

The first challenge enunciated above, pairing zeolites and OSDAs, has been recently studied in depth using ML methods. In the group of Olivetti, two important studies provide simplified recipes of synthesis that have been automatically collected from the scientific literature using advanced AI tools based on natural language processing. First, a random forest regression model allowed prediction of the germanium-containing framework density of synthesized zeolites and predicting experimental conditions that may lead to the discovery of new SiGe-zeolite topologies. The model suggested the Si/Ge, H_2O/(Si+Ge), and OSDA volume, as the more determinant variables to predict the zeolite framework densities of Ge-containing zeolites. Second, 5663 simple synthesis recipes were collected containing zeolite–OSDA pairs and zeolite chemical composition extracted through NLP and text mining techniques. Using Generative Neural Networks, structural descriptors (volume, charge) of the OSDAs were tied with zeolite descriptors (micropore dimensionality, largest sphere that can be included, framework density, and maximum ring size), with little success, at least on a clear-cut basis. Using principal component analysis (PCA), two-dimensional (2D) plots of two such variables (PCA1 and PCA2), without particular physico-chemical meaning, allow us to map reduced regions for each zeolite--OSDA pair. Hence, for instance, if a given set of experimentally-found OSDAs for the synthesis of CHA is located in certain intervals of PCA1 and PCA2 values, it is predicted that other OSDAs within this interval can be candidates for the synthesis of CHA [63].

Along the same idea of trying zeolite and OSDAs, a smart and large database containing interaction zeo-OSDA energies has been recently created, that is widely available [64], fully described in Chapter 4 by Schwalbe-Koda and Gómez-Bombarelli [65], containing 209 zeolites, 1145 OSDAs, and 112,426 zeolite–OSDA pairs. For each OSDA it is possible to download a list of papers containing the zeolite(s) obtained in each case. And for each zeolite, it is possible to download a list of papers containing the OSDA(s) employed in each case, as well as the list of zeolite atom types. Interactive plots also allow us to visually explore calculated zeo-OSDA energies, not only for those experimentally tested, but also for many other combinations.

De novo design of OSDAs for the synthesis of zeolites has been shown to be a viable approach. Success of this approach relies on the ability to accurately predict how well a putative OSDA will direct zeolite synthesis. This is discussed in Chapters 1 [66] and 2 [67]. Novel molecule-generating algorithms based upon deep learning are emerging. Either accurate, but computationally expensive atomistic modeling techniques or rapid, trained, and validated QSAR models using appropriate molecular descriptors and fast machine-learning algorithms, can accurately predict OSDA–zeolite interactions otherwise obtained with computationally expensive atomistic modeling techniques, as described by Daeyaert and Deem in Chapter 2 [67]. These methods in turn can be deployed in large-scale virtual screening in search of novel OSDAs for given target zeolites. Several zeolites have now been synthesized using computationally-designed OSDAs. Remaining challenges include an even more accurate description of constraints on synthetic feasibility of OSDAs and incorporation of additional synthetic parameters, such as pH, inorganic cations, and other solution conditions.

Data science techniques can be used to model and predict synthesis–structure and structure–property relationships and to explore and to optimize molecular structures. Further optimizing these techniques and combining them with automated and high-throughput experiments paves the way to significantly accelerate the discovery of new OSDAs for the synthesis of zeolites [68, 69].

Since mostly supervised ML methods are used for zeolites, definition of descriptors as well as establishing relations between them becomes a crucial point. Decades of synthesis efforts have been beautifully summarized in Chapter 1 [66]. Lew et al. explore synthesis procedures that give different zeolite outcomes when making subtle changes in the synthesis conditions, in particular regarding the choice of OSDA. The wide information given in the chapter contains synthesis variables that are still far away from entering the list of descriptors included in current AI studies, such as the source of silica. In one particular case, an unexpected boron-containing zeolite phase appeared when using an Si,B source from calcined B-beta, and this approach was used to generate novel borosilicate zeolites. Similarly, the joint presence of organic and inorganic (alkali) cations, and their role in the pH of the synthesis, has been controlled as another variable driving the synthesis of new zeolites. These examples shed light on the effect of all variables in the synthesis outcome and provide a wealth of data for future AI-guided studies, not only regarding the amount of experiments but mainly in their interpretation, helped by a careful design of synthesis variables.

A less explored area is co-templating, although a number of cases have been described of zeolites synthesized with two OSDAs. One of the early examples is the silica version of MCM-22 (ITQ-1) using trimethyladamantylammonium and hexamethyleneimine, "templating" respectively the external 12-ring hemi-cavities and the sinusoidal 10-ring channels [70].

The computational modeling-aided application of more than one OSDA to direct the synthesis of zeolite structures containing multiple types of cavities has been shown for SAPO zeotypes, for mixed inorganic-organic SDAs for Al-Si zeolites, and for intergrowths of multiple zeolite frameworks, as systematically described in Chapter 5 [71]. Structure direction of a single-phase aluminosilicate zeolite containing multiple cavity types by the design of multiple well-fitting OSDAs as templates has not yet been unequivocally observed.

Hypothetical zeolites became an exciting research topic after a mathematical breakthrough in tiling theory [72]. Although they had been announced earlier [73], a point arrived when computer algorithms were ready to produce them massively [74]. The closely related problems of zeolite structure determination and zeolite structure prediction can be addressed by either Monte Carlo (MC)- or Evolutionary Algorithm (EA)-based algorithms [15]. The advent of programmable GPUs presents exciting opportunities to speed up calculations, with efficient parallelization schemes being developed. The computationally guided synthesis of a predicted hypothetical zeolite, by selecting appropriate candidate OSDAs, remains an important goal in zeolite science [75–77].

Data-driven approaches sustained by data obtained from high-throughput computation and data collection and curation of historical experimental results can effectively aid the template-based synthesis of zeolites. However, the first-principles synthesis of hypothetical zeolite frameworks hitherto has not experienced major improvements in the last decade and the number of freely available databases remains relatively small. Of those, only one contains a search engine using descriptors [78].

More effective exploration of the OSDA chemical space using AI-based generative models, accelerating simulations using machine-learning algorithms, and methods for computational discovery of template-free synthesis routes are other challenges that will soon be addressed more in-depth.

15.6 Conclusion

The vastness of the field of nanoporous materials makes it ideal for the application of big-data sciences. Development of these methods has already enabled and continues to enable us with new ways to approach our curiosity-driven science, as well as to develop novel technologies. In this book, we aimed to gather the latest developments in the field. As the authors of each chapter have described, AI is empowering the future of nanoporous materials development. The full stack of these methods might enable the lab of the future, where in a synthetic lab, robots driven by AI do chemistry and explore materials spaces that have been narrowed by predictions from machine learning. The success of this approach is yet to be seen. However, we have now indications that this is no longer science fiction. Particularly, as the amount of data generated by the high-throughput labs of the future grows, the quality of the AI model will increase.

References

1 Haber, F. and Bosch, C. (1932). "... we were able to conduct the many separate examinations necessary for the discovery of the catalysts and for development to maximum efficiency, which over the years grew to the number of 20,000." Carl Bosch, Nobel Lecture, 21 May 1932.

2 Juan, Y., Dai, Y., Yang, Y. et al. (2021). Accelerating materials discovery using machine learning. *J. Mater. Sci. Technol.* 79: 178–190.

3 Moosavi, S.M., Jablonka, K.M., and Smit, B. (2020). The role of machine learning in the understanding and design of materials. *J. Am. Chem. Soc.* 142: 20273–20287.

4 Agrawala, A. and Choudhary, A. (2016). Perspective: materials informatics and big data: realization of the "fourth paradigm" of science in materials science. *APL Mater* 4: 053208.

5 Wang, A.Y.-T., Murdock, R.J., Kauwe, S.K. et al. (2020). Machine learning for materials scientists: an introductory guide toward best practices. *Chem. Mater.* 32: 4954–4965.

6 Zhou, T., Song, Z., and Sundmacher, K. (2019). Big data creates new opportunities for materials research: a review on methods and applications of machine learning for materials design. *Engineering* 5: 1017–1026.

7 Li, A., Bueno-Perez, R., and Fairen-Jimenez, D. (2022). Extracting metal–organic frameworks data from the Cambridge Structural Database. In: *AI-Guided Design and Property Prediction for Zeolites and Nanoporous Materials* (ed. G. Sastre and F. Daeyaert), 201–232. John Wiley & Sons.

8 Chung, Y.G., Haldoupis, E., Bucior, B.J. et al. (2019). Advances, updates, and analytics for the computation-ready, experimental metal–organic framework database: CoRE MOF 2019. *J. Chem. Eng. Data* 64 (12): 5985–5998.

9 Ongari, D., Yakutovich, A.V., Talirz, L. et al. (2021). Building a consistent and reproducible database for adsorption evaluation in covalent-organic frameworks. *Materials Cloud Archive* 2021 (100). doi: 10.24435/materialscloud:z6-jn; https://archive.materialscloud.org/record/2021.100

10 Baerlocher, C. and McCusker, L.B. (2022). *Database of Zeolite Structures.* http://www.iza-structure.org/databases

11 Boyd, P., Lee, Y., and Smit, B. (2017). Computational development of the nanoporous materials genome. *Nat. Rev. Mater.* 2: 17037.

12 Earl, D.J. and Deem, M.W. (2006). Toward a database of hypothetical zeolite structures. *Ind. Eng. Chem. Res.* 45: 5449–5454.

13 Treacy, M.M.J., Rao, S., and Rivin, I. (1993). A combinatorial method for generating new zeolite frameworks. In: *Proceedings of the 9th International Zeolite Conference Montreal 1992* (ed. R. Von Ballmoos, J.B. Higgins, and M.M.J. Treacy), 381–388. Stoneham, MA: Butterworth-Heinemann.

14 Barthel, S., Alexandrov, E.V., Proserpio, D.M. et al. (2018). Distinguishing metal–organic frameworks. *Cryst. Growth Des.* 18(3): 1738–1747.

15 Pophale, R., Cheeseman, P.A., and Deem, M.W. (2011). A database of new zeolite-like materials. *Phys. Chem. Chem. Phys.* 13: 12407–12417.

16 Siderius, D., Shen, V., Johnson, R., III. et al. ed. (2014). *NIST/ARPA-E Database of Novel and Emerging Adsorbent Materials*. Gaithersburg, MD: National Institute of Standards and Technology.

17 Ongari, D., Talirz, L., Jablonka, K.M. et al. (2022). Data-driven matching of experimental crystal structures and gas adsorption isotherms of metal–organic frameworks. *J. Chem. Eng. Data* 67: 1743–1756. https://doi.org/10.1021/acs.jced.1c00958.

18 Luo, Y., Bag, S., Zaremba, O. et al. (2022). MOF synthesis prediction enabled by automatic data mining and machine learning. *Chem. Int.* 61: e202200242.

19 Park, H., Kang, Y., Choe, W. et al. (2022). Mining insights on metal–organic framework synthesis from scientific literature texts. *J. Chem. Inf. Model.* 62 (5): 1190–1198.

20 Glasby, L.T., Whaites, E.H., and Moghadam, P.Z. (2022). Machine learning and digital manufacturing approaches for solid-state materials development. In: *AI-Guided Design and Property Prediction for Zeolites and Nanoporous Materials* (ed. G. Sastre and F. Daeyaert), 377–409. John Wiley & Sons.

21 Moosavi, S.M., Chidambaram, A., Talirz, L. et al. (2019). Capturing chemical intuition in synthesis of metal-organic frameworks. *Nat. Commun.* 10: 539.

22 Raccuglia, P., Elbert, K.C., Adler, P.D.F.et al. (2016). Machine-learning-assisted materials discovery using failed experiments. *Nature* 533: 73–76.

23 Jablonka, K.M., Patiny, L., and Smit, B. (2022). Making the collective knowledge of chemistry open and machine actionable. *Nature Chem.* 14: 365–376.

24 Yano, J., Gaffney, K.J., Gregoire, J.et al. (2022). The case for data science in experimental chemistry: examples and recommendations. *Nature Rev. Chem.* 6: 357–370.

25 Willems, T.F., Rycroft, C.H., Kazi, M.et al. (2012). Algorithms and tools for high-throughput geometry-based analysis of crystalline porous materials. *Micropor. Mesopor. Mater.* 149: 134–141.

26 Sarkisov, L., Bueno-Perez, R., Sutharson, M. et al. (2020). Materials informatics with poreblazer v4.0 and the CSD MOF database. *Chem. Mater.* 32: 9849–9867.

27 Lee, Y., Barthel, S.D., Dłotko, P. et al. (2017). Quantifying similarity of pore-geometry in nanoporous materials. *Nature Commun.* 8: 15396.

28 Lee, Y., Barthel, S.D., Dłotko, P. et al. (2018). High-throughput screening approach for nanoporous materials genome using topological data analysis: application to zeolites. *J. Chem. Theory Comput.* 14: 4427–4437.

29 Moosavi, S.M., Xu, H., Chen, L. et al. (2020). Geometric landscapes for material discovery within energy-structure-function maps. *Chem. Sci.* 11: 5423–5433.

30 Hung, T.-H., Xu, Z.-X., Kang, D.-Y. et al. (2022). Chemistry-encoded convolutional neural networks for predicting gaseous adsorption in porous materials. *J. Phys. Chem. C* 126: 2813–2822.

31 Jablonka, K.M., Ongari, D., Moosavi, S.M. et al. (2020). Big-data science in porous materials: materials genomics and machine learning. *Chem. Rev.* 120: 8066–8129.

32 Musil, F., Grisafi, A., Bartók, A.P. et al. (2021). Physics-inspired structural representations for molecules and materials. *Chem. Rev.* 121: 9759–9815.

33 Na, G.S. (2022). Numerical representations of chemical data for structure-based machine learning. In: *AI-Guided Design and Property Prediction for Zeolites and Nanoporous Materials* (ed. G. Sastre and F. Daeyaert), 173–200. John Wiley & Sons.

34 Kim, B., Lee, S., and Kim, J. (2020). Inverse design of porous materials using artificial neural networks. *Sci. Adv.* 6: eaax9324.

35 Yao, Z., Sánchez-Lengeling, B., Bobbitt, N.S. et al. (2021). Inverse design of nanoporous crystalline reticular materials with deep generative models. *Nat. Mach. Intell.* 3: 76–86.

36 Häse, F., Roch, L.M., and Aspuru-Guzik, A. (2019). Next-generation experimentation with self-driving laboratories. *Trends Chem.* 1: 282–291.

37 Haranczyk, M. and Dico, G.L. (2022). Big data science in nanoporous materials: datasets and descriptors. In: *AI-Guided Design and Property Prediction for Zeolites and Nanoporous Materials* (ed. G. Sastre and F. Daeyaert), 319–342. John Wiley & Sons.

38 Rahimi, M., Moosavi, S.M., Smit, B. et al. (2021). Toward smart carbon capture with machine learning. *Cell Rep. Phys. Sci.* 2: 100396.

39 Erdem Günay, M. and Yıldırım, R. (2021). Recent advances in knowledge discovery for heterogeneous catalysis using machine learning. *Catal. Rev. Sci. Eng.* 63: 120–164.

40 Rosen, S., Notestein, J.M., and Snurr, R.Q. (2022). Realizing the data-driven, computational discovery of metal–organic framework catalysts. *Curr. Opin. Chem. Eng.* 35: 100760.

41 Moosavi, S.M., Novotny, B.A., Ongari, D. et al. (2022). A data-science approach to predict the heat capacity of nanoporous materials. *ChemRxiv* Cambridge, UK: Cambridge Open Engage. doi: 10.26434/chemrxiv-2022-5xt71.

42 Datar, A., Lyu, Q., and Lin, L.-C. (2022). Machine-learning-aided discovery of nanoporous materials for energy- and environmental-related applications. In: *AI-Guided Design and Property Prediction for Zeolites and Nanoporous Materials* (ed. G. Sastre and F. Daeyaert) 283–318. John Wiley & Sons.

43 Erlebach, A., Nachtigall, P., and Grajciar, L. (2022). Accurate large-scale simulations of siliceous zeolites by neural network potentials. doi.org/10.48550/arXiv.2102.12404.

44 Heard, C.J., Erlebach, A., Hou, D. et al. (2022). Insights into the dynamics of zeolite encapsulated Pt clusters via machine learning. *Book of Abstracts: 20th International Zeolite Conference*. www.izc2022.com.

45 Gómez-Gualdrón, D.A., Simon, C.M., and Colón, Y.J. (2022). Efficient data utilization in training machine learning models for nanoporous materials screening. In: *AI-Guided Design and Property Prediction for Zeolites and Nanoporous Materials* (ed. G. Sastre and F. Daeyaert), 343–376. John Wiley & Sons.

46 Anderson, R., Biong, A., and Gómez-Gualdrón, D.A. (2020). Adsorption isotherm predictions for multiple molecules in MOFs using the same deep learning model. *J. Chem. Theory Comput.* 16: 1271–1283.

47 Moosavi, S.M., Nandy, A., Jablonka, K.M. et al. (2020). Understanding the diversity of the metal–organic framework ecosystem. *Nat. Commun.* 11: 4068.

48 Daglar, H., Gulbalkan, H.C., Avci, G. et al. (2021). Effect of metal–organic framework (MOF) database selection on the assessment of gas storage and separation potentials of MOFs. *Angew. Chem. Int. Ed.* 60: 7828–7837.

49 Majumdar, S., Moosavi, S.M., Jablonka, K.M. et al. (2021). Diversifying databases of metal organic frameworks for high-throughput computational screening. *ACS Appl. Mater. Interfaces* 13: 61004–61014.

50 Chen, T. and Manz, T.A. (2020). Identifying misbonded atoms in the 2019 CoRE metal–organic framework database. *RSC Adv.* 2020 (10): 26944–26951.

51 Rosi, N.L., Eckert, J., Eddaoudi, M. et al. (2003). Hydrogen storage in microporous metal–organic frameworks. *Science* 300: 1127–1129.

52 Ahmed, A. and Siegel, D.J. (2021). Predicting hydrogen storage in MOFs via machine learning. *Patterns* 2: 100291.

53 Little, M.A. and Cooper, A.I. (2020). The chemistry of porous organic molecular materials. *Adv. Funct. Mater.* 30: 1909842.

54 Curtis, F., Li, X., Rose, T., Vázquez-Mayagoitia, A. et al. (2018). GAtor: a first-principles genetic algorithm for molecular crystal structure prediction. *J. Chem. Theory Comput.* 14: 2246–2264.

55 Toyao, T., Maeno, Z., Takakusagi, S. et al. (2020). Machine learning for catalysis informatics: recent applications and prospects. *ACS Catalysis* 10: 2260–2297.

56 Chen, C. and Ciucci, F. (2016). Designing Fe-based oxygen catalysts by density functional theory calculations. *Chem. Mater.* 28: 7058–7065.

57 Zhang, N., Yang, B., Liu, K. et al. (2021). Machine learning in screening high performance electrocatalysts for CO_2 reduction. *Small Methods* 5: 2100987.

58 Bennett, S. and Jelfs, K.E. (2022). Porous molecular materials: exploring structure and property space with software and artificial intelligence. In: *AI-Guided Design and Property Prediction for Zeolites and Nanoporous Materials* (ed. G. Sastre and F. Daeyaert), 251–282. John Wiley & Sons.

59 Masters, A.F. and Maschmeyer, T. (2011). Zeolites: from curiosity to cornerstone. *Micropor. Mesopor. Mater.* 142: 423–438.

60 Dědeček, J., Sobalík, Z., and Wichterlová, B. (2012). Siting and distribution of framework aluminium atoms in silicon-rich zeolites and impact on catalysis. *Catal. Rev. Sci. Eng.* 54: 135–223.

61 Medeiros-Costa, I.C., Dib, E., Dath, J.-P.et al. (2021). Silanol defect engineering and healing in zeolites: opportunities to fine-tune their properties and performances. *Chem. Soc. Rev.* 50: 11156–11179.

62 Argente, E., Valero, S., Misturini, A.et al. (2022). Computer generation of hypothetical zeolites. In: *AI-Guided Design and Property Prediction for Zeolites and Nanoporous Materials* (ed. G. Sastre and F. Daeyaert), 145–172. John Wiley & Sons.

63 Jensen, Z., Kwon, S., Schwalbe-Koda, D. et al. (2021). Discovering relationships between OSDAs and zeolites through data mining and generative neural networks. *ACS Cent. Sci.* 7: 858–867.

64 https://zeodb.mit.edu

65 Schwalbe-Koda, D. and Gómez-Bombarelli, R. (2022). Generating, managing, and mining big data in zeolite simulations. In: *AI-Guided Design and Property Prediction for Zeolites and Nanoporous Materials* (ed. G. Sastre and F. Daeyaert), 81–111. John Wiley & Sons.

66 Lew, C.M., Xie, D., Schmidt, J.E. et al. (2022).The confluence of organo-cations, inorganic species, and molecular modeling on the discovery of new zeolite structures and compositions. In: *AI-Guided Design and Property Prediction for Zeolites and Nanoporous Materials* (ed. G. Sastre and F. Daeyaert), 1–32. John Wiley & Sons.

67 Daeyaert, F. and Deem, M.W. (2022). *De novo* design of organic structure directing agents for the synthesis of zeolites. In: *AI-Guided Design and Property Prediction for Zeolites and Nanoporous Materials* (ed. G. Sastre and F. Daeyaert) 33–59. John Wiley & Sons.

68 Pophale, R., Daeyaert, F., and Deem, M.W. (2013). Computational prediction of chemically synthesizableor ganic structure directing agents for zeolites. *J. Mater. Chem. A* 1: 6750–6760.

69 Daeyaert, F., Deem, M.W., and Pareto, A. (2017). A Pareto Algorithm for efficient *de Novo* design of multi-functional molecules. *Mol. Inform.* 36: 1600044.

70 Camblor, M.A., Corma, A., Díaz-Cabañas, M.J. et al. (1998). Synthesis and structural characterization of MWW type zeolite ITQ-1, the pure silica analog of MCM-22 and SSZ-25. *J. Phys. Chem. B* 102: 44–51.

71 Chitac, R.G., Shannon, M.D., Cox, P.A. et al. (2022). Co-templating in the designed synthesis of small pore zeolite catalysts. In: *AI-Guided Design and Property Prediction for Zeolites and Nanoporous Materials* (ed. G. Sastre and F. Daeyaert), 113–143. John Wiley & Sons.

72 Delgado Friedrichs, O., Dress, A.W.M. et al. (1999). Systematic enumeration of crystalline networks. *Nature* 400: 644–647.

73 Brunner, G.O. (1990). Criteria for the evaluation of hypothetical zeolite frameworks. *Zeolites* 10: 612–614.

74 Treacy, M.M.J., Randall, K.H., Rao, S. et al. (1997). Enumeration of periodic tetrahedral frameworks. *Z. Kristallogr.* 212: 768–791.

75 Schmidt, J.E., Deem, M.W. et al. (2014). Synthesis of a specified, silica molecular sieve using computationally predicted organic structure directing agents. *Angew. Chem. Int. Ed.* 53: 8372–8374.

76 Gálvez-Llompart, M. and Sastre, G. (2022). Machine learning search for suitable structure directing agents for the synthesis of Beta (BEA) zeolite using molecular topology and Monte Carlo techniques. In: / *AI-Guided Design and Property Prediction for Zeolites and Nanoporous Materials* (ed. G. Sastre and F. Daeyaert), 61–80. John Wiley & Sons.

77 Chaikittisilp, W. (2022). Data-driven approach for rational synthesis of zeolites and other nanoporous materials. In: *AI-Guided Design and Property Prediction for Zeolites and Nanoporous Materials* (ed. G. Sastre and F. Daeyaert), 233–250. John Wiley & Sons.

78 Foster, M.D. and Treacy, M.M.J. (2004). *Atlas of Prospective Zeolite Structures*. http://www.hypotheticalzeolites.net

Index

AI-Guided Design and Property Prediction for Zeolites and Nanoporous Materials, First Edition. Edited by German Sastre and Frits Daeyaert.

g

h

i

k

l

m

s

t

u